Plate 1. Some of the many uses of tall fescue. a. A residential lawn. b. Pasture for cattle. c. A seed crop at sunset. d. A golf course. e. Pasture for horses. (Photograph credit: Larry Kassell, Kassell Concepts, Silverton, Oregon, USA. Sponsor: Oregon Tall Fescue Commission, Salem, Oregon, USA.)

Plate 2. Don and Hazel Pedersen's farm in 1982: note: 'seed rings' around shelters for leafcutting bees that were introduced during crop bloom; crop yielded 890 kg ha^{-1}. (Photograph credit: Daphne Fairey, AAFC, Beaverlodge, Alberta. Sponsors: Peace River Agriculture Strategic Planning Society & Peace Region Forage Seed Association.)

Forage Seed Production
Volume 1: Temperate Species

FORAGE SEED PRODUCTION
Volume 1: Temperate Species

Edited by

D.T. Fairey

Agriculture and Agri-Food Canada,
Alberta,
Canada

and

J.G. Hampton

Seed Technology Centre,
Department of Plant Science,
Massey University,
Palmerston North,
New Zealand

CAB INTERNATIONAL

CAB INTERNATIONAL
Wallingford
Oxon OX10 8DE
UK

CAB INTERNATIONAL
198 Madison Avenue
New York NY 10016-4314
USA

Tel: +44 (0)1491 832111
Fax: +44 (0)1491 833508
E-mail: cabi@cabi.org

Tel: +1 212 726 6490
Fax: +1 212 686 7993
E-mail: cabi-nao@cabi.org

© CAB INTERNATIONAL 1997. All rights reserved. No part of this publication may be reproduced in any form or by any means, electronically, mechanically, by photocopying, recording or otherwise, without the prior permission of the copyright owners.

A catalogue record for this book is available from the British Library, London, UK.

Library of Congress Cataloging-in-Publication Data
Forage seed production / edited by D. T. Fairey, J. G. Hampton
    p. cm.
    Includes index.
    Contents: v. 1. Temperate species.
    ISBN 0-85199-190-4 (alk. paper)
    1. Forage plants—Seeds. 2. Seed technology. I. Fairey, D. T. (Daphne T.). II. Hampton, J. G. (John G.)
SB193.55.F67 1998
633.2′0821—dc21   97-30071
                                                                             CIP

ISBN 0 85199 190 4

Typeset in Photina by AMA Graphics Ltd
Printed and bound in the UK at the University Press, Cambridge

Contents

Contributors vii

Preface xi

Part One

1 **Temperate Forage Seeds: An Introduction** 1
  D.T. Fairey and J.G. Hampton

2 **Reproductive Development and the Establishment of Potential Seed Yield in Grasses and Legumes** 9
  T.S. Aamlid, O.M. Heide, B.R. Christie and R.L. McGraw

3 **Components of Seed Yield in Grasses and Legumes** 45
  J.G. Hampton and D.T. Fairey

4 **Maturation of Grass and Legume Seed** 71
  P. Coolbear, M.J. Hill and Win Pe

5 **Grass Seed Crop Management** 105
  M.P. Rolston, J.S. Rowarth, W.C. Young III and G.W. Mueller-Warrant

6 **Legume Seed Crop Management** 127
  A.H. Marshall, J.J. Steiner, O. Niemeläinen and J. Hacquet

7 **Pollination, Fertilization and Pollinating Mechanisms in Grasses and Legumes** 153
  D.T. Fairey, S.M. Griffith and P.T.P. Clifford

8	Harvest and Postharvest Management of Forage Seed Crops U. Simon, M.D. Hare, B. Kjaersgaard, P.T.P. Clifford, J.G. Hampton and M.J. Hill	181
9	Seed Quality of Grasses and Legumes M.J. Hill, J.G. Hampton and K.A. Hill	219
10	Breeding for Higher Seed Yields in Grasses and Forage Legumes A. Elgersma and A.J.P van Wijk	243
11	The Forage Seed Trade A. Burgon, O.B. Bondesen, W.H. Verburgt, A.G. Hall, N.S. Bark, M. Robinson and G. Timm	271

Part Two: Case Histories

12	*Festuca arundinacea* Schreb. (Tall Fescue) in the USA W.C. Young III	287
13	*Festuca rubra* L. (Creeping Red Fescue) in Canada N.A. Fairey	297
14	*Lolium multiflorum* Lam. (Italian Ryegrass) in Germany W. Schöberlein and E. Lütke Entrup	311
15	*Lolium perenne* L. (Perennial Ryegrass) in Denmark K. Svensson and B. Boelt	321
16	*Poa pratensis* L. (Smooth-stalked Meadowgrass/ Kentucky Bluegrass) in The Netherlands D.A. Donner and G.E.L. Borm	329
17	*Dactylis glomerata* L. (Cocksfoot) in New Zealand M.J. Hill	339
18	*Lotus corniculatus* L. (Birdsfoot Trefoil) in North America P.R. Beuselinck	351
19	*Medicago sativa* L. (Lucerne/Alfalfa) in Canada D.T. Fairey and N.A. Fairey	361
20	*Trifolium pratense* L. (Red Clover) in France S. Bouet and G. Sicard	377
21	*Trifolium repens* L. (White Clover) in New Zealand P.T.P. Clifford	385
22	*Trifolium subterraneum* L. (Subterranean Clover) in Australia K.G. Boyce	395

Index 403

Contributors

Trygve S. Aamlid
The Norwegian Crop Research
    Institute
Apelsvoll Research Centre
Division Landvik, N-4890 Grimstad
NORWAY

Neville S. Bark
British Seed Houses Ltd
Portview Road
Avonmouth, Bristol BS11 9JH
UNITED KINGDOM

Paul R. Beuselinck
USDA-ARS, Plant Genetics
    Research Unit, Columbia
Missouri 65211, USA

Birte Boelt
Danish Institute of Plant and Soil
    Science
Department of Cereals, Seeds and
    Industrial Crops
Roskilde
DENMARK

Ole Bech Bondeson
Danish Seed Council
WA Vestrbrogade
DK-1020 Copenhagen
DENMARK

Gerard E.L. Borm
Applied Research for Arable Farming
    and Field Production of
    Vegetables (PAV)
PO Box 430
8200 AK, Lelystad
HOLLAND

Serge Bouet
FNAMS
Le Verger
49800 Brain sur l'Authion
FRANCE

Kevin G. Boyce
Seed Services Centre,
Primary Industries, South Australia
GPO Box 1617, Adelaide
SOUTH AUSTRALIA 5001

Alex Burgon
British Seed Houses Ltd
Portview Road
Avonmouth, Bristol BS11 9JH
UNITED KINGDOM

Bertram R. Christie
Agriculture and Agri-Food Canada
PO Box 1210
Charlottetown, Prince Edward Island
CANADA

Peter T.P. Clifford
AgResearch
PO Box 60
Lincoln
NEW ZEALAND

Peter Coolbear
The Association of Polytechnics in
    New Zealand
PO Box 10344, Wellington
NEW ZEALAND

Dingena A. Donner
ECAF
Blaauwe Kamer 12
6702 PA, Wageningen
THE NETHERLANDS

Anjo Elgersma
Department of Agronomy
Wageningen Agricultural University
Haraweg 333
6709 RZ Wageningen
THE NETHERLANDS

Ernst Lütke Entrup
Deutsche Saatveredelung Breeding
    Station, D-33154 Thüle
GERMANY

Daphne T. Fairey
Northern Agriculture Research
    Centre
Agriculture and Agri-Food Canada
Beaverlodge, Alberta T0H 0C0
CANADA

Nigel A. Fairey
Northern Agriculture Research
    Centre
Agriculture and Agri-Food Canada
Beaverlodge, Alberta T0H 0C0
CANADA

Stephen M. Griffith
NFSPRC, USDA-ARS, Oregon State
    University
Corvallis, Oregon 97331
USA

Jacques Hacquet
FNAMS
Centre INRA, 8660 Lusignan
FRANCE

A. Graeme Hall
New Zealand Grain and Seed Trade
    Association Ltd
PO Box 1208, Wellington
NEW ZEALAND

John G. Hampton
Seed Technology Centre
Department of Plant Science
Massey University
Palmerston North
NEW ZEALAND

Michael D. Hare
Faculty of Agriculture
Ubon Ratchathani University
Ubon Ratchathani 34190
THAILAND

Ola M. Heide
Department of Biology and Nature
    Conservation
Agricultural University of Norway
N-1432 Ås
NORWAY

Karen A. Hill
Seed Technology Centre
Department of Plant Science
Massey University
Palmerston North
NEW ZEALAND

Murray J. Hill
Seed Technology Centre
Department of Plant Science
Massey University
Palmerston North
NEW ZEALAND

Birthe Kjaersgaard
DLF-Trifolium A/S
4000 Roskilde
DENMARK

Athole H. Marshall
Institute of Grassland and
    Environmental Research
Plas Gogerddan
Aberystwyth SY23 3EB
UNITED KINGDOM

Robert L. McGraw
Department of Agronomy
University of Missouri
Columbia, MO 65211
USA

George W. Mueller-Warrant
USDA-ARS, Oregon State University
Corvallis, Oregon 97331
USA

Oiva T. Niemeläinen
Agricultural Research Centre of
    Finland
Institute of Crop and Soil Science
Plant Breeding Section
FIN-31600, Jokionen
FINLAND

Win Pe
Myanmar Agriculture Service
Gyogon
Insein PO 11011
Yangon
MYANMAR

Michael Robinson
Seed Research of Oregon Inc.
PO Box 1416
Corvallis, Oregon 97339
USA

M. Phil Rolston
AgResearch Grasslands
PO Box 60
Lincoln
NEW ZEALAND

Jacqueline S. Rowarth
Plant Science Department
Lincoln University
Canterbury
NEW ZEALAND

Werner Schöberlein
Landwirtschaftliche Fakultät
Martin-Luther-Universität
D-06108 Halle
GERMANY

Georges Sicard
FNAMS, Le Verger
49800 Brain sur l'Authion
FRANCE

Uwe Simon
Technische Universität München
85350 Freising-Weihenstephan
GERMANY

Jeffrey J. Steiner
National Forage Seed Production
    Research Centre
USDA-ARS, Oregon State University
Corvallis, Oregon 97331, USA

Kenneth Svensson
Federation of Danish Seed Grower's
    Associations
Roskilde
DENMARK

Gary Timm
Research Seeds Inc.
St Joseph, Missouri
USA

Arnold J.P. van Wijk
VanderHave Grasses BV
PO Box 127
5250 AC Vlijmen
THE NETHERLANDS

W.H. Verburgt
Mommersteeg International BV
Postbus 135250 AA
Vlijmen
THE NETHERLANDS

William C. Young III
Department of Crop and Soil Science
Oregon State University
Corvallis, Oregon 97331
USA

Preface

These two volumes on Forage Seed Production (Volume 1: Temperate Species; Volume 2: Tropical and Subtropical Species) have evolved as a result of considerable international collaboration between personnel from CAB INTERNATIONAL (CABI) and the International Herbage Seed Production Research Group (IHSPRG). This collaboration was facilitated greatly by our ability to communicate electronically, which has been particularly important for the editors because it allowed them to strive for more multi-authored contributions in the relatively short time during which these volumes were produced. Hopefully, this has enhanced the utility and scope of the information for all 'students' of this branch of agricultural science.

IHSPRG is a professional association with members in 57 different countries spanning temperate, tropical and subtropical climatic zones, with a major interest in an often neglected aspect of forage crops – seed production. The widely diverging interests and research activities of IHSPRG members include the physiological processes that play a role in seed production, the breeding of cultivars with high potential seed production, the culture and management of the crop and pollinator to optimize seed production, the establishment of standards to measure seed quality, and the marketing of forage seeds in a global economy. All these many aspects are included in these two volumes on forage seed production.

Each volume comprises a number of chapters which attempt to assemble the chronological sequence that documents the establishment of the crop in the field, the determination and realization of its yield potential, and ultimately the sale of the final product – the seed. Experts in various fields and often from different countries and/or continents have collaborated in writing each chapter. All this would not have been possible without the hard work, dedication and enthusiasm of the authors who generously gave of their time and effort. The authors come from both the public (government research laboratories and universities) and private sectors (small research laboratories or multinational seed companies) and, despite their many other commitments, have managed to work together and meet the deadlines. IHSPRG thanks them all for their contribution. The editors would also like to thank the Peace River Agriculture Strategic Planning Society (Fort St John, British Columbia, Canada) and the Oregon Tall Fescue Commission (Salem, Oregon, USA) for contributions towards the cost of printing the colour plates.

Daphne T. Fairey
John G. Hampton

Temperate Forage Seeds: An Introduction

D.T. Fairey[1] and J.G. Hampton[2]

[1]Northern Agriculture Research Centre, Agriculture and Agri-Food Canada, Beaverlodge, Alberta T0H 0C0, Canada; [2]Seed Technology Centre, Department of Plant Science, Massey University, Palmerston North, New Zealand

1.1 THE MULTIPLE USES OF FORAGE SEED

The selected title for this volume, 'Forage Seed Production', does not fully convey the relevance of the subject matter throughout the world. The plants that we refer to as forage or herbage grasses and legumes are now actually utilized and exploited in a multiplicity of ways, and these horizons are likely to continue to expand as we address the challenges of the future. In the narrow sense, the terms 'forage' and 'herbage' simply refer to crops grown and fed to livestock. However, a much broader interpretation of these terms is becoming increasingly necessary as the intended use for many of these grasses and legumes has often little to do with the feeding of animals. Many of these crops are grown specifically for amenity/turf purposes on sports fields, golf courses, lawns, or for landscaping, control of soil erosion and soil amelioration, while others are being used for land reclamation, revegetation or ecological repair following various types of industrial use of the land. New economic uses are emerging as sources of fibre for the manufacture of paper and building materials, or as direct or indirect sources of fuel (biomass energy, ethanol), human food (alfalfa sprouts, protein extracts), and medical/pharmaceutical products (hormone therapy, health preparations). Perhaps it is now time for the scientific community to develop an alternative terminology to replace our somewhat restrictive, current descriptors of 'forage' and 'herbage' for plants whose seed and vegetative growth are utilized for such a diversity of purposes. Such an expanded terminology may eventually help to convey the 'real importance and value' of these plants to the daily lives of mankind, something that has been long overshadowed scientifically and politically by the attention given to annual cereal and oilseed crops.

The forage seed industry is relatively small and quite specialized, but it is of critical importance to mankind in many direct and indirect ways. The vast number of species and the multiplicity of end-uses for forage crops makes the production of their seeds an important component of the global agricultural economy. Many of the benefits of growing forage seed cash crops do not emanate directly from the sale of the seed, but from the byproducts that must be removed from the production fields, such as the forage seed straw that can be used for industrial processing or for livestock feed/bedding, and from the residual effects on soil health and nutrition for subsequent annual grain/oilseed crops.

When consideration is given to all these direct and indirect benefits of growing forage seed crops, it is reasonable to suggest that this sector of the agricultural economy has been neglected, from a research and development viewpoint, in many countries for many years; this is, to a large extent, because of the great diversity of forage-based commodities (e.g. meat, milk, wool, etc.) that have little direct connection to the forage seed industry in the eyes of the public. This will be apparent from the paucity of scientific information available on many topics addressed in this book; many challenges remain for improving our understanding of the seed producing processes in so-called forage or herbage crops. Hopefully, this book will provide a stepping stone for future endeavours that will enable mankind to more fully exploit the characteristics and capabilities of forage seed crops.

1.2 TEMPERATE FORAGE SPECIES

For forage seed production, genera usually accepted as temperate plants include *Agropyron, Agrostis, Arrhenatherum, Bromus, Cynosurus, Dactylis, Festuca, Holcus, Lolium, Phalaris, Phleum* and *Poa* amongst the grasses, and *Coronilla, Hedysarum, Lotus, Medicago, Onobrychis, Ornithopus, Trifolium* and *Vicia* amongst the forage legumes (Kelly, 1988; Anon., 1995). For example, New Zealands' AgResearch Grasslands currently has available cultivars of 15 species from within 10 of these grass genera, and 11 species from within seven of these legume genera, either for forage or amenity purposes.

Internationally however, the forage seed trade is dominated by only a few species. Of the 157,400 t of forage seed produced in the European Union (EU) in 1994 (Kley, 1996), 50% was of ryegrasses (*Lolium* spp.), 17% of red fescue (*Festuca rubra* L.), 13% of legumes (mostly lucerne (*Medicago sativa* L.)) and 20% of other grasses (*Poa* spp., cocksfoot (*Dactylis glomerata* L.) and other *Festuca* spp.). This production was in response to demand; 91,000 t of ryegrass seed was purchased in the EU in 1994, 70% of the EU's total grass seed sales (Kley, 1996). In both Oregon (USA) and New Zealand, ryegrasses account for 55–70% of annual forage seed production (Hampton, 1991), with grasses such as tall fescue (*Festuca arundinacea* Schreb.), cocksfoot and Kentucky bluegrass (*Poa pratensis* L.) also important in Oregon, and white clover (*Trifolium repens* L.) of significance in New Zealand (see Chapters 11, 12 and 21).

While many of the 12 grass and eight forage legume genera that are included as temperate forages are important in some countries, or within regions of one country, it is obvious that they are not all of international significance. Therefore, this book concentrates primarily on *Dactylis, Festuca, Lolium* and *Poa* spp. among the grasses, and *Lotus, Medicago* and *Trifolium* spp. among the forage legumes.

1.3 THE BEGINNING OF A FORAGE SEED INDUSTRY

The history of temperate forage seed production can be traced back to the nineteenth century when farmers required more seed of grass and legume species to renew the extensive livestock pastures that formed the backbone of industrial development. In Europe, shattered seed was collected from grazing stands or hay barns, or from road sides and other public areas by 'stem cutters'; people who cut ripe ears of different species and kept the seed of each species separate. However, there was little commercial exchange of forage seeds, most re-sowings using seeds from local ecotypes (Hampton, 1991).

The realization of the need for improved forage production and quality led to the beginning of cultivar improvement programmes in many countries in the 1920s (Hides and Desroches, 1989; Rolston and Clifford, 1989), and therefore the need to produce seed true to its original description. Kley (1996) describes this as the period of birth of the majority of Europe's forage plant breeding and seed production companies, and the establishment of the first standards for seed quality which finally led to Seed Acts and seed certification.

Developments in Europe were paralleled in North America and Australasia, and forage seed production became an important industry in many countries (see Part Two for case histories of 11 forage species). Whyte (1937) edited Bulletin 19 of the British Imperial Bureau of Plant Genetics herbage publication series, and described it as 'an international exchange of opinions and experiences on the technique of producing seed of Gramineous herbage and forage plants'. Average seed yields were then around 400–500 kg ha^{-1} (Table 1.1), but as the upper ranges of yields demonstrate, in the 1930s there were obviously some seed producers with much expertise (Hampton, 1991).

The benefits of 60 years of plant breeding, seed production research and new technologies have, depending on the species, lead to the doubling or tripling of the average seed yields presented in Table 1.1. However among the seed production problems discussed by Whyte and his colleagues in 1937 were: cover crops versus no cover crops; time of sowing; fertilizer type, rate and timing; weed control; insect

Table 1.1. Seed yields (average and range) for some temperate forage grasses (From Whyte, 1937).

Species	Country	Seed yield (kg ha^{-1})	
		Average	Range
Lolium perenne L.	Northern Ireland	560	400–800
Dactylis glomerata L.	Germany	400	200–700
Phleum pratense L.	Scotland	560	370–1004
Festuca pratensis Huds.	Germany	400	200–900
Festuca rubra L.	Germany	350	200–700
Agropyron cristatum (L.) Gaertn.	USA	500	330–1120
Cynosurus cristatus L.	Northern Ireland	450	300–700
Bromus inermis Leyss.	Canada	250	112–448

and fungal damage and control; seed shedding; method and time of harvesting; meeting seed certification standards; instability of prices for seed produced – all issues still being investigated today and discussed in the relevant chapters of this book.

The original localized activity of seed collection has now become international in scope and has gradually evolved into a multi-billion dollar industry of selected cultivars of a number of species that have been bred for specific end uses. International agreements/associations such as the International Union for the Protection of New Varieties of Plants/Union Internationale pour la Protection des Obtentions Végétales (UPOV) and the International Association of Plant Breeders for the Protection of Varieties (ASSINSEL) provide a framework for the protection and ownership of germplasm. In many instances, improved cultivars bred in one region may be sent to another region for seed multiplication, and the seed progeny is returned to areas where the cultivar is adapted for its intended use.

1.4 REGIONS OF PRODUCTION

1.4.1 The 'Natural' Advantage

While most grass and legume species become reproductive and can produce some seed in most temperate regions, large areas of forage seed crops have been concentrated within certain regions where nature provides ideal conditions for optimal seed yield, harvesting and/or storage. In Europe (Kley, 1996), the main grass seed producer with 40% of the total area is Denmark, followed by Germany (20%) and the Netherlands (16%). Italy has the largest area in legumes (38%), and France (28%) and Spain (18%) are the other major producers. Furthermore, specialized production areas are recognized within each country. For instance, in France, specialized production areas are recognized and ecotypes of many species have been named after the region of production, such as 'trefle d'Issoudun' clover from the central region in Issoudun where most of that country's clover seed is grown (see Chapter 20).

In North America, it is no accident that a majority of the multi-billion dollar forage seed industry of the USA is located in the Pacific Northwest. A significant portion of both the grass (see Chapter 12) and legume seed industries (see Chapter 6) moved to the western regions of North America during the 1940s to take advantage of the more dependable climatic conditions that favour flowering, pollination, seed development and harvest. In Canada, almost half the total annual production of about 44,000 t of forage seeds is from the Peace River region of Alberta and British Columbia (see Chapters 13 and 19). In New Zealand, of the approximately 35,000 ha that are committed to herbage seed production annually, a majority of the country's seed is produced on the Canterbury Plains of the South Island (see Chapter 11). Other small areas of seed production are on the lower east coast of the North Island and in the southern areas of the South Island. Production in Australia is concentrated in the three southern states of New South Wales, Victoria and South Australia.

1.4.2 The 'Other' Advantage

There are a number of other factors such as government subsidies, demands for seed of other crops, urban encroachment, and regulations governing farming practices that determine where and when forage seeds are produced. The financial remuneration under the current European Union Common Agricultural Policy (CAP) makes it more attractive for farmers to grow arable crops such as cereals and oilseed rape rather than forage seeds (see Chapter 11). The CAP can have considerable influence over which crops are grown for seed within its member states, with any changes in crop support, quota systems, livestock reductions, etc., having a significant effect within a relatively short time. In both Canada and the USA, wheat prices in 1996 were at their highest levels in 20 years and there was a reduction in the number of hectares sown to forage seed crops.

Until the early 1990s, the western European countries found it difficult to compete for seed contracts with the Comecon whose member countries (in the former USSR and eastern Europe) received government subsidies for their forage seed production inputs (see Chapter 15). Multinational seed companies located some of their production in Comecon countries, because of the subsidized production costs (see Chapter 14). The immediate effect of the disbanding of the Comecon was a disastrous drop in prices, because of the release of a large surplus in some countries such as the former GDR into the EU (Kley, 1996).

Forage seed production in some countries such as Canada, New Zealand, Australia and the USA is not subsidized, and farmers in these countries are often handicapped when they compete for contracts with their EU counterparts. It is then that other factors come into play. One big advantage in North America is the large farm sizes that, in addition to providing the required isolation from other contaminating crops, permit seed companies to centralize their contracts with a few specialized farmers. In addition, in some areas, the relatively lower land costs, as compared to those in western Europe, lower the production costs. The New Zealand and Australian advantages are those of island nations, well removed from the pests and diseases from other land masses, their mild climates, and their 'out-of-season' production capability for countries in the northern hemisphere.

The benefits of the culture of forage crops to the soil makes forage seed a valuable commodity in the implementation of government-subsidized conservation programmes. Canadian seed producers in the Peace River region of Canada have the climatic conditions that provide for inexpensive long-term storage of forage seed that has often catered to sudden demands brought about by the implementation of conservation programmes (see Chapters 13 and 19).

1.4.3 Other Areas of Production

While four areas of the world (North America, Europe, New Zealand, Australia) produce nearly all the temperate forage seeds that are marketed (Hides and Desroches, 1989), the production of temperate forage seeds is also important to farmers and agro-industry in other regions of the world. There are established temperate forage seed industries in Argentina, Brazil, Turkey, Iran, India, Japan, China and Russia for example (Rolston *et al.*, 1993; Sinizyna, 1996;

J.G. Hampton, Palmerston North, 1996, personal communication), producing primarily perennial ryegrass, tall fescue, cocksfoot, white clover, red clover and lucerne seed lots.

1.5 CHALLENGES FOR THE FUTURE

Seed crop management is an important factor determining both the potential seed yield of a species/cultivar and the realization of that potential; an important component of such management is the selection of suitable sites for production. In most instances, the areas best suited for fodder production are often less suitable for seed production. For successful seed production, special attention has to be given to the crops cultivated previously on the same site, as buried seeds can pose a serious threat to seed purity. This is especially true with legumes where the hard seed of previous crops can remain viable within the soil for many years.

The present standards for seed purity necessitate the use of herbicides. In most countries, the total area devoted to forage seed production is too small for chemical companies to undertake efficacy studies for recommendation of herbicides specifically for forage species, because of the low probability of a pay back on their investment. Very often, potentially useful herbicides must be identified in more widely grown crops where their effectiveness can form the basis of advice for forage seed crops. Unfortunately, herbicides recommended in one country or region within a country may not be recommended for use in another and, as legislation to minimize environmental hazards becomes more restrictive, the chemical methods available for controlling weeds are likely to become more limited.

The stage in forage seed production at which reasonable quantities of good quality seed can be obtained for commonly grown species most of the time has now been reached. However, the chapters of this monograph reveal the lack of a basic understanding of the processes involved in maximizing seed production. Most of the success to date is based on experiments with a relatively arbitrary combination of treatments, predominantly changes in plant management such as plant spacing or fertilizer applications that have been chosen on the basis of previous literature, and evaluated in different environments and with different species and cultivars. A major challenge is to produce an integrated understanding of the physiological responses of grass and legume seed crops and how they might be amenable to modification.

A further challenge for any industry is the ability to meet market demands. While the bulk of present world demand is for ryegrasses, fescues and white clover, there is increasing interest in plant species adapted to low rainfall areas, low to moderate fertility soils, and which require minimal fertilizer inputs (Hampton et al., 1990). Legumes such as *Medicago* spp., *Trifolium* spp. other than red and white clover, *Vicia* spp., *Melilotus* spp. and *Lotus* spp., and grasses such as *Holcus*, *Agropyron*, *Arrenatherum* and *Phalaris* spp. among others, have particular features appropriate for these environments (Hampton, 1991). In conjunction with improving our understanding of the physiological responses of grass and legume seed crops, there is a need to overcome production constraints for the new species and cultivars required by the consumer, and importantly to link seed production goals with environmental policy goals, so that future developments are in tandem, and not in conflict.

REFERENCES

Anon. (1995) *The Grasslands Range of Forage and Conservations Plants.* AgResearch Grasslands, Palmerston North, New Zealand.

Hampton, J.G. (1991) Temperate herbage seed production: an overview. *Journal of Applied Seed Production* 9 (Suppl.), 2–13.

Hampton, J.G., Hill, M.J. and Rolston, M.P. (1990) Potential for seed production of non-traditional herbage species in New Zealand. *Proceedings of the New Zealand Grassland Association* 51, 65–70.

Hides, D.H. and Desroches, R. (1989) The role of seeds in forage production – factors limiting optimal utilization. *Proceedings of the XVI International Grassland Congress,* pp. 1777–1784.

Kelly, A.F. (1988) *Seed Production of Agricultural Crops.* Longman Group UK, Essex.

Kley, G. (1996) Seed production in grass and clover species in Europe. In: Schoberlein, W. and Forster, K. (eds) *Proceedings of the Third International Herbage Seed Conference,* June 18–23, Martin-Luther-Universitat, Halle-Wittenberg, Germany, pp. 12–22.

Rolston, M.P. and Clifford, P.T.P. (1989) Herbage seed production and research – a review of 50 years. *Proceedings of the New Zealand Grassland Association* 50, 47–53.

Rolston, M.P., Lill, G.W., Brougham, R.W. and Jishi, X. (1993) Model herbage seed farm establishment in south-west China, Guizhou Province. *Proceedings of the XVII International Grassland Congress,* pp. 1763–1764.

Sinizyna, S. (1996) The herbage seed growing in Russia. In: Schoberlein, W. and Forster, K. (eds) *Proceedings of the Third International Herbage Seed Conference,* Martin-Luther-Universitat, Halle-Wittenburg, Germany, pp. 197–201.

Whyte, R.O. (1937) *Production of Grass Seed.* Imperial Bureau of Plant Genetics, Bulletin 19, Herbage Publication Series, Aberystwyth.

Reproductive Development and the Establishment of Potential Seed Yield in Grasses and Legumes

T.S. Aamlid[1], O.M. Heide[2], B.R. Christie[3] and R.L. McGraw[4]

[1]*The Norwegian Crop Research Institute, Apelsvoll Research Centre, Division Landvik, N-4890 Grimstad, Norway;* [2]*Department of Biology and Nature Conservation, Agricultural University of Norway, N-1432 Ås, Norway;* [3]*Agriculture and Agri-Food Canada, PO Box 1210, Charlottetown, Prince Edward Island, Canada;* [4]*Department of Agronomy, University of Missouri, Columbia, MO 65211, USA*

2.1 INTRODUCTION

The development of a forage seed crop is commonly divided into two stages – the establishment of the seed yield potential and the utilization of this potential (Hebblethwaite *et al.*, 1980; Hampton, 1990). This chapter covers the first of these stages, following the seed crop through shoot formation, floral induction, floral initiation and differentiation. In the concluding section some implications for the location and management of seed crops are drawn. By highlighting the morphological structures and physiological responses, the objective is to provide the reader with a basis for interpretation of results from applied seed production research.

2.2 SHOOT FORMATION

2.2.1 Shoot Formation in Grasses

While the population density in dicotyledonous canopies is usually recorded as plant number per unit area, the density of grass swards can only be described adequately by shoot number. Lateral shoots always originate from axillary buds which are laid down in acropetal succession from the apex. Leaf and shoot development are highly synchronized, buds being initiated at the same rate as leaf primordia but from deeper, subhypodermal tissues at the opposite side of the apex, and usually two or three phytomers further down (Langer, 1979; Jewiss, 1981). Whereas shoot formation in ryegrasses (*Lolium* spp.), fescues (*Festuca* spp.) and cocksfoot (*Dactylis glomerata* L.) often commences at the coleoptile node, timothy (*Phleum pratense* L.) shoots seldom emerge until the first or second leaf node (Patel and Cooper, 1961; Ryle, 1964a). Grass shoots mostly arise from the nodes of unelongated, vegetative stems. However, especially in lodged seed crops, it is not uncommon to find subsidiary shoots emerging from the elevated culm nodes of elongated, reproductive tillers (e.g. Minderhoud, 1978).

Shoot types: tillers, rhizomes and stolons

Buds in leaf axils of tussock-forming grasses always develop aerial, photosynthesizing shoots. Such shoots are referred to as tillers. The tufted growth habit of some species is due to the fact that tillers develop intravaginally, i.e. within the sheaths of their accompanying leaves, and do not appear externally until near the base of the subtending lamina (Langer, 1979).

Rhizomatous or stoloniferous grasses also develop photosynthesizing, more or less intravaginal tillers, but in addition, axillary buds, especially at the lower apex nodes, tend to produce creeping shoots which eventually become important both for the vegetative expansion of the plant and as storage organs. Such shoots may either grow on the soil surface, like the stolons of rough-stalked meadowgrass (*Poa trivialis* L.) and creeping bent (*Agrostis stolonifera* L. var. *palustris* (Huds.) Farw. (syn. *A. palustris* Huds.)), or they grow underground, such as the rhizomes of smooth-stalked meadowgrass (*Poa pratensis* L.), common bent (*Agrostis capillaris* L. syn. *A. tenuis* Sibth.), smooth bromegrass (*Bromus inermis* Leyss.), reed canarygrass (*Phalaris arundinacea* L.) and the creeping subspecies of red fescue (*F. rubra* L. ssp. *rubra* and *trichophylla*). In either case the creeping shoots display a horizontal growth habit for a certain period of time before they turn upward and develop aerial tillers. The ability of rhizomes to form aerial tillers seems to be stimulated by short photoperiods (Aamlid, 1992), an ample supply of nitrogen, the exposure of rhizome tips to light and decapitation of main tillers, but not by ordinary defoliation treatments (Nyahoza *et al.*, 1974).

Tiller hierarchies and interdependence

Tillers originating from the axillary buds on the main shoot are commonly classified as primary tillers. Secondary tillers arise from buds on the primary tillers, tertiary tillers from buds on the secondary tillers, and hence, a complex system of tillers of various orders develops on the same plant. The rate of tiller production in a grass plant is usually exponential until restricted by some limiting environmental factor or flowering (Langer, 1979).

A number of studies using radioactive carbon have shown that tillers on the same plant maintain close vascular connections throughout their life cycle (e.g. Marshall and Sagar, 1968; Ryle, 1970; Clemence and Hebblethwaite, 1984). Timothy tillers import assimilates from their originating shoots until they have produced four to five leaves and some axillary roots (Williams, 1964), and they do not themselves subtend tillers until they have reached this stage (Ryle, 1964a).

Environmental control of tillering

Low light intensities reduce tillering, presumably as a result of lower assimilate availability (e.g. Patel and Cooper, 1961; Ryle, 1961; Auda *et al.*, 1966). Light intensity also seems to be the most important factor governing tiller mortality (Ong, 1978). However, before light intensity is severely reduced in a grass canopy, phytochrome pigments perceive a reduction in the ratio of red (R) to far red (FR) light at plant bases, and this light quality response causes reduced tillering in a closing sward (Casal *et al.*, 1985, 1987).

Most experimental results show a negative effect of daylength extension on tillering (e.g. Newell, 1951; Auda *et al.*, 1966; Heide, 1982; Aamlid, 1992). The reduction is more pronounced in ecotypes originating at low than at high latitudes (Håbjorg, 1976; Hay and Pedersen, 1986), and there are also fewer negative effects of long photoperiods on tillering at low energy levels (Templeton *et al.*, 1961) or in cold environments (Håbjorg, 1976; Heide, 1984; Aamlid, 1992). Under field conditions it is often difficult to distinguish between a direct effect of photoperiod on tillering and indirect effects through reproductive development and stem elongation in late spring and early summer.

Langer (1979) indicated temperature optima for tillering ranging from 18 to 24°C in perennial ryegrass (*Lolium perenne* L.) to 24–29°C in cocksfoot. The latter may well be correct for Mediterranean cocksfoot, but not for high latitude cultivars which produced more tillers at 15 than at 25°C (Ostgård and Eagles, 1971). Similarly, Heide (1982) noted an increase in tillering for Norwegian and British timothy cultivars as temperature was raised from 12 to 18°C, but there was no further increase at 21°C. In any case, the optimum temperature seems to be lower for tillering than for leaf production (Langer, 1979; Ryle, 1964a; Fig. 2.1). High night temperatures are particularly depressive to tillering (Alberda, 1957; Robson, 1973; Håbjorg, 1976), possibly because of higher respiration losses.

Drought caused a more dramatic decline in tillering than in leaf appearance in field experiments with several grasses (Norris, 1982). Trials in meadow fescue (*Festuca pratensis* Huds.) showed a significant reduction in tillering due to water deficiency in late summer (late August and early September), but this was not reflected in seed yield in the subsequent year (Jonassen, 1992).

Nitrogen application increases tillering both in spaced plants and in swards (e.g. Auda *et al.*, 1966; Lambert, 1967b; Aamlid, 1993). An ample supply of nitrogen reduces tiller death of individual plants (Ong, 1978), but this may be different in dense stands where nitrogen seldom influences the proportion of dead tillers (Langer and Lambert, 1959). Supraoptimal application of nitrogen to grass seed crops often leads to vigorous tillering in early spring, but this is commonly followed by a higher mortality of both vegetative and reproductive tillers during stem elongation (Hebblethwaite and Ivins, 1977).

Defoliation, decapitation and internal control of tillering

Repeated defoliation of single grass plants usually reduces tillering because of carbohydrate exhaustion (e.g. Alberda, 1957; Hume, 1991). This situation is often reversed in dense grass swards where cutting stimulates tiller outgrowth because of more light and higher R/FR ratios at plant bases. Hare (1993) reported an increase in tiller number of New Zealand tall fescue (*Festuca arundinacea* Schreb.) after grazing or mowing in late autumn, but Aamlid (1993) observed no positive effect of defoliation on tiller numbers in Norwegian smooth-stalked meadowgrass (*Poa pratensis* L.), except in the third successive seed-harvest year when plants had acquired a large storage pool of reserve carbohydrates.

If, in mild winter climates, grazing continues for too long into early spring, grass plants are not only defoliated, but also decapitated. From observations that spring grazing (beyond mid-March) caused apical removal in seed crops of Italian ryegrass (*Lolium multiflorum* Lam. var. *italicum*), Young (1980) explained the resultant increase in tiller number as due to release of apical control. It has, however, not been unequivocally established whether apical dominance, primarily mediated by the plant hormone auxin (Thimann and Skoog, 1934), has the same important role in regulating tillering in grasses as it has for branching in dicotyledons (Skinner and Nelson, 1992).

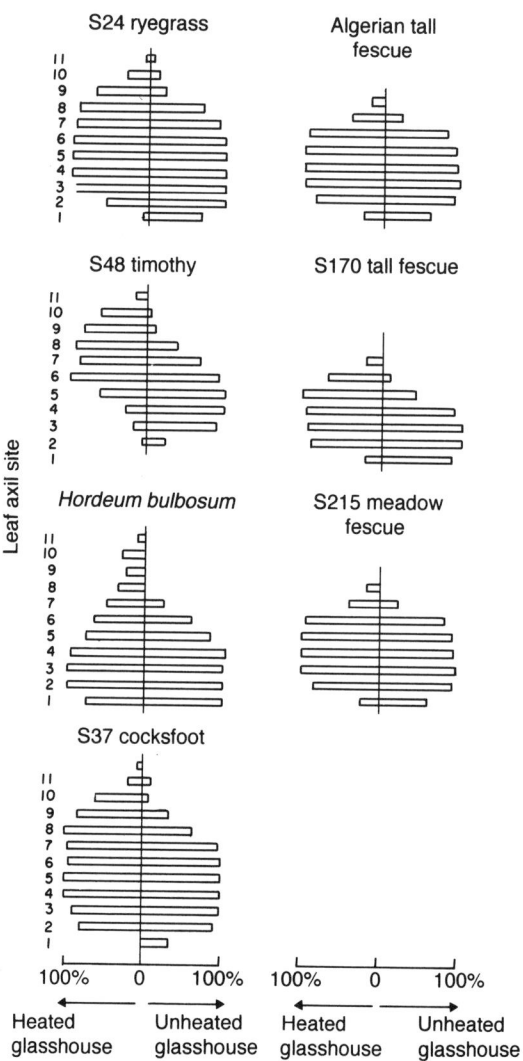

Fig. 2.1. Effect of cultivar and temperature on tillering from successive leaf axil positions in seven perennial grasses in a heated (minimum temperature 15°C) and an unheated glasshouse during a British winter. (From Ryle, 1964a, with permission of the author and Blackwell Science Ltd.)

Tiller longevity and tiller demography

While spring-formed tillers of perennial grasses often die before the end of the growing season, autumn-formed tillers usually survive the winter and make a strong contribution to seed yield in the following year. Tillers of timothy seldom live for more than 1 year, but tillers of meadow fescue and perennial ryegrass may stay alive for 2 or 3 years if the stand is kept open (Jewiss, 1966; Colvill and Marshall, 1984). Occasionally, tillers of meadow fescue have been reported to produce panicles up to 33 months after their formation (Lambert and Jewiss, 1970). Tillers of more than 16 months of age, however, seldom contribute significantly to seed yield.

Many seed production areas, such as Western Oregon, New Zealand, The Netherlands, and to some extent Denmark, have mild winter climates, allowing tiller numbers in seed crops to increase continuously from planting throughout the following winter (Langer *et al.*, 1964; Meijer, 1984; Colvill and Marshall, 1984). In other areas, e.g. Canada, Minnesota and most of Scandinavia, tillering is usually terminated by frost in late autumn (Aamlid, 1996). In either case, early spring appears to be the period with most abundant tillering. Stem extension may be more important than changes at the stem apex in restricting assimilate availability to lateral buds at this stage (Jewiss, 1972) but, for late-flowering timothy, it has also been observed that suppression of tillering occurs before double-ridge formation (Langer *et al.*, 1964). This is compatible with the view that a rapidly expanding leaf canopy reduces the R/FR ratio and the light intensity (in this order) around plant bases before stem internodes start to elongate (Simon and Lemaire, 1987).

The suppressive effect of apical development and stem elongation on tillering gradually dissipates from ear emergence until seed ripening. An abundant supply of water and nitrogen often result in lodging, facilitating light penetration to lower buds and consequently a flush of new tillers during this period (e.g. Lamp, 1952; Hebblethwaite and Ivins, 1977). Although some of these tillers may produce inflorescences in a subsequent year, they are mostly unwanted in seed production because they compete with developing seeds for assimilate (Clemence and Hebblethwaite, 1984) and complicate harvesting operations.

Tiller elongation

In most perennial grasses tiller elongation only occurs in association with reproductive development. Notable exceptions are smooth bromegrass and reed canary-grass in which vegetative tillers often elongate in long photoperiods without previous exposure to inductive conditions (Newell, 1951; Klebesadel, 1970; Heide, 1984, 1994b; Aamlid *et al.*, 1995). Such elongated tillers cannot be induced at a later stage and are therefore deleterious to seed production. Elongation of vegetative timothy tillers in the sowing year is usually indicative of high temperature and/or low light intensities during floral induction (Langer, 1956; Heide, 1982).

2.2.2 Shoot Formation in Forage Legumes

Legume shoots always originate in buds which, except for the primary shoot, are located in leaf axils. Legume shoots vary greatly among species in length, diameter, amount of branching, and woodiness (Nelson and Moser, 1995). Branching of a primary shoot is always preceded by crown formation.

Crowns, aerial shoots, stolons and rhizomes

Crowns are formed by the process of contractile growth. About 6–8 weeks after emergence, lateral growth of the hypocotyl and upper primary root causes these structures to thicken and shorten. Contractile growth draws the first node and the developing crown about 0.5 cm below the surface for white clover (*Trifolium repens* L.), about 1 cm for red clover (*Trifolium pratense* L.) and birdsfoot trefoil (*Lotus corniculatus* L.) and about 2 cm for lucerne (*Medicago sativa* L.) (Nelson and Moser, 1995). The crown and tap root store carbohydrates, proteins and other compounds for shoot growth and for winter survival (Smith, 1962; Hendershot and Volenic, 1993).

Aerial shoots, stolons and/or rhizomes emerge from axillary buds on the crown. Aerial shoots consist of nodes and internodes with leaves attached at the nodes by petioles. In lucerne both the primary and axillary shoots elongate, thus elevating the terminal buds to the upper part of the canopy (Nelson and Moser, 1995). By contrast, according to Taylor and Smith (1995), the primary shoot of red clover does not elongate, but secondary shoots exhibit internode elongation either in the sowing year (mainly early cultivars) or in the subsequent year (mainly late, winter-hardy cultivars).

Stolons are shoots growing along the surface of the soil. Occasionally, stolons have been observed in red clover (Smith and Bishop, 1993) but, in general, their development is far more prominent in white clover than in any other temperate forage legume. In this species, stolons begin to form from axillary buds as growth of the primary stem slows some 6–8 weeks after seedling emergence (Gibson and Cope, 1985). Stolons then become the unit for vegetative growth, spread, and survival of white clover (Pederson, 1995).

The apical bud of a white clover stolon normally contains six or seven leaf primordia. In a vegetative stolon the youngest axillary bud is usually discernible just distal to the third youngest leaf primordium (Fig. 2.2), the two youngest leaf primordia subtending no buds. Such axillary buds either remain dormant or develop into branch stolons. With the transition to reproductive growth, a precocious bud forms in the axil of the youngest leaf primordium, and this eventually develops into an inflorescence (Thomas, 1980b, 1987a,b).

Rhizomes are subterranean shoots with nodes, internodes, scale leaves and adventitious roots. Branch rhizomes can develop from rhizome nodes. Rhizomes of big trefoil (*Lotus uliginosus* Schkuhr. syn. *Lotus pedunculatus* Cav.) develop from the lower crown in autumn (Wedderburn and Gwynne, 1981). In spring, aerial shoots are formed from buds on the rhizomes. Crown development is weak in big trefoil, and rhizomes provide the regrowth sites, carbohydrate storage, and colonization potential (Sheath, 1975). Rhizomes also occur in some cultivars of birdsfoot trefoil (Li and Beuselinck, 1996) and lucerne (Teuber and Brick, 1988).

Environmental control of shoot growth

Most forage legumes are 'sun' species. Although red clover is usually regarded as more shade tolerant than lucerne and birdsfoot trefoil (Rhykerd *et al.*, 1959), low

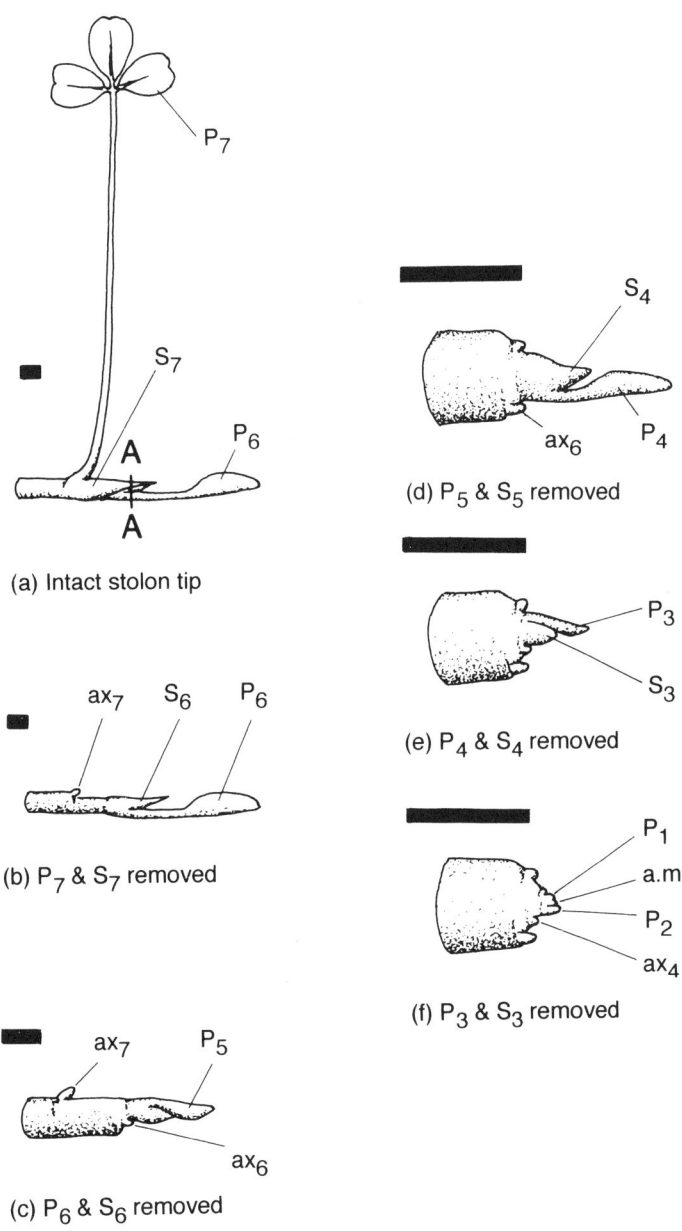

Fig. 2.2. a–f. Structure of the tip of a white clover stolon as revealed by successive stages of dissection. Leaf primordia and leaves are labelled P_1–P_7 in sequence from youngest to oldest, their axillary buds are labelled ax_1–ax_7, respectively, their stipules S_1–S_7 and the apical meristem a.m. The scale is indicated by horizontal bars, each of which represents 4 mm. (From Thomas, 1987a.)

light intensity resulted in etiolated red clover seedlings with low numbers of buds and shoots (Ludwig *et al.*, 1953; Cumming, 1959).

Increasing photoperiods normally decrease leaf number but increase leaf area, shoot number and stem length in red clover (Ludwig *et al.*, 1953; Cumming, 1959; Bowley *et al.*, 1987). Due to stronger suppression from the main shoot, shoot numbers may nevertheless be lower with a 24 rather than a 16 h photoperiod (Pulli, 1988). The critical photoperiod for elongation of vegetative stems in Japanese cultivars of lucerne was only 9–11 h (Susuki *et al.*, 1975), which is considerably shorter than the requirement for stem elongation in most cultivars of red clover. Stem elongation of red clover is normally associated with flower initiation in long days, but even this species can develop elongated vegetative stems if photoperiods are slightly shorter than those critical for flower initiation (e.g. 12–13 h – Bowley *et al.*, 1987).

Each legume species has an optimum temperature range for shoot growth. Shoot numbers decrease with falling temperature from 27 to 15°C in lucerne (Leach, 1971) but increase with temperatures declining from 21 to 9°C in high-latitude red clover (Lunnan, 1989). In the latter study, the number of stem branches per plant increased concomitantly from 2.1 at 21°C to 17.2 at 9°C. Higher temperature results in fewer stolon branches in white clover (Kendall and Stringer, 1985).

Internal control of shoot development in legumes

Apical meristems on white clover stolons do not inhibit buds from being initiated in leaf primordia axils on the same stolon. However, from around the time when the leaflets of their subtending leaves unfold, growth of axillary buds is repressed by the presence of an apical bud. Since excision of the youngest leaf primordia from the stolon apex removes this repression, it is possible that the apical dominance resides in the leaf primordia rather than in the apical meristem itself (Thomas, 1987a).

Longevity and shoot demography of forage legumes

Maintaining stands of forage legumes can be a problem because diseases, pests and frost often cause high plant mortality. Forage legumes can perennate by:

- Individual plant survival, like lucerne and red clover.
- Vegetative plant reproduction via rhizomes or stolons, like big trefoil and white clover.
- Natural reseeding, like common lespedeza (*Lespedeza striata* L. (*Kummerowia striata* (Thunb.) Schindler)) (Beuselinck *et al.*, 1994).

In a birdsfoot trefoil seed crop planted at 44 plants m^{-2}, Li and Hill (1988) found that the shoot population at any one time was mainly composed of shoots less than 3 months old. Shoots rarely survived more than 6 months, and plants exhibited a 'continuous replacement' growth habit as old shoots died and new shoots emerged; similar results have been reported in big trefoil (Tabora and Hill, 1991).

2.3 FLORAL INDUCTION

2.3.1 Terminology

The principal factors controlling the transition from vegetative to reproductive growth in grasses and legumes are photoperiod and temperature. Induction can be defined as the perception of the environmental signals, either directly by the apices, as for temperature, or by leaves, as for photoperiod. In the latter case, perception has to be followed by synthesis and transport of a florigenic signal(s) to the shoot apices; hence, these processes are also included in the definition (Heide, 1994a).

In many perennial species flowering has a dual-induction requirement; plants must undergo a winter influence, involving short days and/or low temperatures (step 1), before they will respond to increasing daylength and flower (step 2). Rather than referring to the first step as (true) low-temperature vernalization or 'short-day vernalization' (Evans, 1964), and the second step as initiation or realization (Calder, 1966), this chapter uses the terms primary and secondary induction to describe the two steps. This terminology, originally proposed by Blondon (1972), was adopted by Heide (1980) and has been used in induction studies with a number of grasses (Heide, 1994a). Although induction requirements are mostly less stringent in forage legumes, the same terminology is useful.

It should be emphasized that the terms primary and secondary induction, as used here, do not relate to whether or not visible changes occur in shoot apices. While concentrating on the environmental signals and their perception in this section, the nature and onset of morphological changes will be returned to later in this chapter.

2.3.2 Juvenility

Definition

Calder (1966) defined juvenility as a phase during which 'plants are insensitive to environmental conditions which later, in the mature or adult phase, promote flowering'. Applied to grasses, this definition only has relevance for seedlings, yet, Calder (1963, 1964, 1966) also admitted that it is uncertain whether juvenility in grasses is a property of every individual tiller, only of main shoots, or of the plant as a whole. Thirty years later this is still an open question and despite the work of Ikegaya (1984) and Nordestgaard (1988), it remains to be resolved whether tillers arising from established plants, e.g. after seed harvest, can be induced at a younger age or smaller size than tillers emerging from seedlings. While there is general agreement that flower induction in perennial seed crops needs to be repeated every year (Cooper, 1952; Evans, 1964; Heide, 1994a), there is also a need for clarification as to whether, and to what extent, the flowering signal(s) produced in one tiller can be transmitted to other tillers on the same plant.

Length of juvenility in various species

A few perennial grasses can be primary induced as germinating seeds. The most prominent example is perennial ryegrass (Bommer, 1961; McCown and Peterson, 1964). Among the forage legumes, 'seed vernalization' at 3°C for 15 days resulted in earlier flowering of white clover (Haggar, 1961), and similar results have been reported in red clover (Fejer, 1960).

Most investigations into the length of juvenility have been conducted with plants at the seedling stage. Calder (1963, 1964), Ikegaya (1984) and Heide (1987) found that juvenility in cocksfoot lasted for about 5 weeks from seedling emergence, and Heichel et al. (1980) reported a 4-week juvenile phase in reed canarygrass. Similar durations were suggested by Bean (1970) for tall fescue and meadow fescue. According to Meijer (1984) the unresponsive phase lasts for approximately 2 and 5 weeks in individual tillers of smooth-stalked meadowgrass and red fescue, respectively. Nittler and Kenny (1964) were able to induce flowering or visible flower buds in most plants of red clover, lucerne and birdsfoot trefoil within 5 weeks of seeding.

Juvenility may be more adequately described by leaf number, or shoot size, than by plant age. Kozumplik and Christie (1972a) reported that cocksfoot seedlings were not receptive to induction until eight leaves had appeared on the main shoot; this stage was reached within 3–7 weeks after emergence depending on temperature and photoperiod. In red clover, Jones (1974) found that an early-flowering cultivar could be induced when two to three leaves were present on the seedling, while a late cultivar needed 12–13 leaves before induction could be successful.

Unless information on the dates of shoot or tiller emergence is provided, results concerning the length of the juvenile phase must always be interpreted in light of the subsequent induction treatment. For example, the surprisingly high leaf number required in Kozumplik and Christie's (1972a) experiments with cocksfoot was probably due to the fact that plants were exposed to primary induction for 6 weeks only. Any extension of the primary induction period is likely to have shown that smaller tillers entered a generative development, as demonstrated by Heide (1987). A similar interaction between plant age and length of primary induction treatment was documented in meadow fescue (Havstad, 1996).

Underlying mechanisms for juvenility

If grasses and legumes need a certain leaf number or leaf area to overcome juvenility, this may well be interpreted as a requirement for a minimum pool of carbohydrates in order to respond to inductive conditions. The fact that only large-seeded grasses are able to respond as germinating seeds may well be taken in support of such a hypothesis (Heide, 1994a); there are, however, indications that green leaf area *per se*, rather than the storage pools of carbohydrates in stubble or underground organs, controls the responsiveness of grass tillers (Aamlid, 1996). An alternative hypothesis is that the termination of the juvenile phase relies more on the appearance of a particularly sensitive (upper) leaf than on the total leaf area

(Evans, 1969), or as shown by King *et al.* (1993) in darnel (*Lolium temulentum* L.) that juvenility is related not only to leaf perceptiveness but to apical size and sensitivity.

2.3.3 Classification of Forage Species According to Induction Requirement

Species with a single-induction requirement

A few temperate grasses flower readily in the sowing year without any exposure to low temperature or short photoperiod. This group includes the long-day annual species, such as darnel and Westerwolds ryegrass (*L. multiflorum* Lam. var. *westerwoldicum* Mansholt ex Wittmack) (Cooper, 1960; Halligan *et al.*, 1991).

One commercially important perennial species which requires long days only for flowering is timothy (e.g. Evans and Allard, 1934; Cooper, 1958; Ryle and Langer, 1963a; Heide, 1982). Critical daylengths for flower induction in this species range from 16 h in northern Norwegian cultivars (67–69°N) to about 14 h in cultivars from south Norway (59–60°N) and the UK (52–53°N) (Heide, 1982). Early-heading cultivars of American origin have shorter critical daylengths (down to 10 h in some cases) than late-heading cultivars from north Europe (Evans and Allard, 1934). High temperatures have an inhibitory effect on induction; this inhibition is strongly aggravated by even modest reductions in light intensity and may, at least in cultivars of high latitude origin, be evident at temperatures as low as 12–15°C (Heide, 1982).

Most temperate forage legumes flower in the sowing year, if the photoperiod is adequate after plants have passed the juvenile stage. Not surprisingly, this group includes annual species such as crimson clover (*T. incarnatum* L.), (Knight, 1985) and arrowleaf clover (*T. vesiculosum* Savi.), (Ball *et al.*, 1974), but also commercially important perennial legumes such as lucerne (Nittler and Kenny, 1964; Fick *et al.*, 1988), red clover (e.g. Aitken, 1964; Bula, 1960, 1969; Bowley *et al.*, 1987; Lunnan, 1989), alsike clover (*T. hybridum* L.), birdsfoot trefoil (Joffe, 1958; McKee, 1963) and vetches (*Vicia* spp.). Admittedly, a period of exposure to low temperatures and/or short days may to some extent hasten flowering and increase the proportion of flowering plants in these legumes (Bula, 1969; Lunnan, 1989), but this does not qualify them as species with double-induction requirements. These legumes usually have critical photoperiods of around 14 h, although some species and cultivars flower with shorter daylengths. High latitude cultivars of red clover usually require 16 h (Schulze, 1957), and even in cultivars from lower latitudes, more flowers were produced at 16 than at 14 h (Ludwig *et al.*, 1953). Furthermore, birdsfoot trefoil grown at 14 h eventually produced some flowers, but the most flowers, and the minimum time to flowering, occurred at 16 h (McKee, 1963).

High temperature and increasing daylength accelerate reproductive development in forage legumes (e.g. Fick *et al.*, 1988). In most species the final number of inflorescences also increases with daylength up to 18 h or more, and with increasing temperature up to at least 20°C (e.g. Puri and Laidlaw, 1984). There is, however, also one report indicating that flowering intensity may diminish with increasing temperatures in high-latitude ecotypes of red clover (Lunnan, 1989).

Species with a dual-induction requirement

Most temperate, perennial grasses have both a primary-induction requirement for low temperatures and/or short days, enabling the plant to initiate floral primordia either directly in short days or after transition to long days, and a secondary-induction requirement for long days allowing for inflorescence development and culm elongation (Fig. 2.3). Such plants may correctly be referred to as short-long-day plants, or as long-day plants with a vernalization requirement (Heide, 1994a).

Primary induction. Temperature and daylength are highly interactive in the primary-induction process. While most species are day-neutral at low temperatures (0–6°C), photoperiod becomes increasingly important as temperature rises up to a certain critical level, above which primary induction is inhibited irrespective of photoperiod and time of exposure (Table 2.1).

Among dual-induction grasses, northern cultivars of smooth-stalked meadowgrass exhibit a minimal requirement for primary induction. In continuous light, 8 weeks' exposure to 3°C sufficed to produce 90% ear emergence in the subarctic cultivar, Holt, from northern Norway (Heide, 1980). The corresponding

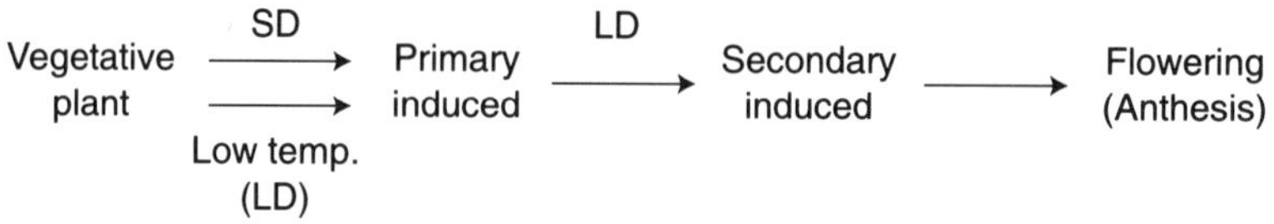

Fig. 2.3. Dual induction pathway for perennial grasses. SD = short days, LD = long days.

Table 2.1. Primary induction requirements for flowering in some temperate perennial grasses. Ranges of exposure time requirements indicate variation with temperature and geographic origin of ecotype/cultivar. (From Heide, 1994a.)

Species	In short days (< 12 h)		In long days (> 16 h)	
	Temperature (°C)	Exposure (weeks)	Temperature (°C)	Exposure (weeks)
Poa pratensis	3–18	6–10	3–12	8–12
Alopecurus pratensis	6–18	6	6–15	6–8
Bromus inermis	6–24	4–6	no induction	
Dactylis glomerata	9–21	8–10	0–3	> 20
Phleum alpinum	3–15	9–12	3–12	12–14
Agrostis capillaris	3–12	15	3–6	15
Phalaris arundinacea	3–15	12–18	no induction	
Lolium perenne	3	12–16	3	12–16
Festuca pratensis	3–15	16–20	3–12	18–20
Festuca rubra	6–15	12–20	3–12	20

requirement in the Scandinavian cultivars, Norma and Atlas, was 12 weeks, while the Swedish cultivar, Fylking, did not flower even after 20 weeks of low temperature vernalization. Although subarctic and temperate cultivars were primary induced in short days at temperatures up to 18 and 12°C, respectively, too long an exposure to short days and moderate temperatures reduced flowering in the former group.

Unlike most other species with dual induction, smooth bromegrass has a specific short-day requirement which cannot be replaced by low temperature (Heide, 1984). Critical photoperiods for primary induction at 15 and 24°C, respectively, were 13.5 and 12 h in the North American cultivar, Manchar, and 14.5 and 13 h in the Norwegian cultivar, Løfar (Fig. 2.4). Within the optimal temperature range (15–21°C), 4–6 weeks of a 10 h photoperiod sufficed for primary induction.

While earlier reports (e.g. Gardner and Loomis, 1953; Jutras, 1965) stated that low temperature was obligatory for primary induction of cocksfoot, Calder (1963, 1964) and Heide (1987) documented induction in short days at 18–21°C. While 10 weeks at 15°C and an 8 h photoperiod were necessary for flowering in Scandinavian cultivars (Heide, 1987), ecotypes originating at lower latitudes had somewhat shorter requirements for primary induction (Fejer, 1966; Kozumplik and Christie, 1972b).

Italian and perennial ryegrass, and their hybrids, are normally classified as dual-induction plants, although their primary-induction requirements vary from obligatory and large in the perennial species to facultative and intermediate in Italian ryegrass and the hybrids (Cooper, 1960; Evans, 1960; Halligan *et al.*, 1991). Cooper and Calder (1964) stated that low temperatures were more important than short days for primary induction in ryegrasses but critical temperature limits for short-day and long-day induction have not yet been established.

The most extreme requirements for primary induction are found among fescue species (Bean, 1970). Although high-latitude ecotypes of red fescue may have a

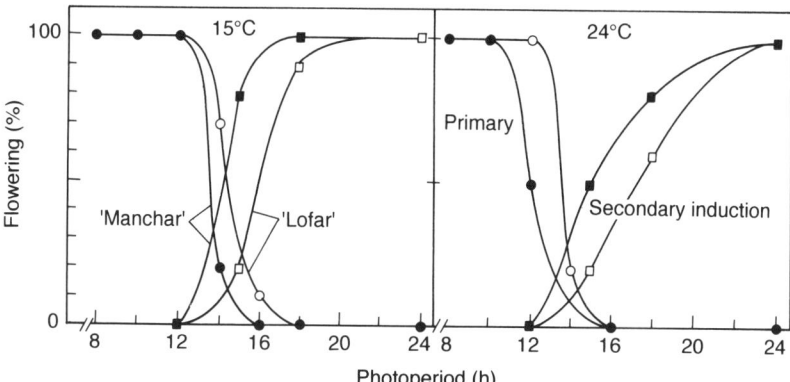

Fig. 2.4. Photoperiodic requirements for primary and secondary induction of flowering at 15°C and 24°C in smooth bromegrass cultivars Løfar and Manchar from Norway and Idaho, USA, respectively. (From Heide, 1984, with permission of the author and *Physiologia Plantarum*.)

somewhat shorter requirement, 16–20 weeks of exposure to short days at 6–12°C was usually necessary for full response in Scandinavian cultivars (Heide, 1990) and in cultivars of meadow fescue (Heide, 1988).

Flowering requirements in white clover are complex and highly diversified (see Thomas, 1987b). Although many cultivars flower in the sowing year without any winter exposure, the conclusion that white clover is a long-day plant is, at best, an oversimplification. This can be illustrated in Table 2.2 where the flowering responses of nine lines of white clover are presented (Thomas, 1982).

Based on reported responses (Gibson, 1957; Thomas, 1961, 1979, 1980a,b, 1981, 1982, 1987b; Norris, 1989) temperate white clover cultivars can broadly be grouped into a high-latitude, summer-growing type (in Table 2.2 typically the Russian ecotypes Jygeva and Kalinin and the British cultivar Kent Wild White) and a low-latitude (Mediterranean), winter-growing type (in Table 2.2 typically 'Tamar' and the Spanish ecotype). Deviating lines occur, such as Louisiana which behaved like a Mediterranean type in autumn but showed a clear long-day response in spring and early summer. In the commercially important cultivar, Grasslands Huia, which possibly originated from natural crossings between plants from various latitudes, both types of reactions have been documented (Thomas, 1979).

Thomas (1981) based his definition of white clover as a short-long-day plant on the fact that inflorescence initiation stopped 3–4 weeks after transfer from short to long days, and he did not interpret the short-day response at high temperature, or the low temperature responses, as a winter requirement similar to that in grasses. However, Thomas (1980b) noted that inflorescences which have initiated during winter in response to cool temperatures (i.e. primary induction) do not emerge until plants have been exposed to long photoperiods and increasing temperatures (secondary induction) in spring and early summer. It is our view that most lines of white clover have a double-induction requirement.

Table 2.2. Floral induction in response to environmental conditions in white clover populations originating at various latitudes. Strength of the response is indicated by the number of pluses, more indicating a greater response, and vice versa. (From Thomas, 1982.)

Population	Latitude (°N/S)	Earliness of direct response to low temp.[1]	Response to natural long photoperiods	Cessation of initiation in long days	Indirect response to low temp.[2]
Jygeva	59	±	+++++	++++	±
Kalinin	57	+	+++++	+++	±
Kent Wild White	51	+	+++++	+++	±
Ladino	45	+	+++++	±	++
Portuguese mat-type	40	+	+++++	±	–
Grasslands Huia	40	++	+++++	+++++	+
Spanish	39	+++	–	+++++	++
Tamar	32	+++++	±	+++++	+++++
Louisiana	31	+++++	++++	++++	+++++

[1]Induced after exposure to cool, short days.
[2]Induced in warm, short days after cool pretreatment.

Secondary induction. After the primary-induction requirement has been met, whether by short days and/or low temperatures, secondary induction by long days is usually necessary for normal flowering in dual-induction species. Critical photoperiods during this phase vary from 9 to 10 h in Mediterranean to about 16 h in high-latitude ecotypes (Heide, 1994a – cf. Fig 2.4). In continuous light the number of inductive cycles required for flowering in all plants previously exposed to primary induction was eight in smooth bromegrass of North American origin, 12 in cocksfoot and meadow fescue of Danish origin, and 16 in Norwegian cultivars of the same species (Heide, 1984, 1987, 1988).

Cocksfoot and meadow fescue, even of high-latitude origin, began ear emergence earlier but produced fewer panicles when the daylength during secondary induction was prolonged from 18 to 24 h (Heide, 1987, 1988). Apparently, continuous light during this period augments reproductive development and the demand for assimilates in the main stem and first primary tillers to such an extent that flowering in weaker tillers is sacrificed. High temperature during secondary induction also has a negative impact on panicle number (e.g. Wilson, 1959; Broué and Nicholls, 1973), and the optimal temperature during this phase is therefore moderate, usually in the range 12–15°C (Heide, 1994a). With the exception of grasses initiating floral primordia during primary induction, increasing temperature normally prolongs the critical photoperiod during secondary induction (Heide, 1984, 1994a; cf. Fig. 2.4).

Marginal secondary induction often leads to stunted culms with small inflorescences and, in many cases, vivipary (e.g. Karlsen, 1988; Heide, 1987, 1990). Viviparous proliferation is most common in grasses originating from high latitude or high altitude and occurs also after marginal induction in single-induction species such as timothy (Heide, 1982; Junttila, 1985; Fig. 2.5).

The requirement for long-day exposure for secondary induction in white clover is shorter than in most dual-induction grasses. For example, plants of the cultivar Grasslands Huia, otherwise kept in warm short days, only required 2 days of a 20 h photoperiod or 3 days of a 16 h photoperiod in order to initiate a minimum of 0.5 inflorescences per stolon (Thomas, 1981). On the other hand, the critical daylength for secondary induction of white clover is mostly the same as in grasses originating at the same latitude (Kendall and Stringer, 1985). While a daylength extension from 14 to 16 h greatly enhanced flowering intensity in cultivar Milkanova from Denmark, and tended to suppress flowering in the Israeli line Tamar, French cultivars flowered equally well at the two photoperiods (Norris, 1989). High temperatures usually increase the daylength at which maximum inflorescence number is obtained (Thomas, 1961) but, perhaps with the exception of high-latitude ecotypes, high temperatures during secondary induction do not reduce inflorescence number to the same extent as in dual-induction grasses. For example, European cultivars produced three times more inflorescences at 22°C than at 17°C (Norris, 1989).

2.3.4 Nonspecific Factors Involved in Flower Induction

In addition to the main control by temperature and photoperiod, both primary and secondary induction may be modified by other external factors. Low light intensities, with concomitant reductions in plant carbohydrate status, reduced flowering in timothy (Ryle, 1961; Heide, 1982), red clover (Ludwig *et al.*, 1953; Bula, 1960), white clover (Thomas, 1981), lucerne and birdsfoot trefoil (Nittler and Kenny, 1964) and the dual-induction grasses meadow fescue, cocksfoot, and perennial ryegrass (Ryle, 1966, 1967; Spiertz and Ellen, 1972; Kleinendorst, 1974). Reed canarygrass is especially sensitive to low light intensity/low carbohydrate status during primary induction (Heichel *et al.*, 1980, Heide, 1994b). These examples indicate that a favourable carbohydrate status resulting from sufficient light intensity is a prerequisite for normal flower induction but, unlike temperature and photoperiod, light intensity does not directly control the transition from vegetative to reproductive growth.

2.3.5 Contribution of Shoots Arising During Various Periods to Seed Yield

Because tillers of most perennial grasses have a juvenile phase during which they are nonreceptive to inductive stimuli, and require a certain exposure to short photoperiod and/or low temperature during primary induction, a logical

Fig. 2.5. Viviparous proliferation of the inflorescence of timothy. Note the growth disturbance with pigtail-like curling of the upper part of the culm. (Photography credit: O.M. Heide)

consequence is that only early-formed tillers have the capacity to produce seed (see Table 2.3).

Since timothy has no obligatory requirement for low-temperature or short-day induction, spring-emerged tillers often produce panicles in this species. However, even in timothy, floret fertility usually declines with delayed tiller emergence. Langer (1956) found tillers emerging in May and later had a negligible impact on seed production of single plants sown in September; while the greatest contribution to seed yield was made by a flush of new tillers in April, tillers arising before or during winter are usually most important in timothy (Lambert, 1966; Hill and Watkin, 1975; cf. Table 2.3).

Most forage legumes have an indeterminate flowering pattern, the apical meristem always remaining vegetative. In such species it makes little sense to calculate the proportion of shoots produced during various periods which 'become reproductive'. Nevertheless, it may be useful to determine the contribution of various shoot categories to the total number of inflorescences at seed harvest. Such an approach was taken by Li and Hill (1988) in birdsfoot trefoil and by Tabora and

Table 2.3. The time of emergence of grass tillers that contribute to the final viable inflorescence count.

Source	Species	Seed harvest year	Date of stand establishment	Contribution (%) to final inflorescence numbers by tillers formed before the given dates			
				Date	%	Date	%
Langer and Lambert, 1959	Meadow fescue	4	(Spring)[1]	15 Dec	72	15 Mar	97
	Cocksfoot	1		15 Nov	82	15 Mar	99
Lambert, 1963	Cocksfoot	2		Late Nov	82	Late Mar	97
Hill and Watkin, 1975	Perennial ryegrass	1	Apr (Oct)	30 Jun (30 Dec)	41	30 Aug (28 Feb)	98
	Timothy	1	Apr (Oct)	30 Jun (30 Dec)	35	30 Aug (28 Feb)	89
	Perennial ryegrass	2		30 Apr (30 Oct)	77	31 Jul (31 Jan)	96
	Timothy	2		30 Apr (30 Oct)	52	31 Jul (31 Jan)	85
Meijer, 1984	Smooth meadowgrass	1	22 May	Oct	85	Mar	100
		1	18 Jul	Oct	58	Mar	100
		1	29 Aug	Oct	25	Mar	100
	Red fescue	1	22 May	Oct	99	Mar	100
		1	18 Jul	Oct	91	Mar	100
		1	29 Aug	Oct	67	Mar	100
Hebblethwaite et al., 1993	Perennial ryegrass	1	4 Sep	17 Dec	93	—	—
Hare, 1994	Tall fescue	3		22 Feb (22 Aug)	72	22 Jul (22 Jan)	99

[1]Experiment in the southern hemisphere; corresponding date in the northern hemisphere given in parentheses.

Hill (1991) in big trefoil. Shoot longevity is seldom more than 6 months in these species, and thus it is hardly surprising that Tabora and Hill (1991) found that 78% of the podbearing umbels at seed harvest in February came from shoots formed during September to December; the contributions of main shoots, primary lateral shoots and secondary lateral shoots to the number of umbels available at harvest were 38, 53 and 9%, respectively.

In white clover the formation of a reproductive bud in a leaf axil is typically followed by two or three leaf primordia with no reproductive buds before a new reproductive bud is formed (Thomas, 1980b). Attention must therefore be focused on the individual bud rather than on the entire shoot (stolon). The fate of each bud (reproductive vs. vegetative) depends directly on temperature and photoperiod at the time when it is laid down and not, as in dual-induction grasses, on subsequent exposure to environmental conditions (Thomas, 1980b).

2.4 FLORAL INITIATION AND DIFFERENTIATION

Floral initiation can be defined as the morphological change of the apex by which the first floral primordia are laid down. In both grasses and legumes the term is often directly connected to the double-ridge stage (see later). Floral differentiation comprises the morphological development of the apex from initiation until all florets/ovules are laid down in the last-formed spikelet/pod. In grasses this apical development is normally associated with stem elongation and ear emergence.

2.4.1 Structures

There is an extensive body of work on the morphological changes in the shoot apices of grasses with the onset of reproductive development (see Barnard, 1964). The appearance of double ridges in the apical mid-region is usually taken as the first visible criterion that the reproductive process has begun. The traditional interpretation of this morphological event is that buds in the axils of the leaf primordia swell and form double protuberances consisting of primordia of a spikelet or branch with a subtending leaf (Jeater, 1956; Latting, 1972). However, recent results from scanning electron microscopy studies of cocksfoot primordia suggest that initiation of leaf primordia culminates well before the swelling of floral protuberances, and that the double ridges are composed of primary and secondary reproductive primordia (Fraser and Kokko, 1993).

After starting in the mid-region of the apex, floral initiation in grasses continues both basipetally and acropetally along the main axis (Jeater, 1956). Depending on species, the protuberances of the double ridge stage are further differentiated into spikelet or branch primordia. In ryegrasses and other species with spike-type inflorescences, first-order lateral protuberances develop directly into spikelets, and the terminal spikelet is borne on the main axis (rachis) itself. In grasses with panicle-type inflorescences, the initiation of lateral branches of continuously higher orders gives rise to clusters and subclusters (Fig. 2.6).

Spikelet initiation can usually be distinguished by the appearance of two concave bract primordia which develop into inner and outer glumes. The lemma

Fig. 2.6. Completion of formation of several spikelets within a cluster (in box) arising from one of the uppermost nodes on a cocksfoot panicle. Both glumes (G1 and G2) are present on the lowermost spikelet that has begun to produce the first florets. (From Fraser and Kokko, 1993, with permission from the authors and the Canadian Journal of Botany. Photograph credit: Agriculture and Agri-Food Canada, Lethbridge Research Centre.)

of the first floret is then initiated within each spikelet, and the rachilla (spikelet axis) gradually starts to elongate. Meanwhile, glumes also elongate and form awns. In ryegrass, growth of the upper glume (next to the rachis) is normally suppressed (Jeater, 1956).

The differentiation of florets within a spikelet always occurs acropetally. Within each floret the stamens develop as three protuberances more or less surrounded by the lemma. These male organs quickly elongate, eventually obscuring the development of the gynoecium (female organ) in their centre. At the same time the palea develops opposite the lemma (Fraser and Kokko, 1993), and lodicules develop at the base of the ovary. The uppermost florets in each spikelet often remain rudimentary or abort during differentiation (Latting, 1972). Concomitant with the differentiation of spikelets and florets, the main axis (rachis and stem), and in panicle grasses also the lateral branches, elongate. Main axis elongation normally starts shortly after initiation of the terminal spikelet (Martin and Field, 1993) and is not finished until anthesis.

Although the developmental pattern described here may appear uniform and straightforward, it is important to realize that within an inflorescence, all stages of differentiation may be found simultaneously. It is, nonetheless, useful to describe apical development by means of stage systems, in which a given number corresponds to a particular stage of development. The system in Table 2.4 is according to Martin and Field (1993), but similar systems have been suggested (Jeater, 1956; Bommer, 1959; Latting, 1972; Sweet et al., 1991).

In legumes the first evidence of floral initiation is the development of meristematic tissue in the axil of the leaf primordium adjacent to the stem apex. In lucerne this development usually occurs at the 10th to 14th node from the crown in the first spring growth, and at the 6th to 10th node during summer growth (Viands et al., 1988). The stems of lucerne are indeterminate, and the stem continues to produce both leaves and flowers (Teuber and Brick, 1988).

Retallack (1987) reported that, in red clover, the first inflorescence arose from a bud in the axil of the second or third leaf. Inflorescences are first produced terminally, and others appear on shoots from axillary buds in a basipetal pattern. The transition from a vegetative to a reproductive meristem can be distinguished as a lateral enlargement in the axil of a trifoliate leaf primordium. The inflorescence dome enlarges becoming convex, and bract primordia are initiated. The floret initials appear proximal to the penultimate bract primordium. The more mature florets are more basal and more lateral (Retallack, 1987). A typical clover floret consists of a tubular calyx with five sepals, a corolla with five petals, 10 diadelphous stamens (nine joined and one free) and a simple carpel. The order of development is sepals, stamens, petals and carpel.

When a white clover plant is transferred into conditions promoting reproduction (e.g. a transfer from short to long days), a bud will develop in the axil of the

Table 2.4. Stages of apical development in perennial ryegrass and corresponding mean stem length in one particular experiment. (From Martin and Field, 1993.)

Stage	Description	Easily recognized features	Main stem length (mm)
0	Vegetative	Small dome	—
1	Elongation	Elongating dome with few primordia	0.3
2	Double ridge	The two sets of primordia often give the dome a 'spiral' appearance	3.6
3	Spikelet primordium	The spikelet primordia	6.1
4	Glume primordium	The second primordium on each spikelet	10.7
5	Floret initials	The three sets of single primordia on the glume	14.2
6	Floret lemma initials	Three distinct sets of two primordia on the glume	23.3
7	Stamen and carpel initials	The spikelet has the shape of an equilateral triangle	25.6
8	Elongation of the spike and spikelet	Elongation of the spikelet	43.2
9	Elongation and development of photosynthetic tissue in the rachis	Photosynthetic rachis	76.0
10	Glume encloses spikelet	Glume encloses spikelet	166.3
11	Ear emergence	Ear emergence	436.8

next leaf primordium to form at the apex of the stolon. This youngest leaf primordium and its subtended bud together form a 'double ridge' (Thomas, 1987b; Fig. 2.7). The apical meristem of a white clover stolon is not affected by the change to reproductive growth.

In red clover, the axillary bud gives rise to a lateral stem or shoot with a terminal inflorescence. In lucerne, the axillary bud gives rise to a raceme. In white clover, the bud gives rise to a flower head on the peduncle. In birdsfoot trefoil, the axillary bud gives rise to an umbel.

As in the grasses, there will always be a range of developmental stages within a forage legume plant and certainly within a canopy. In a crop like lucerne, there can be ripe seed and flowers on the same plant. Kalu and Fick (1981) developed a classification system for lucerne (Table 2.5). Contrary to the grass systems

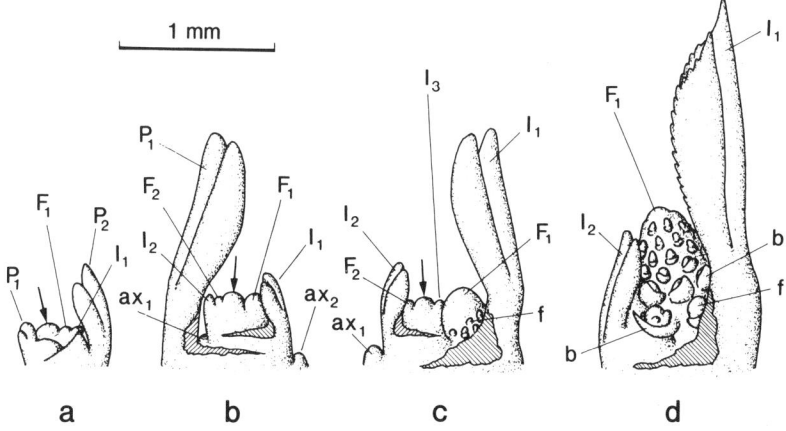

Fig. 2.7. Early stages of development of inflorescence primordia at main stolon apices of white clover. a, b, c and d show apical buds 6, 9, 11 and 13 days respectively after the start of reproductive growth (SRG). P_1 and P_2 are the youngest and next youngest leaf primordium prior to SRG, while ax_1 and ax_2 are subtending axillary buds. I_1, I_2 and I_3 are new leaf primordia successively formed after SRG; F_1, F_2 and F_3 are subtending inflorescence primordia; b are bracts subtending floret primordia; and f are the floret primordia. Arrows point to the meristematic dome at each stolon apex. (From Thomas, 1987b.)

Table 2.5. Stages of development of individual stems of lucerne. (From Kalu and Fick, 1981.)

Stage	Description	Features	Stem length (cm)
0	Early vegetative	No buds, flowers, or seed pods	
1	Mid vegetative	No buds, flowers, or seed pods	16–30
2	Late vegetative	No buds, flowers or seed pods	> 30
3	Early bud	1 or 2 nodes with buds, no flowers, no seed pods	
4	Late bud	> 3 nodes with buds, no flowers, no seed pods	
5	Early flower	1 node with one open flower, no seed pods	
6	Late flower	> 2 nodes with open flowers, no seed pods	
7	Early seed pod	1–3 nodes with green seed pods	
8	Late seed pod	> 4 nodes with green seed pods	
9	Ripe seed pod	Nodes with mostly brown mature seed pods	

described earlier, this system is based more on phenological than on apical development, and it also differs in that it follows the individual stem not only until flower emergence but through to seed maturation.

2.4.2 Timing

Experiments in controlled environments have revealed that only a small group of northern grasses differentiate floral primordia during primary induction. Among species of commercial interest, this group includes meadow foxtail (Heide, 1986) and high-latitude ecotypes of smooth-stalked meadowgrass (Håbjorg, 1978; Heide, 1980). By contrast, perennial ryegrass (Evans, 1960), smooth bromegrass (Heide, 1984), cocksfoot (Heide, 1987), meadow fescue (Heide, 1988), red fescue (Heide, 1990), reed canarygrass (Heide, 1994b) and most cultivars of smooth-stalked meadowgrass (Heide, 1980) do not initiate floral primordia before transfer to long days. In these species the apex elongates and reaches a transitional stage towards the end of the primary induction period but double ridges do not appear before long-day exposure. On the other hand, initiation and differentiation often take place very rapidly after transition to long days and, in some cases, the first floral primordia can be distinguished after only three to four long-day cycles (Gardner and Loomis, 1953).

The timing of inflorescence initiation and differentiation in grasses has also been studied in the field. In Germany, Bommer (1959) found that meadow foxtail initiated floral primordia in September/October; smooth-stalked meadowgrass, sheep's fescue (*Festuca ovina* L.) and smooth bromegrass in January or early February; red fescue and cocksfoot in late February or early March; and meadow fescue and perennial ryegrass from mid March to mid April. The last species to initiate primordia in this investigation were timothy, reed canarygrass and redtop (*Agrostis alba* L.). Finnish experiments with cocksfoot revealed no differentiation before winter in field-sown crops but initiation had occurred by late November in an early-sown pot trial (Niemeläinen, 1990b). Norwegian ecotypes and cultivars of smooth-stalked meadowgrass varied markedly in initiation time depending on their latitude of origin, subarctic cultivars often passing the double-ridge stage by mid August (Håbjorg, 1979). Autumn differentiation of floral primordia was also documented in native Alaskan grasses (Hodgson, 1966), indicating that this is an adaptive strategy to a short and cool growing season (Heide, 1994a).

Under British conditions, it usually takes about 2 months from initiation of double ridges to 50% ear emergence in perennial ryegrass (Hebblethwaite, 1977; Hebblethwaite and Ivins, 1978). In northern environments with late snowmelt, long days and an abrupt rise in temperatures in late April and May, the differentiation period is usually shorter, sometimes only 4–5 weeks. Similarly, in Canada, Fraser and Kokko (1993) reported that differentiation of cocksfoot lasted for a period of at least 6 weeks.

Thomas (1980b, 1982) studied the timing of inflorescence initiation in various white clover cultivars in New Zealand; while low-latitude lines initiated inflorescences in April (corresponding to October in the northern hemisphere) the first initiation of high-latitude ecotypes was not discovered until daylengths had started to increase in July. However, in all ecotypes, the main period of initiation was October/November with flower heads emerging about 6 weeks later. In cold environments, it appears unlikely that native strains of white clover initiate inflorescences before growth cessation in autumn (Thomas, 1980a, 1987b).

2.4.3 Factors Affecting the Number of Spikelets/Florets/Ovules per Inflorescence

In grasses the total number of florets per inflorescence depends on the number of primary branches, and on the number of florets produced per primary branch. While there is only one spikelet per primary branch in ryegrasses, basal branches of panicle grasses usually develop considerably more spikelets, and thus florets, than the terminal ones (Jeater, 1956; Ryle, 1966). The total number of primary branches in an inflorescence is equivalent to the sum of axillary sites available on the meristem at spikelet initiation and the number of floral primordia produced between initiation and the time when the tip of the meristem itself is converted into a primordium (Ryle and Langer, 1963b). The first, and probably most important of these addenda depends on the size of the apex at the time of initiation. As there is comparatively little variation among tillers arising during various periods in their time of initiation (e.g. Wilson, 1959; Ryle, 1963), the number of primary branches normally increases with tiller age (Langer, 1956; Colvill and Marshall, 1984; Ryle, 1964b; Fig 2.8). Old tillers also develop more florets per primary branch than younger ones (Ryle, 1966; Hill and Watkin, 1975; Colvill and Marshall, 1984).

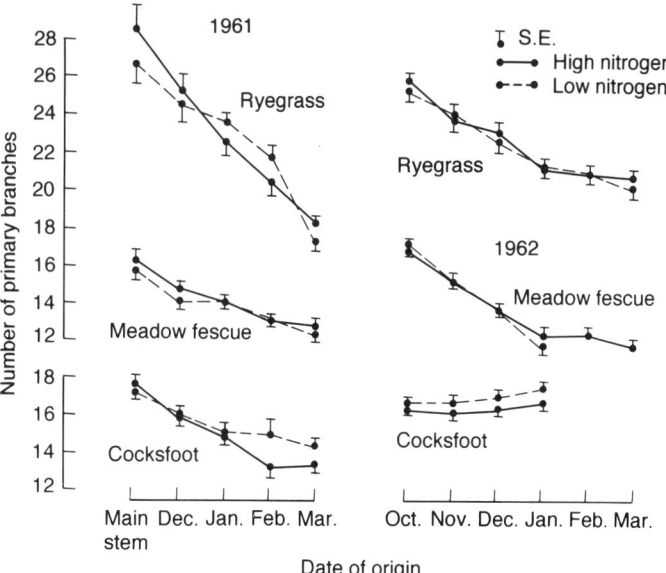

Fig. 2.8. Influence of date of origin of tillers on numbers of primary branches in ears of ryegrass, meadow fescue and cocksfoot. (From Ryle 1964b.)

Increasing temperature and photoperiod during inflorescence development reduced floret number per inflorescence in timothy, perennial ryegrass and darnel (Ryle and Langer, 1963b; Ryle, 1965). Common to these species, and probably to most other grasses as well, is that the greater relative growth rate of the apex/inflorescence at high temperatures, or long days, cannot compensate for the reduction in the time period from initiation to anthesis under such conditions; in the ryegrasses, both spikelet number and florets per spikelet are affected (Ryle, 1965).

Long days enhanced both the number of florets per flower head and the number of ovules per floret in white clover (Thomas, 1961). Temperatures within the range 12–20°C had only a minor effect on the number of florets per inflorescence in red clover (Puri and Laidlaw, 1984). Based on regression lines derived from field experiments with a Japanese line of white clover, Pasumarty *et al.* (1995) recently concluded that a temperature of 7–11°C was optimal for inflorescence size, whereas 16–19°C was optimal for pollen and ovule fertility. In white clover, there is a marked depression not only in inflorescence number, but also in the numbers of florets per inflorescence and ovules per floret, after midsummer (Thomas, 1987b).

Light intensity and nitrogen also affect inflorescence size in many species (Ryle, 1964b, 1966). In grasses the number of spikelets or primary branches is seldom affected unless these factors are very restricted, and the effects are therefore primarily mediated by floret numbers per spikelet (Hill and Watkin, 1975; Ryle, 1966; Table 2.6). In Italian ryegrass, timothy and cocksfoot, nitrogen has been shown to advance the date of inflorescence initiation by up to 2 weeks without similarly advancing the date of ear emergence, thus allowing more time for inflorescence differentiation (Wilson, 1959; Lambert, 1966, 1967b).

2.5 CONCLUSION: IMPLICATIONS FOR LOCATION AND AGRONOMY OF SEED PRODUCTION

The first and most decisive component governing seed yield potential in grasses and herbage legumes is inflorescence number (e.g. Nordestgaard and Andersen, 1991). In order to maximize this component, it is important:

- That the crop develops an adequate number of strong, inducible shoots before the appropriate daylength and/or temperature are encountered.
- That these shoots are exposed to an appropriate induction environment, i.e. inductive conditions of sufficient, but not too long, a duration.
- That the highest possible number of the induced shoots survive and develop inflorescences in spring.

During the latter period, environmental conditions should also be conducive to the development and retention of a high number of florets or ovules per inflorescence.

Table 2.6. Effect of different light intensities (100% = full daylight) on shoot development in perennial ryegrass, meadow fescue, cocksfoot and timothy. Data for timothy were drawn from a separate experiment and are therefore not directly comparable to those for other species. (From Ryle, 1961, 1966.)

Species/light intensity (%)	% fertility of main tiller	Inflorescences per plant	Florets per inflorescence	Primary branches per inflorescence
Perennial ryegrass				
100	100	14.5 ± 0.95	140 ± 7.1	19.4 ± 0.46
50	100	12.1 ± 1.27	142 ± 7.9	20.0 ± 0.92
25	100	9.7 ± 0.77	138 ± 7.2	20.4 ± 0.80
5–10	88	3.2 ± 0.34	45 ± 3.8	18.0 ± 0.99
Meadow fescue				
100	100	4.3 ± 0.33	194 ± 9.4	12.0 ± 0.31
50	80	1.7 ± 0.26	197 ± 7.7	12.8 ± 0.37
25	79	1.5 ± 0.28	180 ± 12.4	12.6 ± 0.31
5–10	0	0		
Cocksfoot				
100	67	1.9 ± 0.35	520 ± 47.2	14.5 ± 0.42
50	40	1.0 ± 0.0	581	14.8 ± 0.49
25	13	1.0 ± 0.0	534	13.5
5–10	0	0		
Timothy			Inflorescence length (cm)	
100	100	12.2 ± 1.01	9.6 ± 0.33	
60	100	8.4 ± 0.76	9.2 ± 0.39	
51	100	6.1 ± 0.52	8.5 ± 0.34	
33	67	1.9 ± 0.44	6.2 ± 0.45	

2.5.1 Adequate Number of Inducible Shoots

Only in the first seed harvest year are seed yields of perennial grasses or legumes sometimes limited by suboptimal shoot densities. This is especially the case when grasses with slow germination (e.g. smooth-stalked meadowgrass), or with a long juvenile period (e.g. red fescue), are sown late in the season or undersown in a vigorous cover crop. Even then, however, more constraints are often imposed by tiller size than by tiller number at the start of primary induction (e.g. Meijer, 1984; Hare and de Ruiter, 1993). The reductions in first year seed yield due to delayed sowing or use of a cover crop normally increase with decreasing length of the growing season, i.e. with increasing latitude and altitude (Klebesadel, 1970; Jonassen and Hillestad, 1990; Table 2.7).

After the first harvest has been taken, and often in the sowing year as well, the shoot density of grasses and legumes in autumn is too high for maximal production of inflorescences. Although sowing rates are always lower and row spacings often wider for seed production than for the production of forage or turf, the wasteful process of self thinning can hardly be avoided. As time goes by, the plant number in such stands gradually decreases, whereas tiller number per plant shows the opposite tendency (Langer *et al.*, 1964). Since seed yields are generally more limited by interplant than by intraplant competition (Lambert, 1967a), it is always advantageous to start the seed crop with a low plant number per unit area.

As in the grasses, the achievement of a high number of inflorescences (flower heads, umbels, racemes) in forage legumes is only possible with an adequate number of aerial shoots, rhizomes or stolon apices at the time when daylength and temperature become suitable for flower induction (e.g. Thomas, 1987b; Li and Hill, 1988; Table 2.8). Forage legumes are often liable to overcrowding of shoots in the seeding year or in the first harvest year.

2.5.2 Appropriate Induction Environment

In most areas of significance for seed production of grasses, the specific requirements for temperature and photoperiod during primary induction (Table 2.1) do not impose serious limitations on seed yields (Heide, 1994a). There are, however, noteworthy exceptions that should be taken into consideration when evaluating the optimal location for seed production of various species and cultivars (Aamlid, 1990; Heide, 1994a).

In winter-cold zones, especially at high latitudes, the short-day primary-induction period in autumn is short and characterized by low temperatures (Heide, 1994a). Since the temperature optimum for short-day induction is relatively high, this restricts seed production, especially of species where the low temperature/long-day route of induction is absent (e.g. smooth bromegrass) or seriously

Table 2.7. Per cent seed yield reduction in the first harvest year at three locations in Norway after sowing various grasses in a spring barley cover crop as compared to sowing on the same date without a cover crop (figures in parentheses indicate the number of trials). (From Jonassen and Hillestad, 1990.)

	Location		
Species	Landvik (58°N)	Hellerud (60°N)	Trondheim (63°N)
Meadow fescue	24 (3)	59 (6)	79 (1)
Cocksfoot	38 (3)	70 (3)	87 (1)
Red fescue	84 (2)	86 (3)	—
Smooth stalked meadowgrass	83 (3)	95 (2)	—

Table 2.8. Optimal conditions for components of the seed yield potential in white clover. (From Thomas, 1987b.)

Component of seed yield potential	Contributing factors	Environmental conditions favouring component
Inflorescences (no. m^{-2})	(1) No. of active stem apices	High light and nutrients
	(2) Strength of flowering stimulus	High light, warm (> 25°C) long days or cool (< 10°C)
	(3) Plant vigour	High nutrients
Rate of inflorescence emergence	Rate of leaf emergence	Warm, moist, long days
Floret no. per infloresence	Inflorescence primordium size (?)	Cool, moist, long days
Ovule no. per floret	Inflorescence primordium size (?)	Cool-mild, moist

restricted (e.g. cocksfoot). This appears to be a major reason for the low seed yields of smooth bromegrass in Alaska (Klebesadel, 1970) and in inland areas of Norway (Aamlid *et al.*, 1995). Under Finnish conditions, Niemeläinen (1990a) estimated that the primary inductive period in cocksfoot lasted for only 7 weeks beginning in mid September.

While seed yields of smooth bromegrass and other species may be limited at high latitudes because of 'under-induction', the opposite situation occurs when species with a short primary-induction requirement and initiation of flower primordia in the autumn are grown for seed in mild winter climates at lower latitudes. In such cases, prolonged exposure to short days and nonfreezing temperatures often lead to 'over-induction' causing abortion of floral primordia (Håbjørg, 1979; Aamlid, 1996). Occasional periods of severe frost in such otherwise mild areas aggravate this situation (Heide, 1980; Niemeläinen, 1990b).

The importance of finding the right environment for seed production can be illustrated by an excerpt from a series of trials with cultivars of smooth-stalked meadowgrass and red fescue at various locations in Scandinavia (Nordestgaard, 1983; Table 2.9). Results for panicle numbers and seed yields clearly indicate that the reproductive capacity of the subarctic cultivar Holt was favoured in the continental, fairly northern climate at Uppsala, Sweden, whereas seed production of the Danish cultivar Norma was most successful in its native environment at Roskilde, Denmark.

For legumes the long-day requirement for flowering imposes clear limitations as to how far south seed production can be carried out successfully (e.g. Beuselinck and McGraw, 1988). In a comparison of flowering of Finnish white clover at two locations in western USA (40 and 32°N), flower initiation was very much reduced at the lower compared to the higher latitude (Valle, 1963).

2.5.3 Survival and Optimal Development of Induced Shoots

Induced apices in grass seed crops are usually exposed to intense competition from spikelet initiation until anthesis. Examination of a large number of tiller apices in perennial ryegrass indicated that of the fertile tillers present in early May, almost 50% died before maximum ear emergence (Hebblethwaite *et al.*, 1980). While edaphic competition often imposes the major constraint on total phytomass production, inflorescence production and survival are usually restricted by

Table 2.9. Panicle number and seed yield of smooth-stalked meadowgrass cv. Holt (origin 69°N) and cv. Norma (origin 56°N) at Uppsala, Sweden (60°N) and Roskilde, Denmark (56°N) together with annual rainfall, mean seasonal temperature and photoperiod at summer solstice at the two locations. (From Nordestgaard, 1983.)

	Panicle number (no. m^{-2})		Seed yield (kg ha^{-1})		Annual rainfall (mm)	Mean seasonal temperature (°C)				Photoperiod on 21 Jun
	'Holt'	'Norma'	'Holt'	'Norma'		Dec–Feb	Mar–May	Jun–Aug	Sep–Nov	
Uppsala	1246	1037	746	573	554	−3.3	4.6	15.9	6.3	18.7 h
Roskilde	920	1863	415	942	553	0.4	6.4	16.1	8.9	17.5 h

competition for light in spring and early summer (Lambert, 1967a). Timothy, meadow fescue and, most notably, cocksfoot are even more vulnerable than perennial ryegrass during this period (Ryle, 1961, 1966; Table 2.9). Under practical field conditions, the amount of light intercepted by individual tillers in spring depends, among other things, on stubble from the previous season and on mutual shading among tillers; hence the positive effect of stubble removal or stand thinning in autumn is probably mostly mediated in spring. Other factors which enhance fertile tiller survival and inflorescence size in grasses include moderate temperatures (e.g. Ryle and Langer, 1963a,b; Ryle, 1965; Heide, 1982) and photoperiods slightly longer than the critical ones (Heide, 1987, 1988; Ryle and Langer, 1963b; Ryle, 1965).

REFERENCES

Aamlid, T.S. (1990) [Important considerations in the location of seed production in Scandinavian cultivars of perennial grasses.] *Norsk Landbruksforsking* 4, 259–277. (In Norwegian, with English summary.)

Aamlid, T.S. (1992) Effects of temperature and photoperiod on growth and development of tillers and rhizomes in *Poa pratensis* L. ecotypes. *Annals of Botany* 69, 289–296.

Aamlid, T.S. (1993) [Autumn treatment in smooth meadowgrass (*Poa pratensis* L.) for seed production.] *Norsk Landbruksforsking* 7, 117–138. (In Norwegian, with English summary.)

Aamlid, T.S. (1996) Second year yield depression in seed production of subarctic *Poa pratensis* in south Norway. *Norwegian Journal of Agricultural Sciences* 10, 113–124.

Aamlid, T.S., Skuterud, R., Heide, O.M. and Torskens, E. (1995) Autumn application of chlormequat chloride (CCC) in seed production of *Bromus inermis* Leyss. *Norwegian Journal of Agricultural Sciences* 9, 271–279.

Aitken, Y. (1964) Flower initiation in pasture legumes. IV. Flower initiation in *Trifolium pratense* L. *Australian Journal of Agricultural Research* 15, 21–36.

Alberda, T. (1957) The effects of cutting, light intensity and night temperature on growth and soluble carbohydrate content of *Lolium perenne* L. *Plant and Soil* 8, 199–230.

Auda, H., Blaser, R.E. and Brown, R.H. (1966) Tillering and carbohydrate content of orchardgrass as influenced by environmental conditions. *Crop Science* 6, 139–143.

Ball, D.M., Hoveland, C.S. and Buchanan, G.A. (1974) Flower and seed production in Yuchi arrowleaf clover. *Agronomy Journal* 66, 581–583.

Barnard, C. (1964) *Grasses and Grasslands*. MacMillan, London.

Bean, E.W. (1970) Short-day and low-temperature control of floral induction in *Festuca*. *Annals of Botany* 34, 57–66.

Beuselinck, P.R. and McGraw, R.L. (1988) Indeterminate flowering and reproductive success in birdsfoot trefoil. *Crop Science* 28, 842–845.

Beuselinck, P.R., Bouton, J.H., Lamp, W.O., Matches, A.G., McCaslin, M.H., Nelson, C.J., Rhodes, L.H., Sheaffer, C.C. and Volenec, J.J. (1994) Improving legume persistence in forage crop systems. *Journal of Production Agriculture* 7, 311–322.

Blondon, F. (1972) Facteurs externes déterminisme floral d'un clone de *Dactylis glomerata* L. In: Chouard, P. and de Bildering, N. (eds) *Phytotronique et Perspective Horticole*. GauthierVillard, Paris, pp. 135–181.

Bommer, D. (1959) Über Zeitpunkt und Verlauf der Blütendifferenzierung bei perennierenden Gräsern. *Zeitschrift für Acker und Pflanzenbau* 109, 95–118.

Bommer, D. (1961) 'Samen'–Vernalisation perennierenden Gräserarten. *Zeitschrift für Pflanzenzüchtung* 46, 105–111.

Bowley, S.R., Taylor, N.L. and Dougerthy, C.T. (1987) Photoperiodic response and heritability of the preflowering interval of two red clover (*Trifolium pratense*) populations. *Annals of Applied Biology* 111, 455–461.

Broué, P. and Nicholls, G.H. (1973) Flowering in *Dactylis glomerata* II. Interaction of temperature and photoperiod. *Australian Journal of Agricultural Research* 24, 685–692.

Bula, R.J. (1960) Vegetative and floral development in red clover as affected by duration and intensity of illumination. *Agronomy Journal* 52, 74–77.

Bula, R.J. (1969) Role of low temperature exposure in floral development of red clover ecotypes. *Crop Science* 9, 82–84.

Calder, D.M. (1963) Environmental control of flowering in *Dactylis glomerata* L. *Nature* 197, 882–883.

Calder, D.M. (1964) Stage of development and flowering in *Dactylis glomerata* L. *Annals of Botany* 28, 187–206.

Calder, D.M. (1966) Inflorescence induction and initiation in the Graminae. In: Milthorpe, F.L. and Ivins, J.D. (eds) *The Growth of Cereals and Grasses*. Butterworths, London, pp. 59–73.

Casal, J.J., Deregibus, V.A. and Sanchez, R.A. (1985) Variations in tiller dynamics and morphology in *Lolium multiflorum* Lam. vegetative and reproductive plants as affected by differences in red/far-red irradiation. *Annals of Botany* 56, 553–559.

Casal, J.J., Sanchez, R.A. and Deregibus, V.A. (1987) Tillering responses of *Lolium multiflorum* plants to changes of red/far-red ratio typical of sparse canopies. *Journal of Experimental Botany* 38, 1432–1439.

Clemence, T.G.A. and Hebblethwaite, P.D. (1984) An appraisal of ear, leaf and stem $^{14}CO_2$ assimilation, ^{14}C-assimilate distribution and growth in a reproductive seed crop of amenity *Lolium perenne*. *Annals of Applied Biology* 105, 319–327.

Colvill, K.E. and Marshall, C. (1984) Tiller dynamics and assimilate partitioning in *Lolium perenne* with particular reference to flowering. *Annals of Applied Biology* 104, 543–557.

Cooper, J.P. (1952) Studies on the growth and development in *Lolium*. III. Influence of season and latitude on ear-emergence. *Journal of Ecology* 40, 352–379.

Cooper, J.P. (1958) The effect of temperature and photoperiod on inflorescence development in strains of timothy (*Phleum* spp.). *Journal of the British Grassland Society* 1, 81–91.

Cooper, J.P. (1960) Short-day and low-temperature induction in *Lolium*. *Annals of Botany* 24, 232–246.

Cooper, J.P. and Calder, D.M. (1964) The inductive requirement for flowering of some temperate grasses. *Journal of the British Grassland Society* 19, 6–14.

Cumming, B.G. (1959) The control of growth and development in red clover. II. Light, temperature, and influence of growth regulators. *Canadian Journal of Botany* 37, 1027–1048.

Evans, L.T. (1960) The influence of temperature on flowering in species of *Lolium* and *Poa pratensis*. *Journal of Agricultural Science (Cambridge)* 54, 410–417.

Evans, L.T. (1964) Reproduction. In: Barnard, C. (ed.) *Grasses and Grasslands*. Macmillan, London, pp. 126–153.

Evans, L.T. (1969) *Lolium temulentum* L. In: Evans, L.T. (ed.) *The Induction of Flowering*. Macmillan, London, pp. 328–349.

Evans, M.W. and Allard, H.A. (1934) Relation of length of day to growth of timothy. *Journal of Agricultural Research* 48, 571–587.

Fejer, S.O. (1960) Response of some New Zealand pasture species to vernalisation. *New Zealand Journal of Agricultural Research* 3, 656–662.

Fejer, S.O. (1966) Growth and reproduction of New Zealand, Mediterranean, and hybrid *Dactylis glomerata* after short day and temperature treatments. *Canadian Journal of Plant Science* 46, 233–241.

Fick, G.W., Holt, D.A. and Lugg, D.G. (1988) Environmental physiology and crop growth. In: Hanson, A.A. (ed.) *Alfalfa and Alfalfa Improvement*. American Society of Agronomy, Madison, Wisconsin, pp. 163–194.

Fraser, J. and Kokko, E.G. (1993) Panicle, spikelet, and floret development in orchardgrass (*Dactylis glomerata*). *Canadian Journal of Botany* 71, 523–532.

Gardner, F.P. and Loomis, W.E. (1953) Floral induction and development in orchardgrass. *Plant Physiology* 28, 201–217.

Gibson, P.B. (1957) Effect of flowering on the persistence of white clover. *Agronomy Journal* 49, 213–215.

Gibson, P.B. and Cope, W.A. (1985) White clover. In: Taylor, N.L. (ed.) *Clover Science and Technology*. American Society of Agronomy, Madison, Wisconsin, pp. 471–490.

Håbjorg, A. (1976) Effects of photoperiod and temperature on vegetative growth of different Norwegian ecotypes of *Poa pratensis* L. *Meldinger fra Norges Landbrukshøgskole* 55(16), 1–26.

Håbjorg, A. (1978) Climatic control of floral differentiation and development in selected latitudinal and altitudinal ecotypes of *Poa pratensis* L. *Meldinger fra Norges Landbrukshøgskole* 57(7), 1–21.

Håbjorg, A. (1979) Floral differentiation and development of selected ecotypes of *Poa pratensis* L. cultivated at six localities in Norway. *Meldinger fra Norges Landbrukshøgskole* 58(4), 1–19.

Haggar, R.J. (1961) Flower initiation in Kent Wild white clover (*Trifolium repens* L.) under controlled environmental conditions. *Nature* 191, 1120–1121.

Halligan, E.A., Forde, M.B. and Warrington, I.J. (1991) Discrimination of ryegrasses by heading date under various combinations of vernalization and daylength: Westerwolds, Italian and hybrid ryegrass varieties. *Plant Varieties and Seeds* 4, 115–123.

Hampton, J.G. (1990) Genetic variability and climatic factors affecting herbage legume seed production: an introduction. *Journal of Applied Seed Production* 8, 45–51.

Hare, M.D. (1993) Postharvest and autumn management of tall fescue seed fields. *New Zealand Journal of Agricultural Research* 36, 407–418.

Hare, M.D. (1994) Effect of vernalisation and tiller age on seed production in tall fescue. *Journal of Applied Seed Production* 12, 77–82.

Hare, M.D. and de Ruiter, J.M. (1993) Seed production of tall fescue (*Festuca arundinacea* Schreb.) established under a barley (*Hordeum vulgare* L.) cover crop. *New Zealand Journal of Agricultural Research* 36, 419–428.

Havstad, L.T. (1996) The effect of plant age on the receptiveness for primary induction in *Festuca pratensis* Huds. In: *Yield and Quality in Herbage Seed Production*. Third International Herbage Seed Conference, 18–23 June 1995. Halle (Saale), Germany, pp. 74–78.

Hay, R.K.M. and Pedersen, K. (1986) Influence of long photoperiods on the growth of timothy (*Phleum pratense* L.) varieties from different latitudes in northern Europe. *Grass and Forage Science* 41, 311–317.

Hebblethwaite, P.D. (1977) Irrigation and nitrogen studies in S.23 ryegrass grown for seed. 1. Growth, development, and seed yield components and seed yield. *Journal of Agricultural Science (Cambridge)* 88, 605–614.

Hebblethwaite, P.D. and Ivins, J.D. (1977) Nitrogen studies in *Lolium perenne* grown for seed. I. Level of application. *Journal of the British Grassland Society* 32, 195–204.

Hebblethwaite, P.D. and Ivins, J.D. (1978) Nitrogen studies in *Lolium perenne* grown for seed. II. Timing of nitrogen application. *Journal of the British Grassland Society* 33, 159–166.

Hebblethwaite, P.D., Wright, D. and Noble, A. (1980) Some physiological aspects of seed yield in *Lolium perenne* L. (perennial ryegrass). In: Hebblethwaite, P.D. (ed.) *Seed Production*. Butterworths, London, pp. 71–90.

Hebblethwaite, P.D., Clemence, T.G.A. and Wiltshire, J.J.J. (1993) Effect of tiller emergence time in relation to autumn and spring defoliation on contribution to seed yield in *Lolium perenne* L. In: *Proceedings of the XVII International Grassland Congress*, pp. 1643–1645.

Heichel, G.H., Hovin, A.W. and Henjum, K.I. (1980) Seedling age and cold treatment effects on induction of panicle production in reed canarygrass. *Crop Science* 20, 683–687.

Heide, O.M. (1980) Studies on flowering in *Poa pratensis* L. ecotypes and cultivars. *Meldinger fra Norges Landbrukshøgskole* 59(14), 1–27.

Heide, O.M. (1982) Effects of photoperiod and temperature on growth and flowering in Norwegian and British timothy cultivars (*Phleum pratense* L.). *Acta Agriculturae Scandinavica* 32, 241–252.

Heide, O.M. (1984) Flowering requirements in *Bromus inermis*, a short-long-day plant. *Physiologia Plantarum* 62, 59–64.

Heide, O.M. (1986) Primary and secondary induction requirements for flowering in *Alopecurus pratensis*. *Physiologia Plantarum* 66, 251–256.

Heide, O.M. (1987) Photoperiodic control of flowering in *Dactylis glomerata*, a true short-long-day plant. *Physiologia Plantarum* 70, 523–529.

Heide, O.M. (1988) Flowering requirements of Scandinavian *Festuca pratensis*. *Physiologia Plantarum* 74, 487–492.

Heide, O.M. (1990) Primary and secondary induction requirements for flowering of *Festuca rubra*. *Physiologia Plantarum* 79, 51–56.

Heide, O.M. (1994a) Control of flowering and reproduction in temperate grasses. *New Phytologist* 128, 347–362.

Heide, O.M. (1994b) Control of flowering in *Phalaris arundinacea*. *Norwegian Journal of Agricultural Sciences* 8, 259–276.

Hendershot, K.L. and Volenic, J.J. (1993) Nitrogen pools in taproots of alfalfa (*Medicago sativa* L.) after defoliation. *Journal of Plant Physiology* 141, 129–135.

Hill, M.J. and Watkin, B.R. (1975) Seed production studies on perennial ryegrass, timothy and prairie grass. 1. Effect of tiller age on tiller survival, ear emergence and seedhead components. *Journal of the British Grassland Society* 30, 63–71.

Hodgson, H.J. (1966) Floral initiation in Alaskan gramineae. *Botanical Gazette* 127, 64–70.

Hume, D.E. (1991) Effect of cutting on production and tillering in prairie grass (*Bromus willdenowii* Kunth.) compared with two ryegrass (*Lolium*) species. 1. Vegetative plants. *Annals of Botany* 67, 533–541.

Ikegaya, F. (1984) Flowering control on orchardgrass (*Dactylis glomerata* L.). *Japan Agricultural Research Quarterly* 17, 260–268.

Jeater, R.S.L. (1956) A method for determining developmental stages in grasses. *Journal of the British Grassland Society* 11, 139–146.

Jewiss, O.R. (1966) Morphological and physiological aspects of growth of grasses during the vegetative phase. In: Milthorpe, F.L. and Ivins, J.D. (eds) *The Growth of Cereals and Grasses*. Butterworths, London, pp. 39–54.

Jewiss, O.R. (1972) Tillering in grasses – its significance and control. *Journal of the British Grassland Society* 27, 65–82.

Jewiss, O.R. (1981) Shoot development and number. In: *Sward Measurement Handbook*. The British Grassland Society, pp. 93–114.

Joffe, A. (1958) The effect of photoperiod and temperature on the growth and flowering of birdsfoot trefoil. *South African Journal of Agricultural Science* 1, 435–448.

Jonassen, G.H. (1992) [Effects of experimental drought periods, irrigation and N-fertilization in meadow fescue (*Festuca pratensis* Huds.) seed meadows.] *Norsk Landbruksforsking* 6, 245–260. (In Norwegian with English summary.)

Jonassen, G.H. and Hillestad, R. (1990) Etablering av froeng uten dekkvekst. In: *NJF seminarium nr 241*. Froavl. Tune, Denmark, 18–20 June 1990, pp. 84–93. (In Norwegian.)

Jones, T.W.A. (1974) The effect of leaf number on sensitivity of red clover seedlings to photoperiodic induction. *Journal of the British Grassland Society* 29, 25–28.

Junttila, O. (1985) Experimental control of flowering and vivipary in timothy (*Phleum pratense*). *Physiologia Plantarum* 63, 35–42.

Jutras, M.W. (1965) Photothermoperiodic responses of orchardgrass. *Dissertation Abstracts* 25, 4895.

Kalu, B.A. and Fick, G.W. (1981) Quantifying morphological development of alfalfa for studies of herbage quality. *Crop Science* 21, 267–271.

Karlsen, Å.K. (1988) Primary and secondary induction requirements for flower initiation in four populations of *Agrostis capillaris* L. *Norwegian Journal of Agricultural Sciences* 2, 97–108.

Kendall, W.A. and Stringer, W.C. (1985) Physiological aspects of clover. In: Taylor, N.L. (ed.) *Clover Science and Technology*. American Society of Agronomy, Madison, Wisconsin, pp. 111–159.

King, R.W., Blundell, C. and Evans, L.T. (1993) The behaviour of shoot apices of *Lolium temulentum in vitro* as the basis of an assay system for florigenic extracts. *Australian Journal of Plant Physiology* 20, 337–348.

Klebesadel, L.J. (1970) Influence of planting date and latitudinal provenance on winter survival, heading and seed production of bromegrass and timothy in the Subarctic. *Crop Science* 10, 594–598.

Kleinendorst, A. (1974) Some effects of vernalization on the reproductive capacity of *Lolium perenne* L. *Netherlands Journal of Agricultural Science* 22, 6–21.

Knight, W.E. (1985) Crimson clover. In: Taylor, N.L. (ed.) *Clover Science and Technology*. American Society of Agronomy, Madison, Wisconsin, pp. 491–502.

Kozumplik, V. and Christie, B.R. (1972a) Completion of the juvenile stage in orchardgrass. *Canadian Journal of Plant Science* 52, 203–207.

Kozumplik, V. and Christie, B.R. (1972b) Heading response of orchardgrass seedlings to photoperiod and temperature. *Canadian Journal of Plant Science* 52, 369–373.

Lambert, D.A. (1963) The influence of density and nitrogen in seed production stands of S 37 cocksfoot (*Dactylis glomerata* L.). *Journal of Agricultural Science (Cambridge)* 61, 361–373.

Lambert, D.A. (1966) The effect of cutting timothy (*Phleum pratense* L.) grown for production of seed. *Journal of the British Grassland Society* 21, 208–213.

Lambert, D.A. (1967a) Competition between plants of cocksfoot (*Dactylis glomerata*) grown for seed. *Journal of the British Grassland Society* 23, 274–279.

Lambert, D.A. (1967b) The effects of nitrogen and irrigation on timothy (*Phleum pratense*) for production of seed. 1. Vegetative growth. *Journal of Agricultural Science (Cambridge)* 69, 225–239.

Lambert, D.A. and Jewiss, O.R. (1970) The position in the plant and the date of origin of tillers which produce inflorescences. *Journal of the British Grassland Society* 25, 107–112.

Lamp, H.F. (1952) Reproductive activity in *Bromus inermis* in relation to phases of tiller development. *Botanical Gazette* 113, 413–438.

Langer, R.H.M. (1956) Growth and nutrition of timothy (*Phleum pratense*) I. The life history of individual tillers. *Annals of Applied Biology* 44, 166–187.

Langer, R.H.M. (1979) *How Grasses Grow*. The Institute of Biology's Studies in Biology no. 34, 2nd edn. Edward Arnold, London.

Langer, R.H.M. and Lambert, D.A. (1959) Ear-bearing capacity of tillers arising at different times in herbage grasses grown for seed. *Journal of the British Grassland Society* 14, 137–140.

Langer, R.H.M., Ryle, G.J.A. and Jewiss, O.R. (1964) The changing plant and tiller populations of timothy and meadow fescue swards. I. Plant survival and the pattern of tillering. *Journal of Applied Ecology* 1, 197–208.

Latting, J. (1972) Differentiation in the grass inflorescence. In: Youngner, V.B. and McKell, C.M. (eds) *The Biology and Utilization of Grasses*. Academic Press, New York, pp. 365–399.

Leach, G.J. (1971) The relation between lucerne shoot growth and temperature. *Australian Journal of Agricultural Research* 22, 49–59.

Li, B. and Beuselinck, P.R. (1996) Rhizomatous *Lotus corniculatus* L. II. Morphology and anatomy of rhizomes. *Crop Science* 36, 407–411.

Li, Q. and Hill, M.J. (1988) An examination of different shoot age groups and their contribution to the protracted flowering pattern in birdsfoot trefoil (*Lotus corniculatus*). *Journal of Applied Seed Production* 6, 54–61.

Ludwig, R.A., Barrales, H.G. and Steppler, H. (1953) Studies on the effect of light on the growth and development of red clover. *Canadian Journal of Agricultural Science* 33, 274–287.

Lunnan, T. (1989) Effects of photoperiod, temperature and vernalization on flowering and growth in high-latitude populations of red clover. *Norwegian Journal of Agricultural Sciences* 3, 201–210.

Marshall, C. and Sagar, G.R. (1968) The interdependence of tillers in *Lolium multiflorum* Lam. – a quantitative assessment. *Journal of Experimental Botany* 19, 785–794.

Martin, M.L. and Field, R.J. (1993) A development scale for perennial ryegrass. In: *Proceedings of the XVII International Grassland Congress*, pp. 1679–1680.

McCown, R.L. and Peterson, M.L. (1964) Effects of low temperature and age of plant on flowering in *Lolium perenne* L. *Crop Science* 4, 388–391.

McKee, G.W. (1963) Influence of day length on flowering and plant distribution in birdsfoot trefoil. *Crop Science* 3, 205–220.

Meijer, W.J.M. (1984) Inflorescence production in plants and in seed crops of *Poa pratensis* L. and *Festuca rubra* L. as affected by juvenility of tillers and tiller density. *Netherlands Journal of Agricultural Science* 32, 119–136.

Minderhoud, J.W. (1978) Pseudostolons and aerial tillers: morphological phenomena of *Lolium perenne* L. In: *Proceedings of the 7th General Meeting of the European Grassland Federation*, 5–9 July 1978, pp. 10.31–10.39.

Nelson, C.J. and Moser, L.E. (1995) Morphology and systematics. In: Barnes, R.F., Miller, D.A. and Nelson, C.J. (eds) *Forages. Volume I: An Introduction to Grassland Agriculture*. Iowa State University Press, Ames, Iowa, pp. 15–30.

Newell, L.C. (1951) Controlled life cycles of bromegrass, *Bromus inermis* Leyss., used in improvement. *Agronomy Journal* 43, 417–424.

Niemeläinen, O. (1990a) Factors affecting panicle production of cocksfoot (*Dactylis glomerata* L.) in Finland. I. Development of panicle production ability and time of floral initiation in Jokioinen. *Annales Agriculturae Fenniae* 29, 217–230.

Niemeläinen, O. (1990b) Factors affecting panicle production of cocksfoot (*Dactylis glomerata* L.) in Finland. II. Effect of juvenile phase, sowing date, and tillering. *Annales Agriculturae Fenniae* 29, 231–239.

Nittler, L.W. and Kenny, T.J. (1964) Induction of flowering in alfalfa, birdsfoot trefoil, and red clover as an aid in testing for variety purity. *Crop Science*, 4, 187–190.

Nordestgaard, A. (1983) [Joint Northern trials on seed production of varieties of smoothstalked meadow grass (*Poa pratensis*) and red fescue (*Festuca rubra*).] *Tidsskrift for Planteavl* 87, 429–444. (In Danish with English summary.)

Nordestgaard, A. (1988) [Inquiries on seed development in red fescue (*Festuca rubra* L.) and cocksfoot (*Dactylis glomerata* L.) for seed production.] *Tidsskrift for Planteavls Specialserie, beretning S. 1937.* (In Danish with English summary.)

Nordestgaard, A. and Andersen, S. (1991) Stability of high production efficiency in perennial herbage seed crops. *Journal of Applied Seed Production (Supplement)* 9, 27–32.

Norris, I.B. (1982) Soil moisture and growth of contrasting varieties of *Lolium, Dactylis* and *Festuca* species. *Grass and Forage Science* 37, 273–283.

Norris, I.B. (1989) *Trifolium repens*. In: Halevy, A.H. (ed.) *Handbook of Flowering*, vol. VI. CRC Press, Boca Raton, Florida, pp. 630–635.

Nyahoza, F., Marshall, C. and Sagar, G.R. (1974) Assimilate distribution in *Poa pratensis* L. – a quantitative study. *Weed Research* 14, 251–256.

Ong, C.K. (1978) The physiology of tiller death in grasses. 1. The influence of tiller age, size and position. *Journal of the British Grassland Society* 33, 197–203.

Ostgård, O. and Eagles, C.F. (1971) Variation in growth and development in natural populations of *Dactylis glomerata* from Norway and Portugal. II. Leaf development and tillering. *Journal of Applied Ecology* 8, 383–391.

Pasumarty, S.V., Higuchi, S. and Murata, T. (1995) Environmental influences on seed yield components of white clover. *Journal of Applied Seed Production* 13, 25–31.

Patel, A.S. and Cooper, J.P. (1961) The influence of seasonal changes in light energy on leaf and tiller development in ryegrass, timothy and meadow fescue. *Journal of the British Grassland Society* 16, 299–308.

Pederson, G.A. (1995) White clover and other perennial clovers. In: Barnes, R.F., Miller, D.A. and Nelson, C.J. (eds) *Forages. Volume I: An Introduction to Grassland Agriculture.* Iowa State University Press, Ames, Iowa, pp. 277–236.

Pulli, S. (1988) Adaptation of red clover to the long-day environment. *Journal of Agricultural Science in Finland* 60, 210–214.

Puri, K.P. and Laidlaw, A.S. (1984) The effect of temperature on components of seed yield and on hardseededness in three cultivars of red clover (*Trifolium pratense* L.). *Journal of Applied Seed Production* 2, 18–23.

Retallack, B. (1987) *Cellular Changes Associated with and used for Selection of Winterhardiness in Red Clover.* Biology Department, Dalhousie University, Halifax, Nova Scotia, Canada.

Rhykerd, C.L., Langston, R. and Peterson, J.B. (1959) Influence of light on the foliar growth of alfalfa, red clover and birdsfoot trefoil. *Agronomy Journal* 51, 199–201.

Robson, M.J. (1973) The effect of temperature on the growth of S.170 tall fescue (*Festuca arundinacea*). II. Independent variation of day and night temperatures. *Journal of Applied Ecology* 10, 93–105.

Ryle, G.J.A. (1961) Effects of light intensity on reproduction in S48 timothy (*Phleum pratense* L.) *Nature* 9, 176–197.

Ryle, G.J.A. (1963) Studies in the physiology of flowering of timothy (*Phleum pratense* L.) III. Effects of shoot age and nitrogen level on timing of inflorescence production. *Annals of Botany* 27, 453–465.

Ryle, G.J.A. (1964a) A comparison of leaf and tiller growth in seven perennial grasses as influenced by nitrogen and temperature. *Journal of the British Grassland Society* 19, 281–290.

Ryle, G.J.A. (1964b) The influence of date of origin of the shoot and level of nitrogen on ear size in three perennial grasses. *Annals of Applied Biology* 53, 311–323.

Ryle, G.J.A. (1965) Effects of daylength and temperature on ear size in S.24 perennial ryegrass. *Annals of Botany* 55, 107–114.

Ryle, G.J.A. (1966) Physiological aspects of seed yield in grasses. In: Milthorpe, F.L. and Ivins, J.D. (eds) *The Growth of Cereals and Grasses*. Butterworths, London, pp. 106–120.

Ryle, G.J.A. (1967) Effects of shading on inflorescence size and development in temperate perennial grasses. *Annals of Applied Biology* 59, 297–308.

Ryle, G.J.A. (1970) Partition of assimilates in an annual and a perennial grass. *Journal of Applied Ecology* 7, 217–227.

Ryle, G.J.A. and Langer, R.H.M. (1963a) Studies on the physiology of flowering of timothy (*Phleum pratense* L.) I. Influence of daylength and temperature on initiation and differentiation of the inflorescence. *Annals of Botany* 27, 213–229.

Ryle, G.J.A. and Langer, R.H.M. (1963b) Studies on the physiology of flowering of timothy (*Phleum pratense* L.) II. Influence of daylength and temperature on size of the inflorescence. *Annals of Botany* 27, 233–244.

Schulze, E. (1957) Photoperiodische Versuche an mehrjährigen Futterpflanzen. *Zeitschrift für Acker und Pflanzenbau* 103, 198–216.

Sheath, G.W. (1975) A descriptive note on the growth habit of *Lotus pedunculatus* Cav. *Proceedings of the New Zealand Grassland Association* 37, 215–220.

Simon, J.C. and Lemaire, G. (1987) Tillering and leaf area index in grasses in the vegetative phase. *Grass and Forage Science* 42, 373–380.

Skinner, R.H. and Nelson, C.J. (1992) Estimation of potential tiller production and site usage during tall fescue canopy development. *Annals of Botany* 70, 493–499.

Smith, D. (1962) Carbohydrate root reserves in alfalfa, red clover, and birdsfoot trefoil under several management schedules. *Crop Science* 2, 75–78.

Smith, R.S. and Bishop, D.J. (1993) Astred – a stoloniferous red clover. In: *Proceedings of the XVII International Grassland Congress*, pp. 421–423.

Spiertz, J.H.J. and Ellen, J. (1972) The effect of light intensity on some morphological and physiological aspects of the crop perennial ryegrass (*Lolium perenne* L. var. 'Cropper') and its effect on seed production. *Netherlands Journal of Agricultural Science* 20, 232–246.

Susuki, S., Inami, S. and Sakurai, Y. (1975) Influence of daylength and temperature on plant growth in the classified groups of lucerne cultivars. *Journal of the Japanese Society for Grassland Science* 21, 245–251.

Sweet, N., Wiltshire, J.J.J. and Baker, C.K. (1991) A new descriptive scale for early reproductive development in perennial ryegrass. *Grass and Forage Science* 46, 201–206.

Tabora, R.S. and Hill, M.J. (1991) An examination of vegetative and reproductive growth habits and their contribution to seed yield in 'Grasslands Maku' Lotus (*Lotus uliginosus* Schk.). *Journal of Applied Seed Production* 9, 7–15.

Taylor, N.L. and Smith, R.R. (1995) Red clover. In: Barnes, R.F., Miller, D.A. and Nelson, C.J. (eds) *Forages. Volume I: An Introduction to Grassland Agriculture*. Iowa State University Press, Ames, Iowa, pp. 217–226.

Templeton, W.C., Mott, G.O. and Bula, R.J. (1961) Some effects of temperature and light on growth and flowering of tall fescue, *Festuca arundinacea* Schreb. I. Vegetative development. *Crop Science* 1, 216–219.

Teuber, L.R. and Brick, M.A. (1988). Morphology and anatomy. In: Hanson, A.A. (ed.) *Alfalfa and Alfalfa Improvement*. American Society of Agronomy, Madison, Wisconsin, pp. 125–162.

Thimann, K.V. and Skoog, F. (1934) On the inhibition of bud development and other functions of growth substance in *Vicia faba*. *Proceedings of the Royal Society, London*, Series B 114, 317–339.

Thomas, R.G. (1961) The influence of environment on seed production capacity in white clover (*Trifolium repens* L.). 1. Controlled environment studies. *Australian Journal of Agricultural Research* 12, 227–238.

Thomas, R.G. (1979) Inflorescence initiation in *Trifolium repens* L.: influence of natural photoperiods and temperatures. *New Zealand Journal of Botany* 17, 287–299.

Thomas, R.G. (1980a) Flowering responses of white clover (*Trifolium repens* L.) growing in Arctic Norway. *Acta Agriculturæ Scandinavica* 30, 51–58.

Thomas, R.G. (1980b) Growth of the white clover plant in relation to seed production. In: *Herbage Seed Production*. Grassland Research and Practice Series No. 1. New Zealand Grassland Association, Palmerston North, New Zealand, pp. 56–63.

Thomas, R.G. (1981) Studies on inflorescence initiation in *Trifolium repens* L. The short-long day reaction. *New Zealand Journal of Botany* 19, 361–369.

Thomas, R.G. (1982) A comparison of the effects of environment on inflorescence initiation in nine lines of white clover (*Trifolium repens* L.). *New Zealand Journal of Botany* 20, 151–162.

Thomas, R.G. (1987a) Vegetative growth and development. In: Baker, M.J. and Williams, W.M. (eds) *White Clover*. CAB International, Wallingford, UK, pp. 31–62.

Thomas, R.G. (1987b) Reproductive development. In: Baker, M.J. and Williams, W.M. (eds) *White Clover*. CAB International, Wallingford, UK, pp. 63–123.

Valle, O. (1963) Breeding and seed production experiences with Finnish white clover. *Annales Agriculturae Fennicae* 2, 51–58.

Viands, D.R., Sun, P. and Barnes, D.K. (1988) Pollination control: mechanical and sterility. In: Hanson, A.A. (ed.) *Alfalfa and Alfalfa Improvement*. American Society of Agronomy, Madison, Wisconsin, pp. 931–960.

Wedderburn, M.E. and Gwynne, D.C. (1981) Seasonality of rhizome and shoot production and nitrogen fixation in *Lotus uliginosus* under upland conditions in southwest Scotland. *Annals of Botany* 48, 5–13.

Williams, R.D. (1964) Assimilation and translocation in perennial grasses. *Annals of Botany* 28, 419–426.

Wilson, J.R. (1959) Influence of time of tiller origin and of nitrogen level on floral initiation and ear emergence of four pasture species. *New Zealand Journal of Agricultural Research* 2, 915–932.

Young, W.C. (1980) Grazing duration effects on seed yield of annual ryegrass (*Lolium multiflorum* L.). MSc Thesis, Oregon State University.

Components of Seed Yield in Grasses and Legumes

3

J.G. Hampton[1] and D.T. Fairey[2]

[1] *Seed Technology Centre, Department of Plant Science, Massey University, Palmerston North, New Zealand;* [2] *Northern Agriculture Research Centre, Agriculture and Agri-Food Canada, Beaverlodge, Alberta TOH OCO, Canada*

3.1 GRASSES

3.1.1 Introduction

Seed yield in forage grasses is the product of seeds per unit area and individual seed weight (Elgersma, 1991). Usually the greatest seed yields are obtained by maximizing seed number; e.g. Hebblethwaite and Hampton (1982) reported that seed number per unit area accounted for between 63 and 98% of the yield variance for two perennial ryegrass (*Lolium perenne* L.) cultivars over a 10-year period at one site. Seed number depends on the number of fertile tillers per unit area, the number of spikelets per tiller, the number of florets per spikelet and floret site utilization (the proportion of florets present at anthesis that produce a seed; Elgersma, 1991). In this section of this chapter, the influence of each of these components on the seed yield of forage grasses and the factors affecting them is discussed.

3.1.2 Fertile Tillers

Relationship to seed yield

Conflicting evidence for the relationship between fertile tiller number and seed yield has apparently been reported. Langer (1980) suggested that in perennial ryegrass, seed yield depended primarily on fertile tiller number, and cited a correlation of $r = 0.90$, $P = 0.01$ between these two factors for a set of New Zealand data. In contrast, Hebblethwaite *et al.* (1981) reported that in two seasons in England, perennial ryegrass tiller numbers accounted for only 7% of the variance in seed yield recorded. The explanation of these apparent contradictions is that this relationship should be considered in two steps.

Initially, seed yield increases as fertile tiller numbers increase (Table 3.1), and the relationship as reported by Langer (1980) can produce positive and highly significant correlations (see Spiertz and Ellen, 1972; Falkowski *et al.*, 1987; Hare, 1992a for perennial ryegrass, Kentucky bluegrass (*Poa pratensis* L.) and tall fescue (*Festuca arundinacea* Schreb.), respectively). However continuing to increase fertile tiller number does not necessarily further increase seed yield (Table 3.2), and may decrease it (Hampton *et al.*, 1983); i.e. in common with other graminaceous crops, seed yield in forage grasses increases with increasing fertile tiller number up to a point where yield plateaus, and the relationship ceases (Fig. 3.1). This means that similar seed yields may be obtained from a relatively wide range of fertile tiller

Table 3.1. Relationship between fertile tiller number and seed yield in four temperate grass species.

Species	Fertile tillers (no. m^{-2})	Seed yield (kg ha^{-1})	Reference
Dactylis glomerata	195	570	Rolston (1991)
	360	960	
	445	1310	
Festuca pratensis	436	488	Falkowski *et al.* (1987)
	824	684	
	1174	1235	
Lolium perenne	891	548	Hampton (1987)
	922	737	
	1303	1127	
Poa pratensis	531	453	Falkowski *et al.* (1987)
	600	563	
	650	753	

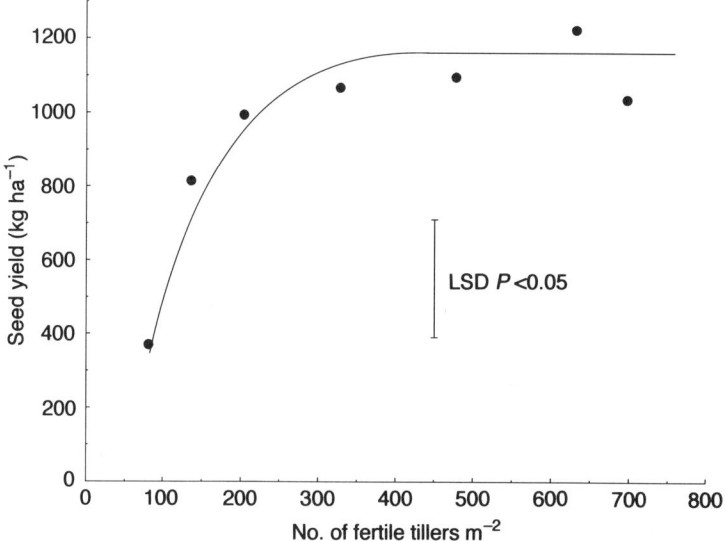

Fig. 3.1. Relationship between fertile tiller number and seed yield in tall fescue. (Data from Hare, 1992a.) LSD, least significant difference.

Components of Seed Yield

Table 3.2. Relationship between fertile tiller numbers and seed yield in two perennial ryegrass cultivars.

cv. Morenne, England[1]		cv. Grasslands Nui, NZ[2]	
Fertile tillers (no. m^{-2})	Seed yield (kg ha^{-1})	Fertile tillers (no. m^{-2})	Seed yield (kg ha^{-1})
2460	1170	1110	1038
2625	1295	1249	1118
2921	1039	1308	936

[1]Hampton *et al.* (1983).
[2]Hampton (1987).

Table 3.3. Number of fertile tillers m^{-2} required for seed production of some temperate grass species.

Grass species	Fertile tillers (no. m^{-2})
Dactylis glomerata	600–850
Festuca arundinacea	600–900
Festuca rubra	1500–3000
Lolium perenne	
agricultural	1800–3000
amenity	2000–3500
Lolium multiflorum	1000–2000
Poa pratensis	800–2000

populations. For example, Hampton and Hebblethwaite (1983) proposed that for perennial ryegrass cultivar S24, this plateau in seed yield was reached at a fertile tiller population of between 2000 and 4000 m^{-2}, and that providing these numbers were achieved, seed numbers and therefore seed yield would depend primarily on conditions governing floret site utilization.

Crop management should therefore be aimed initially at ensuring that fertile tiller numbers are not a limiting factor. The population required will vary with species and cultivar. Some guidelines are presented in Table 3.3.

The origin of tillers, tiller hierarchies and environmental factors affecting tillering have been reviewed in Chapter 2. The grass plant consists of a collection of tillers forming an integrated system with a certain amount of nutritional interdependence (Langer, 1979), and positional effects are very important in terms of the ability to flower and size of the inflorescence (see Chapter 2). However, what are the factors which affect fertile tiller number at harvest?

Factors affecting fertile tiller number at harvest

Sowing date and rate. For species capable of establishing rapidly under a range of environments (e.g. annual and perennial ryegrasses), sowing date has little effect on fertile tiller number at harvest (Hebblethwaite and Peirson, 1983). However, for temperature-sensitive species such as cocksfoot (*Dactylis glomerata* L.) and tall fescue, delaying autumn sowing reduces fertile tiller number because fewer are produced (Hare, 1994; Wilson et al., 1994; Chapter 2).

Fertile tiller numbers are usually not affected by sowing rate, unless extremely high or low rates are used. Meijer (1984) recorded no significant differences in fertile tiller number of Kentucky bluegrass and red fescue (*Festuca rubra* L.) at sowing rates ranging from 3 to 24 kg ha^{-1}. Generally as tiller production increases as a result of increasing sowing rate or row spacing, the percentage of tillers which become fertile reduces (Meijer, 1984; Hampton, 1989).

Nitrogen fertilizer. Provided that other elements are not limiting, nitrogen is the element which has the largest influence on seed yield in forage grasses, primarily through its influence on tillering and fertile tiller numbers (see Chapters 2 and 5). In most species delaying nitrogen application in the spring decreases fertile tiller number (Nordestgaard, 1983), but different species responses have been reported for autumn-applied nitrogen; in Kentucky bluegrass, cocksfoot and red fescue, splitting nitrogen application between autumn and spring produced more fertile tillers than spring application alone (Nordestgaard, 1986), but this did not occur in annual or perennial ryegrass (Hampton, 1987).

Increasing nitrogen rate increases fertile tiller production. However this response appears to be linear only up to a certain point (e.g. around 130 kg N ha^{-1} for perennial ryegrass, Hampton et al., 1983), and further increases may either fail to further increase (Hampton, 1987) or actually decrease (Nordestgaard, 1986) fertile tiller numbers because tillers either die in a lodged canopy or fail to compete with increased numbers of vegetative tillers (see Chapter 5).

Defoliation. The effect of defoliation (cutting or grazing) on fertile tiller numbers depends on the species and the defoliation height. Winter grazing increases fertile tiller number in cocksfoot and tall fescue, but has little effect on red fescue (Green and Evans, 1957). In perennial ryegrass lax (to 7 cm) spring defoliation does not reduce fertile tiller number because of the rapid replacement of autumn-formed tillers by spring-formed tillers (Griffiths et al., 1967), while severe (to 2.5 cm) spring defoliation does reduce fertile tiller numbers (Hebblethwaite and Clemence, 1983). In general close cutting or hard grazing is detrimental for fertile tiller numbers, and particularly in species such as cocksfoot, timothy (*Phleum pratense* L.) and prairie grass (*Bromus willdenowii* Kunth) where few spring-formed tillers contribute to seed yield (Langer, 1980).

Lodging. Fertile tillers die between anthesis and final harvest, and Hebblethwaite *et al.* (1980) showed that in perennial ryegrass, prevention of lodging decreased fertile tiller death by 13–22% depending on cultivar. However lodging does not always reduce fertile tiller number, as Hampton and Hebblethwaite (1985) showed that fertile tiller survival does not always differ between lodged and non-lodged perennial ryegrass plots. The effect of lodging will depend on its severity and time of onset (see Chapter 5). The application of plant growth regulators which reduce apical dominance can modify tillering patterns, and a product such as paclobutrazol can increase fertile tiller numbers (Hampton and Hebblethwaite,

1985). However these responses do not always occur and are dependent on application time, species and season (see Chapter 5).

Weeds. Brown *et al.* (1983) and Oswald (1985) reported that the presence of grass weeds (*Poa annua* L. and *Bromus sterilis* L.) reduced fertile tiller number in cocksfoot and perennial ryegrass respectively. Where seed yields are substantially reduced by weed competition (see Chapter 5), there is usually a positive linear relationship between fertile tiller number and seed yield (Rolston and Hare, 1986).

Irrigation. Moisture stress can significantly reduce fertile tiller numbers, particularly in late flowering cultivars or species (Hebblethwaite *et al.*, 1980). However responses to irrigation will differ from season to season unless the predictability of soil moisture deficits is known and consistent. In an early flowering cultivar of perennial ryegrass, Rolston *et al.* (1994) reported that irrigation to avoid moisture stress between spikelet initiation and anthesis reduced fertile tiller numbers cf. dryland plots, while Hebblethwaite *et al.* (1980) found that irrigating to avoid moisture stress over this same crop development period promoted vegetative tillering and not fertile tiller numbers (see Chapter 5).

Postharvest management. Postharvest treatments such as burning, close mechanical defoliation plus removal of residues, and grazing (see Chapter 8) may increase fertile tiller numbers in the succeeding crop, because residues can restrict tiller development (Chilcote *et al.*, 1980). Meijer (1987) suggested that the effect of residue removal was more likely to be associated with better tiller survival because of decreased competition, particularly for light, rather than solely to the stimulation of tillering.

Cultivar. Within a species, cultivars can differ significantly in their ability to tiller, and hence their ability to produce and sustain fertile tillers (Nordestgaard and Anderson, 1991). For example, in perennial ryegrass Anderson (1982) was able to place 36 Danish cultivars into three seed-yield groups based on their ability to produce fertile tillers when grown in the same environment and under the same management (see also Chapter 10).

3.1.3 Spikelets per Tiller

Relationship to seed yield

Provided a seed crop is managed to ensure maximum survival of an adequate number of autumn-formed tillers (see Chapter 5), the number of spikelets per tiller is reasonably constant within a crop, but may vary between sites, seasons, species and cultivars. Usually, therefore, seed yield is not influenced by the number of spikelets per tiller. Elgersma (1990) found that the number of spikelets per tiller was not correlated with the seed yield of nine perennial ryegrass cultivars grown in different environments, even though this yield component did differ among cultivars, and a similar result was reported by Hare (1994) for tall fescue. Similarly, Hampton and Hebblethwaite (1983) found no relationship between spikelets per tiller and seed number in perennial ryegrass. However if for environmental or poor management reasons autumn-formed tillers are not present in sufficient numbers or are removed, then spikelet number may affect seed yield (Hare *et al.*, 1988).

Factors affecting spikelets per tiller at harvest

Tiller age and number. The size of an individual grass inflorescence is determined by the environment, particularly the external conditions prevailing soon after floral initiation (see Chapter 2). Spikelets arise from buds developing in the axils of the leaf primordia accumulated at the shoot apex, and the number of spikelets depends on the number of primordia accumulated (Ryle, 1964). Tillers formed in autumn and winter have accumulated more unexpanded leaf primordia than those arising in early spring, and therefore have a higher spikelet number per tiller. Because of the relationship between tiller age and inflorescence size, any treatment which affects the contribution of different tiller age groups to the seed-head population can affect the number of spikelets per tiller. For example, in perennial ryegrass delaying sowing in the autumn (Hebblethwaite and Peirson, 1983; Table 3.4), spring defoliation (Hebblethwaite and Clemence, 1983; Table 3.4) and increasing nitrogen application rate (Hampton *et al.*, 1983) all decreased spikelets per tiller because these treatments either decreased autumn-formed tillers (sowing date, defoliation) and/or encouraged the growth of spring-formed tillers (defoliation, nitrogen).

Cultivar. Within a species, significant differences for spikelets per tiller exist among cultivars; Elgersma (1990) reported a range from 19.7 to 22.3 among nine perennial ryegrass cultivars in one season, and from 21.7 to 24.8 for the same nine cultivars in the following season. This between-season variation can be extensive. Hampton *et al.* (1983) recorded a difference of nearly five spikelets per tiller from the same perennial ryegrass cultivar at the same site over two consecutive seasons. While the maximum potential number of spikelets is determined by the genetics of the cultivar, the expression of that potential depends to a certain extent on the environment (see Chapter 2).

Table 3.4. Effect of autumn sowing date[1] and defoliation[2] on spikelets per tiller in perennial ryegrass.

Sowing date	Spikelets per tiller	Defoliation date	Spikelets per tiller
Mid August	20.5	Autumn	20.2
Early September	17.8	Floret initiation	17.4
Mid September	16.2	Ear emergence	16.4
Early October	14.6		

[1]Hebblethwaite and Peirson (1983).
[2]Hebblethwaite and Clemence (1983).

3.1.4 Florets per Spikelet

Relationship to seed yield

As with spikelets per tiller, florets per spikelet are not usually correlated with seed yield (Elgersma, 1990; Hare, 1994) unless management practices favour the production of large numbers of spring-formed tillers.

Factors affecting florets per spikelet at harvest

Tiller age and number. The number of florets per spikelet is influenced by the environment (Ryle, 1966) and by tiller age and number. Hill and Watkin (1975) reported a reduction from 6.6 to 4.6 florets per spikelet in perennial ryegrass as inflorescence emergence was delayed in the spring and similar results have been reported for other species (Ryle, 1964). Similarly, treatments which reduce the number of older tillers, such as spring defoliation, will reduce the number of florets per spikelet (Hebblethwaite and Clemence, 1983). In comparisons between seasons (Hare, 1994) or sowing times (Wilson *et al.*, 1994), florets per spikelet decrease as fertile tiller numbers increase.

Spikelet position. The number of florets per spikelet may vary with spikelet position on the inflorescence and is usually greatest in the mid-region (Table 3.5).

Cultivar. Significant differences in florets per spikelet among cultivars have been recorded within species; e.g. Hare (1994) reported a range from 6.4 to 6.9 florets per spikelet among three tall fescue cultivars for a first year seed crop, and from 5.5 to 6.0 for the same three cultivars at the second year harvest.

Shedding. Florets are shed from the spikelet between anthesis and harvest. However the extent and therefore significance may vary with species. Meijer (1985), for example, recorded a 25–30% reduction in florets per fertile tiller in perennial ryegrass, and Mares Martins and Gamble (1993b) found that from 5 to 15% of these florets had been shed by 10 days after peak anthesis. In contrast, Hare (1992a) recorded a reduction in florets of less than 5% in tall fescue. Mares Martins and Gamble (1993b) found that growth regulator application significantly reduced floret shedding, and suggested that these gibberellic acid (GA) biosynthesis inhibitors reduce floret abscission.

Table 3.5. Effect of spikelet position on the inflorescence on floret number in perennial ryegrass[1] and tall fescue[2].

	Florets per spikelet	
Spikelet position	Perennial ryegrass	Tall fescue
Basal	6.3	5.4
Intermediate	6.8	6.5
Penultimate	5.2	6.3

[1] Hampton (1987).
[2] Hare (1992a).

3.1.5 Floret Site Utilization (FSU)

Relationship to seed yield

Hebblethwaite *et al.* (1980) divided seed yield in grasses into two stages, establishment of the yield potential and utilization of the yield potential. If yield potential is defined as the number of florets per unit area at anthesis (Hebblethwaite and Hampton, 1982), then the yield components discussed so far, i.e. fertile tillers, spikelets per tiller and florets per spikelet, determine that potential. Utilization of that potential is determined by the events at and after anthesis – pollination, fertilization, seed set and seed development (Elgersma, 1985), and excluding the final seed size, this therefore depends on FSU.

Low FSU is the most important limiting factor for high seed yield in many grass species (Elgersma, 1985). However reported correlations between FSU and seed yield range from highly significant (Hampton and Hebblethwaite, 1983) to non-significant (Elgersma, 1990), and percentage FSU in perennial ryegrass for example has been reported to range from 8% to over 90% (see Elgersma and Sniezko, 1988). One possible explanation for this apparent wide variation is that authors are not necessarily reporting (or calculating) the same data because of differences in definition. Elgersma (1985) pointed out the importance of distinguishing between 'biological' and 'economic' FSU, the former being the percentage of florets present at anthesis resulting in a viable seed, while the latter is defined as the percentage of florets present at anthesis resulting in a cleaned pure seed (Elgersma, 1991).

Factors affecting floret site utilization

Losses during seed development. Elgersma (1991) reviewed floret site utilization (FSU), and a full discussion of the relationship between the mother plant, the environment and the developing seed is presented in Chapter 4. In brief the key factors are that around 20% of florets may not be successfully pollinated and fertilized (Mares Martins and Gamble, 1993a,b), and that ovule degeneration/seed abortion can occur within a few days after fertilization (Hill, 1980; Elgersma and Sniezko, 1988); e.g. in perennial ryegrass, around 30% of seeds initially set had aborted within 10 days of fertilization (Mares Martins and Gamble, 1993a), and most of this loss occurred within the first week following fertilization. Abortion after this time is believed to be no more than 5% (Elgersma and Sniezko, 1988).

Both Elgersma and Sniezko (1988) and Mares Martins and Gamble (1993a) considered that seed abortion appears to be independent of assimilate supply, and genetic or cytological factors may be responsible for the failure of seed development (Marshall and Ludlam, 1989). However, Hampton (1988) considered that a major reason for poor floret utilization was abortion of developing seeds because insufficient assimilate was available to satisfy the demands of seed growth at all pollinated sites, an hypothesis originally proposed by Burbidge *et al.* (1978) and indirectly supported by results from many field trials with perennial ryegrass (see Hebblethwaite, 1987; Hampton, 1991a). Whatever the reason(s), the net result is that between 20 and 50% of florets are biologically unproductive, i.e. they do not produce a viable seed.

Management practices which avoid nitrogen stress, prolong photosynthetic tissue duration and/or control disease, and prevent lodging and/or increase fertile tiller survival (see Chapter 5) can increase FSU (Tables 3.6, 3.7), although often these data have been reported as increases in seeds per spikelet (see Hampton, 1988, 1991a). However seed yields can be increased without any change or even a decrease in FSU; e.g. in cultivar Royal (Table 3.6), the highest rate of nitrogen applied decreased FSU but increased seed yield because more fertile tillers were maintained, and a similar reason explained the 1981 seed yield increase in cultivar 'S24' following paclobutrazol application at 2.0 kg a.i. ha^{-1} at spikelet initiation (Table 3.7). Interestingly, cultivars may have different FSU responses to the same treatments at the same site (Table 3.7; Wilson et al., 1994).

Table 3.6. Effect of nitrogen and cultivar on economic floret size utilization (FSU) in perennial ryegrass grown at the same site in the same season[1].

Cultivar	Nitrogen applied (kg ha^{-1})	Seed yield (g m^{-2})	Fertile tillers (no. m^{-2})	TSW (g)[2]	Florets per tiller	% Economic FSU
'Royal'	0	132	2509	2.06	210	12.1
	80	176	2743	2.03	182	17.3
	160	160	3715	2.11	255	8.0
'Morenne'	0	148	2027	1.93	201	18.8
	80	185	2420	1.92	190	20.9
	160	195	2512	1.89	206	19.9

[1]Data calculated from Hampton et al. (1983).
[2]Thousand seed weight.

Table 3.7. Effect of paclobutrazol application to perennial ryegrass[1] and fungicide application to tall fescue[2] on economic floret size utilization (FSU).

	Seed yield (g m^{-2})	Fertile tillers (no. m^{-2})	TSW (g)	Florets per tiller	% Economic FSU
Paclobutrazol					
1981					
0	139	2777	1.85	176	15.3
2[3]	180	3611	1.83	191	14.2
1982					
0	111	2468	1.85	155	15.6
2[3]	256	3285	1.76	153	28.9
Propiconazole applied at					
Nil	97	399	2.72	488	18.3
Ear emergence	201	390	2.92	465	37.9
Anthesis	152	385	2.87	463	29.6

[1]Data calculated from Hampton (1983) and Hampton and Hebblethwaite (1985) for perennial ryegrass cultivar 'S24'.
[2]Data calculated from Hare (1992a) for tall fescue cultivar 'Grasslands Roa'. Propiconazole was applied at 250 g a.i. ha^{-1} at either ear emergence or anthesis.
[3]Applied at spikelet initiation at a rate of 2 kg a.i. ha^{-1}.

Losses during harvest and cleaning. Elgersma (1991) estimated for every 100 florets counted at anthesis, 65 florets produced a seed available to harvest, but that after processing only 25 seeds would be left. While seed shedding or shattering can be an important source of floret and seed loss in some species (Meijer, 1985; Chapters 4 and 8), Elgersma (1991) considered that harvesting and cleaning losses were responsible for the majority of the 'lost' seeds, either because they were immature (too small or light), or had been damaged in some way. However, Roy *et al.* (1994) reported that cleaning losses (small, light or double seeds) in perennial ryegrass accounted for only 6–10% of the florets produced, and concluded that the major reason for poor FSU was the failure of up to 75% of the florets to produce a seed. This is obviously an area of grass seed production which requires further detailed research.

3.1.6 Seed Weight

Relationship to seed yield

The final weight of an individual seed depends mainly on its position within a spikelet (Anslow, 1964), and before cleaning there can be a wide range of individual seed weights within a seed lot. However the cleaning process removes small and light seeds, so that much of this variation is reduced, and the minimum seed weight for the lot depends on the cleaning efficiency and intensity. As seed weight is traditionally expressed as thousand seed weight (TSW) which is recorded after cleaning, TSW tends to be relatively constant and not significantly related to either seed number or yield (Hampton and Hebblethwaite, 1983).

Factors affecting seed weight

Seed weight is often negatively correlated with the number of fertile tillers and seeds per spikelet (Hampton *et al.*, 1985) and floret and spikelet position on the inflorescence (Anslow, 1964). These patterns reflect the relative activities of different sinks within the inflorescence (see Chapter 4). Seed weight can be increased through the application of nitrogen, growth regulators and fungicide (see Chapter 5), but often seed weight does not respond to these treatments, or may be reduced (Hampton *et al.*, 1985). Small variations in seed weight from season to season reflect the flexibility of the grass plant in its adjustment of the yield components to various environmental factors.

3.2 LEGUMES

3.2.1 Introduction

While there are over 4000 temperate legume species, only lucerne (*Medicago sativa* L.), white clover (*Trifolium repens* L.), red clover (*T. pratense* L.), subterranean clover (*T. subterraneum* L.) and birdsfoot trefoil *(Lotus corniculatus* L.) are used extensively in the agriculture of the temperate regions of the world (Lorenzetti, 1993). Many of the remaining legume species are used as sources of germplasm for breeding programmes or for particular end uses in specific climatic and edaphic conditions (Kelly, 1988; Hampton, 1991a). Consequently, most studies on forage legume seed yield components have been conducted with these few species in northerly latitudes, to accommodate the long-day requirement for flowering and the indeterminate growth habit. Although winter survival, insects for pollination, and weather at harvest alter yield components and seed yield, and are critical factors that define where legumes can be grown for seed (Hides *et al.*, 1984; Miller and Steiner, 1995), a major limitation is imposed by latitude. For instance, Valle (1963) observed a different pattern of flowering in Finnish white clover when grown in the USA at two latitudes (40 and 32°N), with flowering being greater at the higher latitude. In red clover, Taylor *et al.* (1990) suggested that cultivars originating between latitude 37 and 38°N should not be grown for seed at latitudes below 40°N to prevent genetic shift emanating from unequal flowering of constituent germplasm. In birdsfoot trefoil grown for seed at low latitudes, flowering is irregular, and occurs over a long period of time, resulting in excessive bloom, extended pollination and low seed set (Beuselinck and McGraw, 1988). Latitude plays a major role in induction and initiation (see Chapter 2), as does crop management (see Chapter 6), in determining both the size of each yield component and its contribution to seed yield. A major limitation in understanding the role of yield components in the determination of seed yield is the relative lack of data on the role that environmental conditions (daylength, temperature, precipitation, heat units . . .) play in seed development (see Hacquet, 1990; Hampton, 1990).

Traditionally, yield components have been measured to explain differences in seed yield between cultivars, to develop selection criteria for breeding higher-yielding cultivars, or for selection of germplasm from different sources for a breeding programme. Some of these aspects are dealt with in detail in Chapter 10.

Measurements of yield components are usually made at harvest when the final seed yield has been determined, and are therefore an end-result of processes that have been altered by inter- or intra-plant interactions at various stages during ontogeny. These changes during ontogeny are influenced by a number of factors and may vary considerably. In most instances, measurements on individual components serve to illustrate the point that differences in final seed yield are the result of previous dynamic processes and events such as the extent of pollination or the control of assimilate partitioning. In most studies, measurements of yield components are often made on the same crop sample that is used for the measurement of total yield. Time and labour constraints, and the growth habit of many legumes, do not always permit an accurate measurement of the individual number of shoots on a per-plant basis, and this is often replaced by a count of the number of fertile shoots per unit area. Inflorescence numbers may be assessed on a per-shoot, per-plant or per-unit area basis. Seed numbers per pod are often averaged numbers for a plant or shoot, as is the weight of the seeds in a pod. Yield component counts are made with greater ease and accuracy in spaced populations; however, this does occur at the cost of some compensation due to inter- and intra-plant competition, and may not agronomically parallel spaced row or sward situations.

3.2.2 Potential Seed Yield and Floral Structure

Potential seed yield is governed by the number of ovules per unit area at anthesis (Lorenzetti, 1993) but the proportion of ovules that develop into mature seed may be quite low (Table 3.8). Seed-yield components represent the size of the reproductive system, which is dictated, for example, by the number of plants per unit area, fertile shoots per plant, inflorescences/racemes per shoot, florets per inflorescence and ovules per floret, and the efficiency of seed setting and maturation. In forage legumes, the fruit is a pod that encloses one or several seeds. The nature and influence of each of these components on seed yield of forage legumes is discussed in this section.

The type of inflorescence determines the number that are borne on a fertile shoot (Gill and Vear, 1966). The flowers in lucerne are in an oblong axillary raceme with 5–40 flowers. Pods are 3–9 mm in diameter with two or three spirals and two to nine seeds per pod.

The flowers of white clover are in globular umbel-like heads, without bracts, borne singly on long peduncles in the axils of the stolon leaves; each head has 20–40 flowers and the pods are sessile with three or four seeds per pod and 75–100 seeds per head. In one study (Clifford and Baird, 1993), five flowering types were described, ranging from mainly crown flowering to mainly stolon-tip flowering. The dominantly crown-flowering types produced the highest seed yield.

The flowers in red clover are densely crowded in mostly terminal, ovoid, racemes subtended by a leaf-like bract. The pods are very short, and oblong to ovoid in shape, single seeded, and open transversely as a pyxidium.

Subterranean clover flowers are in small axillary racemes with two to five fertile flowers; the numerous upper flowers become sterile after anthesis and consist of only a small stiffly toothed calyx. The sterile flowers become reflexed and completely enclose the pods, acting as barbs that help bury the fruit in the soil. The burr-like fruiting heads are raked up and threshed to free the seed from the single-seeded pods that are enclosed within the calyx of the fertile flowers.

The inflorescence in birdsfoot trefoil is an axillary, long-stalked, umbel of three to six flowers. The fruit is a slender cylindrical straight pod about 2.5 cm long, 3 mm wide, and has 5–20 seeds attached to the ventral suture; at maturity, the pods split along both sutures and twist spirally to release the seeds.

Table 3.8. Estimates of the potential and actual seed yield of some forage legumes (Lorenzetti, 1993).

Species	Inflo-rescences (no. m^{-2})	Flowers per inflo-rescence	Ovules per flower	TSW (g)	PSY (t ha^{-1})	OSU (%)	ARSP t ha^{-1}	ARSP % of PSY
Medicago sativa	3750	16	10	2.0	12.0	8	0.5	4
Trifolium repens	600	100	6	0.5	1.8	50	0.4	22
Trifolium pratense	750	110	2	1.6	2.6	25	0.6	23
Lotus corniculatus	400	6	40	1.2	1.2	40	0.2	17

ARSP, agriculturally realized seed potential; OSU, ovule site utilization; PSY, potential seed yield; TSW, thousand seed weight.

3.2.3 Plants per Unit Area

Seed stands are managed to minimize excessive vegetative growth (e.g. by adjusting time of closing, the use of growth regulators, timing of irrigation, cutting, grazing etc.; see Chapter 6). In addition, sowing rates for seed crops are usually lower than those for forage production. These rates vary between 1 and 3 kg ha^{-1} for most major species such as lucerne, red clover, white clover and birdsfoot trefoil (Canadian Seed Growers Association, 1980). Sowing rates are generally higher when establishing with a companion crop (Clifford and Anderson, 1980) and/or the sowing rate of the companion crop is often halved to reduce competition to the legume (Fairey and Lefkovitch, 1991). At these low sowing rates appropriate plant densities for maximum seed production can be achieved. However, similar seed yields can be obtained over a considerable range of plant populations (Kowithayakorn and Hill, 1982; McGraw *et al.*, 1986; Marshall and James, 1988), because of complex compensatory growth. The location and management inputs have a major effect (see Chapter 6).

3.2.4 Shoots per Plant or per Unit Area

In a number of studies, the relationship between the number of fertile shoots per unit area rather than the number of fertile shoots per plant has been correlated with seed yield. The number of fertile shoots per unit area does not appear to have a predominant effect on seed yield of lucerne, where stem production is indeterminate and each stem continues to produce both leaves and flowers (Teuber and Brick, 1988). In a comparison of the yield differences among three lucerne cultivars, Unita (high yield), Ranger (intermediate yield) and Lahontan (low yield), Pedersen *et al.* (1956) found that the number of fertile shoots per unit area did not account for significant differences in seed yield. However, any discussion on the contribution of shoot numbers to seed yield cannot be made in absolute terms, but must allow for the age of the seed stand. Thus, while differences in shoot number may not account for differences in seed yield in any one year, shoot numbers can have a significant effect on seed yield between years. For instance, shoot numbers in Canadian lucerne cultivars dropped from 99 in the first seed harvest to 74 per 1-m row (0.6 m spacing) in the fifth seed harvest while, over the same period, seed yields declined from 429 to 239 kg ha^{-1} (Fairey and Lefkovitch, 1992). Thinning of older stands with consequent declines in seed yield have also been observed in red clover in Norway, where a 40% reduction in the yield of tetraploid cultivars was observed between the first and second harvest year (Aamlid, 1996). On the other hand, those legumes that spread by stolons (Lewis *et al.*, 1983) or rhizomes (Hare, 1992b) are more likely to produce an excessive number of shoots that usually make a greater contribution to vegetative growth as compared to reproductive growth and, consequently, seed stands require thinning by chemical or mechanical means. Defoliation, irrigation, application of growth regulators, and timing of closing by removal of grazing animals at different times are some factors that have been manipulated to promote flowering at the expense of vegetative growth (see Chapter 6).

New shoots are produced during most of the growing season in legumes because the apical meristem remains vegetative. In birdsfoot trefoil, Li and Hill (1989) found that seed yield was governed mainly by the number of shoots. However, optimal shoot numbers had to be produced before peak flowering. New shoots formed after flowering probably competed with the developing seed for assimilates and had a negative effect on seed yield. A similar trend was observed in big trefoil (*L. uliginosis* Schk.) (Tabora and Hill, 1991), where 78% of the pod-bearing umbels at seed harvest came from shoots formed early in the growing season. Of all the pod-bearing umbels, 38% were from main shoots, 53% were from primary lateral shoots, and 9% were from secondary lateral shoots.

Other studies indicate that the number of shoots is relatively unimportant because, within limits, there is a compensation between the number and yield of each shoot depending on plant density. For example, in the establishment year of a field experiment with two lucerne cultivars sown at two locations and 15 rates from 2.2 to 33.6 kg ha^{-1} pure live seed (Kephart *et al.*, 1992), plant density increased linearly with sowing rate. Plant density continued to be affected by sowing rate 4 years after establishment, but the greater plant mortality at high sowing rates caused the response to deviate from linearity. Shoots per plant were negatively correlated with plant density ($r = -0.73$, $P = 0.01$).

3.2.5 Inflorescences per Shoot or per Unit Area

Inflorescence numbers have a significant impact on seed yield of most lucerne cultivars (Table 3.9). Positive correlations between inflorescence numbers and seed yield for lucerne of 0.87 (Pedersen *et al.*, 1959) and 0.73 (Hurst and Pedersen, 1964) have been reported. An increase in the number of racemes following application of the growth regulator paclobutrazol resulted in increased seed yield (Askarian *et al.*, 1994).

In examining 63 agronomically superior, spaced, plants of white clover, Clifford and Baird (1993) found that the number of inflorescences per unit area was

Table 3.9. Correlations between lucerne seed yield and yield components. (Adapted from Hacquet, 1990.)

Country	Racemes/stem or/plant	Pods/raceme	Seeds/pod	Seed weight
USA	0.54*	0.68*	0.57*	
USA		S*	S*	–S*
USA	0.13	0.10	0.48**	–0.10
France			0.62	–0.18
Italy		S*	S*	
USA	S*	S*		
USA	S**		0.52**	0.06
Poland	S***		0.65**	
Russia			S**	
Australia	0.84**		0.77**	
Hungary			0.58**	

*, **, ***, $P < 0.05, 0.01, 0.001$, respectively; S, significant correlation coefficient not given.

the major determinant of seed yield. Maximizing inflorescence production generally increases seed yield despite the fact that plants have some ability to compensate for a reduced inflorescence number. While increasing inflorescence number, by delaying closing, has resulted in higher seed yields (Clifford, 1985), other studies (Clifford, 1987; Marshall and Hides, 1990) have shown that delaying the closing of seed crops of this species often results in a decrease in the number of florets per inflorescence and seed weight. Similar results have been found with moisture stress where the deleterious effects (reduced number of florets or ovules per inflorescence) in white clover are minimized by increases in inflorescence density (Clifford, 1987; Danyach-Deschamps and Wery, 1988). On the other hand, Binek (1983) found that removal of up to 48% of the inflorescences in the white clover cultivar Podkowa did not reduce seed yield because of compensatory increases in other reproductive components in the canopy.

In a study on the seed-producing capability of 15 white clover cultivars, Evans *et al.* (1986) observed that the high yields of cultivars Olwen and Menna were accounted for by the total number of inflorescences produced over a narrow period of time; the result being a high proportion of ripe inflorescences at final harvest. A low-yielding cultivar, AC 20, flowered late and produced few inflorescences. Another cultivar, Anna, had a prolonged flowering period, with a low percentage of ripe inflorescences at harvest and a low seed yield.

In a New Zealand study (Li and Hill, 1989), changes in yield components and their relationship to final seed yield were monitored in birdsfoot trefoil during the 3-month flowering period using a tagged population of approximately 2000 inflorescences on shoots originating in different months and at different positions on the plant. The number of inflorescences per unit area was the most important single yield component, and was positively correlated ($r = 0.97$ at $P < 0.01$) with final seed yield. Similar findings were obtained for other cultivars (Stephenson, 1984; McGraw *et al.*, 1986) and for several local and foreign agroecotypes of birdsfoot trefoil evaluated over a 3-year period in Romania (Dragomir, 1981).

An increase in inflorescence numbers with growth regulators has been shown to increase seed yields. Paclobutrazol increased inflorescence production per unit area in white clover (Hampton, 1991b; Budhianto *et al.*, 1994) but not in all experiments (Boelt and Nordestgaard, 1993). Hampton (1991b) reported that paclobutrazol advanced inflorescence appearance in white clover and Budhianto *et al.* (1994) observed that inflorescences from individual plants treated with paclobutrazol ripened earlier than those from untreated plants.

3.2.6 Florets per Inflorescence

In red clover, floret number per inflorescence increased from 99 at 12°C to 103 and 102 at 16 and 20°C, respectively (Puri and Laidlaw, 1984). A similar observation was recorded with white clover where the number of florets per inflorescence increased between temperatures of 7 and 11°C (Zaleski, 1964; Norris, 1985; Pasumarty *et al.*, 1995). However, optimum temperatures vary among white clover cultivars and a detailed analysis of the effect of temperature on the components of seed yield has not been made. Moisture stress over the reproductive phase favours nutrient partitioning to vegetative growth in preference to reproductive growth, and both inflorescence and floret numbers are reduced (Clifford, 1987). Conversely, unlimited soil moisture favours increased leaf size and reduced seed yield.

In birdsfoot trefoil, the changing pattern of each seed yield component during the extended 3-month flowering period was monitored (Li and Hill, 1989). The number of florets per inflorescence declined with time but was not significantly correlated with seed yield. Total inflorescence number at harvest was the single most important yield component. The flower-carrying ability (inflorescences per shoot and florets per inflorescence) of shoots gradually declined with age. However, as the stand aged, a compensatory increase in shoot numbers occurred and individual shoots bore approximately three inflorescences with about six florets per inflorescence.

Potential seed yield is lost as a result of abortion of florets between development of flower buds to mature pods (Seany and Hansen, 1970; Stephenson, 1984). In some instances, this has been attributed to self-incompatibility (Seaney, 1964), but in others a large proportion of florets in outcrossed flowers may also abort (Stephenson, 1984) and the exact cause is still undetermined.

3.2.7 Pods per Raceme or Flower

A positive correlation between pods per raceme and seed yield has been reported (see Hacquet, 1990; Table 3.9) among lucerne cultivars. In birdsfoot trefoil, only one in three flowers produces a mature pod and three of the five pods initiated subsequently abort (Stephenson, 1984; Li and Hill, 1989). Stephenson (1984) dismissed the possibility of abortion caused by lack of pollination and concluded that a lack of assimilate supply was the major factor. Li and Hill (1989) monitored changes in yield components in tagged shoots during the entire growing season and found that while external factors such as photoperiod and temperature varied during the growing season, the abortion rate remained constant at a one-pod loss per three to five inflorescences that were initiated throughout the flowering period, suggesting a possible 'internal regulating mechanism'.

3.2.8 Seeds per Pod and Seed Weight

In lucerne, the number of seeds per pod (two to nine) is a strongly heritable character and does not depend on the initial number of ovules per ovary (Demarly and Chesneaux, 1966; Dattée, 1972). While low pollen fertility (Blondon et al., 1979) and short pollen tubes (Rice et al., 1970) contributed to low fertilization levels, ovule abortion after fertilization was the major cause of low fertility and reduced seed numbers (Sayers and Murphy, 1966).

Similar findings were reported for white clover where no significant differences between ovule numbers were observed between high- and low-yielding cultivars (Parsumarty et al., 1993). However, the high-yielding cultivar had more fertile ovules and more seeds per pod. The major loss of seeds per pod was due to ovule sterility (30%) and seed abortion (33%) during the first week after pollination. In another study van Bogaert (1977) showed that in cultivars with fertilization levels of 42% and 65%, the average number of seeds per pod was 2.5 and 3.9, respectively. In spaced plants of white clover, Clifford and Baird (1993) found that a high harvest dry matter was associated with reduced numbers of seeds per pod and total seed yield, indicating the possible role of assimilates in determining seed number.

In tetraploid red clover where there are usually problems with pollination (see Chapter 7), selections for single characters such as pollen grain fertility, meiosis regularity, and corolla tube length were not very promising, and better results were later obtained by simply selecting for a high number of seeds per head. Seed set in cultivar Sally was improved by this technique and this was the first Swedish tetraploid red clover cultivar that, under practical field conditions, gave an acceptable seed yield (Sjödin, 1981). Seed set in red clover has been shown to increase with the application of the growth regulator daminozide applied early in the season (Puri and Laidlaw, 1983; Christie and Choo, 1990); however seed weight was reduced (Mela, 1969).

In birdsfoot trefoil, only 40% of the 20–70 ovules in an ovary develop into mature seeds and, in different studies, seed numbers varying between 8 and 20 per pod have been reported (see Seaney and Hanson, 1970). The number of seeds per pod remain relatively constant with only minor fluctuations during the entire flowering period and various factors such as boron treatment, photoperiod, temperature or assimilate limitations do not appear to affect this yield component (Li and Hill, 1989).

The relationship between total flowering and the extent of pollination has an important bearing on seed weight. In many instances, enhanced pollination increases seed set and total seed yield despite decreases in seed weight. High levels of pollination result in higher numbers of seeds and account for the higher seed yield of lucerne, despite a reduced seed weight (Pedersen *et al.*, 1956).

Pedersen and Nye (1962) found that high lucerne seed yield was accounted for by a high number of seeds per pod despite reduced seed and pod weight. Rosellini *et al.* (1994) measured the yield components and seed yield in four lucerne progenies for two successive years and concluded that yield components could not be used to predict seed yield. There was no significant correlation between yield components and seed yield in one year of the study; in the other, seed yield and seeds per pod were significantly correlated. This is in agreement with another study on lucerne where the highest-yielding cultivar had a high number of seeds per pod (Pedersen *et al.*, 1956). Seeds per pod has been shown to be correlated with the degree of crossing (Pedersen, 1968), or bee activity, and is consequently some measure of pollination activity.

In situations where high seed weights are obtained at the expense of seed numbers and a high seed number is necessary, such as in annual self-seeding legumes in pastures, the selection for a high seed weight is undesirable (Cocks, 1990). Lorenzetti (1993) suggested that selection for high seed weight could be justified only when it is considered a factor of seed quality or in obtaining good crop establishment (Evans, 1965; McKersie and Tomes, 1982).

In white clover, seed size varies considerably but is not considered an important agronomic character; its inheritance has apparently not been studied (Williams, 1987). Furthermore, large seed does not necessarily establish better than small seed and any gain in unit seed weight is usually at the expense of yield per inflorescence (Clifford and Baird, 1993). Management practices that enhance seed weight reduce seed yield (Hampton *et al.*, 1987). In cultivar Podkowa removal of about one half of the inflorescences did not affect seed yield because of compensatory increases of between 58 and 78% in both the weight and the number of seeds per pod (Binek, 1983).

Changes in seed weight have been observed with the use of growth regulators but the effects are often inconsistent. Paclobutrazol increased seed weight in some studies (Marshall and Hides, 1991) but reduced it in others (Budhianto *et al.*, 1994). In red clover, the primary effect of the growth regulator daminozide was to improve pollination by delaying flowering (Mela, 1969) and reducing both the height of the nectar column (Puri and Laidlaw, 1983) and corolla tube length (Mela, 1969; Puri and Laidlaw, 1983; Christie and Choo, 1990). Both bumble bees and honey bees were approximately 34–40% more numerous on plots treated with daminozide than on untreated plots (Mela, 1969). While seed set was increased (Puri and Laidlaw, 1983; Christie and Choo, 1990), seed weight was reduced (Mela, 1969).

From a practical viewpoint, despite the wide range in individual seed weights within a seed lot, the thousand seed weight of a processed seed lot tends to be relatively constant over a range of conditions (Marshall, 1985). This is because very small and light seeds are usually eliminated during the cleaning process and thousand seed weight is generally a function of the cleaning intensity.

3.2.9 Other Factors

There are a number of other factors that have a direct impact on the components of yield that operate outside of the compensatory effects between the components themselves. These include a range of factors that affect the floral and pollination processes. For example, boron probably plays a role in pollen tube elongation (Pilbeam and Kirby, 1983) which is essential for fertilization and seed set. That boron is essential in seed production of lucerne (Grizzard and Matthews, 1942), and red and white clovers (Sherrel, 1983), amongst other legumes, has been demonstrated. In Ladino white clover clones, selection based on seed set often increased ovule number and generally improved efficiency of pollination (Dessureaux, 1951; Cebrat *et al.*, 1982). Temperature has an effect on pollen fertility and the number of fertilized ovules increased as temperature increased between 17 and 27°C (Blondon *et al.*, 1979).

The measurement of the individual components of yield often illustrates that performance between cultivars cannot always be explained by a single yield component and that other processes such as the extent of pollination or the duration of seed ripening play an important role. Negative correlations or no correlation between seed yield and yield components have also been reported (see Table 3.9). An added complication in selecting for yield components is that crop management and inter- and intra-plant management alter components drastically. Yield components can, in specific conditions, be useful parameters to help determine management packages or assist in the selection of improved germplasm; these measurements do not, however, explain the physiological mechanisms that govern seed production.

REFERENCES

Aamlid, T.S. (1996) Frøavl av Kolpo rødkløver. In: Abrahamsen, U. (ed.) *Jord og plantekultur 1996. Korn – potet – miljø. Forsøksresultater 1995*. Planteforsk Apelsvoll forskingssenter, Norway, pp. 157–162.

Anderson, S. (1982) Frogivende engenskaber hos rajgres. *Dansk Froavl* 65, 160–164.

Anslow, R.C. (1964) Seed formation in perennial ryegrass. II. Maturation of seed. *Journal of the British Grassland Society* 19, 349–357.

Askarian M., Hampton, J.G. and Hill, M.J. (1994) Effect of paclobutrazol on seed yield of lucerne *(Medicago sativa* L.) cv. Grasslands Oranga. *Journal of Applied Seed Production* 12, 9–13.

Beuselinck, P.R. and McGraw, R.L. (1988) Indeterminate flowering and reproductive success in birdsfoot trefoil. *Crop Science* 28, 842–845.

Binek, A. (1983) [The structure of seed yield in clones of white clover (*Trifolium repens* L.) after the reduction of inflorescences to a standard number per plant.] *Acta Agraria et sylvestria, Agraria* 22, 21–29. (In Polish.)

Blondon, F., Cambier, B., Dattée, Y. and Guy, P. (1979) Influence de la temperature sur la fertilite male et femelle de luzerne: temoins, male-steriles et mainteneurs. *Annales Amélior Plant* 29, 89–96.

Boelt, B. and Nordestgaard, A. (1993) Growth regulation in white clover *(Trifolium repens* L.) grown for seed production. *Journal of Applied Seed Production* 11, 1–5.

Brown, K.R., Rolston, M.P. and Archie, W.J. (1983) 'Grassland Wana' cocksfoot seed production. *Proceedings of the New Zealand Grassland Association* 44, 24–29.

Budhianto, B., Hampton, J.G. and Hill, M.J. (1994) Effect of plant growth regulators on a white clover (*Trifolium repens* L.) seed crop. II. Seed yield components and seed yield. *Journal of Applied Seed Production* 12, 53–58.

Burbidge, A., Hebblethwaite, P.D. and Ivins, J.D. (1978) Lodging studies in *Lolium perenne* grown for seed. 2. Floret site utilization. *Journal of Agricultural Science, Cambridge* 90, 269–274.

Canadian Seed Growers Association (1980) *Pedigreed Forage Seed Production*. Canadian Seed Growers' Association, Box 8455, Ottawa, Ontario, Canada.

Cebrat, J., Kobierzynska-Golab, Z. and Ramenda, S. (1982) [The variability of quantitative characters which affect fertility in five varieties of white clover (*Trifolium repens* L.)] *Hodowla Foslin, Aklimatyzacjai Nasiennictwo* 26, 11–34. (In Polish.)

Chilcote, D.O., Youngberg, H.W., Stanwood, P.C. and Kim, S. (1980) Postharvest residue burning effects on perennial grass development and seed yield. In: Hebblethwaite, P.D. (ed.) *Seed Production*. Butterworths, London, pp. 91–103.

Christie, B.R. and Choo, T.M. (1990) Effect of harvest time and Alar-85 on seed yield of red clover. *Canadian Journal of Plant Science* 70, 869–871.

Clifford, P.T.P. (1985) Effect of cultural practice on potential seed yield of 'Grasslands Huia' and 'Grasslands Pitau' white clover. *New Zealand Journal of Experimental Agriculture* 13, 301–306.

Clifford, P.T.P. (1987) Producing high seed yields from high forage producing white clover cultivars. *Journal of Applied Seed Production* 5, 1–9.

Clifford, P.T.P. and Anderson, A.C. (1980) Red clover seed production – research and practice. In: Hare, M.D. and Brock, J.L. (eds) *Herbage Seed Production*. Grassland Research and Practice Series No 1. New Zealand Grassland Association, Palmerston North, pp. 76–79.

Clifford, P.T.P. and Baird, I.J. (1993) Seed yield potential of white clover: characteristics, components and compromise. *Proceedings of the XVII International Grassland Congress*, pp. 1678–1679.

Cocks, P.S. (1990) Dynamics of flower and pod production in annual medics (*Medicago* spp.) I. In spaced plants. *Australian Journal of Agricultural Research* 41, 911–921.

Danyach-Deschamps, M. and Wery, J. (1988) Effect of drought stress and mineral nitrogen supply on growth and seed yield of white clover in Mediterranean conditions. *Journal of Applied Seed Production* 6, 14–19.

Datté, Y. (1972) Analyse quantitative de l'auto et de l'interfertilité chez quelques familles de luzerne. *Annales Amélior Plant* 22(1), 5–21.

Demarly, Y. and Cheneaux, M.T. (1966) La culture de la luzerne en conditions artificielles. *Annales Amélior Plant* 16, 299–305.

Dessureaux, L. (1951) Ovule formation as a factor influencing seed setting of Ladino white clover. *Science in Agriculture* 31, 373–382.

Dragomir, N. (1981) Seed production by some agroecotypes of birdsfoot trefoil (*Lotus corniculatus* L.). *Analele Institutului de Cercetari pentru Cereale si Plante Tehnice, Fundulea* 48, 151–158. (In Romanian, English abstract.)

Elgersma, A. (1985) Floret site utilization in grasses: definitions, breeding perspectives and methodology. *Journal of Applied Seed Production* 3, 50–54.

Elgersma, A. (1990) Seed yield related to crop development and to yield components in nine cultivars of perennial ryegrass (*Lolium perenne* L.). *Euphytica* 49, 141–154.

Elgersma, A. (1991) Floret site utilization in perennial ryegrass (*Lolium perenne* L.). *Journal of Applied Seed Production* 9 (Suppl.), 38–43.

Elgersma, A. and Sniezko, R. (1988) Cytology of seed development related to floret position in perennial ryegrass (*Lolium perenne* L.) *Euphytica* S, 59–68.

Evans, A.M. (1965) Investigation on diploid and polyploid white clover. The significance of seed size and weight in white clover. In: *Welsh Plant Breeding Station Report for 1964*. Welsh Plant Breeding Station, Aberystwyth, pp. 42–46.

Evans, D.R., Williams, T.A. and Davies, W. Ellis (1986) Potential seed yield of white clover varieties. *Grass and Forage Science* 41, 221–227.

Fairey, D.T. and Lefkovitch, L.P. (1991) Establishing perennial legume seed stands with annual companion crops. *Journal of Applied Seed Production* 9, 49–54.

Fairey, D.T. and Lefkovitch, L.P. (1992) Seed yields of consecutive harvests from legume stands. *Journal of Applied Seed Production* 10, 25–30.

Falkowski, M., Kukulka, I. and Kozlowski, S. (1987) Relationship between the number of generative shoots and the yield of seed grasses. *Journal of Applied Seed Production* 5, 62 (abstract).

Gill, N.T. and Vear, K.C. (1966) *Agricultural Botany*. Gerald Duckworth, London.

Green, J.O. and Evans, T.A. (1957) Grazing management for seed production in leafy strains of grasses. *Journal of the British Grassland Society* 12, 4–9.

Griffiths, D.J., Roberts, H.M., Lewis, J., Stoddart, J.L. and Bean, E.W. (1967) *Principles of Herbage Seed Production*. Welsh Plant Breeding Station, Aberystwyth.

Grizzard, A.L. and Matthews, E.M. (1942) The effect of boron on seed production of alfalfa. *Journal of the American Society of Agronomy* 34, 365–368.

Hacquet, J. (1990) Genetic variability and climatic factors affecting lucerne seed production. *Journal of Applied Seed Production* 8, 59–67.

Hampton, J.G. (1983) Chemical manipulation of *Lolium perenne* grown for seed production. PhD thesis, University of Nottingham.

Hampton, J.G. (1987) Effect of nitrogen rate and time of application on seed yield in perennial ryegrass cv. Grasslands Nui. *New Zealand Journal of Experimental Agriculture* 15, 9–16.

Hampton, J.G. (1988) Herbage seed production. *Advances in Research and Technology of Seeds* 11, 1–28.

Hampton, J.G. (1989) The effect of row spacing, method and time of sowing on seed production of prairie grass (*Bromus willdenowii* Kunth.) cv. Grasslands Matua. *Plant Varieties and Seeds* 2, 171–178.

Hampton, J.G. (1990) Genetic variability and climatic factors affecting herbage legume seed production: an introduction. *Journal of Applied Seed Production* 8, 45–51.

Hampton, J.G. (1991a) Temperate herbage seed production: an overview. *Journal of Applied Seed Production* 9 (Suppl.), 2–13.

Hampton, J.G. (1991b) Effect of paclobutrazol on inflorescence production and seed yield in four white clover (*Trifolium repens* L.) cultivars. *New Zealand Journal of Agricultural Research* 34, 367–373.

Hampton, J.G. and Hebblethwaite, P.D. (1983) Yield components of the perennial ryegrass (*Lolium perenne* L.) seed crop. *Journal of Applied Seed Production* 1, 23–25.

Hampton, J.G. and Hebblethwaite, P.D. (1985) The effect of the growth regulator paclobutrazol (PP333) on the growth, development and yield of *Lolium perenne* grown for seed. *Grass and Forage Science* 40, 93–101.

Hampton, J.G., Clemence, T.G.A. and Hebblethwaite, P.D. (1983) Nitrogen studies in *Lolium perenne* grown for seed. IV. Response of amenity types and influence of a growth regulator. *Grass and Forage Science* 38, 97–105.

Hampton, J.G., Clemence, T.G.A. and McCloy, B.L. (1985) Chemical manipulation of grass seed crops. In: Hare, M.D. and Brock, J.L. (eds) *Producing Herbage Seeds*. Grassland Research and Practice Series No. 2. New Zealand Grassland Association, Palmerston North, pp. 9–14.

Hampton, J.G., Clifford P.T.P. and Rolston, M.P. (1987) Quality factors in white clover seed production. *Journal of Applied Seed Production* 5, 32–40.

Hare, M.D. (1992a) Seed production in tall fescue (*Festuca arundinacea* Schreb). Unpublished PhD thesis, Massey University.

Hare, M.D. (1992b) Inter- cross-row cultivation, atrazine application and band spraying effects on 'Grasslands Maku' Lotus (*Lotus uliginosus* Schk.) seed production. *Journal of Applied Seed Production* 10, 78–83.

Hare, M.D. (1994) Autumn establishment of three New Zealand cultivars of tall fescue (*Festuca arundinacea* Schreb.) for seed production. *New Zealand Journal of Agricultural Research* 37, 11–17.

Hare, M.D., Brown, K.R., Archie, W.J. and Rolston, M.P. (1988) Sowing method, growth regulators and time of nitrogen effects on seed production of spring sown prairie grass. *Journal of Applied Seed Production* 6, 46–54.

Hebblethwaite, P.D. (1987) A review of the chemical control of growth, development and yield in *Lolium perenne* grown for seed. *Journal of Applied Seed Production* 5, 54–59.

Hebblethwaite, P.D. and Clemence, T.G.A. (1983) Effect of autumn and spring defoliation and defoliation method on seed yield of *Lolium perenne*. *Proceedings of the XIV International Grassland Congress*, pp. 257–260.

Hebblethwaite, P.D. and Hampton, J.G. (1982) Physiological aspects of seed production in perennial ryegrass. In: van Bogaert, G. (ed.) *Breeding High Yielding Forage Varieties Combined with High Seed Yield*. Eucarpia, Belgium, pp. 17–32.

Hebblethwaite, P.D. and Peirson, S.D. (1983) The effects of method and time of sowing on seed production in perennial ryegrass. *Journal of Applied Seed Production* 1, 30–33.

Hebblethwaite, P.D., Wright, D. and Noble, A. (1980) Some physiological aspects of seed yield in *Lolium perenne* L. In: Hebblethwaite, P.D. (ed.) *Seed Production*. Butterworths, London, pp. 71–90.

Hebblethwaite, P.D., Hampton, J.G. and McLaren, J.S. (1981) The chemical control of growth, development and yield of *Lolium perenne* grown for seed. In: McLaren, J.S. (ed.) *Chemical Manipulation of Crop Growth and Development.* Butterworths, London, pp. 505–523.

Hides, H.H., Lewis, J. and Marshall, A. (1984) Prospects for white clover seed production in the United Kingdom. In: Thomson, D.J. (ed.) *Forage Legumes, Occasional Symposium No. 16.* British Grassland Society, UK, pp. 36–39.

Hill, M.J. (1980) Temperate pasture grass seed crops: formative factors. In: Hebblethwaite, P.D. (ed.) *Seed Production.* Butterworths, London, pp. 137–149.

Hill, M.J. and Watkin, B.R. (1975) Seed production studies on perennial ryegrass, timothy and prairie grass. I. Effect of tiller age on tiller survival, ear emergence and seedhead components. *Journal of the British Grassland Society* 30, 63–71.

Hurst, R.L. and Pedersen, M.W. (1964) Alfalfa seed production as a function of genetic and environmental characteristics. *Advancing Frontiers in Plant Science* 8, 41–54.

Kelly, A.F. (1988) *Seed Production in Agricultural Crops.* Longman, Essex, UK.

Kephart, K.D., Twidwell, E.K., Bortnem, R. and Boe, A. (1992) Alfalfa yield component responses to seeding rate several years after establishment. *Journal of the American Society of Agronomy* 84(5), 827–831.

Kowithayakorn, L. and Hill, M.J. (1982) A study of seed production of lucerne (*Medicago sativa*) under different plant spacing and cutting treatments in the seedling year. *Seed Science and Technology* 10, 3–12.

Langer, R.H.M. (1979) *How Grasses Grow.* The Institute of Biology's Studies in Biology No. 34. Edward Arnold, London.

Langer, R.H.M. (1980) Growth of the grass plant in relation to seed production. In: Lancashire J.A. (ed.) *Herbage Seed Production.* Grassland Research and Practice Series No. 1., New Zealand Grassland Association, Palmerston North, pp. 6–11.

Lewis, J., James, I.R. and Marshall, A. (1983) Mechanical gapping of second years crops. In: *Report of the Welsh Plant Breeding Station 1983.* Welsh Plant Breeding Station, Aberystwyth, p. 106.

Li, Q. and M.J. Hill. (1989) A study on post-peak flowering shoot manipulation for seed production of *Lotus corniculatus* L. *Journal of Applied Seed Production* 7, 71–75.

Lorenzetti, F. (1993) Achieving potential herbage seed yields in species of temperate regions. *Proceedings of the XVII International Grassland Congress,* pp. 1621–1628.

Mares Martins, V.M. and Gamble, E.E. (1993a) Seed abortion and yield in perennial ryegrass following selective pre-anthesis defoliation of reproductive and vegetative tillers. *Journal of Applied Seed Production* 11, 20–25.

Mares Martins, V.M. and Gamble, E.E. (1993b) Floret dynamics in perennial ryegrass in response to chemical manipulation of the seed crop. *Journal of Applied Seed Production* 11, 39–47.

Marshall, C. (1985) Developmental and physiological aspects of seed production in herbage grasses. *Journal of Applied Seed Production* 3, 43–49.

Marshall, C. and Ludlam, D. (1989) The pattern of abortion of developing seeds in *Lolium perenne* L. *Annals of Botany* 63, 19–27.

Marshall, A.H. and Hides, D.H. (1990) White clover seed production from mixed swards: effect of sheep grazing on stolon density and on seed yield components of two contrasting white clover varieties. *Grass and Forage Science* 45, 35–42.

Marshall, A.H. and Hides, D.H. (1991) Effect of the plant growth regulator Parlay on the seed production of the white clover cvs. Menna and Olwen. II Yield components and potential seed yield. *Journal of Applied Seed Production* 9, 81–86.

Marshall, A.H. and James, I.R. (1988) Effect of plant density on stolon growth and development of contrasting white clover (*Trifolium repens* L.) varieties and its influence on the components of seed yield. *Grass and Forage Science* 43, 313–318.

McGraw, R.L., Beuselinck, P.R. and Ingram, K.T. (1986) Plant population density effects on seed yield of birdsfoot trefoil. *Agronomy Journal* 78, 201-205.

McKersie, B.D. and Tomes, D.T. (1982) A comparison of seed quality and seedling vigour in birdsfoot trefoil. *Crop Science* 22, 1239–1241.

Meijer, W.J.M. (1984) Inflorescence production in plants and in seed crops of *Poa pratensis* L. and *Festuca rubra* L. as affected by juvenility of tillers and tiller density. *Netherlands Journal of Agricultural Science* 32, 119–136.

Meijer, W.J.M. (1985) The effect of uneven ripening on floret site utilization in perennial ryegrass seed crops. *Journal of Applied Seed Production* 3, 55–57.

Meijer, W.J.M. (1987) The influence of autumn cutting on crop structure and seed production of first year crops of *Poa pratensis* L. and *Festuca rubra* L. (Abstract). *Journal of Applied Seed Production* 5, 60–61.

Mela, T. (1969) The effect of N-dimethylaminosuccinamic acid (B-995) on the seed cultivation characteristics of late flowering red clover. *Acta Agralia Fennica* (115), 114.

Miller, D.A. and Steiner, J.J. (1995) Seed production principles. In: Barnes, R.F., Miller, D.A. and Nelson, C.J. (eds) *Forages*. Iowa State University Press, Ames, Iowa, pp. 127–140.

Nordestgaard, A. (1983) Trials on time of nitrogen application in the spring to various grasses grown for seed production. *Proceedings of the XIV International Grassland Congress*, pp. 252–253.

Nordestgaard, A. (1986) Investigations on the interaction between level of nitrogen application in the autumn and time of nitrogen application in the spring to various grasses grown for seed. *Journal of Applied Seed Production* 4, 16–25.

Nordestgaard, A. and Andersen, S. (1991) Stability of high production efficiency in perennial herbage seed crops. *Journal of Applied Seed Production* 9, (Suppl.), 27–32.

Norris, I.B. (1985) Temperature response and flowering of white clover (*Trifolium repens* L.) varieties in controlled environments and the field. *Annals of Applied Biology* 108, 659–665.

Oswald, A.K. (1985) The use of asulam to control *Bromus sterilis* in ryegrass crops grown for seed. *Crop Protection* 4, 329–336.

Pasumarty, S.V., Matsumura, T., Higuchi, S. and Yamada, T. (1993) Ovule fertility – a tool for selecting high-fertility populations of white clover. *Proceedings of the XVII International Grassland Congress*, pp. 1648–1649.

Pasumarty, S.V., Higuchi, S. and Murata, T. (1995) Environmental influences on seed yield components of white clover. *Journal of Applied Seed Production* 13, 25–31.

Pedersen, M.W. (1968) An interaction between pollinators and reciprocals in an alfalfa cross. *Crop Science* 8, 107–109.

Pedersen, M.W. and Nye, W.P. (1962) Alfalfa seed production studies. *Utah Agricultural Experimental Station Bulletin* 436.

Pedersen, M.W., Petersen, H.L., Bohart, G.E. and Levin, M.D. (1956) A comparison of the effect of complete and partial cross-pollination of alfalfa on pod set, seeds per pod, and pod and seed weight. *Agronomy Journal* 48, 177–180.

Pedersen, M.W., Bohart, G.E., Levin, M.D., Nye, W.P., Taylor, S.A. and Haddock, J.L. (1959) Cultural practices for alfalfa seed production. *Utah Agricultural Experimental Station Bulletin* 408.

Pilbeam, D.J. and Kirby, E.A. (1983) The physiological role of boron in plants. *Journal of Plant Nutrition* 6, 563–582.

Puri, K. P. and Laidlaw, A.S. (1983) The effect of cutting in spring and application of Alar on red clover (*Trifolium pratense* L.) seed production. *Journal of Applied Seed Production* 1, 12–18.

Puri, K.P. and Laidlaw, A.S. (1984) The effect of temperature on components of seed yield and on hard seededness in three cultivars of red clover (*Trifolium pratense* L.). *Journal of Applied Seed Production* 2, 18–23.

Rice, J.S., Wang, C.L. and Gray, E. (1970) Relationship of pollen and pistil characteristics with self and cross compatibility in alfalfa. *Crop Science* 10, 59.

Rolston, M.P. (1991) Cocksfoot seed crop tolerance to herbicides applied to seedling and established stands. *Journal of Applied Seed Production* 9, 63–68.

Rolston, M.P. and Hare, M.D. (1986) Competitive effects of weeds on seed yield of first year grass seed crops. *Journal of Applied Seed Production* 4, 34–36.

Rolston, M.P., Rowarth, J.S., DeFilippi, J.M. and Archie, W.J. (1994) Effects of water and nitrogen on lodging, head numbers and seed yield of high and nil endophyte perennial ryegrass. *Proceedings of the Agronomy Society of New Zealand* 24, 91–94.

Rosellini, D., Veronesi, F. and Falcinelli, M. (1994) Seed yield components in lucerne (*Medicago sativa* L.) materials selected for seed yield. *Rivista di Agronomia*, 28(1), 43–49. (In Italian, English summary.)

Roy, S.K., Rolston, M.P. and Rowarth, J.S. (1994) Ryegrass seed yield loss due to undersize seed. *Proceedings of the Agronomy Society of New Zealand* 24, 95–98.

Ryle, G.J.A. (1964) The influence of date of origin of the shoot and level of nitrogen on ear size in three perennial grasses. *Annals of Applied Biology* 53, 311–323.

Ryle, G.J.A. (1966) Physiological aspects of seed yield in grasses. In: Milthorpe, F.L. and Ivins, J.D. (eds) *Growth of Cereals and Grasses*. Butterworths, London, pp. 106–120.

Sayers, E.R. and Murphy, R.P. (1966) Seed set in alfalfa as related to pollen tube growth, fertilization frequency, and post fertilization ovule abortion. *Crop Science* 6, 365–368.

Seaney, R.R. (1964) Cross and self seed in birdsfoot trefoil plants selected for self fertility. *Crop Science* 4, 440–441.

Seaney, R.R. and Hanson, P.R. (1970) Birdsfoot trefoil. *Advances in Agronomy* 22, 119–157.

Sherrel, C.G. (1983) Effect of boron application on seed production of New Zealand herbage legumes. *New Zealand Journal of Experimental Agriculture* 11, 113–117.

Sjödin, J. (1981) Selection for seed setting capacity in tetraploids of clover and grasses. In: *Breeding High Yielding Forage Varieties Combined with High Seed Yield*. Report of the meeting of the Fodder Crops Section of Eucarpia, Belgium, pp. 163–168.

Spiertz, J.H.J. and Ellen, J. (1972) The effect of light intensity on some morphological and physiological aspects of the crop perennial ryegrass (*Lolium perenne* L. var. 'Cropper') and its effects on seed production. *Netherlands Journal of Agricultural Science* 20, 232–246.

Stephenson, A.G. (1984) The regulation of maternal investment in an indeterminate flowering plant (*Lotus corniculatus* L.). *Ecology* 65(1), 113–121.

Tabora, R.S. and Hill, M.J. (1991) An examination of vegetative and reproductive growth habits and their contribution to seed yield in 'Grasslands Maku' Lotus (*Lotus uliginosus* Schk.). *Journal of Applied Seed Production* 9, 7–15.

Taylor, N.L., Rincker, C.M., Garrison, C.S., Smith, R.R. and Cornelius, P.L. (1990) Effect of seed multiplication regimes on genetic stability of red clover. *Journal of Applied Seed Production* 8, 21–27.

Teuber, L.R. and Brick, M.A. (1988) Morphology and anatomy. In: Hanson, A.A. (ed.) *Alfalfa and Alfalfa Improvement*. American Society of Agronomy, Madison, Wisconsin, pp. 125–162.

Valle, O. (1963) Breeding and seed production experiences with Finnish white clover. *Annales Agriculturae Fennicae* 2, 51–58.

van Bogaert, G. (1977) Factors affecting seed yield in white clover. *Euphytica* 26, 233–239.
Williams, W.M. (1987) Genetics and breeding. In: Baker, M.J. and Williams, W.M. (eds) *White Clover.* CAB International, Wallingford, UK, pp. 343–419.
Wilson, S.M., Hampton, J.G., Hill, M.J. and Rolston, M.P. (1994) Effect of cultivar, time of sowing and fungicide application on seed yield in cocksfoot (*Dactylis glomerata* L.). *Proceedings of the Agronomy Society of New Zealand* 24, 103–108.
Zaleski, A. (1964) Effect of density of plant population, photoperiod, temperature and light intensity on inflorescence formation in white clover. *Journal of the British Grassland Society* 19, 237–247.

Maturation of Grass and Legume Seed

4

P. Coolbear,[1] M.J. Hill[2] and Win Pe[3]

[1]*Association of Polytechnics in New Zealand, PO Box 10344, Wellington, New Zealand;* [2]*Seed Technology Centre, Department of Plant Science, Massey University, Palmerston North, New Zealand;* [3]*Myanmar Agricultural Service, Gyogon, Insein PO 11011, Yangon, Myanmar*

4.1 INTRODUCTION

The physiology of seed development and maturation in grasses and legumes is both a fascinating and frustrating topic. Although understanding of the processes involved in seed development has advanced tremendously (especially in terms of description of events), we now know just enough to be aware of the vast range of issues which still remain a mystery. Herein lies the frustration: not only is most detailed knowledge limited to a very few major food crops, but even in these species much information is fragmentary. For the forage species there are still many questions which have yet to be addressed.

Some aspects, such as those relating to the understanding of the establishment and maintenance of cell polarity, or the control of the processes of differentiation and development at a molecular level, are fundamental issues in modern biology and well beyond the scope of this book. Others, however, are of immediate relevance and importance to improved seed production. That we still know so little about the control of relative sink strength during reproductive development, the key environmental factors which affect seed quality, and the physiology of the acquisition of dormancy might be taken as an indictment on the emphasis of much postwar research in this field.

Without a basic understanding of the processes involved in any particular species, most agronomic research in seed production becomes a 'try-it-and-see' affair. Various manipulations, such as plant spacing, closing date or growth regulator treatments, are chosen on the basis of previous literature (plus a little guess work) and conscientiously evaluated. Successes are often re-evaluated with differences in environment, cultivar and scale; failures tend to be forgotten. Too frequently, questions about 'why' are only given summary consideration. Even less frequently is there any strategic plan of research involving an attempt to produce an integrated understanding of the seed crop's physiological responses and how they might be amenable to modification. On the other hand, the very complexities which are revealed by detailed physiological studies put such an integrated approach way beyond the wildest budgetary dreams of most researchers. In the search for publishable research outputs, it is easier to continue short studies on the effects of a relatively arbitrary combination of treatments.

Against this background this chapter attempts to review the current state of knowledge about the processes of seed development and maturation in forage grasses and legumes. It will be quickly evident how little is known and how much scope there is for improving our understanding of some of the basic events in seed development in these species.

4.2 FORAGE GRASSES

Unlike their close relatives, the cereals, forage grasses have not been selected for their capacity for seed production. Although this statement might be regarded as a truism, the level of inefficiency in seed production shown by this group of species may still come as a surprise. For instance, Hebblethwaite et al. (1980) demonstrated that the actual yield of perennial ryegrass (*Lolium perenne* L.) seed in the UK may not reach even 8% of yield potential. Losses are due to a range of factors, but particularly include premature death of fertile tillers, poor seed set, abortion during seed development and shedding of seed prior to harvest. Furthermore, harvested seed may be of variable quality due not only to variation in production conditions between lots, but also to the inherent heterogeneity of the grass seed crop. Nevertheless, the whole strategy of the plant is geared to producing a quantity of reproductive units with a reasonable chance of survival.

It is easy to make general statements about the pattern of grass seed development (summarized in Section 4.2.1), but the search for increased levels of production of higher quality seed needs a more detailed data base. Thus Section 4.2.2 deals at some length with the morphology and physiology of seed development. A detailed review of the biochemistry of seed reserves has not been included, because most of the work has been done from the point of view of the food value of cereal grains. However, an increased understanding of the biochemistry of storage proteins in forage grasses may become necessary in the near future, particularly if electrophoretic methods of cultivar verification continue to be developed at their present rate.

The third section (4.2.3) is devoted to seed ripening and shedding. Firstly, the changing water status of the seed during development – inherently related to assimilate supply and caryopsis growth rates – is discussed. Secondly, we consider the other major event occurring as most pasture grass seeds begin to dry: seed losses by shedding.

Apart from Section 4.2.1, this section on grasses has proved a difficult one to write. Rost and Lersten's (1973) major review and bibliography showed a paucity of published research on forage grass seed development. Collating the literature since then suggests that the gap in our knowledge in relation to the cereals has become even wider. Tremendous advances have been achieved in cereal research in the past 10 years. We now have quite a sophisticated understanding of the control processes involved in the development of cereal grains, although, of course, the story still has a great many gaps. Forage grass seeds are more difficult to work with; they are smaller and crop development is less uniform, but recent work with cereals is offering new ideas and approaches (e.g. Murray, 1995).

Inevitably, therefore, this chapter has had to draw on data from the cereals at many points in the discussion. No apologies are made: this kind of research provides a useful and often exciting context for studies on other members of the *Gramineae*. If nothing else, it is hoped that this chapter might provide new insights and ideas about how this context might be exploited for research into the grasses.

4.2.1 General Pattern of Seed Development in Grasses

Following successful fertilization, three semi-independent processes begin in the developing seed. Two of these are the development of the embryo and the endosperm. The third component is the development of the seed coat, a fused pericarp-testa. It is worth remembering that each of these components has a different genetic history, and that while we can summarize the developmental pattern of the grass seed as a whole, the rates of each process may differ between the embryo, endosperm and seed coverings.

Classically, because these protective seed coverings do not entirely originate from the integuments (which give rise to the testa), but also from the ovary wall which produces the pericarp, the propagules of the *Gramineae* are technically fruits (caryopses), rather than seeds. Practically this distinction hardly matters, and, of course, the term 'seed' remains in common usage. Much more important is the fact that the harvested reproductive unit of grasses is often the caryopsis enclosed in residual floral parts, especially the lemma and palea and sometimes (as in the case of prairie grass (*Bromus willdenowii* Kunth)), the glumes.

Shaw and Loomis (1950) identified three phases of reproductive development in maize (*Zea mays* L.) after seed set. A phase of rapid growth is followed by a period of food reserve accumulation where growth rate is often constant. Maximum seed dry weight is attained at the end of this second phase. The seed loses water in the third ripening stage. Hyde *et al.* (1959) were able to characterize this pattern in perennial ryegrass on the basis of fresh and dry weight changes in the developing caryopses, a sequence of events which is summarized in Table 4.1. This general sequence applies to most seed, including forage legumes (see Section 4.3.1).

This pattern can be recognized in the development and maturation of most grass seed crops, and provides a useful reference point for subsequent discussions on the seed development process. In practice, sample variation in these heterogeneous crops may cause deviation from expected patterns, especially when weather conditions affect the synchrony of reproductive tillering and early maturing seeds are lost by shedding. One example of this is the data presented by Hill and Watkin (1975) where, in an unusually wet growing season, an unexpected continuous increase in the fresh weight of developing ryegrass seed was observed throughout Stage II. Figure 4.1 shows typical changes in components of seed development for timothy (*Phleum pratense* L.) and perennial ryegrass.

These three stages summarize the different morphological and metabolic processes which occur during development and will be used as reference points throughout this chapter where the events will be described in detail. During Stage

Table 4.1. The general pattern of seed development in grasses. (From data obtained by Hyde et al., 1959.)

Stage	Pattern of growth	Seed moisture content (% of fresh weight)	Approximate duration for perennial ryegrass (days)
I	Rapid increase in both fresh and dry weight	Remains constant at around 75–80%	About 10
II	A continued, almost linear increase in dry weight Slower or no increase in fresh weight	Seed moisture content falls to around 40%	10–14
III	Dry weight stays constant, fresh weight decreases	Seed moisture content falls to equilibrium with the relative humidity of the environment	3–7

I, growth is mainly by cell division. Because the embryo has yet to differentiate fully, the seed is not usually an independently viable unit. Grabe (1956) reported that a few seeds of smooth brome (*Bromus inermis* Leyss.) were capable of germination as early as 5 days post-anthesis (DPA). Acquisition of germinability during development is shown in Fig. 4.2 for seed crops of ryegrass and timothy. Seeds usually become fully viable during Stage II.

It is important that seed viability is not confused with seed germinability. A seed population may be fully viable, but have very low germination because of immaturity or dormancy. Similarly, just because a seed is germinable does not mean that development is complete. Germinable seeds at an early stage of development are well known to have poor vigour. Despite the early onset of germination in smooth brome, Grabe (1956) clearly demonstrated that the maximum capacity for seedling growth was not reached until seeds had attained maximum dry weight at the end of Stage II. This point has been variously referred to as morphological, functional or physiological maturity (Anderson, 1944; Grabe, 1956; Shaw and Loomis, 1950; respectively). The last of these three terms is the most common; a little unfortunately perhaps, because the physiological development of a seed may be far from complete at this time, despite the fact that it has gained its final dry weight. This fact may be inferred from the data for timothy presented in Fig. 4.2. Physiological maturity is reached at 35 DPA, but immediate germinability at this stage is less than 55%. Recent work by Richard Ellis and his team at Reading has demonstrated that considerable numbers of changes in seed quality and therefore seed physiology occur during Stage III of seed development (e.g. Ellis and Pieta Filho, 1992). They suggest the use of the term 'mass maturity' to replace 'physiological maturity'.

Very few generalizations can be made about the onset of germinability even within species, since environmental conditions may have considerable impact on this parameter. For example, Anslow (1964) found considerable variation in the development of germinability between ryegrass seeds formed on early heads compared to those emerging more than a week later on the same plant. The seed population from late spikes took 28 DPA before the development of maximum

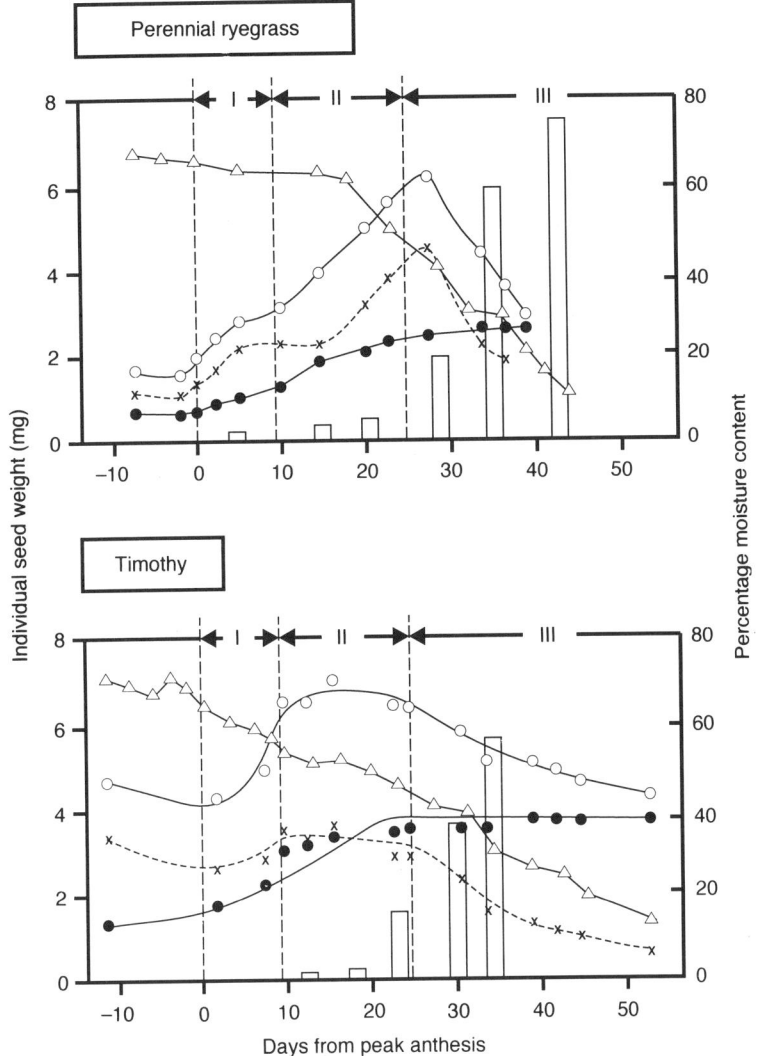

Fig. 4.1. Patterns of seed development in perennial ryegrass cultivar Grasslands Ruanui and timothy cultivar Grasslands Kahu. Changes in mean seed fresh (o) and dry (●) weights and absolute (x) and percentage (△) moisture contents are shown. The histogram bars represent the percentage of heads shedding seed at different times. The approximate stages of seed development are indicated for each species. (Data taken from Hill, 1971.)

germination capacity which was never as high as that from early spikes (18 DPA). He also observed that rates of seed development varied significantly within an individual spike, with significant suppression of development in apical florets. Germination capacity tended to be highest from seeds on the middle of the head. Influences of climate may also be confounded with the action of pathogens; for

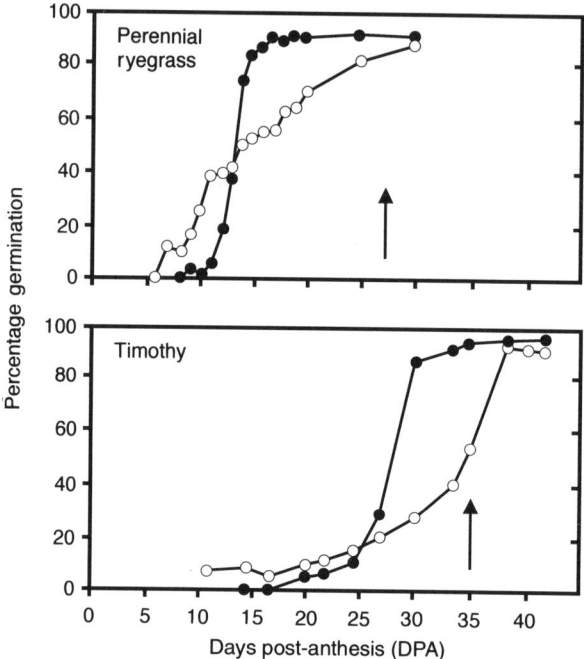

Fig. 4.2. Comparisons of the timing of onset of germinability during seed development between perennial ryegrass cultivar Grasslands Ruanui and timothy cultivar Grasslands Kahu. Both mean percentage germination immediately after removal from the plant (○) and after 3 months' dry storage (●) are shown. The estimated times when Stage II of development ends in each seed crop are indicated (↑). (Data from Hill, 1971.)

example, blind seed disease (*Gloeotinia granigena* (Quel.) T. Schumacher) depresses germinable seed yield of ryegrass under wet conditions (Grant, 1981).

Stage III of seed development is the ripening stage when dry weight is approximately constant and the seed tissues lose moisture as rapidly as ambient environmental conditions permit. Anslow (1964) was one of the first to suggest that the beginning of this stage is determined by the internal development of the seeds themselves rather than the status of the parent plant. Certainly this last stage of development is independent of assimilate uptake from the culm (Stoddart, 1965). In grasses, seed ripening may be associated with considerable loss of seed through shedding (Fig. 4.1), but because the seed is now functionally independent of the parent plant, harvesting may well be timed earlier, even late in Stage II for some species.

4.2.2 Developmental Morphology and Physiology of the Caryopsis

Embryogenesis

The uncertainties that have already been mentioned in the small amount of literature available on the pattern of embryo development in the *Gramineae* are no doubt due to the diversity of the family. Any assumption that the pattern of embryogenesis found in the cereals applies to all grasses is a dangerous one. For instance, in his mammoth survey of mature embryo structure, Reeder (1957, 1962) showed that there were six different grass embryo types. More recently, Bhanwra (1988) has continued this work, having completed an extensive survey of patterns of embryogenesis in NW Indian species, including cultivars of many temperate grasses. Once again the emphasis has been on diversity of developmental patterns within the family. Accordingly, the short account in this section can do no more than provide a general summary of events.

Although the primary endosperm nucleus may begin to divide rapidly after the fusion of gametes (see 'Endosperm development' below), there is usually a delay before the first cell division of the zygote (e.g. Elgersma and Sniezko, 1988). This appears to vary considerably between species, but Hill (1971) found a delay of over 60 h between pollen tube entry through the micropyle and this first division in ryegrass. In prairie grass the interval was only about 20 h.

When the first cell division in the zygote does occur, it is unequal, forming a small, highly meristematic apical cell and a larger basal cell which faces the micropyle (Fig. 4.3a). There is some uncertainty about whether the basal cell undergoes any further divisions at all (Bhatnagar and Johri, 1972; Bewley and Black, 1985). Whatever the case, subsequent cell divisions give rise to an axially symmetrical mass of cells, the pro-embryo (Figs 4.3b, 4.4). In cereals this is supported by a short suspensor region which may have a regulatory function (Bewley and Black, 1985). Hill (1971), however, could not find any trace of a suspensor in either ryegrass or prairie grass, and suggested that the absence of this organ may be associated with the much more rapid development of germinability in these embryos compared to the cereals.

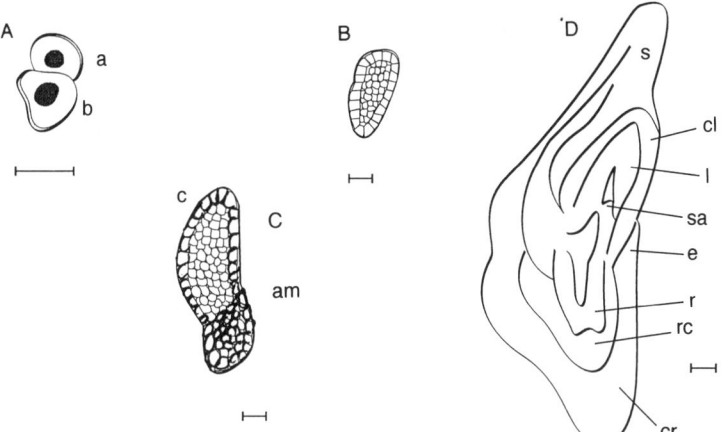

Fig. 4.3. Schematic diagram of embryogenesis in perennial ryegrass (longitudinal section). A. First cell division (≈ 3 DPA); B. Pro-embryo stage (4 DPA); C. Development of axial asymmetry (7 DPA); D. Mature embryo. Bars indicate 100 µm. Key : a, apical cell; b, basal cell (nearest micropyle); c, cotyledon; am, apical meristem; s, scutellum; v, vascular strands; cl, coleoptile; l, first leaf; sa, stem apex, e, epiblast; r, radical; rc, root cap; cr, coleorhiza.

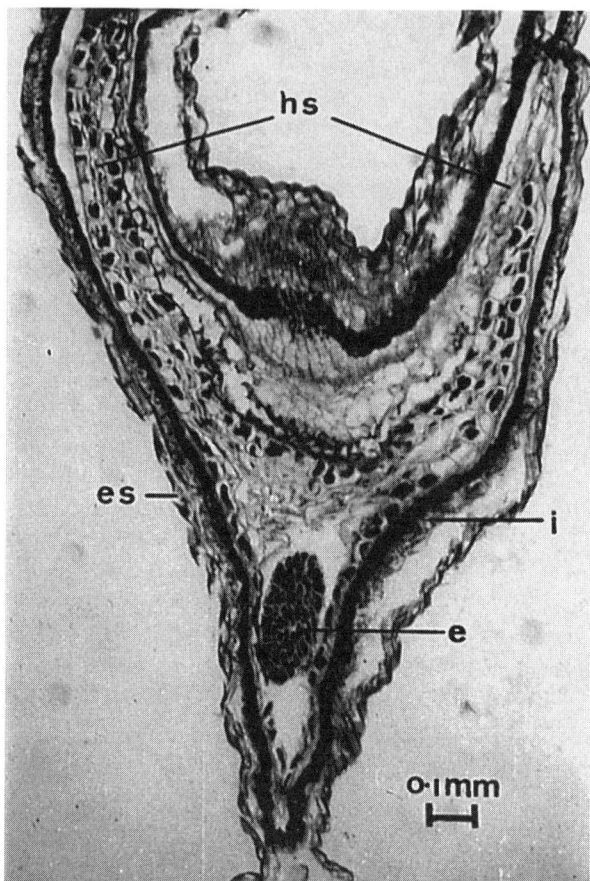

Fig. 4.4. Longitudinal section of developing prairie grass seed, four days after pollination. hs, horns of style; i, integuments; e, embryo; es, endosperm (some shrinkage of tissue occurs with fixing). (From Hill, 1971.)

After the pro-embryo stage, the developing embryo loses its axial symmetry because of the development of organizing centres within the ball of cells. These may result from some meristematic cells becoming quiescent, thus allowing changes in direction of divisions of the active cells around them (Bhatnagar and Johri, 1972) and/or the appearance of oblique cell divisions (Batygina, 1969).

From this stage on, the embryo continues to develop with bilateral symmetry until all the characteristic features of a mature embryo are evident (Fig. 4.3d). Data of Ingle *et al.* (1965) show that the growth of the embryo in maize continues by both cell elongation and cell division right up to the stage of mass maturity. This is in marked contrast to the pattern of development in the endosperm (see 'Endosperm development' below). In the Festucoid grasses the embryo is usually very small relative to the caryopsis, but this is not always the case in Panicoid species where the embryo may constitute 50% of the seed (Reeder, 1957).

The single cotyledon, apparent very rapidly after the pro-embryo stage, extends above and often below the embryo axis, developing into the scutellum which will later act as the intermediary between the germinating embryo and its main food reserve, the endosperm. Although some earlier workers have suggested that the coleoptile, which forms a protective sheath around the plumule, was derived from the lower part of the cotyledon, Guignard and Mestre (1970) consider the coleoptile to be the first leaf produced by the shoot apex, while the coleorhiza is the outer covering of the degenerate embryonic root. The roots that will develop as the seed germinates are often adventitious, their primordia being formed late in embryo development (Barnard, 1964).

Another unexplained feature of cereals and most, but not all of the Festucoid grasses (Reeder, 1957), is the epiblast which again appears late in development and has been regarded as a vestigial cotyledon (Barnard, 1964). However, Guignard and Mestre (1970) demonstrated that this organ's cells are very similar to those of the coleorhiza.

Embryo formation without fertilization (gametophytic apomixis) is common in tropical grasses, but unusual in the temperate Festucoid types except for species in the *Poa* and *Eragrostis* genera (McWilliam, 1964). This phenomenon arises when either a diploid somatic cell substitutes for the embryo sac (apospory) or, less usually, when the embryo sac itself fails to undergo meiosis (diplospory). Both types have been noted in *Poa*. In species where it occurs, apomitic seed formation becomes a major consideration for breeding work (see Chapter 10).

Before concluding this section, it must be emphasized that embryo development is not always a smooth, inevitable process. Early abortion after just a few cell divisions is common (Hill, 1971; Elgersma and Sniezko, 1988). Johnston (1960) found shrivelled caryopses at various stages of development in cocksfoot (*Dactylis glomerata* L.). Additionally, Hill (1971) noted that developing seeds of ryegrass and prairie grass may go through a resting phase in early development after which they may either resume growth or abort. The duration of this resting phase may be as long as 3 weeks. Whether or not abortion will occur seems to be a function of the activity of adjacent reproductive units, rather than overall assimilate availability.

Endosperm development

In most grass species, seed development as a whole very much reflects the development of the largest organ, the endosperm. At 8 DPA, for example, the wheat (*Triticum aestivum* L.) embryo, developing according to its own programme, with cell division destined to continue right through to mass maturity, might be expected to have only 100 cells in contrast to the several thousand already present in the endosperm (Herzog, 1986). Thus Stage I of seed development represents rapid growth of the endosperm by cell division, while Stage II is essentially the result of cell expansion as the endosperm cells accumulate food reserves. In wheat, Stage I is normally completed by 16 DPA, although in barley (*Hordeum vulgare* L.) it may extend beyond 28 DPA, the endosperm of this species having at least twice as many cells as wheat (Evers, 1981).

In the early 1980s, there was some argument about whether or not endosperm cell division was totally complete by the end of Stage I (Cochrane and Duffus, 1981), but this confusion seems to have arisen partly due to false comparisons between wheat and barley, and also because some increases in DNA content of the endosperm had been observed during early Stage II, and it was assumed that this represented cell division. Knowles and Phillips (1985) have shown that central endosperm nuclei in maize show considerable accumulation of DNA after mitotic activity ceases. The precise function of this new DNA is unclear. It may reflect massive amplification of gene coding for storage proteins and the enzymes involved in starch synthesis or deposition, or it may simply represent a means of storing nucleic acid material for later remobilization to the developing embryo.

By the end of the cell division phase the endosperm is a full, flattened structure, but as cell enlargement begins in Stage II, it rapidly becomes folded in transverse section as the lobes hinge back across the thick-walled cells to form the furrow or crease. During this time the average wheat endosperm cell will increase three-fold in volume (Briarty et al., 1979).

The original outer meristematic layer of the endosperm is destined to become the aleurone tissue which may be one, or less commonly, several cells thick. This layer will be chiefly responsible for the mobilization of food reserves contained in the body of the endosperm during germination. However, the functionality of the aleurone tissue in many grass species is still unclear.

In marked contrast to the highly specialized, active aleurone layer, cells in the rest of the endosperm (the starchy endosperm) will be dead by the time the seed ripens. These expanding cells become so filled with food reserves, especially starch and protein bodies, that their cytoplasmic contents become disrupted. Starch begins to accumulate immediately after cell division ceases, grains forming in protoplasts which are already present at the end of the cell division stage. Usually endosperm cell walls are thin, but occasionally they may themselves constitute a significant carbohydrate reserve, e.g. the thickened β-glucan rich walls found in *Bromus* (Macleod and McCorquodale, 1958; MacLeod et al., 1964).

The other major reserve in the starchy endosperm is protein, generally around 10% dry weight. This may be stored in discrete membrane-bound bodies or as a matrix surrounding the starch grains. Because of their nutritional importance, a great deal of information has been collected on the storage proteins in the cereals (reviewed by Bewley and Black, 1978; Murray, 1984), but again, information in grasses is minimal.

In many grasses the process of seed development can be traced by the developing hardness of the endosperm as the amount of insoluble material increases within the seed. In ryegrass, the milky endosperm stage is when the free sugars in the seed are maximal. For this species this may be as early as 12 DPA (Stoddart, 1968). Radiolabelling studies showed that little movement of assimilate to the spike occurs after 12 DPA.

Various workers have used different descriptors for the stages of endosperm hardening (e.g. Zadoks' scale, sections 7 and 8, Zadoks et al., 1974, or that used by Pegler, 1976). This causes added confusion to what is already a very subjective measurement. Pegler (1976) performed an extensive study, in which he monitored the patterns of endosperm development in a range of forage grasses and correlated these data with maximum germinable seed yield. There were clearly differences in the relative timing of development between cultivars, and occasionally between different production years for the same cultivar. Typically, however, seeds of ryegrass, fescue, cocksfoot and timothy were at the stage of maximum germinable seed yield when the embryo was at the firm (cheesy) dough stage. He suggested that endosperm observations might be a useful complement to monitoring moisture contents during seed ripening. Andersen and Andersen (1980) have emphasized the need for large sample sizes to be taken in order to compensate for the huge variation which can occur within a forage grass seed crop. Hill (1971) found endosperm assessment of little value in determining the mass maturity of ryegrass, timothy or prairie grass, where, in each species, the hard dough stage occurred several days before maximum seed dry weight was achieved. This may seem surprising, in that at the hard dough stage the endosperm must be largely inactive, but may be explained by continuing photosynthetic activity of the pericarp and surrounding tissues in some species (see 'Development of the seed coat' below). In several species of Festucoid grasses the endosperm never fully solidifies (Rost and Lersten, 1973).

Development of the seed coat

Up to four different tissues may be involved in the production of the protective coverings of a grass seed, namely, the remnants of the nucellus, the integuments of the ovary, the ovary wall (pericarp) and the two floret bracts (lemma and palea). The apparent variability in seed coverings in the *Gramineae* are permutations of differentiation patterns of these four tissues.

Considerable changes in morphology will occur as the propagule develops. Cochrane and Duffus (1981), for example, showed that a single layer of epidermal nucellus cells were still present in barley 23 DPA, but had degenerated into an inert layer of compressed wall material (the hyaline layer) 10 days later. This layer causes close adhesion between the aleurone cells and the testa in this species. During the same period the two layers of testa cells become extensively differentiated, the outer layer becoming much thinner than the inner one. As the caryopsis matures, the two layers of cells usually become crushed, the inner layer often containing the grain pigment (e.g. Reilly, 1984). One known exception is *Elusine coracana* L. (Rost and Lersten, 1973) where the testa does retain its cellularity in the mature seed. Similarly, the pericarp may degenerate in some species or become highly differentiated. Protective layers of cross and tube cells (running at right angles and parallel to the seed axis, respectively) are a typical feature (Reilly, 1984).

Huang *et al.* (1983) demonstrated that the structural relationships of pericarp and testa in wheat may be one of the multiple factors involved in dormancy and susceptibility to sprouting damage. Sprouting-resistant hard red wheats have a much tighter and more impermeable testa-pericarp structure than related white wheats which show a looser arrangement with a more folded and weaker pericarp. No doubt similar genetic variation in seed-coat structure is an important component of dormancy behaviour in some forage species.

It is well known that photosynthetic activity in the ears of cereals, especially awned varieties, can make a considerable contribution to yield (e.g. Thorne, 1965). The fertilized floret is, of course, photosynthetically active and Cochrane and Duffus (1979) demonstrated that stomata were even present on the exposed pericarp of developing cereal grains, leading them to speculate about the role of photosynthesis by the grain itself in the early stages of development. Floret photosynthesis is probably an important contributor to grain dry weight in forage

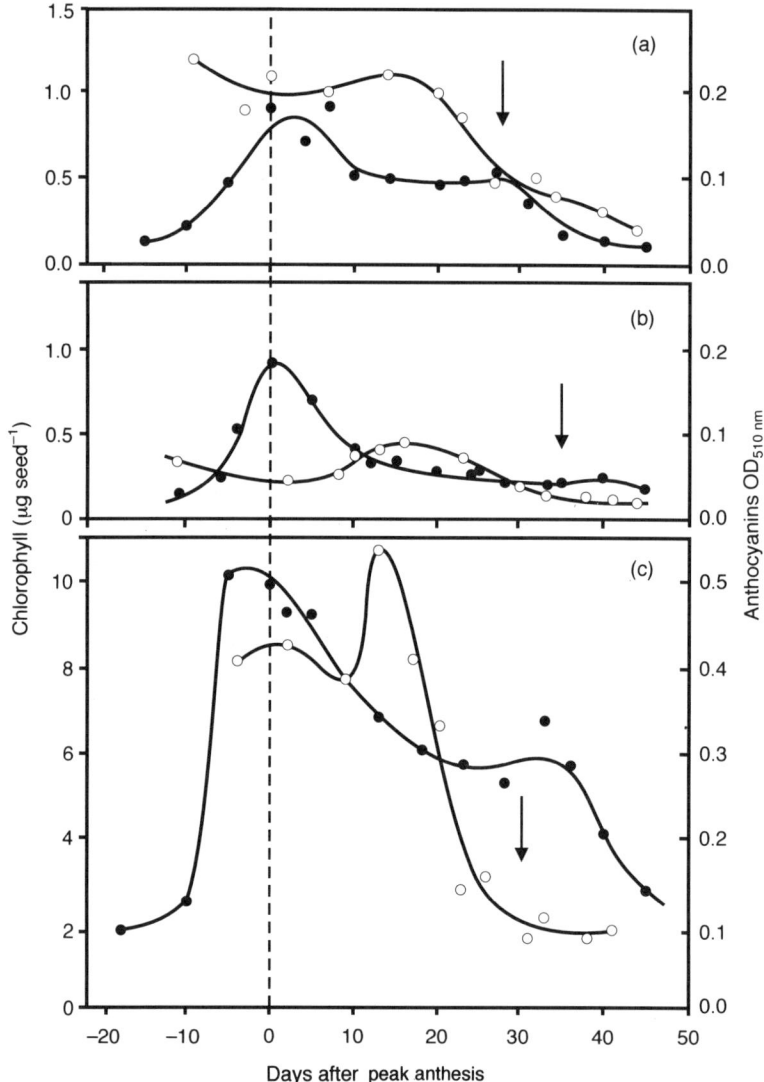

Fig. 4.5. Pigment changes during grass seed development: a. perennial ryegrass cultivar Grasslands Ruanui; b. timothy cultivar Grasslands Kahu; c. prairie grass. o——o, chlorophyll (µg seed^{-1}); ●——●, anthocyanins (OD$_{510nm}$) combined extracts from a single seed made up to 1 ml. ↓, estimated time of completion of Stage II of seed development. (Data from Hill, 1971.)

grasses too. Hill and Watkin (1975) suggested that prolonged photosynthetic activity of the seed head can contribute to late increases in seed weight even when the endosperm has solidified.

Some interest has been shown in losses of chlorophyll and other pigments from the seed heads of grasses, particularly from the point of view of determining the best time to harvest. Stoddart (1964) suggested that extensive chlorophyll degradation in the seed head indicated that abscission layers were fully formed and the seeds functionally separated from the parent plant. However, evidence is accumulating that isolation of the caryopsis may not depend on changes in abscission zones, but on blockage of the chalazal region (see Section 'Water loss in maturing seeds' below).

Hill (1971) studied in detail the losses of chlorophyll and anthocyanin from ryegrass, timothy and prairie grass. Contrary to Stoddart (1964), he found that percentage chlorophyll content began to decline in all three species before anthesis. However, when absolute levels per floret are plotted (Fig. 4.5), it can be seen that total chlorophyll levels remain stable or even increase during Stage I of development. Levels of chlorophyll in ryegrass and timothy were still relatively high at the hard dough stage, losses occurring during late Stage II and in Stage III.

Neither Stoddart (1964), nor Hill and Watkin (1975), are advocates of using pigment changes as an indicator of the correct time to harvest, due to both population variation within the seed crop and difficulties of making objective measurements. Generally, anthocyanin levels in the forage grasses reach a maximum at anthesis, declining slowly thereafter.

4.2.3 Seed Ripening and Shedding

It has been recognized in the forage grasses for at least 50 years (Evans, 1937) that, at the crop level, it is difficult to be precise about the pattern of seed development and especially difficult to select the optimum time for harvest. This then becomes a compromise between the attainment of maximum seed dry weight and seed losses due to shedding. The practicalities of choosing the correct time and method of harvesting are discussed in Chapter 8, but an understanding of the events occurring during the final stages of seed development is an important prerequisite for appropriate decision making in the field. Accordingly, this section will be devoted to a short consideration of the ripening processes of grass seed development with particular attention to, first, the mechanics of water loss from the seed and then the processes involved in seed shedding.

Water loss in maturing seeds

From general discussion in Section 4.2.1 we know that the percentage moisture content of seeds begins to fall from the beginning of Stage II of development. However, as can be seen in Fig. 4.1, the absolute water content remains reasonably constant during this time. If these descriptions are accurate, we would expect the fall in percentage moisture content during this phase to be approximately constant. Surveying the literature on different species of the *Gramineae*, this seems to be the usual case (Jensen, 1976; Raja Harun and Bean, 1979). After the end of Stage II seed drying becomes much more susceptible to the environment, clearly becoming a function of temperature and relative humidity (Andersen and Andersen, 1981).

It is important to recognize that there is no direct vascular connection between the parent plant and the developing seed. Additionally, as found in wheat and barley, there may be no direct connection between xylem elements in the developing pericarp and those in the rachilla (Barlow *et al.*, 1980). Accordingly, much of the water reaching the endosperm cavity in a developing grain is likely to arrive symplastically via the phloem and the xylem parenchyma in the chalaza.

During the reserve accumulation stage of seed development a large amount of assimilate in solution is arriving at the caryopsis. Therefore, a continued increase in water content might be reasonably expected, rather than the absolute amount of water in the seed remaining relatively constant. However, the developing seed has the ability to regulate the rate at which it loses water during Stage II, maintaining a constant rate of water movement through the seed under conditions of water stress and perhaps slowing it down when the supply of water is plentiful. Movement of assimilate to the developing grain can thus be viewed as being regulated via mass flow of water, and the grain filling rate is driven – and stabilized – by a constant rate of water movement.

Lee and Atkey (1984) have calculated that most of the water lost by a developing grain is via evaporation from the pericarp. From an early age the caryopsis of wheat is fully expanded (as it is in *Festuca* also, Jensen, 1976) and they suggest that the pericarp surface provides an essentially fixed resistance to water loss. They have calculated this is sufficient to drive the movement of assimilate into the grain at a constant rate. There may also be, however, an apoplastic return path for water via the cell walls of the chalaza or its eqivalent (Oparka and Gates, 1981; Cochrane, 1983).

Eventually, the ultimate source of water supply to the grain from the rachis xylem dries up because of accumulation of pectins in the elements. This accumulation will be gradual, but can be seen to be complete in many ears of wheat and barley when grain moisture content is down to 50% (Cochrane, 1985). At the same time, through Stage II, polyphenolic compounds, most probably tannins, accumulate in the chalazal cells. Initially these are membrane bound, but when they are released in late development will precipitate out cytoplasmic proteins and complete the isolation of the caryopsis from the parent plant (Cochrane, 1983). Seed drying rate is then entirely at the mercy of the elements and the water permeability of the enveloping structures.

Given that many forage grass seed crops need to be harvested very close to mass maturity to avoid major losses due to shedding, seed moisture content may be expected to be a reasonably objective measure of crop ripeness and has indeed been widely advocated (see Chapter 8). Hill and Watkin (1975), for example, recommended a seed moisture content of around 45% as being a good indicator of swathing time for perennial ryegrass. A problem with this is, of course, that inaccurate measurements can be caused by wet weather, while, given the inherent heterogeneity of the seed crop, careful sampling is absolutely crucial. More basic, however, is the fact that completion of Stage II can occur at different seed moisture contents in different years depending on ambient conditions (e.g. Clarke, 1983). Further, there may be considerable ecotypic variation in seed shedding rates at the same moisture content, as was found for Italian ryegrass by Raja Harun and Bean (1979).

As might be inferred from the data in Fig. 4.2, water loss is often a necessary prerequisite for inducing seed germinability. Kermode *et al.* (1986) have discussed the importance of desiccation as an option for initiating the germinative mode in many seeds, and it seems that some species of the *Gramineae* are no exception. While maturation drying will cause a decrease in all kinds of metabolic activity and eventually cause changes in membrane organization within live tissues (e.g. see Perl, 1987), it may also have an important role to play in transcriptional control (e.g. Finklestein and Crouch, 1986). Such control may involve either terminating metabolic events associated with development or preparing for the new ones of germination to be initiated as soon as the seed rehydrates.

Physiology of seed shedding

As we have already stated, while the cereals have been selected deliberately for seed retention, this has not been the case for the forage grasses and remains a serious problem. Losses due to seed shedding may be severe before the bulk of the population has matured enough to be of acceptable quality (see Fig. 4.1). Harvest timing may be critical; for example, Hebblethwaite *et al.*(1980), have reported around 30% losses of seeds due to shedding in perennial ryegrass (a species generally regarded to have less of a shedding problem than many) when harvesting was delayed from a moisture content of 40 to 30%, and in other species losses of from 20 to 60% have been reported (Jensen, 1976; Raja Harun and Bean, 1979).

Separation of seeds from the parent plant involves either disarticulation of the rachis just below the glume (typical of tropical species) or breakage of the rachilla just above the glumes (McWilliam, 1980). Many grasses, e.g. *Agropyron* spp. (Sharma and Gill, 1982) and cocksfoot (Falcinelli, 1987), break at both positions. In one of the few recent studies on the anatomy of seed shedding in grasses, Burson *et al.*(1983) found abscission zones in both positions in *Panicum* species which tend to shed seed very rapidly (10–14 DPA). The main breaking zone, located at the base of the glumes, was fully developed by anthesis. Falcinelli (1987) reported broadly similar results in cocksfoot. Both high and low shattering cultivars exhibited both abscission zones, differences in shedding being dependent on rates of zone development.

The general pattern of events involved in abscission has been comprehensively reviewed by Sexton and Roberts (1982). At the site of separation a distinct zone of small cells arises across the intended break, vascular connections are plugged with tyloses and cell wall breakdown occurs as a result of rapid increases in cellulase and polygalacturonase activity. The process appears to be subject to hormonal control in that high levels of auxin (as would be present in an actively growing seed) are inhibitory, while abscisic acid (ABA) and, more directly, ethylene are promotory.

The terms 'shattering' and 'shedding' seem to be often used synonymously in discussions on seed loss at harvest. This is not strictly correct, in that shattering is the physical separation of the seed from the parent plant while shedding is its loss to the ground. Where seeds break off, but are still retained within fixed glumes, these may be distinctly separate processes. In phalaris (*Phalaris tuberosa* L.), for instance, shattering occurs because of the swelling of the bases of sterile lemmas under each side of the maturing seed. The pressure created forces the seed away from the parent plant, breaking the fragile rachilla such that the seed is loosely held within the glumes. When the inflorescence dries, the glumes open and only then are the seeds shed (McWilliam, 1980).

McWilliam (1980) recommended selection of non-shedding mutants as the most successful approach to preventing seed losses. Such selections are now being actively sought in a range of forage species (see Chapter 10).

4.3 FORAGE LEGUMES

Once fertile legume flowers have reached maturity, the formation of ripe seeds depends on the successful completion of the steps from pollination through to fertilization, endosperm formation and embryogenesis. Attempts to describe this sequence fully, although of obvious significance for seed production, once again reveal the inadequacies of our knowledge of these aspects, and clearly identify the need for further studies to provide a sound basis for improving legume seed production. Unlike the wind-pollinated grasses with their determinate growth habit, legumes are often indeterminate plants which may require specific insect pollinators for cross pollination. They are often more susceptible to environmental change and may have remarkably precise climatic requirements (particularly daylength and temperature) which must be met before they will deliver adequate seed yield.

As with grasses, the high level of reproductive inefficiency in legume seed production is immediately obvious. In most species the main factor limiting seed production by legumes is low seed set per floret (Thomas, 1987). The causes of this are not yet understood. The development sequence from pollination to seed set and the effect of nutrition and environment on seed development and maturation also need further study.

Seed development and maturation in legumes will be discussed in a similar format to the previous section, beginning with the general pattern of development (Section 4.3.1). Developmental morphology and seed physiology in forage legumes, comparatively neglected in the literature apart from Thomas (1987), is discussed in Section 4.3.2. The final section on legumes (4.3.3) considers seed ripening, seed coat development and, in particular, the valvular function of the hilum and the role of the strophiole in regulating seed moisture levels.

4.3.1 General Pattern of Seed Development in Forage Legumes

The early work by Hyde (Hyde, 1950; Hyde *et al.*, 1959) demonstrated that the sequence already outlined for grasses in Table 4.1, i.e. a growth stage, a food accumulation stage and a ripening stage, could also be defined in legumes. Examples from the tetraploid red clover (*Trifolium pratense* L.) cultivar Grasslands Pawera and the lucerne (*Medicago sativa* L.) cultivar Wairau are shown in Figs 4.6 and 4.7, respectively.

As might be expected, the timing of different stages is widely variable between different species. Thus, while white clover (*Trifolium repens* L.) might reach full seed viability 18 days after pollination (DAP) and maximum seed dry weight after 25 days (Hyde *et al.*, 1959), corresponding times for lucerne are typically longer (e.g. 22 and 40 DAP, Fig. 4.7).

Of course, cultivar and environmental differences will also cause considerable variations in the timing of different events. Thus, for example, although the two

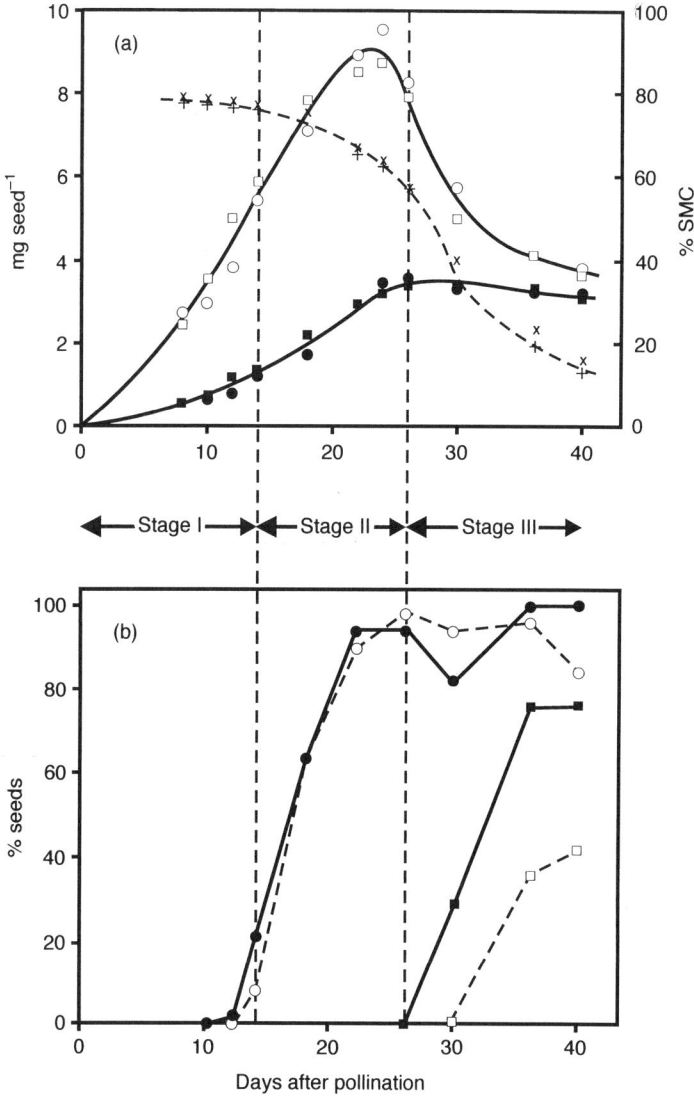

Fig. 4.6. The pattern of seed development in red clover cultivar Grasslands Pawera in two consecutive seasons.

	1976	1977
a. fresh weight changes	○	□
dry weight changes	●	■
% seed moisture content	x	+
b. % seed viability	○	●
% hard seed	□	■

(Replotted from Win Pe, 1978.)

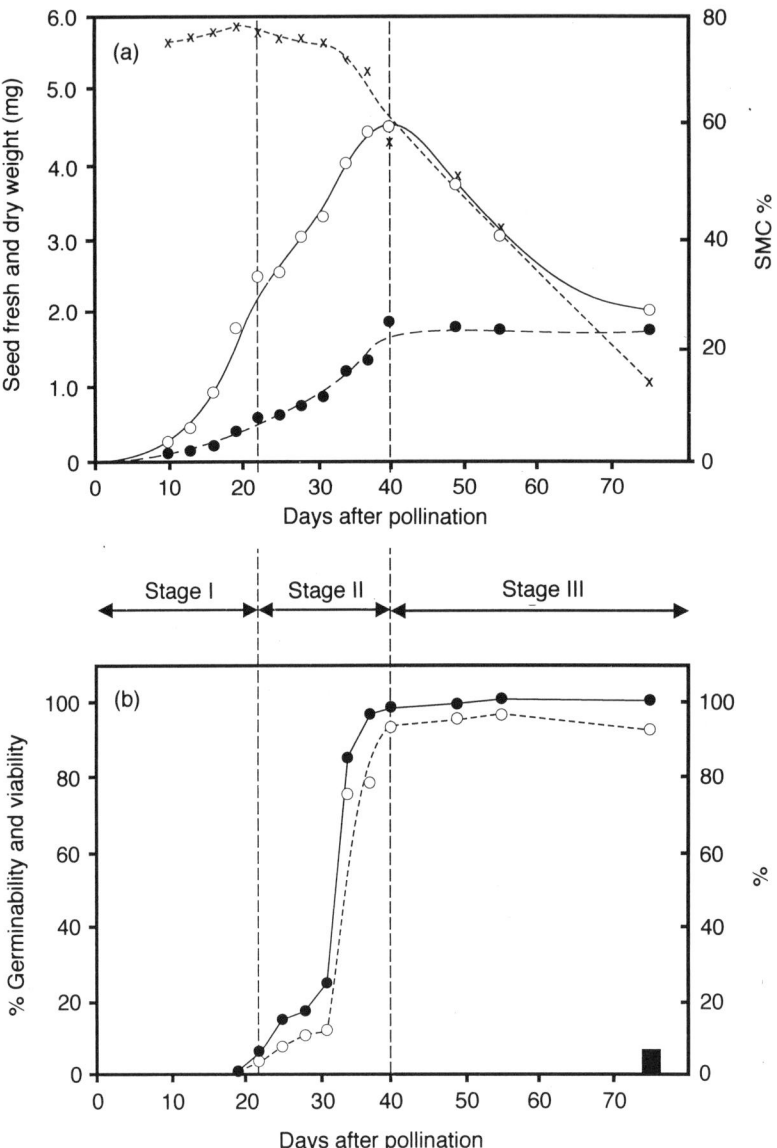

Fig. 4.7. Seed development in lucerne cultivar Wairau. a. o——o, Fresh weight changes; ●---●, dry weight changes; x---x, % seed moisture content. b. ●——●, % seed viability; o---o, % normal seedlings produced. Histogram shows percentage hard seed. (Redrawn from Kowithayakorn and Hill, 1982.)

crops of red clover shown in Fig. 4.6 showed remarkably similar patterns of development in successive years, there were considerable differences in levels of hardseededness. Even position of the ovary on the clover flower may have a considerable impact on seed dry weight changes and the onset of seed germinability (Fig. 4.8).

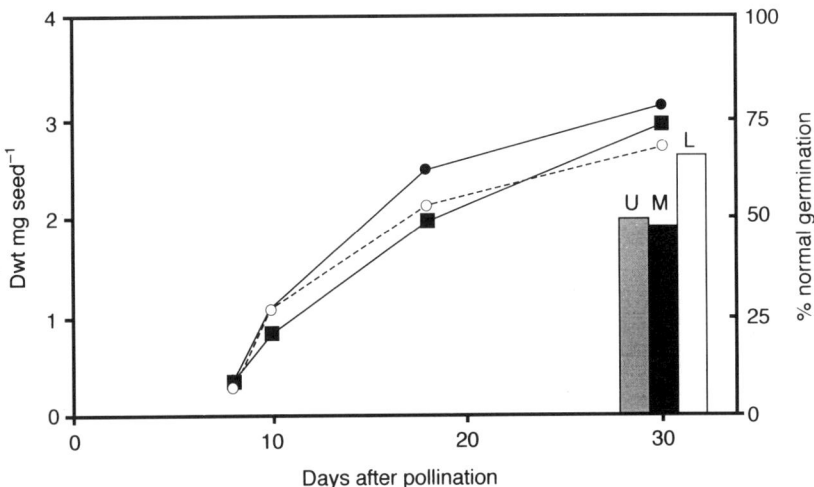

Fig. 4.8. Variation in seed characteristics during development, depending on position of the ovary in open pollinated red clover. Dry weight (Dwt) of seeds from: ■——■, upper; ●——●, middle; ○---○, lower positions on the inflorescence. Histograms show seed germinability 30 DAP from upper (U), middle (M) and lower (L) ovaries. Data from Win Pe (1978).

As with grasses, the inappropriateness of identifying the end of Stage II as physiological maturity is evident, especially because in many legumes hardseededness develops during Stage III (Fig. 4.6) or even in some cases after harvest (e.g. lucerne: Kowithayakorn and Hill, 1982). Similarly, legume species have been shown to lay down storage protein well after the end of Stage II (Millerd and Spencer, 1974).

4.3.2 Developmental Seed Morphology and Physiology

The endosperm and early nutrition of the developing seed

Apart from the notable exceptions in the *Trifoliae* (Bewley and Black, 1978), most mature forage legume seeds tend to be non-endospermic, the endosperm tissues having been digested by the developing embryo. In all species, however, the endosperm is an important source of embryo nutrition at the start of development. Evidence of this is clear from the often reported observation that poor endosperm development seems to be a common cause of embryo failure in crosses between different clover species (e.g. Chen and Gibson, 1971; White and Williams, 1976; Williams and White, 1976).

Initial endosperm development is free nuclear, division usually beginning promptly after fertilization and continuing in a synchronous, or at least partly synchronous fashion depending on species (White and Williams, 1976; Sangduen *et al.*, 1983). In the earliest stages of development endosperm nuclei and cytoplasm tend to remain around the edge of the embryo sac in the region of the embryo (Fig. 4.9a) but an endosperm haustorium develops rapidly, extending to the chalazal

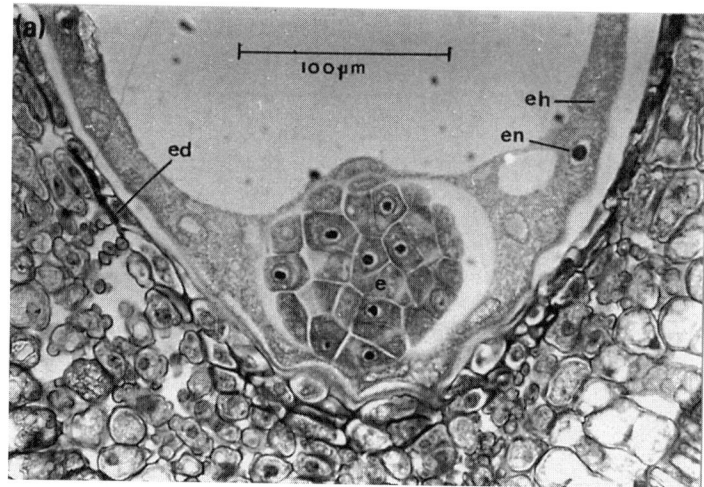

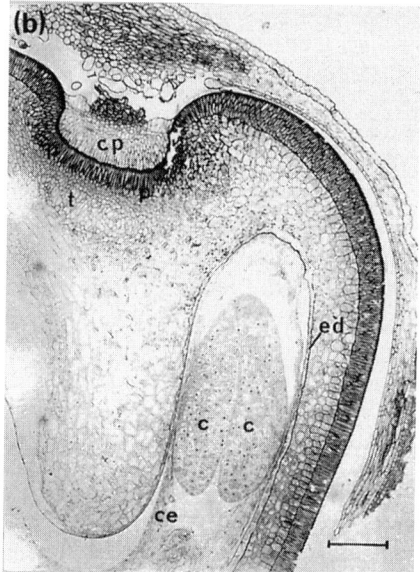

Fig. 4.9. (*and opposite*) Seed development in red clover. a. Six days after pollination: globular embryo stage. e, Embryo; eh, endosperm haustoria; en, endosperm nuclei; ed, endothelium. b. Eight days after pollination: torpedo stage, embryo enveloped by a tube of cellular endosperm. ce, Cellular endosperm; c, cotyledons; cp, counter palisade; p, palisade; t, tracheids; ed, endothelium (scale bar = 100 μm). c. Ten days after pollination: cellular endosperm proliferating ahead of the embryo. ce, Cellular endosperm; e, embryo, eh, endosperm haustoria; t, tracheids (scale bar = 100 μm). d. Eighteen days after pollination: transverse section across the embryo cavity. al, Aleurone layer; enc, cellular endosperm; sc, subcuticular layer; c, cuticle; hf, hilar fissure (scale bar = 100 μm) (Win Pe, 1978). e. Eighteen days after pollination: hilar region. c, Cuticle; hf, hilar fissure; cp, counter palisade; ll, light line; p, palisade; tb, tracheid bar (scale bar = 100 μm) (Win Pe, 1978).

Maturation of Seed

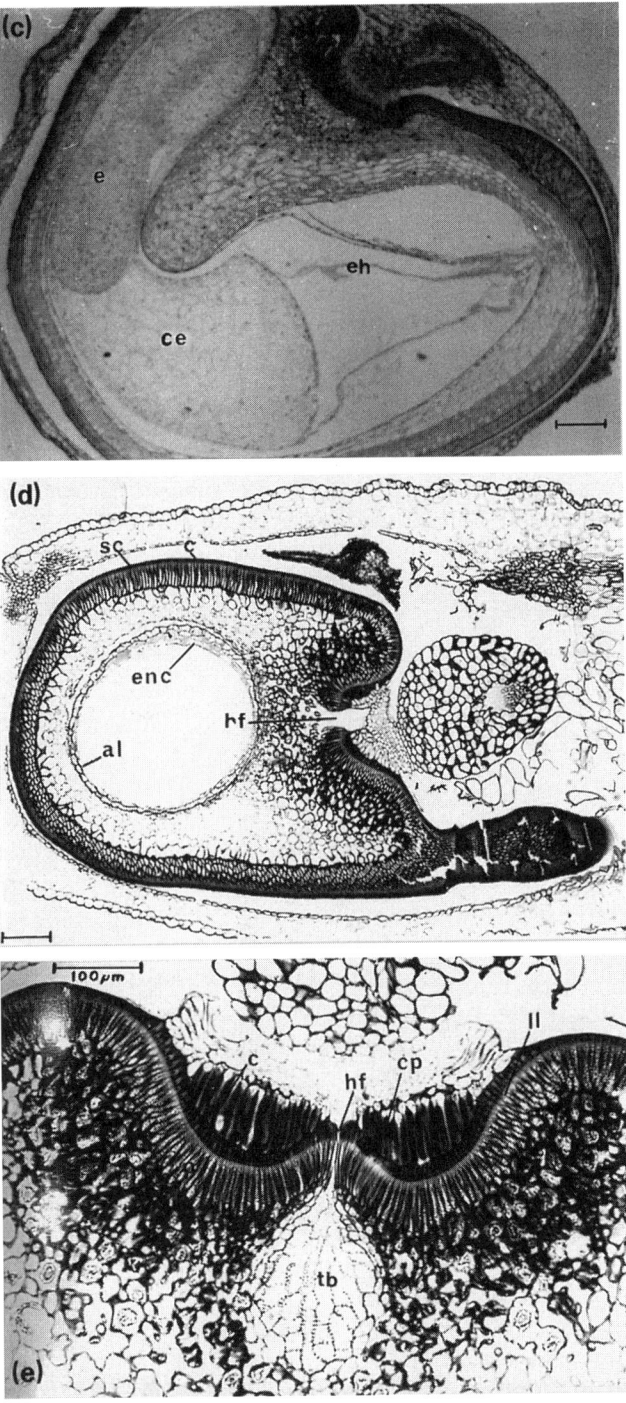

end of the embryo sac (Fig. 4.9b, c). Chen and Gibson (1971, 1973), for example, have pointed out the importance of this haustorium in the absorption and translocation of food reserves in *Trifolium* species. Usually the rest of the endosperm becomes cellular before it is digested by the embryo, but the haustorium may remain multinucleate until it degenerates.

Where the endosperm is persistent, the main stored food reserve is galactomannan which is laid down throughout Stage II of development, beginning with the cells nearest the embryo (possibly to enable the endosperm to resist further digestion). These mucilaginous polysaccharides are formed outside the plasmalemma of the endosperm cells and continue to be laid down until the cytoplasm is completely occluded (Bewley and Black, 1978). The only endosperm cells to escape this fate are those of the aleurone layer, the outermost cells of the endosperm (Watson, 1948). This layer is responsible for the mobilization of these polysaccharide reserves on germination, although there does not appear to be the same tight embryo control of the process that is found in cereal grains (Reid and Meier, 1972). Bewley and Black (1978) point out that this additional endosperm reserve in these seeds is not essential for successful seedling establishment, but may have adaptive significance in allowing seeds of the *Trifoliae* to establish under water stress conditions because of the water-holding capacity of galactomannans. The aleurone layer is clearly visible in *T. pratense* at 18 DAP (Fig. 4.9 d).

Causes of seed loss

There is considerable variation in the number of ovules present in the ovaries of forage legumes, ranging from just two in red clover (Povilaitis and Boyes, 1960) to six in white clover (Chen and Gibson, 1971) to 18 in lucerne (Thomas, 1967) and to over 70 in birdsfoot trefoil (*Lotus corniculatus* L.) (Hill and Supanjani, 1993). However, many of these will not produce mature seed, either through the failure of embryo sac development, poor fertilization (often due to slow pollen tube growth) or, finally, abortion. Thus it is quite usual for large-seeded cultivars of lucerne to produce an average of only three to four seeds per pod (Thomas, 1967).

Accordingly, before going on to discuss embryo development proper, it seems appropriate at this stage to consider the reasons for these losses. Recent data on the pollination of white clover, using a fluorescent staining technique to monitor pollen tube growth, have demonstrated that pollination is unlikely to be a problem for this species. Concurrent shading studies suggest that ovule fertility is a limiting factor pre-pollination, with an additional yield constraint of post-fertilization abortion if the number of fertilized ovules is too high (Pasumarty et al., 1993). It is likely that abortion results from poor competitive ability with adjacent ovules for limited nutrient supply. Povilaitis and Boyes (1960) showed that late fertilized ovules in red clover were the ones most likely to abort. Poor nutrient supply as a consequence of arrested vascular development is cited as a cause of seed abortion in broadbean (*Vicia faba* L.) (White et al., 1984).

Essentially there seem to be three sources of nutrients for the developing embryo in legumes. First, metabolites available from the breakdown of various structures around the embryo sac, such as the synergids, antipodals and some nucellar tissue before fertilization, may be crucially important in determining the success of early development (Chen and Gibson, 1971). Following this stage there is continued absorption of ovule tissues themselves or assimilate from the parent plant which normally occurs via the endosperm, but may in some cases be partly mediated by – or even be directly through – the suspensor (see 'Embryogenesis' below).

Embryogenesis

As noted in grasses, often the fertilized zygote does not divide immediately, but may undergo a resting stage during which time the primary endosperm nucleus begins to divide. Thus in Kenya white clover (*T. semipilosum* L.) there may be four endosperm nuclei before the zygote divides (White and Williams, 1976), while in white clover Chen and Gibson (1971) found that although there may be as many as 14 endosperm nuclei 24 h after fertilization, there are a maximum of only four embryo cells. The physiological basis for this so-called zygote dormancy is, as yet, undetermined (see discussion in Evenari, 1984), nor is it universal in the forage legumes: for example, Chen and Gibson (1971) reported embryos reaching the 20-cell stage before a single endosperm nucleus division in ball clover (*T. nigrescens* Viv.)

Essentially there is little difference in the embryogenesis of forage legumes from the general pattern described by Bhatnagar and Johri (1972). The first division of the zygote is a transverse, unequal one. The smaller apical cell of the resulting proembryo faces into the embryo sac and will give rise to the embryo itself. The larger outer cell, lying near the micropyle, usually divides to produce a multicellular suspensor which functions to locate the embryo in the enlarging embryo sac. In most legume species, the suspensor is persistent, and in some of the larger seeded genera may become multinucleate (Davies and Williams, 1985). Clover species are unusual in this respect in that their suspensor is comparatively poorly developed (Lersten, 1983).

Where the suspensor is persistent, Sangduen *et al.* (1983) have shown that there can be quite subtle differences in suspensor structure between related species. Lucerne shows transfer cells in the chalazal portion of the mature suspensor, complemented by wall ingrowths along the embryo sac from this region to the base of the suspensor. This identifies a role for the suspensor in taking up nutrients from maternal tissue via the endosperm. In the related annual, snail medic (*Medicago scutellata* (L.) Mill), transfer sites seem to be located at the suspensor base, implying direct uptake of nutrients from the embryo sac. Such assumptions about nutrient movement are largely circumstantial, but are supported by ultrastructural features such as proliferation of mitochondria near wall ingrowths. It is also interesting that the apparently more direct assimilate transfer route found in snail medic is associated with a significantly faster rate of embryo development compared to the related perennial. However, this may not be directly nutrient related: studies in Italy on runner bean (*Phaseolus coccineus* L.) (e.g. Lorenzi *et al.*, 1978) show that the suspensor is also an important source of hormones during early embryo development.

The inner apical cell divides to form a sphere of cells at a rate which differs considerably between species. Thus, after 2 DAP, a typical white clover embryo may contain 22 cells, while this stage is unlikely to be reached until 6 days in red clover (Fig. 4.9a), 7 days in broadbean and may take more than 12 days in white lupin (*Lupinus albus* L.) (Chen and Gibson, 1971; Davies and Williams, 1985). Apart from genotypic differences it is clear that raised ambient temperature can increase the rate of the development process, not only decreasing the time interval between pollination and fertilization due to increased pollen tube growth rates, but also speeding up cell division once fertilization has occurred (Chen and Gibson, 1973).

In their comparative study of embryo morphogenesis in grain legumes, Davies and Williams (1985) have demonstrated that cell division in the embryo loses synchrony after the second division. There is no reason to believe that this is any different in the smaller seeded species, and probably represents the first stages of loss of symmetry in the embryo, although morphologically the legume embryo will appear spherical (or at least ovoid) until after the 32-cell stage.

The first indications of the development of cotyledonary outgrowths is typically the flattening of the distal side of the embryo. Cells in this region temporarily lose their meristematic activity and become vacuolate (Win Pe, 1978), acting as an organizing centre around which the two cotyledonary outgrowths develop (Fig. 4.9b). The differentiating embryo takes on a heart-shaped, and then a 'torpedo'-shaped appearance, before attaining its characteristic form. There will generally be little cell differentiation in the embryo axis (Thomas, 1987). In red clover, the embryo attains maximum size very rapidly after this stage (around 14 DAP), even though it has only attained 40% of its final dry weight (Win Pe, 1978).

Smith (1983) distinguished three main types of cotyledons in the legumes, varying from a thin foliar type structured to produce good leaf expansion on epigeal germination, to thick fleshy tissue designed essentially for storage. In the *Trifoliae* and *Loteae* the thin foliar cotyledons have a distinct layer (or layers) of palisade mesophyll cells with spongy mesophyll below. In *Vicia* and *Lathyrus* species the thick, fleshy cotyledon consists of undifferentiated parenchyma. Where germination is epigeal, cotyledon expansion is minimal and, although there may be considerable greening, the net photosynthetic contribution of these tissues is likely to be small.

Chlorophyll is usually retained by developing embryos until after the end of Stage II when the seeds begin to yellow as they ripen. There is some discussion about how much net photosynthate the seeds themselves contribute, but it is likely to be minimal. Cotyledons of white clover, for example, have no stomata (Thomas, 1987). Atkins and Flinn (1978) have demonstrated that developing lupin seeds in Stage II fixed little free CO_2 from the air spaces in the developing pod. Ribulose biphosphate carboxylase (RuBisCO) activity begins to decline during this stage of seed development, but is always present in significant amounts, while phosphoenolypyruvate (PEP) carboxylase increases throughout this time. These observations strongly suggest that, although the seed is most unlikely to make a net contribution to its own demand for assimilate, it can refix the carbon dioxide it loses through its own respiration. Interestingly, both RuBisCO and PEP carboxylase activities increase considerably during seed ripening, yet another illustration (if one was needed) of the fallacy of designating the attainment of maximum dry weight as *physiological* maturity.

Kolloffel and Matthews (1983) reported their findings on respiratory activity in developing pea cotyledons which probably constitutes a pattern representative of most legumes. Activity increases during the first half of Stage II as mitochondrial activity increases. Although it can be demonstrated that the potential respiratory capacity of mitochondria continues to increase up to and a few days beyond the end of Stage II, the respiratory activity of the cotyledons reaches a plateau level during Stage II which is maintained until mass maturity when it begins to decline. This levelling off of respiration is thought to be due to the limited availability of respiratory substrate at this stage. Up until this time respiration is driving the growth processes of seed development. However, as moisture contents begin to fall in Stage III, respiration is still required for maintenance metabolism as the seed undergoes the stresses of desiccation. Under normal ripening conditions loss of mitochondrial competence, and thus respiratory activity, is gradual, allowing these requirements to be fulfilled.

4.3.3. Seedcoat Development, Water Loss and the Development of Hardseededness

During legume seed development the inner of the two integuments largely disappears and it is the outer one which differentiates into a multilayered seed coat (Esau, 1972). The structural details of the testa vary considerably between species, as does the nature of the secondary thickening present in the mature seed. What follows below is an attempt at a generalized description of events.

The epidermis of the outer integument differentiates into an outer, mostly single, layer of elongated cells with domed apices. These macrosclereids are often referred to as either a palisade layer or, less ambiguously, malphigian cells. It is the characteristics of these cells which appear to be largely responsible for the onset of hardseededness in many, but not all legume crops (Rolston, 1978). These cells are overlain with two layers: an outer cuticle which may vary in thickness according to species, supported by a subcuticular matrix of pectins and hemicelluloses. Both these layers begin to become visible in the developing red clover seed at 8 DAP (Fig. 4.9d). Later the macrosclereids undergo extensive secondary thickening, especially at the apex. This process of suberization in peas has been well documented by Spurny (1973), who demonstrated that such structural changes were closely related to a decrease in seed-coat permeability in this species. In other species it is commonly reported that suberin or cutin is laid down in the apical cones of the malphigian cells. These cells remain viable until the lumen of the upper parts of the cells is completely obstructed. It is likely that the so-called 'light line', a conspicuous refraction boundary near the surface of many legume seeds, signifies a change in the pattern of secondary thickening lower down the cell. Secondary thickening in developing red clover can be detected as early as 14 DAP and a clear light line at 18 days (Fig. 4.9e) (early Stage II), although no hardseededness was noted here until the onset of Stage III (Win Pe, 1978).

Below the epidermis there is sometimes an intermediate layer (Watson, 1948) before a layer of osteosclereids deriving from the subepidermal layer of the outer integument. These may vary in shape in different legumes and are variously referred to as pillar cells or hour-glass cells. These cells, dead at maturity, serve to separate the palisade layer from the rest of the seed coat. Typically, there are intercellular spaces between part of the height of these cells, often at the site of the connection with the malphigian cells (see Thomas, 1987, for a detailed description for white clover).

Underneath the asteosclereids are layers of parenchyma which usually collapse in the mature seed. During development these chlorophyll-containing cells may function as an important route, and also a temporary store, for nutrients (Grusak and Minchin, 1988). Assimilate is unloaded from the vascular tissue of the funiculus into the developing seed coat and then apoplastically into the developing endosperm and embryo. Thorne (1965) comments that the extent of phloem vascularization through this layer is not correlated to the rate of seed development, emphasizing the role of this tissue as a whole in regulating the nutrient supply.

As seed development proceeds into Stage III, the seed eventually breaks contact from the parent plant. An abscission layer develops across the funicle leaving indented scar tissue at the hilum. During much of Stage III, the progress of water loss in legume seeds is similar to that found in grasses; however, the hilum appears to play a crucial role in regulating the final water status of the mature seed. The distinct nature of the hilar area is clear at a very early stage of development and well differentiated by mid Stage II (Fig. 4.9d and, in particular, Fig. 4.9e). In the middle of the hilum is a fissure or groove. Around this and above the cuticle in this region is a layer of counter palisade cells, while below the fissure itself is a group of tracheids known as the tracheid bar (Fahn, 1982).

Lersten (1982) conducted an extensive survey of the hilar region of a wide range of species and, commenting on its surprising uniformity, suggested a well-defined function. He suggested that this region may allow the opening and closing of the hilar fissure which acts as a one-way valve, first demonstrated by Hyde (1954) in clover and tree lupin. When the counter palisade cells lose turgor, the hilar fissure is drawn open allowing more rapid water loss from the desiccating seed. In higher relative humidity the counter palisade cells swell and force the fissure closed. Hyde showed that during ripening a typical legume seed loses water rapidly through the testa until it attains a seed moisture content of 20% (fresh weight basis) when the testa begins to become impermeable and water is lost more slowly. At this stage water losses through the hilum of the now detached seed become increasingly important. At around 12% moisture no more water loss can occur through the testa of a hard seed, and any further losses are entirely through the action of the hilum valve. As the fissure is only open to allow continued water loss under conditions of low relative humidity (RH), the final moisture content of the seed reflects the lowest RH with which it has been allowed to equilibrate during ripening.

Two other features visible on the testa of most leguminous seeds are the micropyle (the original site of entry of the pollen tube into the ovule) and the strophiole (or lens – see Tran and Kavanagh, 1984, for a discussion of the terminology of this region). The strophiolar region can be identified before fertilization by virtue of the greatly elongated cells with less dense cytoplasm in this region. In the mature seed this extreme elongation is retained and appears as a ridge on the testa. Typically, osteosclereids are absent in this region and the strophiole is underlain with loosely arranged sclerenchyma.

Taken as a whole, the literature on the impermeability of the legume seed coat is confusing. Watson (1948) could detect no correlation between structural features and impermeability in different legume species. Nearly 50 years later, the situation is little clearer. There is still some debate about the chemical identification of suberin in the secondarily thickened malphigian cells of many species (Werker, 1980) and it is clear that secondary thickening may also comprise quinone complexes and, possibly, some lignin (Rolston, 1978). In particular, a correlation has been reported between hardseededness in peas and catechol oxidase activity (Werker, 1980). This enzyme is a key component of the oxidation pathway of phenols to quinones and thus to pigmented complexes with proteins and/or other quinones. Several studies have been reported on the associations between seed pigmentation and hardseededness. However, in subterranean clover, Slattery *et al.* (1982) found that these events preceded, and were not correlated with, hardseededness.

It is unlikely that the water impermeability causing hardseededness in many seed crops relies on one single structural feature of the testa, but in different species that each component of the tissue system has the potential to present some barrier to water. In crownvetch (*Coronilla varia* L.), for example, McKee *et al.* (1977) reported that the crushed inner integument and possibly the aleurone layer had to be pierced before the embryo was able to take up water. Despite suggestions to the contrary in some of the older literature, it seems fairly clear that it is the strophiole, rather than the micropyle, which offers the main potential route for water uptake by a hard legume seed. Ballard (1976) demonstrated in *Medicago*, *Stylosanthes* and *Trifolium* spp. that only 2% of mildly scarified hard seed commenced water uptake elsewhere. Wounding of the testa transmits stress to the strophiole which is the weakest point on the testa. Interestingly, he found that direct impact damage on the strophiole was less effective at breaking hardseededness than that on other parts of the testa. This is possibly due to the cushioning effect of the radicle pocket just behind this structure. Hopkinson (1993) points out that the legume families *Mimosaceae* and *Caesalpiniaceae* have a strophiole occluded by a plug of tissue which, when induced to erupt, allows water entry.

4.4 SOME CONCLUSIONS

A recurring theme of this chapter is how little we know of the detail of seed development in forage species and how often we can do no more than extrapolate from a very small range of related food crops. If nothing else, this chapter should highlight the dangers of doing that.

A particular deficiency of work on forage seed (in common with other seed crops) is that most seed production studies fall short of gathering information on seed quality and in particular seed vigour. The issue of seed vigour is explored more thoroughly in Chapter 9, but perhaps here we can highlight two prevalent assumptions which should be treated with some caution. The first of these is that maximum seed vigour is usually attained when maximum seed dry weight is achieved. As we have seen at various times in this chapter, there are several important aspects of seed development which occur during Stage III of seed development, after the seed has reached maximum dry weight. The practical implication of this is that if the decision is made to harvest seed at high moisture contents, particular attention must be paid to subsequent handling of harvested material to maximize seed quality. For example, Stoddart (1968) observed that the loss of storability of perennial ryegrass seed caused by premature cutting could be avoided if the swaths were allowed to dry out slowly.

The second potentially dangerous assumption is that seed vigour is always positively correlated with seed size. While Bean (1980), reviewing the small amount of literature available on forage grasses, found that there was very little evidence of effects of water, light and temperature on seed quality which could not be attributed to seed weight, Shimizu et al. (1979) found that smaller seeds produced at higher temperatures were more resistant to temperature stress than those produced under a low temperature regime. No clear picture has yet emerged on the effects of nutrients on seed quality but there is a small amount of evidence in grasses (Bean, 1980) and in legumes (Hadavizadeh and George, 1989) that both the nitrogen and phosphate status of the parent plant may be determinants of seed quality in ways other than via seed size.

Another aspect of seed quality which needs consideration is dormancy. Although seed dormancy bears no direct relation to seed vigour, an unrecognized dormancy problem can severely reduce planting value. The inability of a seed to germinate immediately after harvest under otherwise normal germination conditions may be a function of the seed coat, a metabolic block in the embryo itself, or both. Dormancy has, of course, adaptive significance as the basis of a temporal dispersal mechanism for many seeds, but in many cases simply represents residual mechanisms designed to prevent germination on the parent plant. It is not surprising, therefore, that environmental changes during seed production may have variable effects on natural dormancy levels within seeds. For instance, seeds of perennial ryegrass produced at low temperatures have a more pronounced pre-chilling requirement (Komatsu et al., 1980). Similarly, Juntilla (1977) reported that cocksfoot produced in north Norway showed complex dormancy responses unlike that of other populations produced in Denmark. This may represent genetic segregation, but is equally likely to be due to environmental factors during seed production.

Finally, because of limitations of space, we have made a conscious decision in this Chapter not to address the issues of the physiology of the control of seed production and development. In particular we have neglected to discuss the potential roles of the endogenous plant growth regulators. This may have been a mistake, except that the literature in this field based directly on forage grasses or legumes is next to non-existent. Now that immunoassay techniques for plant hormones such as abscisic acid and cytokinins are becoming routine tools, this kind of study on seeds of major forage species becomes feasible for the first time. It is work which should be done. Only then can we truly address some of the issues about minimal realization of potential seed yield, variable seed size and seed dormancy which bedevil this branch of the seed production industry.

REFERENCES

Andersen, K. and Andersen, S. (1981) Increase in dry matter and decrease in moisture content during ripening of barley. *Acta Agriculturae Scandinavica* 31, 70–74.

Andersen, S. and Andersen, K. (1980) The relationship between seed maturation and seed yield in grasses. In : Hebblethwaite, P.D. (ed.) *Seed Production*. Butterworths, London, pp. 151–172.

Anderson, J.C. (1944) The effect of nitrogen fertilisation on the gross morphology of timothy (*Phleum pratense* L.). *Journal of the American Society of Agronomy* 36, 584–587.

Anslow, R.C. (1964) Seed formation in perennial ryegrass. II. Maturation of seed. *Journal of the British Grassland Society* 19, 349–357.

Atkins, C.A. and Flinn, A.M. (1978) Carbon dioxide fixation in the carbon economy of developing seeds of *Lupinus albus* (L.). *Plant Physiology* 62, 486–490.

Ballard, L.A.T. (1976) Strophilar water conduction in seeds of the *Trifoliae* induced by action on the testa at non-strophiolar sites. *Australian Journal of Plant Physiology* 3, 465–469.

Barnard, C. (1964) Form and structure. In: Barnard, C. (ed.) *Grasses and Grasslands*. Macmillan, London, pp. 47–72.

Barlow, E.W.R., Lee, J.W., Munns, R. and Smart, M.C. (1980) Water relations of the development wheat grain. *Australian Journal of Plant Physiology* 7, 519–520.

Batygina, T.B. (1969) On the possibility of separation of a new type of embryogenesis in Angiospermae. *Review Cytologia and Biologia Vegetale* 32, 335–341.

Bean, E.W. (1980) Factors affecting the quality of herbage seeds. In: Hebblethwaite, P.D. (ed.) *Seed Production*. Butterworths, London, pp. 593–604.

Bewley, J.D. and Black, M. (1978) *Physiology and Biochemistry of Seeds in Relation to Germination. Volume I. Development, Germination and Growth*. Springer-Verlag, Berlin.

Bewley, J.D. and Black, M. (1985) *Seeds: Physiology of Development and Germination*. Plenum Press, New York.

Bhanwra, R.K. (1988) Embryology in relation to systematics of *Gramineae*. *Annals of Botany* 62, 215–233.

Bhatnagar, S.P. and Johri, B.M. (1972) Development of angiosperm seeds. In: Koslowski, T.T. (ed.) *Seed Biology, Volume I. Importance, Development and Germination*. Academic Press, New York, pp. 77–149.

Briarty, L.G., Hughes, C.E. and Evers, A.D. (1979) The developing endosperm of wheat – a stereological analysis. *Annals of Botany* 44, 641–658.

Burson, B.L., Correa, J. and Potts, H.C. (1983) Anatomical basis for seed shattering in kleingrass and guineagrass. *Crop Science* 23, 747–751.

Chen, C.-C. and Gibson, P.B. (1971) Seed development following the mating of *Trifolium repens* × *T. uniflorum*. *Crop Science* 11, 667–672.

Chen, C.-C. and Gibson, P.B. (1973) Effect of temperature on pollen tube growth in *Trifolium repens* after cross- and self-pollination. *Crop Science* 13, 563–566.

Clarke, J.M. (1983) Time of physiological maturity and post-physiological drying rates in wheat. *Crop Science* 23, 1203–1205.

Cochrane, M.P. (1983) Morphology of the crease region in relation to assimilate uptake and water loss during caryopsis development in barley and wheat. *Australian Journal of Plant Physiology* 10, 473–491.

Cochrane, M.P. (1985) Assimilate uptake and water loss in maturing barley grains. *Journal of Experimental Botany* 36, 770–782.

Cochrane, M.P. and Duffus, C.M. (1979) Morphology and ultrastructure of immature cereal grains in relation to transport. *Annals of Botany* 44, 67–72.

Cochrane, M.P. and Duffus, C.M. (1981) Endosperm cell number in barley. *Nature* 289, 399–401.

Davies, S. and Williams, W. (1985) The rate of morphogenesis of embryos and seeds in four species of grain legumes. *Annals of Botany* 56, 429–435.

Elgersma, A. and Sniezko, R. (1988) Cytology of seed development in perennial ryegrass, *Lolium perenne* L. In: Cresti, M., Gori, P. and Pacini, E. (eds) *Sexual Reproduction in Higher Plants*. Springer-Verlag, Berlin, pp. 377–382.

Ellis, R.H. and Pieta Filho, G. (1992) The development of seed quality in spring and winter cultivars of barley and wheat. *Seed Science Research* 2, 9–15.

Esau, K. (1972) *Plant Anatomy*, 3rd edn. Wiley, New York.

Evans, G. (1937) Techniques of grass seed production at the Welsh Plant Breeding Station. *Imperial Bureau of Plant Genetics, Herbage Publication Series*, Bulletin 22. Imperial Bureau of Plant Genetics, Aberystwyth.

Evenari, M. (1984) Seed physiology: from ovule to maturing seed. *Botanical Review* 50, 143–170.

Evers, A.D. (1981) Endosperm cell number in barley. *Nature* 293, 682.

Fahn, A. (1982) *Plant Anatomy*, 3rd edn. Pergamon Press, New York.

Falcinelli, M. (1987) Breeding for seed retention in orchardgrass (*Dactylis glomerata* L.). *Journal of Applied Seed Production* 5, 25–31.

Finkelstein, R.R. and Crouch, M.L. (1986) Rapeseed embryo development in culture on high osmoticum is similar to that in seeds. *Plant Physiology* 81, 907–912.

Grabe, D.F. (1956) Maturity in smooth bromegrass. *Agronomy Journal* 48, 253–256.

Grant, A.S. (1981) A seed development study on the effects of different soil moisture regimes on three perennial ryegrass cultivars (*Lolium perenne* L.). Unpublished DipAgrSc Dissertation, Massey University, New Zealand.

Grusak, M.A. and Minchin, P.E.U. (1988) Seed coat unloading in *Pisum sativum*. Osmotic effects in attached versus excised empty ovules. *Journal of Experimental Botany* 39, 543–559.

Guignard, J.L. and Mestre, J.C. (1970) L'èmbryon des Gramineè. *Phytomorphology* 20, 190–197.

Hadavizadeh, A. and George, R.A.T. (1989) The effect of mother plant nutrition on seed yield and seed vigour in pea *Pisum sativum* cultivar Sprite. *Acta Horticulturae* 253, 55–61.

Hebblethwaite, P.D., Wright, D. and Noble, A. (1980) Some physiogical aspects of seed yield in *Lolium perenne* L. (perennial ryegrass). In: Hebblethwaite, P.D. (ed.) *Seed Production*. Butterworths, London, pp. 71–90.

Herzog, H. (1986) Source and site during the reproductive period of wheat. Development and its regulation with special reference to cytokinins. *Journal of Agronomy and Crop Science* (suppl. 8).

Hill, M.J. (1971) A study of seed production in perennial ryegrass, timothy and prairie grass. PhD Thesis, Massey University, New Zealand.

Hill, M.J. and Supanjani (1993) Reproductive abortion in birdsfoot trefoil, *Lotus corniculatus* L. *Proceedings of the XVII International Grassland Congress*, pp. 1645–1646.

Hill, M.J. and Watkin, B.R. (1975) Seed production studies on perennial ryegrass, timothy and prairie grass. 2. Changes in physiological components during seed development and time and method of harvesting for maximum seed yield. *Journal of the British Grasslands Society* 30, 133–143.

Hopkinson, J.M. (1993) The strophiole in leguminous seeds. *International Herbage Seed Production Research Group Newsletter* 18, 7–8.

Huang, G., McCrate, A.J., Varriano-Marston, E. and Paulsen, G.M. (1983) Caryopsis structural and imbibitional characteristics of some hard red and white wheats. *Cereal Chemistry* 60, 161–165.

Hyde, E.O.C. (1950) Studies on the development of white clover seed. *Proceedings of the 12th Conference of the New Zealand Grassland Association*, pp. 101–107.

Hyde, E.O.C. (1954) The function of the hilum in some *Papilionaceae* in relation to the ripening of the seed and the permeability of the testa. *Annals of Botany* 18, 241–256.

Hyde, E.O.C., McLeavey, M.A. and Harris, G.S. (1959) Seed development in ryegrass and red and white clover. *New Zealand Journal of Agricultural Research* 2, 947–952.

Ingle, J., Bietz, D. and Hageman, R.H. (1965) Changes in composition during development and maturation of maize seeds. *Plant Physiology* 40, 835–839.

Jensen, H.A. (1976) Investigation of anthesis, length of caryopsis, moisture content, seed weight, seed shedding and stripping – ripeness during development and ripening of a *Festuca pratensis* seed crop. *Acta Agriculturae Scandinavica* 26, 264–268.

Johnston, M.E.H. (1960) Investigations into seed setting in cocksfoot seed crops in New Zealand. *New Zealand Journal of Agricultural Research* 3, 345–357.

Juntilla, O. (1977) Dormancy in dispersal units of various *Dactylis glomerata* populations. *Seed Science and Technology* 5, 463–471.

Kermode, A.R., Bewley, J.D., Dasgupta, J. and Misra, S. (1986) The transition from seed development to germination: a key role for desiccation? *HortScience* 21, 1113–1118.

Knowles, R.V. and Phillips, R.L. (1985). DNA amplification patterns in maize endosperm nuclei during kernel development. *Proceedings of the National Academy of Sciences, USA* 82, 7010–7014.

Kolloffel, C. and Matthews, S. (1983) Respiratory activity in pea cotyledons during seed development. *Journal of Experimental Botany* 34, 1026–1036.

Komatsu, T., Shimizu, N. and Suzuki, S. (1980) Studies on seed development and ripening in temperate grasses. 3. Effects of temperature on seed development and ripening, and germination behaviour in perennial ryegrass. *Bulletin of the National Grassland Research Institute* 16, 56–66.

Kowithayakorn, L. and Hill, M.J. (1982) A study of lucerne seed development and some aspects of hard seed content. *Seed Science and Technology* 10, 179–186.

Lee, D.R. and Atkey, P.T. (1984) Water loss from the developing caryopsis of wheat (*Triticum aestivum*). *Canadian Journal of Botany* 62, 1319–1326.

Lersten, N.R. (1982) Tracheid bar and vestured pits in legume seeds (Leguminosae: Papilionoideae). *American Journal of Botany* 69, 98–107.

Lersten, N.R. (1983) Suspensors in Leguminosae. *Botanical Review* 49, 233–257.

Lorenzi, R., Bennici, A., Cionini, P.G., Alpi, A. and D'Amato, F. (1978) Embryo-suspensor relations in *Phaseolus coccineus*: cytokinins during seed development. *Planta* 143, 59–62.

MacLeod, A.M. and McCorquodale, H. (1958) Water-soluble carbohydrates of seeds of the Gramineae. *New Phytologist* 57, 168–182.

MacLeod, A.M., Johnston, C.S. and Duffus, C.M. (1964) Ultrastructure of caryopsis of the Gramineae. I. Aleurone and central endosperm of *Bromus* and barley. *Journal of the Institute of Brewing* 70, 303–307.

McKee, G.W., Peiffer, R.A. and Mohsenin, N.N. (1977) Seed coat structure in *Coronilla varia* L. and its relations to hard seed. *Agronomy Journal* 69, 53–58.

McWilliam, J.R. (1964) Cytogenetics. In: Barnard C. (ed.) *Grasses and Grasslands*. McMillan, London, pp. 154–167.

McWilliam, J.R. (1980) The development and significance of seed retention in grasses. In: Hebblethwaite, P.D. (ed.) *Seed Production*. Butterworths, London, pp. 51–60.

Millerd, A. and Spencer, D. (1974) Changes in RNA-synthesizing activity and template activity in nuclei from cotyledons of developing pea seeds. *Australian Journal of Plant Physiology* 1, 331–341.

Murray, D.R. (1984) Accumulation of seed reserves of nitrogen. In: Murray, D.R. (ed.) *Seed Physiology. Volume I. Development.* Academic Press, Sydney, pp. 83–137.

Murray, P.A. (1995) Seed factors involved in early seedling establishment of *Festuca arundinacea* (Tall fescue). MSc Thesis, Massey University, New Zealand.

Oparka, K.J. and Gates, P. (1981) Transport of assimilate in the developing caryopsis of rice (*Oryza sativa* L.). *Planta* 152, 388–396.

Pasumarty, S., Matsumura T., Higuchi, S., and Yamada, T. (1993) Cultivar variation for seed development in white clover (*Trifolium repens* L.) *Euphytica* 65, 211–217.

Pegler, R.A.D. (1976) Harvest ripeness in grass seed crops. *Journal of the British Grassland Society* 31, 7–13.

Perl, M. (1987) Biochemical aspects of the maturation and germination of seeds. *Advances in Research and Technology of Seeds* 10, 1–27.

Povilaitis, B. and Boyes, J.W. (1960) Ovule development in diploid red clover. *Canadian Journal of Botany* 38, 507–532.

Raja Harun, R.M. and Bean, E.W. (1979) Seed development and seed shedding in north Italian ecotypes of *Lolium multiflorum*. *Grass and Forage Science* 34, 215–220.

Reeder, J.R. (1957) The embryo in grass systematics. *American Journal of Botany* 44, 756–768.

Reeder, J.R. (1962) The bambusoid embryo: a reappraisal. *American Journal of Botany* 49, 639–641.

Reid, J.S.G. and Meier, H. (1972) The function of the aleurone layer during galactomannan mobilisation in germinating seeds of fenugreek (*Trigonella foenum-graecum* L.), crimson clover (*Trifolium incanartum* L.) and lucerne (*Medicago sativa* L.): a correlative biochemical and ultra-structural study. *Planta* 106, 44–60.

Reilly, M.L. (1984) Functional aspects of cereal structure. In: Gallagher, E.J. (ed.) *Cereal Production*. Butterworths, London, pp. 137–160.

Rolston, M.P. (1978) Water impermeable seed dormancy. *Botanical Review* 44, 365–396.

Rost, T.L. and Lersten, N.R. (1973) A synopsis and selected bibliography of grass caryopsis anatomy and fine structure. *Iowa State Journal of Research* 48, 47–87.

Sangduen, N., Kreitner, G.L. and Sorensen, E.L. (1983) Light and electron microscopy of embryo development in perennial and annual *Medicago* species. *Canadian Journal of Botany* 61, 837–849.

Sexton, R. and Roberts, J.A. (1982) Cell biology of abscission. *Annual Review of Plant Physiology* 33, 133–162.

Sharma, H.C. and Gill, B.S. (1982) Variability in spikelet disarticulation in *Agropyron* species. *Canadian Journal of Botany* 60, 1771–1775.

Shaw, R.H. and Loomis, W.E. (1950) Basis for the prediction of corn yields. *Plant Physiology* 25, 225–244.

Shimizu, N., Komatsu, T. and Ikegaya, F. (1979) Studies on development and ripening of seeds in temperate grasses. 2. Effects of temperature on seed development and ripening, and germination behaviour in cocksfoot and Italian ryegrass. *Bulletin of the National Grassland Research Institute* 15, 70–87.

Slattery, H.D., Atwell, B.J. and Kuo, J. (1982) Relationship between colour, phenolic content and impermeability in the seed count of various *Trifolium subterraneum* L. genotypes. *Annals of Botany* 50, 373–378.

Smith, D.L. (1983) Cotyledon anatomy in the Leguminosae. *Botanical Journal of the Linnean Society* 86, 325–335.

Spurny, M. (1973) The imbibition process. In: Heydecker, W. (ed.) *Seed Ecology.* Butterworths, London, pp. 367–389.

Stoddart, J.L. (1964) Seed ripening in grasses. III Changes in chlorophyll and anthocyanin content. *Journal of Agricultural Science* 63, 397–402.

Stoddart, J.L. (1965) Post-harvest changes in seed of *Lolium* species. *Journal of Agricultural Science* 65, 365–370.

Stoddart, J.L. (1968) Biochemical aspects of seed ripening in grasses. *Journal of the National Institute of Agricultural Botany* 11, 370–377.

Thomas, R.G. (1967) Flowering and reproductive development in lucerne. In: Langer, R.H.M. (ed.) *The Lucerne Crop.* A.H. & A.W. Reed, Wellington, New Zealand, pp. 195–204.

Thomas, R.G. (1987) Reproductive development. In: Baker, M.J. and Williams W.M. (eds) *White Clover.* CAB International, Wallingford, pp. 63–123.

Thorne, G.N. (1965) Photosynthesis of ears and flag leaves of wheat and barley. *Annals of Botany* 29, 317–329.

Tran, V.N. and Kavanagh, A.K. (1984) Structural aspects of dormancy. In: Murray, D.A. (ed.) *Seed Physiology Volume 2. Germination and Reserve Mobilisation.* Academic Press, Sydney, pp. 1–44.

Watson, D.P. (1948) Structure of the testa and its relation to germination in the Papilionaceae tribes *Trifoliae* and *Loteae. Annals of Botany* 12, 385–409.

Werker, E. (1980) Seed dormancy as explained by the anatomy of embryo envelopes. *Israel Journal of Botany* 29, 22–44.

White, D.W.R. and Williams, E. (1976) Early seed development after crossing *Trifolium semipilosum* and *T. repens. New Zealand Journal of Botany* 14, 161–168.

White, G., Gates, P. and Boulter, D. (1984) Vascular development in the reproductive tissues of *Vicia faba* L. In: Hebblethwaite, P.D., Dawkins, T.C.K., Heath, M.C. and Lockwood, G. (eds) *Vicia faba.* Martinus Nijhoff/Dr W. Junk, Netherlands, pp. 15–22.

Williams, E. and White, D.W.R. (1976) Early seed development after crossing of *Trifolium ambiguum* and *T. repens. New Zealand Journal of Botany* 14, 307–314.

Win Pe (1978) A study of seed development, seed coat structure and seed longevity in 'Grasslands Pawera' red clover (*Trifolium pratense* L.). PhD Thesis, Massey University, New Zealand.

Zadoks, J.C., Chang, T.T. and Konzak, C.F. (1974) A decimal code for the growth stages of cereals. *Weed Research* 14, 415–421.

Grass Seed Crop Management 5

M.P. Rolston,[1] J.S. Rowarth,[2] W.C. Young III[3] and
G.W. Mueller-Warrant[4]

[1] AgResearch Grasslands, PO Box 60, Lincoln, New Zealand; [2] Plant Science
Department, Lincoln University, Canterbury, New Zealand; [3] Department of
Crop and Soil Science, Oregon State University, Corvallis, Oregon 97331, USA;
[4] USDA-ARS, Oregon State University, Corvallis, Oregon 97331, USA

5.1 INTRODUCTION

Temperate grass seed production involves a wide range of species grown for forage or turf in countries ranging from 45° South (New Zealand) to 61° North (Norway and Canada), excluding the tropics but including high altitude subtropical regions from latitudes of 26° (China). Management systems in these countries are affected by climate and soil type, by the availability of resources such as fertilizer and chemical sprays, and by local government regulations. In all developed countries a compromise is being effected between the desires of the seed grower for achieving maximum yields and profits, and what government regulations will allow in terms of fertilizer application, use of chemical sprays and residue burning. It is the challenge of the seed scientist to work within the climate, resources and regulations to help the seed grower produce optimum seed yield and quality from cultivars generally bred for vegetative performance.

This chapter reviews the effects of management on seed yield of the major temperate grass species.

5.2 SITE AND CULTIVAR/SPECIES SELECTION

5.2.1 Adaptation for Production, Projected Marketing

Most grass seed, particularly seed of new cultivars, is grown under certification programmes in specialized seed-growing areas far from the regions of consumption, and often in regions of different latitudes from those where the breeding was undertaken. Information on the cultivars' seed yield potential, species and cultivar production, carryover, and future demand should be considered by both the producer and the seed company when selecting a species or cultivar for seed increase.

5.2.2 Physical and Environmental Constraints

The selection of species (and cultivar) will be influenced by the physical and environmental constraints that include drainage, availability of irrigation, time of flowering (and therefore harvest) in relation to moisture availability, nutrient requirements or tolerances of the species, weed problems and herbicide options available.

5.3 CROP ESTABLISHMENT

The most common problem in grass seed crop establishment is failure to observe basic principles of sowing. These include:

- Removal or destruction of the previous vegetation, and control of weed species.
- Preparation of a seedbed suitable for rapid establishment of the sown species.
- Provision of adequate nutrients.
- Appropriate time and sowing method.
- Careful management during seedling establishment.

Establishment of a new seed field is expensive and often seed is limited. Therefore, it is important that the most favourable environment be provided to overcome adverse conditions during the critical germination and seedling development period.

5.3.1 Site Preparation

Certification rules should be consulted for field history requirements for production of certified seed. Grasses may follow grain or cultivated crops in areas well supplied with moisture. In dry areas it may be necessary to fallow prior to the sowing year. A fallow period allows time for preparation of a good seedbed, provides an opportunity for weed control (see Section 5.5), and allows for accumulation of moisture and soil nutrients for the seedlings during the establishment year. Fallow land should be left in a rough condition during the winter months to reduce the danger of winter and early spring erosion.

The primary objective of soil preparation for any crop is to provide a seedbed which will enhance germination and rapid emergence of the seedlings. Existing vegetation should be killed with cultivation or herbicides. Any remaining plant residue must be buried well below the seedbed. The seedbed should be firm to allow seeds to absorb moisture rapidly, with a moist zone covered by well-worked soil, free from large clods and air pockets.

Fertilizer application should be made during seedbed preparation or immediately before sowing. The amount of nitrogen (N) fertilizer to be applied is determined by the residual N from the previous crop, and the species to be planted. Soil tests should be used to determine the amount of phosphorus or potassium to be applied. Lime, where needed, should be applied and thoroughly mixed with the soil prior to final seedbed preparation.

The chances of establishing good and vigorous stands of grasses are usually much better where they are sown alone rather than with a nurse or companion crop (which competes and decreases herbicide options). However, the advisability of using a nurse or companion crop will depend on the individual situation. A companion crop may provide additional income during the establishment year in spite of reducing the seed yield in the first harvest year. Meadow fescue (*Festuca pratensis* Huds.) and cocksfoot (*Dactylis glomerata* L.) have been found to respond favourably to companion cropping; tall fescue (*Festuca arundinacea* Schreb.) has not (Chastain and Grabe, 1989a, b). If a companion crop is used, the sowing rate of the companion crop should not be higher than one-half the usual rate per hectare.

5.3.2 Time of Sowing

Spring sowings must be sufficiently early for seedlings to become established before the onset of dry, hot summer conditions. Late summer sowings often encounter high soil temperatures, dry soil and summer weed competition; irrigation may be needed. Autumn sowings must be sufficiently early for plants to become established and tiller before cold temperatures inhibit growth, freeze soft tissue and cause frost heave. Species such as tall fescue, Kentucky bluegrass (*Poa pratensis* L.) and cocksfoot are slow to establish and to develop a plant of adequate size to allow floral induction in the first production season or to develop enough tillers to produce a good seed crop. These species are normally sown in the spring or late summer for harvest in 12–15 months. Ryegrasses (*Lolium* spp.) are commonly autumn sown for harvest in 9–11 months, while non-vernalizing grasses (e.g. Westerwolds type annual ryegrass) (*L. multiflorum* Lam.) can be sown in spring for harvest in 3–4 months.

5.3.3 Sowing Method

Several types of drills give good results if they are adjusted properly. The ordinary grass drills and fluted grain drills will handle most grass seeds. Alternate coulters on the drill may be closed to obtain the desired row spacing. An inert carrier such as clean rice hulls or sawdust can be mixed uniformly with the seed to provide additional bulk to obtain more uniform distribution of seed.

The carbon-banding, broadcast-diuron system was developed to control seedling weeds between the rows in new plantings of grasses grown for seed (Lee, 1973; Mueller-Warrant *et al.*, 1991). The system has provided good, but not total, control of all grass weeds; weeds can germinate within the carbon-band and weeds germinating from more than 20 mm below the soil surface are often poorly controlled. Control is improved when good soil moisture prior to planting is available to germinate a major flush of weeds, which can then be killed with paraquat or glyphosate. Ethofumesate is often applied after rainfall has dispersed the activated charcoal band in order to control annual poa (*Poa annua* L.) seedlings within the row.

Table 5.1. Effect of sowing rate on seed yield in Kentucky bluegrass and red fescue. (After Meijer, 1984.)

	kg ha^{-1}			
Kentucky bluegrass				
Sowing rate	3	6	12	24[1]
Seed yield	1658	1708	1696	1570
Red fescue				
Sowing rate	4	8	16	32[1]
Seed yield	1311	1346	1319	1143

[1] For both species, seed yield was significantly decreased at the highest sowing rate.

5.3.4 Spatial Patterns

Recommended sowing rates depend upon a number of factors including season, seedbed condition, sowing equipment, row spacing, cultivation equipment and seed availability. The objective is to obtain an acceptable population of strong and vigorous plants, rather than a dense population of weak plants, and sowing rates for seed production are therefore normally lower than the optimum rates recommended for forage production (Acikcoz and Karagoz, 1989). In practice, seed production of temperate grasses is often little affected by sowing rate, as similar yields can be obtained from a wide sowing rate range (Table 5.1). With good seedbed preparation and attention to sowing depth and row spacing, sowing rates of 5 kg ha^{-1} or lower can be used, but for small seeded species, precision drills may be required.

There is often conflicting experimental evidence for the effects of row spacing on seed yield (Hampton, 1988), but for forage grasses, similar yields can be obtained from row spacings which vary from 15 to 45 cm. Row spacings commonly used are from 10 to 30 cm for perennial ryegrass, from 20 to 35 cm for red fescue (*Festuca rubra* L.), from 30 to 40 cm for cocksfoot, and from 40 to 60 cm for tall fescue. Often the choice of row spacing will depend on equipment design and availability rather than agronomic considerations.

5.4 NUTRIENTS

5.4.1 Nitrogen

Nitrogen (N) forms 3–5% of the dry matter of an herbage seed crop. Availability of N is a major determinant of plant yield; in seed crops N affects tillering, dry matter production, inflorescence determination, yield component dynamics and, ultimately, seed yield and quality (Hebblethwaite and Ivins, 1977; Rolston *et al.*, 1985).

Form of nitrogen

Plants can take up N as nitrate or ammonium ions. In biologically active soils, nitrate ions tend to dominate, even at low temperatures. Recent research has shown that at temperatures below 3°C nitrate can be measured in the soil within 2 h of application of urea (R. Sherlock, Lincoln University, 1995, personal communication).

Cost, ease of application, environmental conditions and requirements for other nutrients determines fertilizer-N choice. Urea is generally the cheapest form, but is also the most mobile. At soil surface temperatures above 15°C volatilization of urea increases exponentially; losses of 20% of N have been recorded (Black *et al.*, 1987). Volatilization can be reduced by irrigation. However irrigation (or rainfall) of 50 mm can wash 60% of the applied N below 200 mm in the soil (Francis and Haynes, 1991), thus removing it from the zone of maximum nutrient uptake.

Rate of nitrogen

Research on the fertilizer-N required by grass seed crops has tended to concentrate on rate and timing of fertilizer. Increasing the amount of N fertilizer applied usually increases seed yield until an optimum application rate is reached (Table 5.2; Hampton, 1988). Estimates of the optimum rate vary widely, e.g. from 60 (Young *et al.*, 1995) to 180 kg N ha^{-1} (J.S. Rowarth, Lincoln, 1996, unpublished) in perennial ryegrass and from 100 (Nordestgaard, 1986) to 320 kg N ha^{-1} (Jiminez-Merino *et al.*, 1993) in cocksfoot. Few reports indicate soil nutrient status (soil nitrogen has been estimated at 10 to 60–105 kg ha^{-1} (Hampton, 1987; White, 1990)), cropping history, soil temperature, rainfall or atmospheric inputs (estimated at 50 kg ha^{-1} year^{-1} in Britain (S. Jarvis, North Wyke, 1995, personal communication)), yet all these factors influence nitrogen availability. In some experiments yields were increased by the highest amount of fertilizer-N applied, implying that the optimum rate may not have been reached. Furthermore, the variation in maximum seed yields recorded implies that factors other than N may have limited yields. Thus generalizations about optimum fertilizer-N requirements

Table 5.2. Effect of nitrogen on seed yield in two cultivars of perennial ryegrass over three seasons at the same site. (Adapted from Hampton *et al.*, 1983.)

		Seed yield (kg ha^{-1})					
		Royal[3]			Morenne[4]		
Applied-N (kg ha^{-1})[1]	Total-N (kg ha^{-1})[2]	1979	1980	1981	1979	1980	1981
0	55	—[5]	—	1318	—	—	1483
80	135	2006	1252	1755	1699	1038	1846
120	175	1997	1127	—	1542	1170	—
160	215	1883	1162	1601	1515	1039	1952
Significance		NS	NS	**	NS	NS	**

[1]Applied as urea.
[2]Applied-N plus measured available soil-N.
[3]Amenity cultivar.
[4]Forage cultivar.
[5]Data not available
NS, Not significantly different at $P < 0.05$.
**Significantly different from the control at $P < 0.01$.

and the extrapolation of results to commercial practice are suspect. Recognition that the traditional experimental approach to estimating optimum fertilizer-N applications for seed crops is temporally and spatially specific has stimulated research on alternative methods for predicting fertilizer-N requirements (Rowarth et al., 1993; Rowarth and Archie, 1994) (see 'Yield prediction' below).

Timing of nitrogen

The physiological age of the plant when fertilizer-N is applied affects subsequent yield response. The optimum time for applying fertilizer-N in spring is thought to be between spikelet initiation (Hampton, 1987) and stem elongation (Brown, 1980). The effect of N application just prior to spikelet initiation, i.e. application in late winter rather than early spring, which is the time of maximum concentration of N in a plant, occurring just before rapid spring growth (Scharrer and Mengel, 1960), has not been reported. Rolston et al. (1985) suggested that fertilizer-N should be applied to make up the difference between soil mineralizable-N and the N requirements of the crop (130 kg ha^{-1}; Hampton et al., 1983). However, mineralization is slow in early spring (when N demand by the plant is high) as soil temperatures are low. Mineralizable-N gives an indication of the amount of N that is likely to become available during the growing season, and might be considered as an indicator of the likely response to additional N later in the season.

The balance between autumn and spring nitrogen is also of concern in considering timing of fertilizer-N application, particularly in those countries where the growing season is short, e.g. Denmark (Nordestgaard, 1986). Results are difficult to compare because of lack of information on developmental stage of the crop, soil nutrient status and large differences in amount of N applied and the yields achieved. In general, seed yields have been shown to respond to fertilizer-N applied one-third in autumn followed by a second application (two-thirds of the total) in early spring (Nordestgaard, 1986; Meijer and Vreeke, 1988). Delaying application in spring reduces number of fertile tillers, regardless of the amount of autumn-N; in some species (e.g. cocksfoot, red fescue and Kentucky bluegrass) the decrease in fertile tillers also decreases seed yield (Nordestgaard, 1986). Delaying spring-N application may increase TSW (thousand seed weight) and number of seeds per fertile tiller. However, late-flowering species such as timothy respond to mid-spring N; Italian ryegrass is relatively insensitive to timing of N application (Nordestgaard, 1986). Early spring-N may be insufficient to fulfil plant requirements throughout the growing season if soil-N is inadequate, in which case N applied at anthesis may have a positive effect on TSW (Green, 1993). When soil-N is more plentiful, plant requirements can be met by a single spring application (Hampton, 1987). However, when large amounts of urea (over 100 kg ha^{-1} of N) are used, splitting the application reduces risk of leaf-burn and loss of N due to extremes of weather which could result in volatilization or leaching.

Seed yield component response

Yield responses to increasing N are generally due to increasing spike size (Langer, 1979), spikelets per tiller (Ryle, 1964; Hill and Watkin, 1975; Hare and Rolston, 1990), florets per spikelet (Hare and Rolston, 1990; Young et al., 1995) and increased seeds per spike (Young et al., 1995). There are few reports of spring-N having a positive effect on fertile tillers (exceptions are Rolston *et al.*, 1994; Young *et al.*, 1995). This is probably because by the time spring-N is applied, most of the tillers large enough to produce a seed head by the traditional harvest date have already become reproductive. Those that are stimulated by N fertilizer to increase in size and become reproductive will not be sufficiently mature to contribute to harvest. Applying fertilizer after stem elongation decreases seed yield (Hebblethwaite and Ivins, 1978) and, particularly in an N-deficient crop where there is little competition for light, stimulates new vegetative tillers (Meijer and Vreeke, 1988) which can cause harvesting difficulties. Differences in reported responses in components of yield to added N probably reflect the fact that crops were at different developmental stages (physiological age) when the fertilizer-N was added; physiological age is rarely indicated (exceptions are Hampton, 1987; Young *et al.*, 1995), making comparisons between experiments difficult.

Yield prediction
The concentration of N in ryegrass herbage cut just before stem elongation (early spring) is related to seed yield at harvest (Rowarth *et al.*, 1993; Rowarth and Archie, 1994). This suggests that the concentration of N in foliage can indicate plant nutrient status and seed yield (*ceteris paribus*) and that maximum seed yields can be achieved with a foliar N concentration of 5–6% (Rowarth and Archie, 1994, 1995). For cocksfoot, 3.75% N is thought to be optimum (Schoberlein and Wahl, 1993). The concentration of N in foliage is a direct measure of plant nutrient status – it integrates N supplied from the soil, atmosphere and fertilizers. Determining plant-N status in early spring allows growers time to correct deficiencies by applying fertilizer-N, and assists them to avoid over- or under-application.

Considerations other than yield
The trend towards increasing use of fertilizer-N has resulted in increased yields (cf. Evans, 1963, 40 kg ha^{-1} N and 1100 kg ha^{-1} seed; Rolston *et al.*, 1994, 120 kg ha^{-1} N and 2150 kg ha^{-1} seed) and decreased incidence of blind seed disease (see Section 5.10.1). However, there is evidence that lodging has also increased, as have concerns about the possibility of environmental contamination, particularly of drinking water; the latter has resulted in increased emphasis on fertilizer recovery efficiency (Rowarth *et al.*, 1994).

5.4.2 Other Major Elements
There has been little research on the fertilizers required for grass seed production for nutrients other than nitrogen. This is partly because nitrogen is highly mobile in the soil, and hence plant-available supplies fluctuate, whereas most other nutrients are retained in the soil to some degree. Furthermore, grass seed crops are often part of a crop rotation, in which they are considered to be less demanding in terms of nutrients than other crops. However, restrictions on field burning (and, therefore, inorganic nutrient return to the soil), particularly in Oregon, have resulted in surveys and the formulation of some recommendations:

Sulphur (S) should be applied early in the growing season (Hart et al., 1989); a foliar concentration of about 4% S at stem elongation was associated with maximum yields in pot trials (J.S. Rowarth, Lincoln, 1996, unpublished).

Soil phosphorus (P) should be above 25 µg g^{-1} Bray in Oregon (Horneck and Hart, 1988). In New Zealand, no response to P has been found above an Olsen of 6 µg g^{-1} (Brown, 1980), but maximum yields achieved were only 1000 kg ha^{-1}, suggesting some other limiting factor (e.g. soil moisture). Top yields (2250 kg ha^{-1}) have been achieved with an Olsen of 17 µg g^{-1} (J.S. Rowarth, Lincoln, 1996, unpublished).

Potassium (K) is recommended to be above 100 µg g^{-1} exchangeable K in Oregon (Horneck et al., 1993). In New Zealand no response was found to K in unirrigated field trials (Brown, 1980) and no relationship was found between soil K and seed yield in a survey of 80 crops (J.S. Rowarth, Lincoln, 1996, unpublished), reflecting the recent nature of many of New Zealand's cropping soils.

5.5 WEED CONTROL AND HYGIENE

5.5.1 Losses Caused by Weeds

Weeds interfere with the establishment of new sowings through the obvious means of intercepting light, depleting soil moisture, and using nutrients; first-year seed yields of cocksfoot were reduced an average of 0.9%, or 7 kg ha^{-1}, by each 1% increase in weed cover during the sowing year (Rolston and Hare, 1986). Weeds also pose serious yield threats to successfully established stands. Yield losses due to weeds can be difficult to measure because treatments that control weeds often damage the crop, and would reduce yield if applied to weed-free stands. As a result, the true magnitude of yield losses due to weeds is often understated.

Paraquat between rows has been reported to increase cocksfoot seed yields by 35–47% (Rolston, 1991), while broadcast application of four different herbicides increased perennial ryegrass yields by an average of 8% (Mueller-Warrant et al., 1994). Atrazine, chlorprophan, diuron, ethofumesate, fenoxaprop, pronamide, dichlobenil + monolinuron and simazine have all been found to control volunteer seedlings in forage grasses; seed yield increases were not always apparent (e.g. Lescar and Bouchet, 1968; Hammond et al., 1976; Johnson et al., 1982; Wright and Hebblethwaite, 1983; Mueller-Warrant and Brewster, 1986; Mueller-Warrant, 1990, 1991).

Awns are found on many weeds common in grass seed fields, including Italian ryegrass, rattail fescue (*Vulpia myuros* (L.) K.C. Gmel.), downy brome (*Bromus tectorum* L.) and California brome (*Bromus carinatus* H. and A.). These awns not only aid in the spread of weeds by machines and livestock, but also hinder seed cleaning. Awns become stuck in screens that sort seed by size, reducing capacity and forcing frequent removal and manual cleaning of the screens. Awns also break off from weed seeds, leaving either a de-awned seed or a naked caryopsis that the seed cleaning equipment may not be able to separate from the crop seed due to similarity in size. Sizes of weed and crop seed often overlap, causing loss of some of the smaller size fraction of the crop seed with the larger size fraction of the weed seed, or *vice versa*. Losses of good perennial ryegrass seed during cleaning to remove common weed seeds have been estimated at 10% during the first pass through an air-screen seed cleaner. Seed laws may prohibit sale of seed containing designated noxious weeds, and specific markets often set maximum tolerance levels for certain weed species including *Rumex* spp. Crops free of *Poa annua* L. typically sell for 4–10% more than contaminated seed lots.

5.5.2 Techniques

Herbicides

Many of the herbicides used to control weeds in grass seed crops are only marginally selective, especially when used in attempts to control genetically similar grasses. Their safety depends on a variety of factors, including stand age, plant health, growing conditions and season of application. Selectivity between seedling volunteer crop or other annual weeds and the established crop plants is often based on the greater ability of the established crop to regrow following destruction of leaf tissue, although differences in rooting patterns may also play a role. Safety margins for use of photosynthesis inhibitors in grass seed crops are often quite small, particularly when crops such as perennial ryegrass begin their autumn regrowth. The carbon-band planting technique (see Section 5.3.3) is used, despite its expense, because it safely controls a wider spectrum of seedling grasses during the establishment year than any other approach. The tolerance of seedling grasses to herbicides normally increases with stand age and stage of plant development (Faulkner, 1974), although the converse has been found in perennial ryegrass. Environmental conditions such as temperature, light intensity, humidity and moisture stress also play a role in the tolerance of crops to herbicides (e.g. crop injury from photosynthesis inhibitors such as diuron, atrazine and metribuzin is greatest under full sunlight, and damage is accentuated when a lack of rainfall after application allows substantial foliar absorption of the chemicals), as do the presence of invertebrate pests and pathogens. Grass seed crops become sensitive to a variety of herbicides in the boot to anthesis stages, including the growth-hormone types such as MCPA, 2,4-D and dicamba (Darwent and Smith, 1988), the wild oat (*Avenua fatua* L.), materials such as MSMA and fenoxaprop (Mueller-Warrant, 1990, 1991), and the sulfonylureas.

The introduction of herbicides in the 1950s caused dramatic changes in grass seed production practices and seed trade expectations for freedom from weed seed contamination. The impact of herbicides on grass seed production is enormous; the economic benefit of oxyfluorfen alone to Oregon grass seed growers was estimated at US $69 million for the 1995–1996 growing season. In 1987, changes in rules about herbicide registration in the USA led to withdrawal of grass seed labels for atrazine, simazine, propham and chlorpropham. Mounting public pressure to reduce field burning in western Oregon led to compromise legislation in 1991, phasing down field burning and imposing much greater expectations on the role of herbicides in the control of weeds. Recent similar restrictions in Europe on the use of field burning and pesticides are expected to increase problems with *Poa annua* in grass seed crops.

Cultural

Cultural methods of controlling weeds were de-emphasized during the widespread adoption of herbicides, although clipping during the sowing year to limit weed seed production has continued. It is commonly done with spring sowings, and with autumn sowings of species other than perennial ryegrass in cases where the crop is thinly tillered or heavily infested with weeds. Recent recognition that the near-total reliance on herbicides for weed control in grass seed production poses a variety of risks, including the potential for chaos if herbicide registrations are suddenly withdrawn, as well as the problem of developing herbicide-resistant weeds, has generated renewed interest in cultural control methods that include clipping, inter-row cultivation, carbon banding, flame sterilization and grazing.

Biological

Biological weed control has generally been ignored in grass seed production, due to the relatively small area involved, perceived needs for near-total weed control, lack of effort in developing specific agents, and success of herbicides. Grazing has sometimes been used as a biological control method, being most useful against weeds that animals find more palatable than the crop. Several bacteria show promise for control of specific weeds, e.g. *Pseudomonas fluorescens* (Trevisan) Migula suppresses downy brome in Kentucky bluegrass, and *Xanthomonas campestris* (Pammel) Dowson controls *Poa annua*; the major factor limiting use is year-to-year inconsistency in control.

5.5.3 Hygiene

Buried seed

Grass seed is usually produced by growers who have specialized for many years. As a result, the same fields have often been used to produce many seed crops and may contain large quantities of buried seed of various species. However, the viability of many commonly grown species declines rapidly in the soil, and a 2-year rotation out of grass seed production is sufficient to eliminate the viability of nearly all perennial ryegrass, tall fescue, cocksfoot and fine fescue seed. Some Italian ryegrass seed remains viable after 6 years of burial in the soil. Dormancy in bluegrass (*Poa* spp.) and bentgrass (*Agrostis* spp.) species is a problem; the minimum time to change cultivars for these species is 3 and 5 years, respectively. Most weedy annual grasses possess some degree of seed dormancy, a factor which contributes to their success.

Deep ploughing can help reduce the density of buried seed near the soil surface, i.e. the potential germination zone, although subsequent ploughing will bring some seed back to the soil surface. Deep ploughing is best used with species with short seed survival in the soil and for which no easy way exists to control them during the rotational crops. Seed of species possessing substantial dormancy and soil longevity may be better controlled by leaving them on or near the soil surface and encouraging germination, followed by use of tillage or non-selective herbicides to destroy them. Preventing weeds from going to seed is the best general hygiene available to farmers. However, nearly all weeds that pose problems in grass seed production shed their seed by the time of grass seed harvest. Indeed, seed from the crop itself is a major weed from the perspective of future sowings.

General hygiene

Weed seed dispersal is aided by many activities. Intakes in irrigation systems should be screened to restrict weed seed dispersal. Animals should be 'emptied' or grazed on weed-free pasture before grazing seed fields as many weed seeds survive passage through the animal's digestive system. Hay should not be fed on grass seed fields, and if it is, the feeding area should be restricted to a small area which will not then be harvested. Weed seeds will also be dispersed throughout the field through faeces as the animals graze.

Thorough cleaning of all equipment moving between grass seed fields is vital to prevent contamination of one seed lot by weed or crop seeds from another field (see Section 5.5.2). Perhaps the most crucial area for general hygiene is the seed used to plant new stands. Most incidents of long-distance weed seed transport occur in the seed lot used to plant new stands; seed analysis certificates should be examined before buying seed lots.

5.6 DEFOLIATION AND CLOSING

Defoliation is practised either to improve crop yields by the removal of excess vegetative growth that may impede light penetration and tiller development or cause early lodging, and/or to increase enterprise profitability by grazing with livestock during autumn/winter and early spring at a time when forage is scarce.

The time to close the crop (cease grazing or cutting) is based on the need for an adequate number of reproductive tillers in the spring. Cutting or grazing prior to internode extension results in the removal of leaf material only; the subsequent regrowth arises from the extension of leaf primordia from the terminal meristem and from the unexpanded leaves that remain (Jewiss, 1972). Defoliation after stem elongation commences will damage the reproductive meristem, resulting in decreased seed yields (Roberts, 1965; Watson and Watson, 1982; Table 5.3).

Table 5.3. The effect of closing data on mean spikelet number and seed yield of perennial ryegrass (Hill, 1971).

Date of closing from grazing	Stage of reproductive development[1]	Number of spikelets per ear	Relative seed yield
18 August	Vegetative	19.4	98
1 September	Spikelet initiation	19.7	97
15 September		19.6	100
29 September		19.1	95
6 October	Floret initiation	17.4	79
13 October		16.1	77
21 October		16.3	70
27 October	Ear emergence	15.8	57

[1] For the plots closed on August 18.

5.7 PLANT GROWTH REGULATORS

Plant growth regulators (PGRs) have been widely evaluated in grass seed crops, especially ryegrass, with their application resulting in seed yield increases that are variable but often large (8–136%). Earlier work on PGRs was reviewed by Hebblethwaite (1987) who concluded that the age of chemical manipulation seemed to be near in ryegrass seed production. A decade later the use of PGRs in seed production is low, with growers still waiting for an acceptable, cost-effective chemical that gives consistent results.

Lodging has been identified as one of the most important factors reducing seed yield in grasses; losses due to lodging have been estimated to be as great as 60% (Hampton et al., 1985; Griffith, 1991). Lodging increases the death of fertile tillers by intensifying competition for light and nutrients; photosynthesis becomes less efficient and the developing seeds may abort or fail to develop fully due to inadequate assimilate supply (Hebblethwaite et al., 1980; Hebblethwaite, 1987; Griffith, 1993). Lodging tends to increase with increasing fertilizer-N, but yield increases in response to fertilizer-N applied at near optimum rates are greater than the yield lost due to lodging (Rolston et al., 1994).

In grass seed production the PGRs that have been most commonly used are those that will reduce lodging and stem length, i.e. gibberellin biosynthesis inhibitors that include paclobutrazol (PP333, Parlay, Cultar) (Hebblethwaite, 1987) and chlormequat chloride (CCC, Cycocel) (Hampton, 1986a), the latter being registered for use in grass seed production in New Zealand. Other triazole PGRs including ancymidol (Wright and Hebblethwaite, 1979), flurprimadol (EL500 Elanco) (Hebblethwaite et al., 1985), triapenthenol (RSW0411 Bayer) and XE-1019 (Chevron Chemical Company) (Young et al., 1985) have also been evaluated; responses are generally similar to paclobutrazol, although some are less residual. None of these products have yet been released for commercial use, either because of concerns about soil residual effects or the inconsistency of results.

5.8 WATER

Water is a constituent in the photosynthetic reaction upon which all life depends; as such it is a plant nutrient. However, it also fills many other roles in plant function and metabolism; for example, support through turgor pressure, nutrient uptake and transport medium, medium for biochemical reactions, participant in redox and hydrolysis reactions. Water is the major limiting factor in crop production around the globe. A water loss of 10–15% can markedly affect plant metabolic processes, reducing cell growth, and closing stomata, which results in a decrease in photosynthesis. These effects occur before there are any visible signs of wilting and will affect the leaf area of a plant, the maintenance of which throughout the growing season is of importance for seed yield (Lorenzetti, 1993).

Despite the importance of water in crop production, there has been little research on the influence of water supply on seed yields. For many years the best available advice for seed growers was to irrigate to avoid moisture stress. Since the mid 1980s in New Zealand, irrigation schedulers have monitored soil moisture using neutron probes, and assisted growers in their attempts to apply water efficiently. The neutron probe measures water in the soil profile below 200 mm (above this depth it is unreliable as the neutrons escape from the soil surface), to whatever depth above 1500 mm is chosen. However, the relationship between seed yield and soil moisture potential has not yet been defined.

In general, irrigation prolongs the period of reproductive development, delays harvest and increases seed yield (Lambert, 1967; Hebblethwaite, 1977; Rolston et al., 1994). Early water deficit (before stem elongation) reduces the number of fertile tillers produced which reduces seed yield (Hebblethwaite, 1977). A soil water deficit of less than 100 mm after stem elongation has little effect on floret site utilization (FSU) (Hebblethwaite, 1977) and seed yields of over 2000 kg ha^{-1} can be obtained (Rolston et al., 1994). However, moisture deficit greater than 100 mm does reduce FSU and prevents response to nitrogen (Rolston et al., 1994). Moisture stress at anthesis decreases seed yields, possibly because the photoassimilate supply, which is vital to the developing ovule, is reduced. Moisture stress after anthesis decreases thousand seed weight (Lambert, 1967), probably because leaf area and photosynthetic capacity are reduced. Excess water, however, can increase vegetative tillers (Hebblethwaite, 1980) which are a stronger assimilate sink than reproductive tillers; assimilate partitioned away from developing seeds results in increased abortion and decreased seed yields (Griffith, 1992). Furthermore, these vegetative tillers create harvesting difficulties – not only do they retain moisture, inhibiting seed drying, but also they can grow through the cut crop.

Further research to define what constitutes moisture stress in herbage seed production and how it varies with species and developmental stage is necessary.

5.9 TEMPERATURE

Temperature affects the rate of all enzymatically mediated reactions and all reactions occurring in a medium. Hence, temperature affects the speed of life. All growth and metabolism is temperature dependent; temperature also affects critical stages in the development of a seed crop, e.g. through the processes of stratification and vernalization (see Chapter 2).

Unseasonally low temperatures during tillering reduce dry matter production, but 'optimum temperature' has not been defined. Between initiation of floral primordia and inflorescence emergence, increasing temperature (range 13–23°C) reduces spikelets per inflorescence and florets per spikelet, thus reducing seed yield potential (Ryle, 1965). Increasing temperature also reduces time to anthesis and reduces the period of seed development, resulting in reduced thousand seed weight (TSW) (Bean, 1971). Air temperatures just before and after anthesis account for over 70% of the variability in seed number; temperatures below 8°C are likely to decrease FSU in perennial ryegrass (Hebblethwaite, 1985). In Kentucky bluegrass, however, there is no effect on seed set of reducing night temperatures to 2°C (Aamlid, 1995). A detailed study on tall fescue (Hare, 1993) indicated sensitivity to frost from ear emergence onwards: a −5°C frost killed all seed heads; a −2°C frost at ear emergence or anthesis decreased seed yield per tiller, decreased TSW and decreased germination; a −2°C frost after anthesis decreased seed yield but did not affect TSW or germination. Applying frost protectant failed to prevent damage. Frosts at anthesis reduce or destroy pollen viability (Nikolaevskaya, 1973), whereas increasing temperature at anthesis (range 14–26°C) enhances pollen tube growth (Elgersma *et al.*, 1989). Reasons for effects at other developmental stages have not been elucidated.

As temperature is a major determinant of plant growth and development, growth data presented as a function of accumulated growing degree days would not only allow comparison of results between experiments in different years and climatic regions, but would also enable increased accuracy of crop models predicting response to fertilizer or water addition (Griffith *et al.*, 1994).

5.10 DISEASES, PESTS AND ENDOPHYTES

5.10.1 Diseases

Establishment diseases

Damping off, caused by soil-borne diseases such as *Fusarium* spp. and *Pythium* spp., can be controlled with a fungicide seed treatment, and may be important when low sowing rates (1–5 kg ha^{-1}) are being used (Falloon and Fletcher, 1983).

Seed/seedhead diseases

Blind-seed disease (*Gloeotinia granigena* (Quelet) Schumacher) is a seed-borne fungus that kills the embryo of many grasses (ryegrass and tall fescue are especially susceptible); periodic epidemics are associated with damp weather at flowering. Infected seed lying close to the soil surface produce apothecia which release ascospores during the flowering period of the grass crop. Primary infection occurs when ascospores land on the unfertilized or immature seed, germinate and ramify throughout the endosperm and ovary, killing the seed. Secondary infection takes place after conidia are spread from infected seeds to other seeds by rain splash or surface moisture. Control of blind seed disease is attributed to field burning in Oregon (Hardison, 1980); increased usage of urea was associated with reduced blind seed disease incidence in ryegrass (Hampton and Scott, 1980; Chapter 9), but will not result in adequate control in epidemic years. Triazole fungicides or benlate applied at flowering have been reported to reduce blind seed disease (M. Mebalds, Melbourne, 1993, personal communication), while seed treatment with benomyl or other fungicides prevents apothecial formation (N. Grbavac, Palmerston North, 1995, personal communication).

Ergot (*Claviceps purpurea* (Fr.) Tul.) is commonly seen as large, cylindrical, purplish-black sclerotia, that replace the seed and serve to carry the fungus over from one year to the next (Latch, 1980). The life cycle of the ergot fungus is similar to that of the blind seed fungus. Kentucky bluegrass and *Agrostis* species are more susceptible to ergot than other grass species, with 52 and 13% respectively of seed heads from Oregon seed fields being infected compared with chewings fescue (3%), ryegrass (2%) and tall fescue (1%) (Alderman, 1988). Triazole fungicides applied with silicone surfactants have given good control of ergot in Kentucky bluegrass (Johnston et al., 1995).

Foliar and stem diseases

Stem rust (*Puccinia graminis* Pers.) can reduce ryegrass seed yields by up to 80% (Kerse and Ballard, 1989) and also infects tall fescue and cocksfoot, cutting off the translocation of nutrients to the developing seed head. Stem rust is readily controlled by triazole fungicides, with one or two applications being required at head emergence and flowering (Table 5.4). Stem rust epidemics are explosive and regular monitoring (2–3-day intervals) of seed fields is required if a full preventive fungicide programme is not being used. Crown rust (*P. coronata* Corda.) and brown rust (*P. recondita* Rob. ex Desun.) are believed to contribute to loss of green leaf area during the reproductive phase, and may require treatment with triazole fungicides. Cocksfoot seed yield is affected by both stem rust and stripe rust (*P. striiformis striiformis* Westend. var. *dactylidus* Mannus).

Drechslera siccans (Drechsl.) Shoemaker and *D. andersenii* Scharifoxham leaf spots can be important in wet cool springs or wet summers, causing damage to ryegrass, tall fescue and cocksfoot. Fungicides used in grass seed crops for rust control generally give poorer control of *Drechslera*.

Rhynchosporium leaf blotch (*Rhynchosporium orthosporum* Caldwell) is a wet weather disease in spring and early summer that causes leaf lesions but has not been reported as causing seed yield loss, except, perhaps, through loss of green leaf area.

Mildew (*Erysiphe graminis* DC.) is a common disease that infects most grass species, particularly ryegrass and fescue. A seed yield increase (26%) following propiconazol fungicide application at the beginning of stem elongation has been attributed to control of an early attack of powdery mildew (Rijckaert, 1995).

Maintenance of green leaf area (GLA). Seed yield increases (from 5 to 43%) after fungicide application either just before or just after anthesis in ryegrass, cocksfoot and prairie grass have been associated with increased GLA or increased

Table 5.4. Effect of triadimefon and time of application on the incidence and severity of stem rust, green leaf area of the flag and penultimate leaf, and seed yield of perennial ryegrass. (Adapted from Hampton, 1986b.)

	Stem rust[1]			% Green leaf area[1]		
	% Infected tillers	% Area with pustules			Penultimate leaf	Seed yield (kg ha^{-1})
Treatment		Stem	Ear	Flag leaf		
Nil	82	38.3	20.6	38	3	648
Head[2]	45	11.8	3.8	59	27	818
Full[3]	3	8.1	0.5	72	37	903

[1]Assessed 4 days prior to harvest.
[2]One application just prior to head emergence.
[3]Two applications, just prior to head emergence and just before anthesis.

reproductive leaf dry matter accumulation in the virtual absence of recognized pathogens (Hampton et al., 1985; Rolston et al., 1989). Seed yield increases were explained by an increase in the number of seeds retained per spikelet.

5.10.2 Pests

Pest problems in grass seed crops occur sporadically. With the exception of aphids, many insect pests are regional problems, reflecting the species that are endemic to the region. Generally the same species cause damage in turf, pasture and cereals in the area. Control measures for insect pests depend on the species but include seed treatment with an insecticide, sowing a granular insecticide with the seed, or broadcast insecticide treatments. Biological control programmes using bacteria or parasitic wasps to control pasture pests that attack seed crops can be expected to reduce specific pests to acceptable levels. Beneficial insects (e.g. ladybird beetles) that are effective in controlling pests such as aphids are often destroyed by non-selective broadcast insecticide applications; these chemicals should be used only if significant seed yield loss is expected.

Vertebrate pests are also regional in distribution, and sporadic in effect. They include those animals that are local pests of pastures and cereals because they burrow and/or feed on forage including rabbits, field mice, moles and gophers which are usually controlled by poison baits, shooting or trapping. Larger herbivores may be controlled with electric fences. Birds are rarely a problem.

5.10.3 Endophyte

The endophyte *Acremonium* fungi of ryegrass and fescues (both *F. arundinacea* and *F. pratensis*) are seed borne and have implications for seed producers. The endophyte–plant interaction produces a number of alkaloids, including peramine, which reduces insect pest damage and thus enhances plant persistence, while other alkaloids (lolitrem B and ergovaline) are associated with animal health problems that include staggers and heat stress, resulting in low animal growth rates. Novel endophytes that have low or nil levels of the detrimental alkaloids can be introduced into new cultivars, and seed producers and Seed Certification authorities need to ensure that production systems protect and maintain the endophytic status of the cultivar. Seed yield of endophyte versus non-endophyte cultivars depends upon insect pest pressure (Rolston et al., 1994; Schoberlein et al., 1995).

Endophyte viability in seed declines rapidly in 1–6 months when stored at high humidity and warm temperatures (Rolston et al., 1986), and declines more rapidly than seed viability. To maintain endophyte viability for 12–24 months, seed moisture content should be less than 11.5%, and seed should be stored in a cool (5–15°C) environment.

The use of fungicides for rust control generally does not result in reduced endophyte in the seed, the only exception being multiple applications of tebuconazole (but not propiconazole) to an artifially inoculated novel endophyte. New fungicides should be screened for activity on endophyte as well as their efficacy for disease control. Many fungicides used for seed treatment can kill endophyte, but benomyl (3.5 g kg^{-1} seed) does not detrimentally affect endophyte (N. Grbavac, Palmerston North, 1995, personal communication).

5.11 POSTHARVEST MANAGEMENT

Management of seed fields during the postharvest period is extremely important for seed production during the following season. Removal of postharvest residue from the plant crown is essential in maintaining a vigorous and productive stand.

5.11.1 Residue Removal

Burning

Open-field burning has been the most effective method of removing stubble and destroying unharvested crop and weed seed. Where practised, burning should be completed as soon after harvest as possible and before autumn regrowth. Heat from the fire and direct combustion of seed can, but rarely does, destroy the viability of all the unharvested seed left on the soil surface after harvest; autumn application of herbicides is generally required. Large-seeded weeds tend to be better controlled by field burning than smaller ones. Furthermore, field burning has only a small effect on soil temperature, and the temperature gradient in the first few centimetres above and below the soil surface can be quite large, allowing small differences in seed position to have large effects on survival.

Two primary factors impair the performance of field burning as a weed-control tool: the perennial crop can itself be 'burned out' by a fire that is too hot, and there are restrictions on scheduling of field burns. Furthermore, soot (carbon) building up in the upper layers of the soil is capable of adsorbing herbicides, forcing growers to increase their chemical rates on fields with prolonged histories of burning.

Fire also has effects on seed germination patterns. Seed exposed to heat but not killed will generally have reduced dormancy compared to seed not exposed to heat. The practice of propane flaming after baling was widely used during the initial shift away from universal field burning in western Oregon. At travel speeds and quantities of propane per hectare commonly used, propane flaming was very effective in removing straw and uncut stems, and relatively ineffective in controlling weeds.

Clipping, grazing and timing

Clipping has already been mentioned as a means to control weed seed production during the establishment year (see Section 5.5.2). Significant differences in seed yield due to the method of postharvest residue removal generally depend upon the efficacy of the removal method; a 'good' burn achieves similar results to grazing or mowing (Hare, 1993). Grazing can be important during the winter and early spring (Section 5.6).

5.12 CONCLUSION AND FUTURE DIRECTIONS

The challenge for seed production researchers has always been to understand the mechanisms and interactions involved, and thereby improve seed yield and quality. Considerable progress has been made over the last two decades, but now further challenges are appearing, including the effects of environmental restrictions on chemical use (for both pesticides and fertilizers) and water use, competition for land use (emphasizing the requirement to increase financial returns) and the general problems associated with climate change. Some of these can be answered by the plant breeders (aiming for increased yield potential and decreased seed shattering) and others by molecular geneticists (creating disease resistance), but the onus is on the seed scientist. None of this work would be useful if outcomes did not reach the grower. Technology transfer is an increasingly important focus for the seed scientist; seeing the uptake of a new technique or management practice is as rewarding as the development process itself.

REFERENCES

Aamlid, T.S. (1995) Do low, non-freezing night temperatures during anthesis affect seed set in *Bromus inermis* Leyss. and *Poa pratensis* L. *Journal of Applied Seed Production* 13, 1–9.

Acikgoz, E. and Karagoz, A. (1989) Effect of row spacing, seeding rate and nitrogen fertilization on seed yield of perennial ryegrass under dryland conditions. *Journal of Applied Seed Production* 7, 50–52.

Alderman, S. (1988) Distribution of *Gloeotinia temulenta*, *Claviceps purpurea*, and *Anguina agrostis* among grasses in the Willamette Valley of Oregon in 1988. *Journal of Applied Seed Production* 6, 6–13.

Bean, E.W. (1971) Temperature effects upon inflorescence and seed development in tall fescue (*Festuca arundinaceae* Schreb.). *Annals of Botany* 35, 891–897.

Black, A.S., Sherlock, R.R. and Smith, N.P. (1987) Effect of timing of simulated rainfall on ammonia volatilization from urea, applied to soil of varying moisture content. *Journal of Soil Science* 38, 679–687.

Brown, K.R. (1980) Seed production in New Zealand ryegrasses. *New Zealand Journal of Experimental Agriculture* 8, 33–39.

Chastain, T.G. and Grabe D.F. (1989a) Spring establishment of orchardgrass seed crops with cereal companion crops. *Crop Science* 29, 466–493.

Chastain, T.G. and Grabe, D.F. (1989b) Spring establishment of turf-type tall fescue seed crops with cereal companion crops. *Agronomy Journal* 81, 488–493.

Darwent, A.L. and Smith, J.H. (1988) Effects of 2,4-D and dicamba time of application on bromegrass seed production. *Canadian Journal of Plant Science* 68, 811–815.

Elgersma, A., Stephenson, A.G. and Den Nijs, A.M.P. (1989) Effects of genotype and temperature on pollen tube growth in perennial ryegrass (*Lolium perenne* L.). *Sexual Plant Reproduction* 2, 225–230.

Evans, G. (1963) Seed rates of grasses for seed production. III. Early perennial ryegrass S.24. *Empire Journal of Experimental Agriculture* 31, 34–40.

Falloon, R.E. and Fletcher, R.H. (1983) Increased herbage production from perennial ryegrass following fungicide seed treatment. *New Zealand Journal of Agricultural Research* 26, 1–5.

Faulkner, J.S. (1974) The effect of dalapon on thirty-five cultivars of *Lolium perenne*. *Weed Research* 14, 405–413.

Francis, G. and Haynes, R.J. (1991) The leaching and chemical transformation of surface applied urea under flood irrigation. *Fertilizer Research* 28, 139–146.

Green, W. (1993) Effects of ethophon and nitrogen rates on Concord seed yield. *New Zealand Herbage Seed Subsection of Federated Farmers Newsletter* 2, November, 5–6.

Griffith, S.M. (1991) Reproductive herbage grass shoot: carbohydrate and nitrogen status as affected by lodging. *Journal of Applied Seed Production* 9, 51 (Abstract).

Griffith, S.M. (1992) Changes in post-anthesis assimilates in stem and spike components of Italian ryegrass (*Lolium multiflorum* Lam.) I. Water soluble carbohydrates. *Annals of Botany* 69, 243–248.

Griffith, S.M. (1993) Tall fescue dry matter partitioning and seed yield components in response to lodging. *Proceedings of the XVII International Grasslands Congress*, pp. 1642–1643.

Griffith, S.M., Alderman, S.C. and Streeter, D.J. (1994) Annual ryegrass fertilization and growth as related to growing degree days in two contrasting climatic years. In: Young III, W.C. (ed.) *Seed Production Research at Oregon State University: USDA-ARS Cooperating*. Oregon State University, Corvallis, pp. 35–38.

Hammond, C.H., Griffiths, W., van Hoogstraten, S.D. and Whiteoak, R.J. (1976) The use of ethofumesate in grass seed crops. *Proceedings 1976 British Crop Protection Conference – Weeds*, pp. 657–663.

Hampton, J.G. (1986a) The effects of chlormequat chloride application on seed yield of perennial ryegrass (*Lolium perenne* L.). *Journal of Applied Seed Production* 4, 8–13.

Hampton, J.G. (1986b) Fungicidal effects on stem rust, green leaf area, and seed yield in 'Grasslands Nui' perennial ryegrass. *New Zealand Journal of Experimental Agriculture* 14, 7–12.

Hampton, J.G. (1987) Effect of nitrogen rate and time of application on seed yield in perennial ryegrass cv. Grasslands Nui. *New Zealand Journal of Experimental Agriculture* 15, 9–16.

Hampton, J.G. (1988) Herbage seed production. *Advances in Research and Technology of Seeds* 11, 1–28.

Hampton, J.G. and Scott, D.J. (1980) Blind seed disease of ryegrass in New Zealand. 1. Occurrence and evidence for the use of nitrogen as a control measure. *New Zealand Journal of Agricultural Research* 23, 143–147.

Hampton, J.G., Clemence, T.G.A. and Hebblethwaite, P.D. (1983) Nitrogen studies in *Lolium perenne* grown for seed. IV. Response of amenity types and influence of a growth regulator. *Grass and Forage Science* 38, 97–105.

Hampton, J.G., Clemence, T.G.A. and McCloy, G.L. (1985) Chemical manipulation of grass seed crops. In: Hare, M.D. and Brock, J.L (eds) *Producing Herbage Seeds*. Grassland Research and Practice Series No. 2. New Zealand Grassland Association, Palmerston North, pp 9–14.

Hardison, J.R. (1980) The role of fire for disease control in grass seed production. *Plant Disease* 64, 641–645.

Hare, M.D. (1993) Seed production in tall fescue (*Festuca arundinacea* Schreb.). PhD Thesis. Massey University, Palmerston North, 186pp.

Hare, M.D. and Rolston, M.P. (1990) Nitrogen effects on tall fescue seed production. *Journal of Applied Seed Production* 8, 28–32.

Hart, J.M., Horneck, D., Peek, D. and Young III, W.C. (1989) Nitrogen and sulfur uptake for cool season forage and turf grass grown for seed. In: Young III, W.C. (ed.) *Seed Production Research at Oregon State University:USDA-ARS Cooperating*. Oregon State University, Corvallis, pp. 15–16.

Hebblethwaite, P.D. (1977) Irrigation and nitrogen studies in S. 23 ryegrass grown for seed 1. Growth, development, seed yield components and seed yield. *Journal of Agricultural Science, Cambridge* 88, 605–614.

Hebblethwaite, P.D. (ed.) (1980) *Seed Production*. Butterworths, London-Boston, 694pp.

Hebblethwaite, P.D. (1985) The influence of environmental and agronomic factors on floret site utilization in perennial ryegrass. *Journal of Applied Seed Production* 3, 57–59.

Hebblethwaite, P.D. (1987) A review of the chemical control of growth, development and yield in *Lolium perenne* grown for seed. *Journal of Applied Seed Production* 5, 54–59.

Hebblethwaite, P.D. and Ivins, J.D. (1977) Nitrogen studies in *Lolium perenne* grown for seed I. Level of application. *Journal of the British Grassland Society* 32, 195–204.

Hebblethwaite, P.D. and Ivins, J.D. (1978) Nitrogen studies in *Lolium perenne* II. Timing of nitrogen application. *Journal of the British Grassland Society* 33, 159–166.

Hebblethwaite, P.D., Hampton, J.G., Batts, G.R. and Barrett, S. (1985) The effects of time of application of the growth retardant flurprimidol (EL500) on seed yields and yield components in *Lolium perenne* L. *Journal of Applied Seed Production* 3, 15–19.

Hebblethwaite, P.D., Wright, D. and Noble, A. (1980) Some physiological aspects of seed yield in *Lolium perenne* L. (perennial ryegrass). In: P.D. Hebblethwaite (ed.) *Seed Production*. Butterworths, London, pp. 71–90.

Hill, M.J. (1971) Closing ryegrass crops for seed production. *New Zealand Journal of Agriculture* 123, 43.

Hill, M.J. and Watkin, B.R. (1975) Seed production studies on perennial ryegrass, timothy and prairie grass. *Journal of the British Grassland Society* 30, 63–71.

Horneck, D. and Hart, J.M. (1988) A survey of nutrient uptake and soil test values in perennial ryegrass and turf type tall fescue fields in the Willamette Valley. In: Youngberg, H.W. and Burcham, J. (eds) *Seed Production Research at Oregon State University:USDA-ARS Cooperating*. Oregon State University, Corvallis, pp. 13–14.

Horneck, D.A., Hart, J.M. and Young III, W.C. (1993) Effect of soil K on perennial ryegrass and tall fescue. In: Young III W.C. (ed.) *Seed Production Research at Oregon State University:USDA-ARS Cooperating*. Oregon State University, Corvallis, pp. 14–17.

Jewiss, O.R. (1972) Tillering in grasses – its significance and control. *Journal of the British Grassland Society* 27, 65–82.

Jimenez-Merino, A., Cadena-Meneses, A. and Castrellon-Montelongo, J. (1993) Seed yield and quality of orchardgrass at five levels of added nitrogen. *Summaries of Proceedings of XVII International Grasslands Congress* 3.

Johnson, J., Scott, J.L., Dibb, C. and Greenwood, M.A. (1982) The effect of controlling volunteer cereals, *Poa trivialis* and ryegrass seedlings on the seed yield of perennial ryegrass. *Proceedings 1982 British Crop Protection Conference – Weeds*, pp. 407–414.

Johnston, W.J., Golob, C.T. and Sitton, J.W. (1995) Control of ergot in Kentucky bluegrass using SDI fungicides and a surfactant. *Journal of Applied Seed Production* 13, 66 (abstract).

Kerse, G.W. and Ballard, D.L. (1989) Cyproconazole – a new DMI fungicide. *Proceedings of the New Zealand Weed and Pest Control Conference* 42, 114–118.

Lambert, D.A. (1967) The effects of nitrogen and irrigation on timothy (*Phleum pratense*) grown for production of seed. *Journal of Agricultural Science, Cambridge* 69, 231–239.

Langer, R.H.M. (1979) *How Grasses Grow*. The Institute of Biology's Studies in Biology no. 34, 2nd edn. Edward Arnold, London.

Latch, G.C.M. (1980) Importance of diseases in herbage seed production. In: Lancashire, J.A. (ed.) *Herbage Seed Production*. Grassland Research and Practice Series No. 1. New Zealand Grasslands Association, Palmerston North, pp. 36–40.

Lee, W.O. (1973) Clean grass seed crops established with activated carbon bands and herbicides. *Weed Science* 21, 537–541.

Lescar, L. and Bouchet, F. (1968) Trials for the control of self-sown seedlings and other weed grasses in grass seed crops. *Proceedings of the 9th British Weed Control Conference*, pp. 527–532.

Lorenzetti, F. (1993) Achieving potential seed yields in species of temperate regions. *Proceedings of the XVII International Grasslands Congress*, pp. 1621–1628.

Meijer, W.J.M. (1984) Inflorescence production in plants and seed crops of *Poa pratensis* L. and *Festuca rubra* L. as affected by juvenility of tillers and tiller density. *Netherlands Journal of Agricultural Science* 32, 119–136.

Meijer, W.J.M. and Vreeke, S. (1988) Nitrogen fertilization of grass seed crops as related to soil mineral nitrogen. *Netherlands Journal of Agricultural Science* 36, 375–385.

Mueller-Warrant, G.W. (1990) Control of roughstalk bluegrass (*Poa trivialis*) with fenoxaprop in perennial ryegrass (*Lolium perenne*) grown for seed. *Weed Technology* 4, 250–257.

Mueller-Warrant, G.W. (1991) Enhanced activity of single-isomer fenoxaprop on cool-season grasses. *Weed Technology* 5, 826–833.

Mueller-Warrant, G.W. and Brewster, B.D. (1986) Control of roughstalk bluegrass (*Poa trivialis*) in perennial ryegrass (*Lolium perenne*) grown for seed. *Journal of Applied Seed Production* 4, 44–51.

Mueller-Warrant, G.W., Mellbye, M.E. and Aldrich-Markham, S. (1991) Pronamide improves weed control in new grass plantings protected by activated charcoal. *Journal of Applied Seed Production* 9, 16–26.

Mueller-Warrant, G.W., Young III, W.C. and Mellbye, M.E. (1994) Influence of residue removal method and herbicides on perennial ryegrass seed production: II. crop tolerance. *Agronomy Journal* 86, 684–690.

Nikolaevskaya, T.S. (1973) The effect of frost on the reproductive organs of *Dactylis glomerata*. *Herbage Abstracts* 43, 3529.

Nordestgaard, A. (1986) Investigations on the interaction between level of nitrogen application in the autumn and time of nitrogen application in the spring to various grasses. *Journal of Applied Seed Production* 4, 16–25.

Rijckaert, G. (1995) Effect of fungicides on seed yield and disease control in perennial ryegrass (*Lolium perenne* L.). *Journal of Applied Seed Production* 13, 57 (abstract).

Roberts, H.M. (1965) The effect of defoliation on the seed-producing capacity of bred varieties of grasses. III. Varieties of perennial ryegrass, cocksfoot, meadow fescue and timothy. *Journal of the British Grasslands Society* 20, 283–289.

Rolston, M.P. (1991) Cocksfoot seed crop tolerance to herbicides applied to seedling and established stands. *Journal of Applied Seed Production* 9, 63–68.

Rolston, M.P. and Hare, M.D. (1986) Competitive effects of weeds on seed yield of first year grass seed crops. *Journal of Applied Seed Production* 4, 34–36.

Rolston, M.P., Brown, K.R., Hare, M.D. and Young, K.A. (1985) Grass seed production: weeds, herbicides and fertilisers. In: Hare, M.D. and Brock J.L. (eds) *Producing Herbage Seeds*. New Zealand Grassland Association, Palmerston North, pp.15–22.

Rolston, M.P., Hare, M.D., Moore, K.K. and Christensen, M.J. (1986) Viability of Lolium endophyte fungus in seed stored at different moisture contents and temperatures. *New Zealand Journal of Experimental Agriculture* 14, 297–300.

Rolston, M.P., Hampton, J.G., Hare, M.D. and Falloon, R.E. (1989) Fungicide effects on seed yield of temperate forage grasses. *Proceedings XVI International Grasslands Congress*, pp. 669–670.

Rolston, M.P., Rowarth, J.S., DeFilippi, J.M. and Archie, W.J. (1994) Effects of water and nitrogen on lodging, head numbers and seed yield of high and nil endophyte perennial ryegrass. *Proceedings of the Agronomy Society of New Zealand* 24, 91–94.

Rowarth, J.S. and Archie, W.J. (1994) The nutrient needs of small seed crops: a new concept in optimising seed yields. *Proceedings of the Agronomy Society of New Zealand* 24, 87–89.

Rowarth, J.S. and Archie, W.J. (1995) A diagnostic method for prediction of seed yield in perennial ryegrass. *Journal of Applied Seed Production* 13, 43 (abstract).

Rowarth, J.S., Archie, W.J. and Baird, D.B. (1993) Nitrogen requirements for Italian ryegrass seed production. In: Barrow, N.J. (ed.) *Plant Nutrition from Genetic Engineering To Field Practice*. Kluwer Academic, Dordrecht, pp. 513–516.

Rowarth, J.S., Jin, Q.F. and Scott, W.R. (1994) Nitrogen recovery efficiency in a browntop seed crop. *Soil News* 42(5), 156.

Ryle, G.J.A. (1964) The influence of date of origin of the shoot and level of nitrogen on ear size in three perennial grasses. *Annals of Applied Biology* 53, 311–323.

Ryle, G.J.A. (1965) Effects of day length and temperature on ear size in S. 24 perennial ryegrass. *Annals of Applied Biology* 59, 297–308.

Scharrer, K. and Mengel, K. (1960) On the transient occurrence of visible magnesium deficiency in oats. *Agrochimica* 4, 3–24.

Schoberlein, W. and Wahl, H.J. (1993) Efficient nitrogen fertilisation in grass seed crops with regard to soil mineral nitrogen content and analysing grass tillers in spring. *Proceedings of the XVII International Grasslands Congress*, pp. 1695–1696.

Schoberlein, W., Eggestein, St. and Pfannmoller, M. (1995) Effects of endophyte-infected varieties of *Festuca pratensis* on seed production. *Journal of Applied Seed Production* 13, 77 (abstract).

Watson, C.E., Jr and Watson, V.H. (1982) Nitrogen and date of defoliation effects on seed yield and seed quality of tall fescue. *Agronomy Journal* 74, 891–893.

White, J.G.H. (1990) Herbage seed production. In: Langer, R.H.M. (ed.) *Pastures*. Oxford University Press, Auckland, pp. 370–408.

Wright, D. and Hebblethwaite, P.D. (1979) Lodging studies in *Lolium perenne* grown for seed. 3. Chemical control of lodging. *Journal of Agricultural Science, Cambridge* 93, 669–679.

Wright, D. and Hebblethwaite, P.D. (1983) Volunteer winter wheat: its effects and control in ryegrass seed production. *Weed Research* 23, 273–281.

Young, W.C., Chilcote, D.O. and Youngberg, H.W. (1985) In: H.W. Youngberg (ed.) *Seed Production Research at Oregon State University: USDA-ARS Cooperating*. Department of Crop Science Extension/CRS 54, Corvallis, Oregon, pp. 17–19.

Young, W.C.III, Chilcote, D.O. and Youngberg, H.W. (1995) Seed yield response of perennial ryegrass to spring applied nitrogen at different rates of paclobutrazol. *Journal of Applied Seed Production* 13, 10–15.

Legume Seed Crop Management 6

A.H. Marshall,[1] J.J. Steiner,[2] O. Niemeläinen[3] and J. Hacquet[4]

[1] *Institute of Grassland and Environmental Research, Plas Gogerddan, Aberystwyth SY23 3EB, UK;* [2] *National Forage Seed Production Research Centre, USDA-ARS, Oregon State University, Corvallis, Oregon 97331, USA;* [3] *Agricultural Research Centre of Finland, Institute of Crop and Soil Science, Plant Breeding Section, FIN-31600, Jokionen, Finland;* [4] *FNAMS, Centre INRA, 8660 Lusignan, France*

6.1 INTRODUCTION

Management of the crop determines the potential seed yield of temperate forage legumes, and can also influence the proportion of the potential yield that is harvested. Many factors, including location and climate, determine the management systems used to maximize seed yield of temperate forage legumes. It is beyond the scope of this chapter to describe in detail the management systems used for all temperate forage legume species in all the diverse seed production areas. Our purpose is therefore to highlight some of the main differences between seed crop management systems utilized for some legume species.

6.2 CROP SITE

6.2.1 Site Selection and Seedbed Preparation

The areas best suited for fodder production of legumes are often less suitable for seed production. Weather conditions can be too wet during the pollination and harvesting period or even throughout the whole season, resulting in excessive vegetative growth. White clover (*Trifolium repens* L.) is a good example of such a crop. Kjaersgaard (1994) reported that in Denmark, white clover requires 110–125 mm rainfall in the April–June period and temperatures of 15–20°C to produce plants of sufficient size for seed production; higher levels of rainfall produce excessive vegetative growth.

Although forage legumes vary greatly in their adaptation to different environments and in the optimal weather conditions required during seed production, dry and warm weather should prevail during the peak-flowering period and harvesting to give a homogeneously ripened crop. However, care is necessary as lack of water during that period can reduce seed number (Kjaersgaard, 1994).

Climatic factors are therefore extremely important in identifying the appropriate regions for seed production of different legume species. Within these regions, edaphic and environmental factors can also be important. For example, forage legumes differ in their reaction to the height of the water table. Deep-rooted lucerne (*Medicago sativa* L.) requires a well-drained soil and will suffer if the water table is high, while white clover can withstand these conditions or even benefit from them. All legume seed crops are sensitive to standing water and therefore the drainage of fields should be good. In areas with hard winter conditions fields with south-facing slopes are most suitable.

Legume crops do well on mineral soils but the fertility of the soil does not need to be high. Depending on the weather conditions and species, coarse or fine textured mineral soils are preferable. In Canada, light gravelly loams are better suited for red clover (*Trifolium pratense* L.) than heavy clay soil (Belzile, 1991). Where irrigation is not available and there is a risk of summer drought, soils of high water-holding capacity are preferable. Where rainfall is abundant or irrigation is possible, coarse soils are also suitable.

Legume seed crops need a fine tilth and a firm seedbed to establish rapidly and evenly. Seedbed preparation should be uniform throughout the whole field to facilitate even development of the stand. On very loose soils it may even be beneficial to firm the soil by rolling before sowing to produce a more uniform sowing depth. Seeds of forage legumes tend to be sown at a relatively shallow depth (between 2 and 3 cm). Sowing should be followed by further harrowing, if necessary, and rolling immediately afterwards.

6.2.2 Place in the Cropping System

For successful seed production, special attention has to be given to the crops cultivated previously on the same site as buried seeds can pose a serious threat to cultivar purity (Hampton *et al.*, 1987). This is especially true with legumes where the hard seed of previous crops can remain viable within the soil for several years. Seed certification regulations stipulate a minimum time that must have elapsed before the same species can be cultivated on the same site. The regulations are distinct for each country. To further minimize the risk to the genetic purity of the crop, appropriate isolation distances between the seed crop and crops or plants of the same species in the proximity of the seed production field are required.

The field should be free from noxious weeds and other species whose seeds are prohibited or restricted under rules for certified seed production. The use of weed-control measures before seed crop establishment is therefore an important management practice.

6.3 CROP ESTABLISHMENT

6.3.1 Place of the Seed Crop in the Rotation

Whilst soil conditions and climate determine the areas where seed of forage legumes can be produced, other factors may influence the place of seed production within the crop rotation. On predominantly arable farms, legumes provide a valuable cash break crop, and an alternative to the ley or root crop in the cereal cropping sequence. Red and white clover often precede perennial grass seed crops in the rotation. The benefits include N-fixation (they may supply 66% of the first-year grass seed crop requirement), and reduced pests and diseases. On mixed livestock-arable farms, forage legume seed production can be integrated into the livestock feed requirements, providing a remedy to the problem of forage removal.

6.3.2 Cover Crops

The seed producer has several options for establishing the seed crop. Seed of white and red clover and other legume species can be broadcast or sown in rows under a cover crop. As forage legumes are generally slow to establish, the cover crop could provide some income for the farmer in the establishment year, help to control weeds, and provide some protection to the developing legume seed crop during adverse weather conditions. Establishing white clover under a cover crop of barley, (*Hordeum vulgare* L.), (Fig. 6.1) or peas (*Pisum sativum* L.), can produce more inflorescences and higher potential seed yields than when sown in pure stands (Hollington *et al.*, 1993). Some seed producers advocate reducing the seed rate of the cereal cover crop to encourage light penetration to the developing legume crop and improve establishment. In northern Canada, Fairey and Lefkovitch (1991) established lucerne, alsike clover (*T. hybridum* L.) and red clover under cover crops of barley, oats (*Avena sativa* L.) and canola (*Brassica campestris* L.). They reported

Fig. 6.1. White clover (*Trifolium repens* L.) seed crop sown at 60-cm row spacing under a cover crop of spring barley (*Hordeum vulgare* L.), in the establishment year after cereal harvest.

no beneficial effect of reducing the sowing rate of the cover crop to 50% of the recommended rate, but obtained a different yield response of the legumes depending on the cover crop used. Oats gave the highest legume seed yield, whilst canola and barley reduced legume seed yields in the first harvest year. Steiner and Snelling (1994) reported that most legume seed crops grown in the western region of the USA are speciality crops that are sown without a cover crop in the autumn or spring, depending upon weather conditions, though some red clover is sown with spring-sown cereals in Oregon and there is an economic benefit for establishing slow-growing Caucasian clover (*T. ambiguum* Bieb.) with wheat (*Triticum aestivum* L.).

6.3.3 Sowing Date

Several factors influence sowing date, such as the vernalization requirements of the legume species and the location of the field. When cover crops are used, seed is normally sown as early as possible in spring. Seedlings of clover species, especially white clover, establish slowly in the autumn and stands in the seedling stage can be easily damaged through frost heaving. However, direct sowing in late summer may prove successful in seasons when conditions are favourable. In the UK, if seed of white clover cannot be sown before late August, it is safer to delay doing so until the following spring. In New Zealand, McCartin (1985) has demonstrated the potential benefit of direct drilling seed of white clover into stubble of wheat or barley in February (August in northern hemisphere). In California, USA, white clover and lucerne are sown in spring. In eastern Oregon they are spring sown with irrigation if a first-year seed crop is required. In western Oregon, seed crops of white and red clover may be autumn sown (particularly if irrigated) but are mainly sown in the spring, while crimson clover (*T. incarnatum* L.), rose clover (*T. hirtum* All.) and vetches (*Vicia* spp.) are only sown in the autumn.

6.3.4 Sowing Rate

Sowing rates for seed production depend upon the species, time of sowing, presence or absence of a companion crop and area of production. For example, in the UK white clover is sown in pure stands at 2–4 kg ha^{-1}. In Denmark, white clover is normally sown at a rate of 1.5 kg ha^{-1} with smooth-stalked meadow grass (*Poa pratensis* L.) at a rate of 7–8 kg ha^{-1} under a cover crop of spring barley. Seed of white clover is harvested in the first harvest year, then removed by spraying with a herbicide and seed of the slower to establish grass harvested in the second and third harvest years. In the UK, red clover is best pure sown at a rate of 8–10 kg ha^{-1} for diploid cultivars and 11–13 kg ha^{-1} for tetraploid cultivars. Traditionally, a companion grass may be sown in wide rows for seed production after the first harvest year, e.g. cocksfoot (*Dactylis glomerata* L.) with broad red types and timothy (*Phleum pratense* L.) with the late flowering types. In New Zealand, seed crops of red clover are sown at a rate of 3 kg ha^{-1} if sown in spring and when seed is to be harvested in the next season, but at 4–5 kg ha^{-1} if sown in autumn (late January in New Zealand) and generally 1–2 kg ha^{-1} higher if sown with a companion grass (Clifford and Anderson, 1980).

Fig. 6.2. White clover (*Trifolium repens* L.) seed crop in spring of the first harvest year after sowing at 30-cm row spacing.

6.3.5 Row Spacing

Forage legumes may be sown in rows (Fig. 6.2) or broadcast through a precise row width depending on species, location and climate. In dry areas of the UK, establishing seed crops of white clover in narrow row spacings (15–18 cm) is preferable to broadcasting. However cultivars of contrasting leaf size respond differently to row spacing (Hollington *et al.*, 1993). Large-leaved types tend to produce most ripe inflorescences and highest seed yields when broadcast, but small-leaved types benefit from wide row spacings (60 cm). In North America, seed crops of forage legumes are mainly sown in rows of 30 cm spacing or above and not broadcast because this may depress seed yields in some species such as white clover. Lucerne, however, may be sown in 50-cm-row spacings or greater. In New Zealand, highest seed yields of first-year seed crops of the white clover cultivars Grasslands Huia and Grasslands Pitau have been achieved from row spacings of 30 cm rather than 15, 45 (Clifford, 1985) or 60 cm (Clifford, 1977). The 30-cm spacing, however, reduced seed yields of both cultivars in the second harvest year compared to other spacings (Clifford, 1977). Hare (1984) showed that seed yields of lotus (*Lotus pedunculatus* Cav.) were significantly higher in 30–45-cm rows at population densities of 66 and 133 plants m^{-2}, than in 15-cm rows at population densities of 66 and 133 plants m^{-2}. Clifford (1974) showed that for red clover 60-cm row spacing and a density of 10 plants per metre length of row produced the highest seed yield compared to 15-cm row spacing and 5 plants per metre row length.

All methods of establishment seek to produce an optimal plant density. For white clover there is some flexibility in plant number as there is considerable compensatory growth at densities between 9 and 100 plants m^{-2} (Marshall and James, 1988). In birdsfoot trefoil (*Lotus corniculatus* L.), McGraw *et al.* (1986) have suggested that at least 19 plants m^{-2} are required to produce maximal seed yields, while for lucerne plant densities of between 4 plants m^{-2} (Rincker, 1977) and 25 plants m^{-2} have been suggested (Kowithayakorn and Hill, 1982). Clearly there is considerable variability in the optimal plant density and a number of factors, including a genetic component, which are important.

6.4 FERTILIZER REQUIREMENTS

All legumes need adequate levels of nitrogen (N), phosphorus (P), potassium (K), calcium (Ca) and other nutrients to ensure satisfactory growth. Although legume seed crops do not require soils of high fertility, local soil tests will determine the amounts of each nutrient that are needed for adequate plant growth to support seed production. The amount of each nutrient required for seed production may be less than that needed for optimal forage production. In some cases, excessive concentrations of specific elements may give greater reductions in seed yield than nutrient deficiencies.

The nitrogen requirements for forage legume seed crops are generally met by the symbiotic relationship of nitrogen-fixing bacteria within nodules attached to the plant roots. The species of rhizobia specific for the crop should be applied to the seeds at the time of planting to help ensure adequate inoculation. However, particularly hot and dry soil conditions at the time of planting may greatly reduce rhizobia viability.

High P levels in white clover seed fields may result in excessive production of stolon buds in deference to floral primordia, thereby reducing seed yield (Clifford, 1987). White clover leaf size is also increased with increasing P availability which results in reduced seed yield (Clifford and Rolston, 1990). Potassium requirements are lower for optimal seed production (Clifford and White, 1986) compared to forage production (Cornforth and Sinclair, 1984).

Boron (B) is required in higher amounts for optimal white clover seed production than for forage production (Johnson and Wear, 1967; Marshall *et al.*, 1991). Boron probably serves a role in facilitating pollen tube elongation (Pilbeam and Kirby, 1983), which is essential for syngamy and the resulting seed set and development to occur. This has been demonstrated in seed production of alsike, red and white clovers (Sherrell, 1983), crimson clovers (Rogers, 1947; Davis, 1949), vetch (*Vicia* spp.) (Rogers, 1947) and lucerne (Grizzard and Matthews, 1942; Sherrell, 1983).

Soil pH may greatly influence seed yield potential. Legume seed crops such as lucerne, crimson clover and strawberry clover (*T. fragiferum* L.) are very sensitive to acid soil conditions. Strawberry clover is best adapted to alkaline conditions. Soil pH can, however, be modified by the addition of lime. The amount of lime required is dependent upon the reactivity of the soil and desired pH. Many nutrients are not readily available to the plant when soil pH is at either extreme (Miller and Reetz, 1995). Major nutrients such as K, sulphur (S), Ca, and magnesium (Mg), and micronutrients such as B and molybdenum (Mo) can be relatively unavailable at

6.5 WEED CONTROL AND HYGIENE

6.5.1 Choice of Clean Fields

Weeds compete with the crop for water, nutrients and light, reduce yields, increase cleaning costs and can reduce the quality of seed lots. To ensure successful seed production, fields which contain large numbers of weed species which would be difficult to destroy or to clean out from the legume seeds must be avoided.

Weeds are easiest to control when they are young and small but are sometimes impossible to destroy after they are established. Usually, pre-emergence and post-emergence herbicides are combined; in some countries, e.g. USA (California), where crop rows are widely spaced, mechanical methods of control such as cultivators and disc harrows are used in conjunction with chemical control (Bell and Marble, 1985). As recommendations for herbicides are specific for each country or even region, a detailed description of the herbicides used is not within the scope of this chapter. However, as an example, in France, the efficacy of numerous herbicides on different legume seed crops is illustrated in Fig. 6.3 (Hacquet and Jouy, 1994).

6.5.2 Herbicides[1]

The area of forage legume seed production is too small for chemical companies to undertake efficacy studies for recommendation of herbicides specifically for legume species. In most instances potentially useful herbicides are identified in more widely grown crops where their effectiveness can form the basis of advice for legume seed crops. Unfortunately, herbicides recommended in one country may not be recommended for use in another, and as new legislation becomes more restrictive, the chemical methods available for controlling weeds are likely to become more limited.

Pre-sowing herbicides must be incorporated into the soil before sowing. In France the main herbicide is benfluralin with an acceptable selectivity for all legumes species. Neburon is the most widely used pre-emergence herbicide in lucerne crops, particularly when sunflower (*Helianthus annuus* L.) is a companion crop. It is also effective for seed crops of sainfoin (*Onobrychis viciifolia* Scop.). However this chemical requires moist soil conditions to be effective.

The older post-emergence herbicides are hormonal, e.g. 2-4 DB for lucerne and MCPB for clover species, and should be applied between the two and eight trifoliate leaf stage at temperatures below 25°C to be most effective. Figure 6.3 shows other post-emergence herbicides used in France.

Table 6.1 shows the effectiveness of weed control obtained by mixtures or programmes of spraying (Jouy and Hacquet, 1992) when lucerne is sown under

---

[1]The inclusion of a chemical (herbicide, pesticide or growth regulator) should not be considered as a recommendation for its use.

CROP STAGE	ACTIVE INGREDIENT AND CONCENTRATION	RATE l or kg ha⁻¹ (or g. a.m.)	SELECTIVITY							EFFICACY										
			Alfalfa	Red Clover	White Clover	Crimson Clover	Sainfoin	Birdsfoot trefoil	Wild oat	Rye grass	Fathen	Crucifers	Picris echioides	Pineappleweed	Black nightshade	Wireweed	Redshank	Speedwells
Presowing incorporated	Benfluralin 180 g l⁻¹	6	●	O	O	*	*	O										
Pre-emergence	E.P.T.C. 360 g l⁻¹	8	*	*	◆	*	*	*										
	Neburon	2400 g	●	◆	◆	◆	*	◆										
2 trifoliate leaves	Pyridate 450 g l⁻¹ /45%	2	●	◆	●	*	*	◆										
	Pendimethalin 400 g l⁻¹	1.5	O	●	◆	O	O	O										
	2–4 DB 300 g l⁻¹	7	●	*	◆	◆	O	◆						?				
3–4 trifoliate leaves	MCPB 400 g l⁻¹	4	◆	◆	◆	O	◆	◆					?	?				
	Bentazone 480 g l⁻¹	2.5	*	*	O	*	*	◆					?	?				
After 2 trifoliate leaves	Cycloxydime 100 g l⁻¹	2	●	●	●	*	*	*						?				
After 3 trifoliate leaves	Carbetamide 70%	3	●	●	●	*	*	*						?				

SELECTIVITY
● satisfactory and officially confirmed herbicides
* satisfactory but not officially confirmed
O low selectivity margin and not officially confirmed
◆ toxic for the crop

EFFICACY
■ Satisfactory
▨ Medium
□ Insufficient
? Information not available

Fig. 6.3. Main recommended herbicides in young legume seed crops. (From Hacquet and Jouy, 1994.)

Legume Seed Crop Management

Table 6.1. Post-emergence herbicide efficacy on young lucerne. (From Jouy and Hacquet, 1992.)

Stages for spraying and active ingredients	Rate of active ingredients (g ha^{-1})	Crop Selectivity notes (variability)	Amaranths	Ammi-majus	Fat hen	Picris echioides	Sowthistle	Black nightshade	Turnipweed	Wild buckwheat	Wirewed	Redshank	Charlock	Sunflower	Speedwells
Postemergence 3–4 trifoliate leaves															
Dinoseb + 2-4 DB	900 + 1500	1 (0–3)	5	4	8	8	7	7	7	6	5	6	9	8	8
Bentazone	1440	1 (0–2)	6	7	1	8	7	9	8	6	0	8	8	7	5
Bentazone + 2-4 DB	720 + 900	2 (0–4)	9	8	9	8	6	10	9	7	6	9	9	8	4
Bentazone + pyridate	720 + 900	2 (1–3)	8	8	9	9	8	10	9	6	2	8	7	8	7
Pyridate + 2-4 DB	900 + 1500	1 (0–3)	9	6	9	8	7	9	9	6	7	8	8	9	4
Programmes 2–3 leaves then 4–5 leaves															
Bentazone then bentazone	720 then 720	1 (0–2)	—	9	5	9	8	—	9	8	0	—	10	7	3
Bentazone then pyridate	720 then 900	2 (1–3)	10	—	10	9	—	10	9	7	0	9	—	9	9
Pyridate then bentazone	900 then 720	1 (0–2)	10	9	9	9	9	10	9	8	1	10	9	8	8
Pyridate then 2-4 DB	900 then 1500	0 (0–1)	10	7	9	9	9	10	9	7	6	8	9	9	8

Efficacy notes (from 0 to 10)

a cereal companion crop. If sown under sunflower, the main companion crop in France, the only known solution is a programme of benfluralin application followed by neburon.

6.5.3 Weed Control in Established Crops

Few active ingredients are selective on established legume seed crops, and they must be sprayed during winter when there is no growth or just after a cut. Figure 6.4 shows the main herbicides recommended in France (Hacquet and Jouy, 1994). Except for lucerne seed crops, the possibilities of weed control in legume seed crops are restricted. In lucerne crops, the best efficiency is obtained by hexazinone, especially against persistent perennial weeds.

Many noxious weeds are forbidden in certified seed lots but it is not always possible to completely destroy them with the usual chemical and cleaning processes. In such circumstances it is important to carry out regular examinations of the crop and if necessary hand weed until the crop is harvested (Bell and Marble, 1985).

CROP STAGE	HERBICIDES TRADENAME (active ingredient and concentration)	RATE l or kg ha^{-1} (or g. a.m.)	SELECTIVITY						EFFICACY									
			Alfalfa	Red Clover	White Clover	Crimson Clover	Sainfoin	Birdsfoot trefoil	Grasses	Quackgrass	Chickweed	Sowthistle	*Picris echioides*	Pineappleweed	Plantain	Dandelion	*Silene alba*	Speedwells
Beginning of winter rest	Hexazinone 90%, paraquat 100 g l^{-1}	0.5 + 1	*	♦	♦	♦	o	♦	Satisfactory	Insufficient	Satisfactory	Satisfactory	Medium	Medium	Medium	Medium	Medium	Satisfactory
	Metsulfuron-methyl 20%	0.02	♦	♦	♦	♦	♦	o	Insufficient	Insufficient	Satisfactory	Satisfactory	? ?	Satisfactory	Medium	Satisfactory	Satisfactory	Medium
	Diuron	2000 g	●	♦	♦	♦	♦	♦	Satisfactory	Insufficient	Satisfactory	Insufficient	Insufficient	Satisfactory	Insufficient	Insufficient	Insufficient	Satisfactory
Winter rest	Paraquat 100 g l^{-1} + diquat 50 g l^{-1}	5	●	♦	♦	♦	♦	♦	Medium	Insufficient	Satisfactory	Insufficient	Insufficient	Insufficient	Insufficient	Insufficient	Insufficient	Medium
	Paraquat 100 g l^{-1}	6	●	●	♦	♦	?	o	Medium	Insufficient	Satisfactory	Insufficient	Insufficient	Insufficient	Insufficient	Insufficient	Insufficient	Medium
	Carbetamide 50% + dimefuron 25%	3.5	*	o	?	♦	*	♦	Satisfactory	Medium	Satisfactory	Insufficient	Insufficient	Medium	Insufficient	Insufficient	? ?	Satisfactory
Beginning of regrowth or after cutting	Hexazinone 90%	1	●	♦	♦	♦	♦	♦	Satisfactory	Medium	Satisfactory	Satisfactory	Satisfactory	Satisfactory	Medium	Medium	Medium	Satisfactory
Beginning of Dock stem elongation	Asulame 400 g l^{-1}	4	●	♦	♦	♦	♦	♦	Dock weed control									

SELECTIVITY
- ● satisfactory and officially confirmed herbicides
- * satisfactory but not officially confirmed
- o low selectivity margin and not officially confirmed
- ♦ toxic for the crop

EFFICACY
- Satisfactory
- Medium
- Insufficient
- ? Information not available

Fig. 6.4. Main recommended herbicides for established legume seed crops. (From Hacquet and Jouy, 1994.)

6.6 DEFOLIATION AND CLOSING

6.6.1 Objective of Forage Removal

Seed crops of some forage legumes may be defoliated several times during the establishment and seed production year, either by grazing, if livestock are available, or mechanically (cutting), or a combination of both. Some annual crops such as crimson clover and the *Vicia* spp. are not defoliated, whilst this is essential in others, e.g. subterranean clover (*T. subterraneum* L.). Defoliation seeks to fulfil a number of functions. If legumes are sown in the spring or autumn in the year before the first seed harvest, defoliation during the establishment year encourages strong plant growth before the onset of winter and helps to prevent winter kill. In the UK, seed crops of both white and red clover sown under a cover crop benefit from grazing or cutting in early autumn when the cover crop has been removed.

During spring of the harvest year, defoliation is often carried out to remove excessive leaf growth, stimulate flowering, improve the synchrony of flowering with the availability of natural pollinators and facilitate easier harvesting. Many legume species are known to benefit from being defoliated in spring, particularly if, like white clover, red clover and lucerne, they are able to recover rapidly from defoliation. In western Oregon, removing a hay crop in late spring is common practice for seed crops of white and red clover (Steiner, 1992) but in eastern Oregon, where red clover is spring sown and irrigated, no forage is removed. Although there is a general consensus of the necessity of spring defoliation for many species, the benefit of spring defoliation is not universal. Recent research has shown that Caucasian clover should not be cut at all in spring as this severely depresses seed yield (Steiner, 1992).

6.6.2 Method of Forage Removal

The method of forage removal depends upon the farming system in which the seed is produced, the species being grown and in many cases the particular cultivar. In many countries, seed crops may be grazed in the autumn and early spring (Fig. 6.5), then cut in late spring if necessary when more control over the extent of defoliation is required. Though mechanical defoliation is probably more usually carried out in spring, there are some reports of the effect of grazing on seed yield. In Australia and western Oregon, subterranean clover must be grazed by sheep in spring to maximize seed yields. In white clover, Marshall and Hides (1990) showed that small-leaved types are more tolerant of sheep grazing than large-leaved types. For these types defoliation by cattle may be less severe and more appropriate.

The optimal height of defoliation may depend upon when, within the growing season, it is imposed. In early spring, seed crops of white clover can be quite severely cut or grazed but as the crop develops, a cutting height of 3 cm is normal to prevent stolon removal. For Caucasian clover and zig-zag (*T. medium* L.) clover cutting heights of > 1 cm have been suggested (Daly *et al.*, 1993).

Fig. 6.5. White clover (*Trifolium repens* L.) seed crop (small-leaved cultivar Aberdale) being grazed by sheep in the spring of the first harvest year.

6.6.3 Timing of Forage Removal

If seed crops are grazed or cut in spring then the timing of defoliation can have a significant effect on seed production. Considerable research has been carried out to determine at what stage of seed crop development in spring, grazing or mechanical defoliation should cease and the crop be left to flower and produce seed. This is often referred to as the 'closing date'. For example, Clifford (1985) showed that in New Zealand, delaying closing of white clover from October to November (from April to May in the northern hemisphere) increased inflorescence number and potential seed yields. The effects of later closing have also been demonstrated. Marshall and Hides (1990) have shown that delaying closing seed crops of white clover reduces inflorescence size and seed weight, though this can be compensated by an initial increase in numbers of inflorescences. Subsequent delays in closing can however be detrimental to seed yield. Seed crops of white clover which were grazed should be closed some 2 weeks earlier than crops which were cut to allow sufficient time for crop recovery (Marshall and Hides, 1990). In other species, Daly *et al.* (1993) have reported reduced seed yields of Caucasian and zig-zag clovers after delaying closing beyond the stage of appearance of the flower buds. In western Oregon, sheep grazing of subterranean clover should cease after early bur fill to avoid reduced seed yields (Steiner and Grabe, 1986).

6.6.4 Consequences of Late Closing

Generally, if high seed yields are to be achieved then seed crops should not normally be defoliated more than 1 or 2 weeks after bud emergence. In the UK, seed crops of white clover begin to produce inflorescences on or about 15 May, though this may be one or two weeks earlier or later depending upon climatic conditions in spring. However, delaying closing may have other indirect consequences for reproductive development and seed yield. In white clover it can delay peak flowering and therefore the stage of optimal crop maturity. This may make harvesting later and at a time when harvesting conditions are less satisfactory (Marshall and Hides, 1990). These are all factors which the seed producer must consider when deciding how and when to defoliate the crop.

6.7 PLANT GROWTH REGULATORS

In spite of extensive research on the use of growth regulators, their current use on forage legume seed crops is limited. Their price is one obstacle, as is concern over possible residual effects. The beneficial effects of growth regulator application in terms of increased seed yield components and harvested yields have often been demonstrated, yet a common feature of much of the research thus far has been the variable effect of growth regulator treatment between years and in differing soil conditions. Some researchers (e.g. Choo, 1983) have suggested that growth regulators may have a specialized role in seed multiplication, rather than routinely in commercial seed production. For example, Choo (1983) proposed that daminozide may be of value in red clover breeding programmes for accelerating seed multiplication. At present, however, their role in forage legume seed production is limited to experimental studies and they are not a common part of crop management. Nevertheless, the potential of plant growth regulators to alter the morphology of legume crops and to increase seed yield is considerable and has been well studied. Beneficial effects on seed yield have been attained by an improvement in a number of seed yield components and changes in the pattern of growth. In the following sections the effects of the most commonly studied growth regulators on the growth of some forage legumes and their potential role in seed production are briefly described.

6.7.1 Paclobutrazol

This is probably the most commonly studied chemical in legume seed crops. Paclobutrazol (PP333, Parlay) is taken up into the xylem through leaves, stems or roots, and translocated to growing subapical meristems. It inhibits gibberellin biosynthesis (Anon., 1991), is soil active and its uptake requires adequate moisture in the upper part of the soil profile (Hampton and Hebblethwaite, 1984).

In glasshouse and field experiments on white clover, paclobutrazol reduced primary stolon length and nodes per stolon but increased the number of axillary buds per stolon as well as the number of reproductive buds per stolon and per unit area (Marshall and Hides, 1987, 1991a). It has been suggested that an increase in the number of nodes per unit area without a corresponding effect on stolon density leads to an increase in the number of harvestable inflorescences and potential seed yield. Results from New Zealand have shown similar responses (Budhianto et al., 1994a), and reflected in increased inflorescence production per unit area (Marshall and Hides, 1987; Hampton, 1991a; Budhianto et al., 1994b) and ripe inflorescences at harvest (Rijckaert, 1991), though not in all experiments (Boelt and Nordestgaard, 1993).

Paclobutrazol has also influenced the canopy structure of white clover seed crops so that inflorescences are located more appropriately for pollination. It has reduced petiole length more effectively than peduncle height and elevated flowers above the leaf canopy so that they are more accessible to pollinating insects (Marshall and Hides, 1986, 1987, 1991a; Rijckaert, 1991; Boelt and Nordestgaard, 1993; Budhianto et al., 1994a). This has often increased seed set per inflorescence (Marshall and Hides, 1991b; Rijckaert, 1991) though not always (Budhianto et al., 1994b). Increases in actual and potential seed yield have also been reported (Hampton, 1991a; Marshall and Hides, 1991b; Rijckaert, 1991; Budhianto et al., 1994b). The beneficial effects on yield have been dramatic though inconsistent. In Belgium (Rijckaert, 1991) it increased seed yield by an average of 271%. Actual seed yield increases were from 124 to 558 kg ha^{-1} in a wet and cold year (1987) and from 909 to 1200 kg ha^{-1} and 886 to 1084 kg ha^{-1} during warmer and drier seasons (1989 and 1990, respectively). This was achieved using a rate of 0.5 kg a.i. ha^{-1} applied at the appearance of the first green flower buds.

In red clover, soil-applied paclobutrazol at a rate of 1.68 kg a.i. ha^{-1} increased seed yield by 14%. Foliar application (1.12 kg a.i. ha^{-1}) reduced canopy height and increased seed yield by 21% in normal conditions due to increased inflorescence production, but in dry conditions it had no effect on yield. In Finland, paclobutrazol increased the seed yield of tetraploid red clover by 38 and 29% in two different years (Niemelainen, 1987).

In New Zealand, paclobutrazol improved seed yields of *Lotus* spp. by 82% and 68% by increasing the number of reproductive stems leading to more umbels per unit area (Clifford and Hare, 1987). Seed yields of lucerne have also been increased through an increase in the number of harvestable racemes, due to the stimulation of primary lateral shoot production and pods per raceme (Askarian et al., 1994). Seed yield increases of between 36 and 150% were achieved.

6.7.2 Daminozide

Daminozide (Alar, B9, B995) is absorbed by the leaves and translocated throughout the plant (Anon., 1991). It has increased seed yields of red clover at rates of 2.5–10.0 kg a.i. ha^{-1}. It has also reduced stem length and lodging of red clover (Puri and Laidlaw, 1983; Christie and Choo, 1990) and reduced plant height (Mela, 1969).

It has had no reported effect on the number of stems per unit area or the number of nodes per stem (Puri and Laidlaw, 1983), but has increased the number of primary branches per stem (Valle and Bergt, 1965) and the number of flower heads per unit area (Puri and Laidlaw, 1983; Christie and Choo, 1990) due to an increase in inflorescences per stem (Puri and Laidlaw, 1983).

Daminozide has reportedly delayed the flowering of red clover (Niemeläinen, 1987) and reduced the height of the nectar column and corolla tube length (Puri and Laidlaw, 1983; Christie and Choo, 1990), resulting in improved pollination by bees. This is confirmed by reports that bumble bees and honey bees were some 34–40% more numerous on plots treated with daminozide than on untreated plots (Puri and Laidlaw, 1983) and that it increased seed set (Puri and Laidlaw, 1983; Christie and Choo, 1990). These beneficial effects have been reflected in increased seed yields of diploid red clover (19–28%) (Mela, 1969) and tetraploid red clover (7–36%, Jakesova and Svetlik, 1984; 13%, Niemeläinen, 1987) and an increase in harvested seed yield from 474 to 604 kg ha^{-1} (Puri and Laidlaw, 1983). In New Zealand it also increased seed yields of *Lotus* spp. by 35 and 50% by promoting stem branching (Clifford and Hare, 1987), although peak flowering was later than normal.

6.7.3 Triapenthenol

Triapenthenol (RSW0411) has increased seed yield of white clover from 124 to 372 kg ha^{-1} in a wet growing season in Belgian conditions (Rijckaert, 1991). During drier seasons, yields increased from 909 to 1018 kg ha^{-1} and from 886 to 989 kg ha^{-1} when triapenthenol was applied at a rate of 2.8 kg a.i. ha^{-1} at the appearance of the first flower buds.

6.7.4 Chlormequat

Chlormequat inhibits cell elongation, hence shortening and strengthening the stem (Anon., 1991). Its applications to seed production plots of white clover have yielded few promising results (Marshall and Hides, 1986; Boelt, 1991; Budhianto *et al.*, 1994a), but in red clover chlormequat at a rate of 2.5 l ha^{-1} at flower emergence had a slight positive effect on seed yield (Niemeläinen, 1987).

Plant growth regulators can therefore have beneficial effects on the seed yields of forage legumes but until approved for use, their role in seed crop management is limited.

6.8 WATER MANAGEMENT

Forage legume seed crops are grown in ecological regions ranging from those limited by moisture in the southwestern USA, southern and western South America, Australia and the Mediterranean basin, to more humid marine environments in the USA Pacific Northwest, northern and central Europe, and New Zealand where more than adequate moisture is often available but thermal energy may be limited. A significant portion of the temperate forage legume seed industry in the USA moved to the western regions of North America during the 1940s to take advantage of the more dependable climatic conditions that favour flowering, pollination, seed development and harvest.

The specific water requirements for optimizing seed production from forage legume seed crops may differ greatly from those for hay or pasture crops. Common to all crops is the need to have adequate soil-water available to facilitate flowering, pollination and seed development. Different crops may express different mechanisms for water stress adjustment, so a single irrigation management strategy is not possible. Successful seed production depends upon the temporal climatic pattern throughout the cropping season, the variability and distribution of the actual amount of precipitation received in a season, and the proper use of supplemental irrigation, where available.

Lucerne is primarily grown for seed in arid regions requiring supplemental irrigation. When soil-water conditions are such that excessive vegetative plant growth occurs, seed yields are reduced (Taylor et al., 1959; Yamada et al., 1973; Hageman et al., 1975; Beukes and Barnard, 1985; Steiner et al., 1992). However, seed yields can also be reduced when the plants are severely water stressed (Hageman et al., 1975; Steiner et al., 1992). Different amounts of water may be required for the establishment and subsequent years of production, and the amount of water stored in the soil profile during the winter can greatly influence seed yield (Steiner et al., 1992). The 3-year average water requirement for a lucerne seed crop was 800 mm.

Flower bud development and flowering increase with increased amounts of applied water. However, the percentage of flowers tripped by pollinators was decreased (Goldman and Dovrat, 1980) and the number of florets aborted increased (Steiner et al., 1992) as water application amounts by irrigation were increased. By maintaining continued slow plant growth that promotes flower production and seed development but which restricts excessive vegetative development, seed yields are maximized (Taylor et al., 1959; Yamada et al., 1973; Hageman et al., 1978; Beukes and Barnard, 1985; Taylor and Marble, 1986; Steiner et al., 1992). The use of water management to reduce vegetative growth may also have the added benefit of facilitating pollinator effectiveness and reducing the amounts of insecticides and desiccants needed to produce the crop (Steiner et al., 1992).

The response of white clover to water management is similar to lucerne in that unrestricted water availability reduces seed yield. Stolon tips can form either secondary stolons or flowers (Thomas, 1961), so when vegetative growth is not restricted, seed yield will be reduced due to fewer flowers being differentiated. Constraining summer foliage growth by properly timed water applications during the flowering period increases flower density and results in higher seed yields (Zaleski, 1966; Clifford, 1986; Danyach-Deschamps and Wery, 1988; Bullitta et al., 1988; Oliva et al., 1994c).

The plant growth response of birdsfoot trefoil grown for seed at varying levels of crop water stress differs greatly from that of lucerne and white clover. When soil-water is not limited to the plant, flower differentiation is reduced. For the climatic conditions found in western Oregon, plants maintained under low-stress conditions have reduced numbers of umbels per unit area, seeds per pod and seed mass compared to plants grown under limited soil-water availability (Garcia-Diaz and Steiner, 1995). In a first-year seed crop, total seed yield increases as the amount of accumulated seasonal water stress increases. Plants grown under conditions where soil-water availability is limited also have lower harvest losses due to seed shattering because the duration of the flowering period is limited, which results in a greater proportion of the seed pods being mature at the same time.

Red clover responds differently to the amount of available soil-water. In Oregon, USA, optimal red clover seed production results when there is no limitation to soil-water availability (Oliva *et al.*, 1994b). The most economical treatment is when a single application of water is applied at the time of peak flowering. Water applied soon after the common practice of spring forage removal is of less benefit to the crop because the soil-water content may still be relatively high (> 80%) from late-spring precipitation. Also less soil-water is available late in the growing season for seed development when irrigation occurs soon after the time of forage removal.

Seed yields are lower in second-year production fields due to the increased severity of the root and crown rot disease complex that restricts water uptake by the crop. As a result, the seed yield water-use efficiency for a second-year crop is generally lower than in the first year of production (Oliva *et al.*, 1994a). Irrigation during the period of late-season reproductive development reduces the severity of root rot incidence and its effect on seed yield.

The overall reduction in red clover seed yield from increased plant water stress is primarily caused by decreased floret fertility. Final seed yield is more the result of a loss in the already established potential than from an initially low potential (Hampton, 1991b). The establishment of potential seed yield depends mainly on the number of flowers per unit area and the number of florets per flower at anthesis. Therefore, the delayed application of water until the time of reproduction should result in higher seed yields. In western Oregon, USA, the annual water requirement for red clover was about 400 mm (Oliva *et al.*, 1994a).

6.9 PESTS AND DISEASES

Several pests and diseases are known to damage legume seed crops but insect pests are the major cause of damage.

6.9.1 Insect Pests

About 60 noxious species are known in Europe but as lucerne is the main legume seed crop, more information is available on this than other species. In the USA, over 500 species have been identified in lucerne seed crops (Johansen *et al.*, 1979), but the majority have no direct influence on seed production and many are parasites or predators.

Using information from various countries (Australia: Berg and Boyd, 1984; USA: Johansen *et al.*, 1979; Rincker *et al.*, 1988), the following are examples of the insect groups considered to be the important pests of legume seed crops.

Coleoptera

Beetles in this family cause extensive damage. The larvae of the lucerne weevil (*Hypera postica* Gyll.) feed on the leaves and buds of the lucerne seed crop in all the western States of the USA, Canada and in southern Europe. The pea leaf weevil (*Sitona lineatus* L.), the adults of which attack young legume crops, is very frequent in Europe and in America. The sitona weevil (*S. discoideus* Gyll.) was introduced to Australia and New Zealand and their larvae cause considerable damage to lucerne roots (Peterson *et al.*, 1982). The larva of the clover root borer (*Hylastinus obscurus* Marsham) bores galleries inside the tap roots of red clover (Bouet, 1993a).

Other beetles cause damage by feeding on the seeds of the legume crop. The lucerne pod weevil (*Tychius aureolus* Keisenwetter), whose larvae cause damage in some regions of France and in eastern Europe, may eat 40% of seeds within pods. In France, clover seed weevils (*Apion trifolii* Bach. and *A. apricans* Ierbst.) are the main pests of red clover.

Lepidoptera

The native budworm (*Heliotis* spp.) in lucerne in Australia and cutworms (various species) are frequently observed on legume crops. The lucerne seed moth (*Cydia medicaginis* Kuzn.) is common in southern Europe, but yield losses do not exceed 10%.

Diptera

This group is mainly represented by gall midges of which each legume species has one or two specific midges. The flower gall midge (*Contarinia medicaginis* Kieff.) damages European lucerne seed crops, particularly during wet springs, but has become much less frequent in the last 10 years. The clover seed midge (*Dasyneura leguminicola* Lint.) is a serious pest on red clover in North America.

Hymenoptera

Chalcids are considered the most damaging pests of legume seed crops. The lucerne seed chalcid (*Bruchophagus roddi* Guss.) causes considerable damage on lucerne in both Australia and the USA. Each larva hatches quickly inside one seed, then destroys it. In France, this damage is generally light due to the climate and numerous parasites of the chalcid. Up to 30% crop losses have been reported for the trefoil seed chalcid (*B. kolobovae* Fedoseeva) on birdsfoot trefoil.

Heteroptera

Bugs are (with chalcids) the most serious pest of lucerne seed crops. They are predominantly mirid bugs whose nymphs and adults feed on lucerne, destroying buds, and causing flower drop and seed blast. The main species are *Lygus hesperus* Knight (in California), *L. elisus* Vand. (in the northwest USA and in Canada) and *Adelphocoris lineolatus* Goeze. (in France, Hungary and Canada).

Homoptera

These are widespread and prolific and damage seed crops as they suck sap and transmit viruses. Aphids cause considerable damage in America and Australia but are not considered dangerous in Europe, except in the drier south (Greece and Spain). The pea aphid (*Acyrthosiphon pisum* Harris) is the most common aphid on lucerne and is observed in most growing regions.

Legume Seed Crop Management

6.9.2 Insect Control

Before treatment, it is necessary to identify the pests, to measure the risks, to verify the efficacy of available insecticides, their selectivity on pollinators and the suitable stage of the crop for successful control. Some pests require preventive and systematic treatments because spraying is ineffectual when damage becomes perceptible. In most other cases, pest populations must be quantified. Sweep netting is the most common method of sampling and counting pest species (Gregory *et al.*, 1966). Sampling requires 25 sweeps in the plot (or more, in several places if the field is large) and each sweep covers a 90 degree arc, striking the top 25 cm of the plants.

The threshold for commencing treatment depends upon the pest. In France (FNAMS and INRA, 1986), treatment is appropriate when 100 lucerne weevil larvae in 25 sweeps, or 75 mirid bugs, or 25 lucerne pod weevils are found. In Washington State, spraying is advised if there are 100 aphids per sweep, or three lygus bugs per sweep (Johansen *et al.*, 1979). The following guidelines are useful for insect control:

- Do not spray a pesticide when it is not justified or when the crop is not at the suitable stage.
- Carefully follow dosage recommendations. Overdosing may kill predators or parasites; it may also produce a selection pressure for development of resistant insects.
- Choose selective insecticides which do not affect pollinators and other beneficial insects. It is advised to spray before the flowering stage when possible, or to treat during late evening or night.

Integration of different control methods

Chemical control remains the principal means for most pests. Figure 6.6 shows the efficiency of insecticides for controlling the main pests of lucerne in France (Hacquet, 1995) where new chemicals have replaced most of the pesticides used

PESTICIDE Active ingredient	RATE Active ingredient (g ha^{-1})	Sitona weevil	Alfalfa weevil	Apion pisi	Flower gall midge	Mirid bugs	Alfalfa moth	Alfalfa pod weevil (2)	SELECTIVITY (3) on leafcutter bees
Deltamethrin	6.25								Good
Alphamethrin	15				?				Medium
Lambda-cyhalothrin	7.5				?				Very good
Tau-Fluvalinate	72				?				?
Esfenvalerate	10			?	?				Very good
Fenvalerate	30								Very good
Bifenthrin	7.5		?	?	?			?	?
Phosalone	1000								Medium
Furathiocarb	800 g q^{-1}	Seed treatment without other efficacy							-

(1) Efficacy (FNAMS trials from 1980 to 1993)
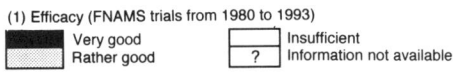
Very good
Rather good
Insufficient
? Information not available

(2) Specific dose for Alfalfa weevil
(3) Selectivity on Leafcutter bees taken as reference pollinators, according to Tasei J.N., INRA Lusignan, France.

Fig. 6.6. Insecticide efficacy in lucerne seed crops and selectivity on pollinators. (From: Hacquet, 1995.)

10 years ago. Some still remain, e.g. parathion which is applied before flowering of red clover (Bouet, 1993b).

Seed treatments offer new possibilities for pest control. In France Hacquet (1994a,b) controlled pea leaf weevil in young legume crops using the systemic chemical furathiocarb (800 g per 100 kg of seed). This is relatively efficient and removes the need for several treatments to the crop.

Biological control is still rare in legume crops, although there are some examples – breeding of parasites against the lucerne seed chalcid (Aeschlimann and Vitou, 1989). Cultural techniques are, at present, the only method of controlling the lucerne seed chalcid. This involves cutting hay to delay its life cycle and reduce population numbers, removing volunteer lucerne plants, cultivation and irrigation in the autumn, burning postharvest debris and burying dropped seeds (Pedersen et al., 1990).

Plant selection can also be used to breed plants for resistance, particularly to aphids, and resistant cultivars are available in the USA, Australia and New Zealand.

6.9.3 Diseases

Although fungi, bacteria, viruses and other diseases are prevalent on legume seed crops, they are much less important than pests. The following diseases can be observed on seed crops (Cooperative Extension Services of Washington State University, 1972; Goplen et al., 1980; Raynal et al., 1989) but differ in the damage they cause.

Fungi

Most fungi are not dangerous in seed crops. Sclerotinia rot (*S. trifoliorum* Erikss) may however cause severe damage when crops are sown in autumn and more frequently on red clover, crimson clover and sometimes white clover in regions with humid and cool winters (UK, Germany, Sweden, France, North America). Some red clover cultivars bred in France and Belgium show good resistance.

Some diseases such as Verticillium wilt (*V. albo-atrum* Reinke & Berth) cause little damage. Lucerne rust (*Uromyces striatus* Schroet.) can damage lucerne in some Mediterranean countries. Resistant cultivars are available in the USA. Sooty blotch (*Polythrincium trifolii* Kunze.) can cause early defoliation on crimson or white clover.

Root diseases

Fusarium wilt (*F. oxysporum* Schlecht.) and Rhizoctonia stem blight (*R. solani* Kühn.) are widespread on lucerne in western Canada and in the USA, sometimes causing considerable damage. Damping off (*Pythium ultimum* Trow.) is frequently recorded in France with damage to young lucerne seed crops.

Viruses, mycoplasma and bacteria

Approximately 36 viruses have been identified on legume crops but the damage they cause is often difficult to measure. Many viruses are transmitted by seed, so sanitary control is necessary, particularly with alfalfa mosaic virus.

Clover phyllody, caused by phytoplasma, can be problematical in white clover crops. The symptoms are spectacular (racemes like leaves, dwarf plants) and rhizobia are unable to fix nitrogen. Control may be acheived by controlling leafhoppers that transmit the phytoplasma (Cousin, 1989).

Bacterial wilt (*Corynebacterium insidiosum* McCull) causes huge losses in lucerne seed crops in North America, Australia, New Zealand and eastern Europe. The bacteria are in the soil and seeds and enter the plants by wounds or lesions caused by nematodes. Plants develop small, cupped leaves followed by death. Resistant cultivars exist in the USA, and strict quarantine rules are applied in Europe.

Nematodes

Stem nematode (*Ditylenchus dipsaci* Kühn.) is the main species causing damage to lucerne and clovers. It is widespread in Europe and America, causing losses, particularly when seed crops are wet. This parasite survives several years in soil or in vegetative debris and in seed lots. Resistant cultivars are available, and it is also possible to fumigate seeds with methyl bromide.

In most legume seed crops, however, no particular management is needed to control disease, but careful observation is recommended so that the appropriate hygiene measures can be adopted when necessary.

6.10 CONCLUSIONS

Considerable care is required in selecting fields with the appropriate soil structure, soil type and crop history to enable the production of high seed yields whilst maintaining genetic purity. Many crop management systems have been developed to accommodate the range of environments in which seeds of forage legume species are produced. As environmental concerns surface, and legislation on herbicide and pesticide usage becomes tighter, alternative methods of weed and pest control are developing and will become more important in the future. The challenge for seed producers is to maintain the high quality standards necessary during seed production, as some of the chemicals currently available are removed from the recommended lists.

ACKNOWLEDGEMENTS

The authors thank Mr D.H. Hides, Institute of Grassland and Environmental Research, Aberystwyth, UK for reviewing this manuscript.

REFERENCES

Aeschlimann, J.P. and Vitou, J. (1989) Observations on the lucerne seed chalcid, *Bruchophagus roddi* (Hym., Eurytomidae), and its parasitoids in mediterranean France: a promising candidate for classical biological control in Australia. *Acta Oecologica Oecologia Applicata* 10(2), 129–133.

Anon. (1991) In: Tomlin, C. (ed.) *The Pesticide Manual. Incorporating the Agrochemical Handbook*, 10th edn. British Crop Protection Council and The Royal Society of Chemistry, Cambridge, 1341 pp.

Askarian, M., Hampton, J.G. and Hill, M.J. (1994) Effect of paclobutrazol on seed yield of lucerne *(Medicago sativa* L.) cv. Grasslands Oranga. *Journal of Applied Seed Production* 12, 9–13.

Bell, A. and Marble, V.L. (1985) Alfalfa seed production in California. In: *National Alfalfa Symposium*. Iowa State University, USA.

Belzile, L. (1991) Effet des types de sol sur la production de semence du trefle rouge. Summary: the effect of soil type on red clover seed yield. *Canadian Journal of Plant Science* 71, 1039–1046.

Berg, G. and Boyd, M. (1984) *Insects in Lucerne: A Guide to Identification and Control of Insects in Lucerne*. The Grassland Society of Victoria, Australia, pp. 5–27.

Beukes, D.J. and Barnard, S.A. (1985) Effects of level and timing of irrigation on growth and water use of lucerne. *South African Journal of Plant Science* 2, 197–202.

Boelt, B. (1991) The influence of growth regulation on seed yield and yield parameters in white clover *(Trifolium repens* L.). *Journal of Applied Seed Production* 9 (Suppl.), 60 (abstract).

Boelt, B. and Nordestgaard, A. (1993) Growth regulation in white clover *(Trifolium repens* L.) grown for seed production. *Journal of Applied Seed Production* 11, 1–5.

Bouet, S. (1993a) Trèfle violet, gare au scolyte. In: *Bulletin Semences* No. 122. FNAMS (Fédération Nationale des Agriculteurs Multiplicateurs de Semences), France.

Bouet, S. (1993b) Trèfle violet, les nouvelles armes anti-apion. In: *Bulletin Semences* No. 123. FNAMS (Fédération Nationale des Agriculteurs Multiplicateurs de Semences), France.

Budhianto, B., Hampton, J.G. and Hill, M.J. (1994a) Effect of plant growth regulators on a white clover *(Trifolium repens* L.) seed crop. I. Plant growth and development. *Journal of Applied Seed Production* 12, 47–52.

Budhianto, B., Hampton, J.G. and Hill, M.J. (1994b) Effect of plant growth regulators on a white clover *(Trifolium repens* L.) seed crop. II. Seed yield components and seed yield. *Journal of Applied Seed Production* 12, 53–58.

Bullitta, P., Bullitta, S. and Roggero, P.P. (1988) Water management for white clover seed production in a Mediterranean environment. *Agricoltura Mediterranea* 118, 354–360.

Choo, T.M. (1983) Effect of Alar on seed yield of red clover. *Forage Notes* 27, 22–23.

Christie, B.R. and Choo, T.M. (1990) Effect of harvest time and Alar85 on seed yield of red clover. *Canadian Journal of Plant Science* 70, 869–871.

Clifford, P.T.P. (1974) Influence of inter- and intra-row spacing on components of seed production of tetraploid red clover 'Grasslands Pawera'. *New Zealand Journal of Experimental Agriculture* 2, 261–263.

Clifford, P.T.P. (1977) Cultural methods for 'Grasslands Pitau' white clover seed crops. *New Zealand Journal of Experimental Agriculture* 5, 147–149.

Clifford, P.T.P. (1985) Effect of cultural practice on potential seed yield of 'Grasslands Huia' and 'Grasslands Pitau' white clover. *New Zealand Journal of Experimental Agriculture* 13, 301–306.

Clifford, P.T.P. (1986) Effect of closing date and irrigation on seed yield and some seed yield components of 'Grasslands Kopu' white clover. *New Zealand Journal of Experimental Agriculture* 14, 271–277.

Clifford, P.T.P. (1987) Producing high seed yields from high forage producing white clover cultivars. *Journal of Applied Seed Production* 5, 1–9.

Clifford, P.T.P. and Anderson, A.C. (1980) Red clover seed production-research and practice. In: Lancashire, J.A. (ed.) *Herbage Seed Production*. Grassland Research and Practice Series No 1. New Zealand Grassland Association, Palmerston North, pp. 76–79.

Clifford, P.T.P. and Hare, M.D. (1987) Improved *Lotus pedunculatus* seed yields from the use of growth regulators. *Journal of Applied Seed Production* 5, 67 (abstract).

Clifford, P.T.P. and Rolston, M.P. (1990) Mineral nutrient requirements for white clover seed production. *Journal of Applied Seed Production* 8, 54–58.

Clifford, P.T.P. and White, S.D. (1986) A sulphur response: white clover seed production. *New Zealand Journal of Experimental Agriculture* 14, 97–99.

Cornforth, I.S. and Sinclair, A.G. (1984) *Fertiliser and Lime Recommendations for Pasture and Crops in New Zealand*. 2nd revised edn. New Zealand Ministry of Agriculture and Fisheries, Wellington.

Cooperative Extension Services of Washington State University (1972) *Alfalfa Analyst, Diseases and Insects, An Aid To Identification*. Washington State, USA.

Cousin, M.T. (1989) La Phyllodie du Trèfle : ni virus, ni vrai mycoplasme. In: *Bulletin Semences* No. 109. FNAMS (Fédération Nationale des Agriculteurs Multiplicateurs de Semences), France.

Daly, G.T., Gurung, J. and Lucas, R.J. (1993) Stand age and closing date effects on seed yield of Caucasian clover and zigzag clover cultivars. *Proceedings of the XVII International Grassland Congress, Nice*, pp. 1664–1666.

Danyach-Deschamps, M. and Wery, J. (1988) Effects of drought stress and mineral nitrogen supply on growth and seed yield of white clover in Mediterranean conditions. *Journal of Applied Seed Production* 6, 14–19.

Davis, F.L. (1949) Effects of liming on responses to minor elements of crimson clover, soybeans, and alyce clover. *Agronomy Journal* 41, 368–374.

Fairey, D.T. and Lefkovitch, L.P. (1991) Establishing perennial legume seed stands with annual companion crops. *Journal of Applied Seed Production* 9, 49–54.

FNAMS and INRA, Lusignan (1986) Les Ravageurs de la luzerne porte-graine, depliant de vulgarisation. In: *Supplement a Bulletin Semences* No.94. FNAMS (Fédération Nationale des Agriculteurs Multiplicateurs de Semences), France.

Garcia-Diaz, C.A. and Steiner, J.J. (1995) Birdsfoot trefoil seed crop water relations. *American Society of Agronomy. Agronomy Abstracts*. ASA, CSSA, SSSA, Madison, WI, USA, p.139.

Goldman, A. and Dovrat, A. (1980) Irrigation regime and honey bee activity as related to seed yields in alfalfa. *Agronomy Journal* 72, 961–965.

Goplen, B.P., Baenziger, H., Bailey, L.D., Gross, A.T.H., Hanna, M.R., Michaud, R., Richards, K.W. and Waddington, J. (1980) *Growing and Managing Alfalfa in Canada*. Agriculture Canada Ottawa Publication 1705. Ministre des Approissionments et Service Canada 1982, Ottawa.

Gregory, E.J., Ferris, C. and Marble, V. (1966) *Alfalfa Seed Production in Fresno County*. Agricultural Extension Service, University of California.

Grizzard, A.L. and Matthews, E.M. (1942) The effect of boron on seed production of alfalfa. *Journal of the American Society of Agronomy* 34, 365–368.

Hacquet, J. (1994a) *Bruchophagus roddi : Ravageur de la luzerne, Etude bibliographique*. FNAMS (Fédération Nationale des Agriculteurs Multiplicateurs de Semences), France.

Hacquet, J. (1994b) Luzerne : vivement des semences traitées. In: *Bulletin Semences* No. 129. FNAMS (Fédération Nationale des Agriculteurs Multiplicateurs de Semences), France.

Hacquet, J. (1995) Luzerne porte-graine, Quels traitements contre Quels ravageurs? In: *Bulletin Semences* No. 132. FNAMS (Fédération Nationale des Agriculteurs Multiplicateurs de Semences), France.

Hacquet, J. and Jouy, L. (1994) Légumineuses à petites graines, l'art du désherbage. In: *Bulletin Semences* No. 126. FNAMS (Fédération Nationale des Agriculteurs Multiplicateurs de Semences), France.

Hageman, R.W., Ehling, C.F., Huber, M.J., Reynoso, R.Y. and Willardson, L.S. (1978) Effect of irrigation frequencies on alfalfa seed yield. *California Agriculture* 32, 17–18.

Hageman, R.W., Willardson, S., Marsh, A.W. and Ehling, C.F. (1975) Irrigating for maximum alfalfa seed yield. *California Agriculture* 29, 14–15.

Hampton, J.G. (1991a) Effect of paclobutrazol on inflorescence production and seed yield in four white clover *(Trifolium repens* L.) cultivars. *New Zealand Journal of Agricultural Research* 34, 367–373.

Hampton, J.G. (1991b) Temperate herbage seed production: an overview. *Journal of Applied Seed Production* 9 (suppl.), 2–13.

Hampton, J.G. and Hebblethwaite, P.D. (1984) The influence of rainfall on paclobutrazol (PP333) response in the perennial ryegrass *(Lolium perenne* L.) seed crop. *Journal of Applied Seed Production* 2, 8–12.

Hampton, J.G., Clifford, P.T.P. and Rolston, M.P. (1987) Quality factors in white clover seed production. *Journal of Applied Seed Production* 5, 32–40.

Hare, M.D. (1984) 'Grassland Maku' Lotus *(Lotus pedunculatus* (Cav.)) seed production. 2. Effect of row spacing and population density on seed yields. *Journal of Applied Seed Production* 2, 65–68.

Hollington, P.A., Marshall, A.H. and Hides, D.H. (1993) The effect of row spacing and cover-crop on stolon development and the seed yield components of white clover cultivars of contrasting leaf size. *Grass and Forage Science* 48, 1–10.

Jakesova, H. and Svetlik, V. (1984) Einfluss von Aussaat, des Einsatzes von Mikronahrstoffen un Morphoregulatoren auf die Ertragshohe und Qualitat des Saatgutes von tetraploidem Rotklee. *Wissenschaftliche Beitrage, Martin-Luther Universitat, Halle-Wittenberg* (54) 378–389 (abstract).

Johansen, C., Baird, C., Bitner, R., Fisher, G., Undurraga, J., Lauderdale, R. and Eves, J. (1979) *Alfalfa Seed Insect Pest Management.* Western Regional Extension Publication. Washington State University, USA.

Johnson, W.C. and Wear, J.I. (1967) Effect of boron on white clover (*T. repens* L.) seed production. *Agronomy Journal* 59, 205–205.

Jouy, J. and Hacquet, J. (1992) Légumineuses à petites graines: sélectivité et efficacité d'herbicides sur jeunes cultures. In: *XV International Conference* COLUMA, ANPP, Versailles, France.

Kjaersgaard, B. (1994) Lokalisering af hvidkloverproduktion i Nordiske lande. *NJF-Seminarium* 241, 47–50.

Kowithayakorn, L. and Hill, M.J. (1982) A study of seed production of lucerne (*Medicago sativa*) under different plant spacing and cutting treatments in the seedling year. *Seed Science and Technology* 10, 3–12.

Kubota, J. and Allaway, W.H. (1972) Geographic distribution of trace element problems. In: Mortvedt, J.J., Giordano, P.M. and Lindsay, W.L. (eds) *Micronutrients in Agriculture.* Soil Science Society of America, Madison, Wisconsin, pp. 525–554.

Marshall, A.H. and Hides, D.H. (1986) The effect of growth regulators on seed yield components of white clover. *Journal of Applied Seed Production* 4, 5–7.

Marshall, A.H. and Hides, D.H. (1987) Modification of stolon growth and development of white clover *(Trifolium repens* L.) by growth regulators, and its influence on flower production. *Journal of Applied Seed Production* 5, 18–25.

Marshall, A.H. and Hides, D.H. (1990) White clover seed production from mixed swards: effect of sheep grazing on stolon density and on seed yield components of two contrasting white clover varieties. *Grass and Forage Science* 45, 35–42.

Marshall, A.H. and Hides, D.H. (1991a) Effect of the plant growth regulator Parlay on the seed production of the white clover cvs. Menna and Olwen. I Stolon growth and development. *Journal of Applied Seed Production* 9, 73–80.

Marshall, A.H. and Hides, D.H. (1991b) Effect of the plant growth regulator Parlay on the seed production of the white clover cvs. Menna and Olwen. II Yield components and potential seed yield. *Journal of Applied Seed Production* 9, 81–86.

Marshall, A.H. and James, I.R. (1988) Effect of plant density on stolon growth and development of contrasting white clover (*Trifolium repens* L.) varieties and its influence on the components of seed yield. *Grass and Forage Science* 43, 313–318.

Marshall, A.H., Khrbeet, H.K. and Hides, D.H. (1991) Influence of boron on the reproductive growth of white clover (*Trifolium repens* L.) cultivars. *Annals of Applied Biology* 119, 541–548.

McCartin, J. (1985) Alternative establishment strategies for white clover seed production. In: Hare, M.D. and Brock, J.L. (eds) *Producing Herbage Seeds*, Grassland Research and Practice Series No 2. New Zealand Grassland Association, Palmerston North, pp.33–35.

McGraw, R.L., Beuselink, P.R. and Ingram, K.T. (1986) Plant population density effects on seed yield of Birdsfoot trefoil. *Agronomy Journal* 78, 201–205.

Mela, T. (1969) The effect of N-dimethylaminosuccinamic acid (B-995) on the seed cultivation characteristic of late flowering red clover. *Acta Agralia Fennica* (115) 152 pp.

Miller, D.A. and Reetz, H.F. Jr (1995) Forage fertilization. In: Barnes, R.F., Miller, D.A. and Nelson, C.J. (eds) 1, *An Introduction to Grassland Agriculture*, vol. 1, *Forage*. Iowa State University Press, Ames, pp. 71–87.

Niemeläinen, O. (1987) The effect of some growth regulators on the seed yield of tetraploid red clover. *Journal of Applied Seed Production* 5, 67 (abstract).

Oliva, R.N., Steiner, J.J. and Young III, W.C. (1994a) Red clover seed production. I. Crop water requirements and irrigation timing. *Crop Science* 34, 178–184.

Oliva, R.N., Steiner, J.J. and Young III, W.C. (1994b) Red clover seed production. II. Plant water status on yield and yield components. *Crop Science* 34, 184–192.

Oliva, R.N., Steiner, J.J. and Young III, W.C. (1994c) White clover seed production. II. Soil and plant water status on yield and yield components. *Crop Science* 34, 768–774.

Pedersen, M.W., Bohart, G.E., Marble, V.L. and Klostermeyer, E.C. (1990) Seed production practices. *Plant Disease* 82, 703–704.

Peterson, J.R., Buchanan, C., Grasser, H.A., Maerae, R. and Holikamp, R. (1982) Lucerne seed production. In: *Agfacts*. Department of Agriculture. New South Wales, Australia.

Pilbeam, D.J. and Kirby, E.A. (1983) The physiological role of boron in plants. *Journal of Plant Nutrition* 6, 563–582.

Puri, K.P. and Laidlaw, A.S. (1983) The effect of cutting in spring and application of Alar on red clover (*Trifolium pratense* L.) seed production. *Journal of Applied Seed Production* 1, 12–18.

Raynal, G., Godran, J., Bournoville, R. and Courtillot, M. (1989) Maladies des légumineuses. In: Raynal, G., Godran, J., Bournoville, R. and Courtillot, M. (eds) *Ennemis et Maladies des Prairies*, Institut National de la Recherche Agronomique, Paris, pp.85–131.

Rijckaert, G. (1991) Application of growth regulators in seed crops of white clover (*Trifolium repens* L.) under Belgian climatic conditions. *Journal of Applied Seed Production* 9, 55–62.

Rincker, C.M. (1977) Alfalfa seed yields from seeded rows vs. spaced transplants. *Crop Science* 16, 268–270.

Rincker, C.M., Marble, D.E., Brown, D.E. and Johansen, C.A. (1988) Seed production practices. In: Hanson, A.A. (ed.) *Alfalfa and Alfalfa Improvement*, Agronomy No. 29. American Society of Agronomy, Madison, Wisconsin.

Rogers, H.T. (1947) Boron response and tolerance of several legumes to borax. *Journal of the American Society of Agronomy* 39, 897–913.

Sherrell, C.G. (1983) Effect of boron application on seed production of New Zealand herbage legumes. *New Zealand Journal of Experimental Agriculture* 11, 113–117.

Steiner, J.J. (1992) Effect of haying on Kura clover (*Trifolium ambiguum*) grown for seed. *Journal of Applied Seed Production* 10, 15–18.

Steiner, J.J. and Grabe, D.F. (1986) Sheep grazing effects on subterranean clover development and seed production in western Oregon. *Crop Science* 26, 367–372.

Steiner, J.J. and Snelling, J.P. (1994) Kura clover seed production when intercropped with wheat. *Crop Science* 34, 1330–1335.

Steiner, J.J., Hutmacher, R.B., Gamble, S.D., Ayars, J.E. and Vail, S.S. (1992) Alfalfa seed water management. I. Crop reproductive development and seed yield. *Crop Science* 32, 476–481.

Taylor, S.A., Haddock, L. and Pedersen, M.V. (1959) Alfalfa irrigation for maximum seed production. *Agronomy Journal* 51, 357–360.

Taylor, A.J. and Marble, V.L. (1986) Lucerne irrigation and soil water use during bloom and seed set on a red-brown earth in southeastern Australia. *Australian Journal of Experimental Agriculture* 26, 577–581.

Thomas, R.G. (1961) The influence of environment on seed production capacity in white clover (*Trifolium repens* L.). I. Controlled environment studies. *Australian Journal of Agricultural Research* 12, 227–238.

Valle, O. and Bergt, K. (1965) Stem-shortening experiments on red clover with growth regulators. *Acta Agralia Fennica* 104(1), 119.

Yamada, H., Henderson, D.W., Miller, R.J. and Hoover, R.M. (1973) Irrigation water management for alfalfa seed production. *California Agriculture* 27, 6–7.

Zaleski, A. (1966) White clover seed investigations. III. Effect of irrigation in seed production. *Journal of Agricultural Science* 67, 249–253.

Pollination, Fertilization and Pollinating Mechanisms in Grasses and Legumes

D.T. Fairey,[1] S.M. Griffith[2] and P.T.P. Clifford[3]

[1]*Northern Agriculture Research Centre, Agriculture and Agri-Food Canada, Beaverlodge, Alberta T0H 0C0, Canada;* [2]*NFSPRC, USDA-ARS, Oregon State University, Corvallis, Oregon 97331, USA;* [3]*AgResearch, PO Box 60, Lincoln, New Zealand*

7.1 INTRODUCTION

There are many important sequences of events, from inflorescence development to seed physiological maturity, which have a significant bearing on the outcome of the final seed harvest. At each stage or sequence, genetic and environmental factors can degrade or enhance this outcome. Some events are more sensitive to these factors than others. In many respects, our knowledge of these factors and the mechanisms by which they exert their influence in forages are poorly understood.

Many aspects of the floral biology of temperate grasses and legumes have been extensively reported elsewhere (Cleveland, 1985; Thomas, 1987; Owens, 1992). This chapter presents information on the pollination and fertilization of grasses and legumes, and the pollinating mechanisms that permit successful seed production in these species.

7.2 POLLINATION

Pollination is the transfer of pollen grains from the anther of a stamen to the receptive part of a carpel, the stigma. In most cases, the agency by which the pollen is transferred is one of two kinds, by insects (for legumes bearing entomophilous flowers) or by wind (for grasses bearing anemophilous flowers). The characteristics that are of importance to pollination in both these types of flowers can be briefly summarized as follows (Vines and Reese, 1964; Gill and Vear, 1966):

Entomophilous flowers are adapted to attract insects. They are usually large, brightly coloured and often scented. If the flowers are small, they are aggregated in numbers to form a broad expanse of colour. The flowers provide a source of food for insects which imbibe sugary nectar from the nectaries or collect the pollen, or both. The situation of the nectaries, anthers and stigmas within the flower is such that if the nectar or pollen has to be reached by the insect, the anthers and stigma are disturbed, usually to effect pollination. Relatively small quantities of pollen are produced but the pollen grains are usually thick walled, sticky and spiny; these are special features to adhere to insect bodies.

In legumes, surplus flowers, which may eventually abort or abscise, may serve as important attractants to pollinators which are obligatory for successful pollination (Wilson and Price, 1977). This, in turn, increases the amount and distribution of pollen. Florets which abort play an important role as pollen donors (Silverton, 1982) particularly in legume species because they are frugal pollen producers.

By comparison, anemophilous flowers are often small and inconspicuous. They are not scented and do not develop nectaries. They produce very large amounts of pollen to maximize the chances of pollination. The stigmas are often large, feathery and very sticky, and extend well outside the flower, making pollen more likely to reach them. The pollen grains are usually very tiny, smooth walled, dry and light.

Once upon the stigma surface, pollen grains germinate and form pollen tubes which penetrate and grow through the stigma tissue and enter the ovary. The pollen tube eventually enters the micropyle and makes contact with the female gametophyte. The resulting differences in turgor pressure between the pollen tube and female gametophyte cytoplasm cause the pollen tube to burst, discharging its nuclei.

In grasses, pollination begins with the two lodicules enlarging and glumes opening (Elgersma and Sniezko, 1988). Stamens and stigmas emerge together and pollen is released from the anthers within minutes after emergence from the glume. This usually occurs in more than one floret within a spikelet in the same day, but is controlled by climatic conditions (see Hill, 1980). In forage legumes, anther dehiscence mainly occurs during the bud stage before the flower petals are fully elongated (Thomas, 1987) but seed set, with some exceptions (Atwood, 1943), requires the aid of a pollinating agent.

7.2.1 Pollen Production and Viability

Grasses generally produce more pollen than legumes; pollen release is cyclic and lasts for defined periods of time (see Faegri and van der Pijl, 1980). Pollen production varies with species, climate and time of day (e.g. Jones and Brown, 1951). Tall fescue (*Festuca arundinacea* Schreb.) grown in Oklahoma in the USA, starts to shed pollen at 1100 h and peaks around 1430 h. Bromegrass (*Bromus inermis* Leyss.) is a species that, under optimal conditions, produces large quantities of pollen. In contrast, switchgrass (*Panicum virgatum* L.) and Johnsongrass (*Sorghum halepense* L.) produce pollen in very sparing amounts.

Under controlled conditions, pollen sterility in white clover (*Trifolium repens* L.) is strongly affected by temperature (Thomas, 1961). Sterile pollen percentages averaged 10.8% at 30°C, 11.2% at 20°C and 42.0% at 10°C. These data support field data collected at Palmerston North, New Zealand, where white clover pollen sterility was highest in the winter and lowest in the summer (Thomas, 1981). Temperature also plays a role in grass pollen viability. Under controlled environment conditions, both the genotype of the maternal plant and temperature affected pollen grain performance in perennial ryegrass (*Lolium perenne* L.) (Elgersma *et al.*, 1989).

Environmental factors, namely temperature, humidity and salinity, can adversely affect pollination processes (Elgersma *et al.*, 1989). Low temperature during pollination can reduce seed yield in perennial ryegrass (Hampton and Hebblethwaite, 1983) and pollen grain performance (Elgersma *et al.*, 1989). These authors concluded that daily fluctuations in temperature during pollination could conceivably alter the occurrence of self-pollen and cross-pollen fertilizations. This would potentially impact both population genetics and the resulting seed yield. The genotypic differences in the degree to which cross-pollination is successful or the degree to which the stigma is receptive may explain the extent to which potential seed yield is realized (Elgersma *et al.*, 1989).

7.3 FERTILIZATION

Fertilization is a fusion of male and female gametes. One male gamete fuses with the egg cell to form the zygote which develops into the embryo and the other male gamete fuses with the central cell to form the endosperm. This method of sexual reproduction (amphimixis) occurs in most species. In some temperate grasses another form of reproduction, apomixis, that does not involve nuclear or cellular fusion, occurs. In apomicts the chromosome number in the female gamete is not reduced and the embryo is formed without fertilization. Apomixis is a form of vegetative reproduction through seed. Dioecious species such as *Poa* spp. primarily use this mode of reproduction (see Chapter 10). Apomixis, like self-fertilization, would be an evolutionary dead-end if it were completely obligate (Berg, 1972). However, most species that were originally thought to be obligately apomictic were subsequently found to be simultaneously facultatively apomictic (Clausen, 1954). True apomixis rarely occurs in nature, if at all. Facultative apomixis occurs when both modes of reproduction, sexual and apomictic, occur; these modes have been shown to occur simultaneously and independently (Sherwood *et al.*, 1980). The advantages and disadvantages of breeding and seed multiplication of apomictic and facultatively apomictic cultivars are discussed in Chapter 10. The type of pollination in some grass species is presented in Table 7.1

7.4 SELF- AND CROSS-POLLINATION

Self-pollination is common in most annual species whereas many perennial species are cross-pollinated (Evans, 1964). Self-pollination and the annual growth habit are considered more evolved or advanced than cross-pollination and the perennial growth habit (Stebbins, 1957). Masses of pollen are shed very close to the stigmatic surface in self-pollinating species; this apparently contributes to selfing.

The rate (t) of outcrossing has been determined in a number of self-incompatible and self-compatible species. Self-incompatible species such as Italian ryegrass (*Lolium multiflorum* Lam.) and perennial ryegrass (Spoor, 1976) both have rates of t = 0.92 (Arcioni and Maritti, 1983), whereas in a self-compatible species such as annual bluegrass (*Poa annua* L.) the rate is t = 0.15 (Ellis, 1974).

Table 7.1. The type of pollination in a number of grasses (Gould, 1968; Canadian Seed Growers Association, 1986).

Species	Pollination[1]
Bluegrass	
(Big) *Poa ampla* Merr.	A
(Canada) *P. compressa* L.	A
(Kentucky) *P. pratensis* L.	A
(Rough) *P. trivialis* L.	A
Bromegrass	
(Downy) *Bromus tectorum* L.	SP
(Meadow) *B. biebersteinii* Roem and Schult.	CP
(Smooth) *B. inermis* Leyss.	CP
Fescue	
(Chewings) *Festuca rubra* var. *commutata* Gand.	CP
(Creeping red) *F. rubra* L.	CP
(Hard) *F. ovina* var. *duriuscula* (L.) Koch	CP
(Meadow) *F. pratensis* Huds.	CP
(Sheep) *F. ovina* L.	CP
(Tall) *F. arundinacea* Schreb.	CP
Foxtail	
(Creeping) *Alopecurus arundinaceus* Poir.	CP
(Meadow) *A. pratensis* L.	CP
Cocksfoot	
Dactylis glomerata L.	CP
Reed Canarygrass	
Phalaris arundinacea L.	CP
(Altai wild) *Leymus angustus* (Trin.) Pilger	CP
Ryegrass	
(Darnel) *Lolium temulentum* L.	SP
(Italian) *L. multiflorum* Lam.	CP
(Perennial) *L. perenne* L.	CP
Psathyrostachys juncea (Fisch.) Nevski	CP
Timothy	
Phleum pratense L.	CP
Wheatgrass	
(Intermediate) *Elytrigia intermedia* (Host) Nevski	CP
(Pubescent) *E. intermedia* subsp. *trichophora* A. and D. Löve	CP
(Slender) *E. trachycanulus* (Link.) Gould ex Shinners	SP
(Tall) *E. pontica* (Podp.) Holub	CP
(Western) *Pascopyrum smithii* (Rydb.)	CP

[1] A, apomictic; CP, cross-pollinated; SP, self-pollinated.

7.4.1 Mechanisms for Self-incompatibility

Most grasses and legumes are cross-pollinated and self-incompatible. Self-fertility occurs at a very low frequency primarily due to inhibition of pollen tube growth (Brink and Cooper, 1938) and lack of fertilization (Cooper and Brink, 1940). The frequency of ovule abortion is greater in zygotes and embryos resulting from self- than from cross-pollination (Cooper and Brink, 1940).

Pollen tube growth is influenced by the pollen receptivity of the stigma. In a self-incompatible species such as perennial ryegrass, pollen tube penetration and growth is more rapid following cross-pollination as compared with selfing (Elgersma *et al.*, 1989). Impregnation of ryegrass pollen tubes into the stigma surface following a self-pollination event is inhibited or blocked due to callose formation (Lundqvist, 1961). Although self-incompatibility is strong in perennial ryegrass, it is not obligate, and thus at low frequency mature seeds result (Foster and Wright, 1970; Cornish *et al.*, 1980). Apparently, force selfing of self-incompatible plants can result in the selection of genotypes with improved selfing performance (Jones and Jenabzadeh, 1981).

The probability of selfing in lucerne is reduced by a cuticular membrane that forms a film over the stigma (Kreitner and Sorensen, 1984). This membrane is broken by the pollen-covered body of a tripping agent (a bee) that transfers pollen from another plant on to the newly receptive stigma. Lucerne pollen grain germination and pollen tube growth are influenced by both environment and genetic factors.

In clovers (*Trifolium* spp.) it appears that pollen will not germinate until mucilaginous secretion begins at the stigma surface (Baker and Williams, 1987). This results following the rupture of the stigma cuticle surface, usually after one or several visits from a pollinating insect. Similar to grass pollen, legume pollen germinates quickly when in contact in the mucilaginous stigma surface (Chubirko, 1965). Concrete evidence for differences in the rates of germination between compatible and incompatible pollen in clover is not available. Compatible pollen has been shown to germinate faster than incompatible pollen (Atwood, 1941) but conflicting evidence exists (Chen and Gibson, 1973). The rate of pollen tube growth is faster with cross-pollination of compatible pollen (Atwood, 1941); incompatible pollen tubes often fail to grow out of the style into the ovary in white clover. Pollen tube growth is temperature dependent. Rates increase with rise in temperature up to 35°C. Temperature also plays a role in grass pollen viability. Under controlled environment conditions, both the genotype of the maternal plant and the temperature affect pollen grain performance in perennial ryegrass (Elgersma *et al.*, 1989).

A large portion of white clovers are self-incompatible, while a very small number are strongly self-compatible (Atwood, 1940, 1941). Thus, cross-pollination by bees is of utmost importance for most white clovers, but is not a prerequisite in some species/genotypes (Atwood, 1943). Selfing of self-incompatible white clover can result in 0–5.7 seeds per flower head (Atwood, 1941; Thomas, 1987). This wide range in frequency may be attributed to differences in temperature and genotype (Baker and Williams, 1987).

While pollination is not a major problem in white clover, it can be in red clover (*Trifolium pratense* L.) and lucerne which are largely self-incompatible, despite some adaptations to facilitate cross-pollination such as the continued receptivity of the stigma for up to 2 weeks if the flower is not visited by a bee (Walton, 1983). To compensate for a single visit of an insect being insufficient, some clovers such as red clover and alsike clover (*T. hybridium* L.) have evolved a mechanism for the floret to be visited more than once. The pistils and stamens of these species protrude only an instant after tripping, after which they move back to their original positions. A second or third visit by an insect will have the same effect, and the chances of the pistil being properly fertilized will last as long as it remains in a condition to receive pollen. This is not the case with lucerne where each floret has only one chance to be fertilized, as the stamens and pistils do not return to their original positions.

On the other hand, some floral characteristics discourage insect visitation. A very high proportion of honey bees visiting lucerne flowers fail to pollinate them (see Free, 1993). The tripping force of the staminal column strikes the underside of the bee's head, momentarily trapping it between the staminal column and the standard petal. To avoid being struck on the head, the bee learns to insert its head between the standard and wing petals, but it then fails to trip the flower on most visits. Honey bees cannot reach the nectaries in the flowers of tetraploid red clover (Dennis, 1980) and learn to avoid them in favour of diploid cultivars or alsike clover (*T. hybridium* L.), which have shorter corolla tubes than tetraploids (Julén, 1975). Preferential pollination by bees can cause shifts in populations and is an important consideration for cultivar maintenance (see Chapter 10).

7.5 SEED SET

Seed yield in grasses and legumes is low and often unpredictable. This has partly resulted from intensive breeding for agronomic performance and not for seed yield *per se* (Hides and Desroches, 1990). Seed set can be defined as the commencement of early growth of the embryo and endosperm. Much of the potential seed yield is lost during this period. A common theme among all species is to produce more potential offspring than can be supported by the mother organism or the environment (Harper, 1977). This is true in grass and legume crops which often show a high potential for seed production but never realize that full potential (see Elgersma, 1985; Bawa and Buckley, 1989). Realized seed yields in ryegrass require less than 30% of the potential seed sites and infertile or unpollinated sites can account for much of this loss (Knowles and Baenziger, 1962; Burbidge *et al.*, 1978). Other factors such as pollination, fertilization, seed set, seed development, harvesting and cleaning also contribute to high losses (Hill, 1980). An explanation of seed losses at each of these steps is discussed in detail in other chapters. No one process contributes more than another in reducing final seed yield but successful or unsuccessful seed set has important implications with regard to the survival and vigour of following generations and for seed quality.

The success of seed set varies from species to species, but is often related to the type of plant life cycle which is inherent in that species (Wiens, 1984; Wiens *et al.*, 1987). For example, annual species have high rates of seed set while perennials do not. In darnel or annual bearded ryegrass (*L. temulentum* L.), seed set success can range from 88 to 98% (Beddows, 1931), whereas in perennial ryegrass seed set has been reported to be less than 60% (Hill, 1980). This demonstrates the characteristic nature of perennial species to abort high numbers of ovules. Several factors have been attributed to cause ovule abortion (Burbidge *et al.*, 1978; Elgersma, 1985). These range from hormonal inhibition and assimilate limitation to deleterious genetic factors (Weins *et al.*, 1987; see Chapters 4 and 10).

7.6 POLLINATING MECHANISMS

7.6.1 Wind Pollination

A large proportion of the grasses are wind pollinated. With a few exceptions, this is a generally wasteful process, as the transfer of pollen is non-directional (Faegri and van der Pijl, 1980). Wind-pollinated plant species compensate for this relatively inefficient method of pollen distribution by producing large quantities of dry, low-density pollen (for some species in excess of 452,000 pollen grains per anther) to ensure fertilization. This is in sharp contrast to the insect-pollinated species which produce small quantities (e.g. up to 20,000 pollen grains per anther) of heavy, sticky pollen (Pohl, 1937). However, aerodynamic studies have shown that wind pollination in grasses is not a completely random process of pollen capture (Niklas, 1985). To capture air-borne pollen grains, grasses have been modified to function in an aerodynamic environment, albeit in a relatively passive way, because the plants are in a fixed position. Grass inflorescences can be described as either closed or tight, where numerous spikelets are densely clustered, or open and loose, where the peduncles are long. The aerodynamic properties of these two types are dramatically different. Closed types create substantial irregular eddies on the leeward side, while open types produce much less leeward turbulence. Preliminary studies indicate that the grass inflorescence shows a remarkable morphological variation in size, shape and the extent to which flowers can move in air currents, e.g. with the aid of large feathery stigmatic surfaces, pendulous stamens and hollow stems which cause flowers to sweep through greater volumes of airspace, enhancing the probability of colliding with airborne pollen (Faegri and van der Pijl, 1980).

Another consideration in wind pollination is the distance over which pollen must be transported, the duration of the viability of the pollen, and the receptivity of the stigma on receipt of the pollen. Grasses have a clonal habit and, when cultivated for seed production, are maintained as large, isolated monocultures. Under these conditions, the supply of pollen is often intermittent and does not always coincide with receptivity of the stigmatic surfaces. For example, when pollination is delayed in perennial ryegrass (Gregor, 1928), the flowers close and, after a certain time has elapsed, they may re-open (or partially re-open) for several days. However, the stigmas are receptive only during the early part of this period. This lack of synchrony between pollen viability and stigma receptivity is often the cause of floret sterility.

7.6.2 Insect Pollination

For most temperate, perennial, legume species, pollination is obligate on the adequate supply of an insect's food requirements. In commercial seed production, that insect is normally a bee. The bee obtains energy in the form of sugars (glucose, sucrose, fructose) from the flowers' nectar, and proteins, fats, vitamins and minerals from the pollen; these major resources are associated with both self-preservation and larval food requirements.

Bee species effectiveness (visitation rate) varies in relation to the product being collected. The length of the visitation is shortest for nectar, intermediate for mixed loads of pollen and nectar, and longest for pollen only (Free, 1965). Among bee genera, visitation rates vary according to the requirements of the bee species. The proportion of flower visits for nectar collection is greater for honey bees (*Apis* spp.) because of the requirement to maintain a colony over the winter. In comparison, bumble bees (*Bombus* spp.) do not require abundant stores for the winter. The workers die at the end of the growing season and the young queens hibernate until nature provides them with a fresh store of provisions the following spring. Solitary bees (*Megachile, Osmia* spp.) collect nectar for self-preservation and for their offspring. In white clover, honey bees and bumble bees have been shown to probe more florets per inflorescence than *Megachile* spp. and their visits produced more seed (Marshall *et al.*, 1995). However, honey bee visitation, as compared to bumble bee visitation, gave more seeds per inflorescence, and per pod, due to improved homogeneity of pollen loads. Thus, species-specific foraging patterns affect pollination efficiency.

In honey bees, crop identification for foraging is performed by scout bees and is communicated to other hive workers by way of an orientation dance (Lindauer, 1971). In contrast, bumble bees appear to have no inherent guidance system; they are initially attracted from afar by 'colour-block' vision (Brian, 1954). Consistency of revisitation will only occur after sampling acceptance is established. The latter appears to be the case with solitary pollinators as well.

The preferences shown by pollinating insects to certain plant genera/species is governed by effort/reward activity (Morse, 1978), which is established by the anatomy of the bees' mouth parts in relation to that of the plants' flowers. Thus, long-tongue bumble bee species prefer the longer corolla tubes of red clover flowers, whereas short-tongue bumble bees prefer alsike clover and lucerne that have short corolla tubes (Fairey *et al.*, 1992). Another factor that appears to assist in keeping a particular bee species on a candidate legume crop is the removal of other flowering species in the vicinity of legume seed fields. In instances where honey bees are used for pollination, care must be taken when growing crops of different genera, but similar flowering patterns, in close proximity. The New Zealand experience (P.T.P. Clifford, AgResearch, Christchurch, New Zealand, 1996, personal communication) indicates that phacelia (*Phacelia tanacetifolia* Benth.) and borage (*Borage officinalis* L.) are preferred to all legumes, while white clover and birdsfoot trefoil are visited to the exclusion of chicory (*Cichorium intybus* L.). For *Megachile* spp., birdsfoot trefoil (*Lotus corniculatus* L.) is preferred over lucerne (Fairey, 1993). These observations indicate that a knowledge of similar flowering pattern/bee species preference is important for successful seed production.

Nectar availability, accessibility and concentration *per se* are a function of plant management within the immediate environment. In red clover, nectar secretion is dependent on the presence of a sugar surplus over that needed for growth, respiration and other concurrent processes (Shuel and Pedersen, 1952). Nectar secretion is directly related to the solar radiation reaching the plant, and its concentration is modified by relative humidity (Shuel, 1952). These responses are similar for white clover and certain other plants (Oertel, 1946). In contrast to both honey bees and solitary bees, bumble bees forage with much less regard for individual thermal cost (lower temperatures) to gain higher nectar rewards (Willmer, 1983). Therefore, adequate temperatures to promote flight-energy cost efficiencies are a prerequisite in the location of the regional sites that are selected for legume crops for pollination by the former two genera.

In the absence of alternative satisfactory food sources, bees will rob (non-floret entry) nectar at the expense of pollination. *B. terrestris* will bite holes at the base of red clover florets, which are subsequently used by honey bees as secondary robbers. Similar happenings occur in *Vicia* spp. (Morse, 1978).

While insect pollination is the paramount step in the evolution of a mature seed capable of recreating its parents, postfertilization ovule-provisioning limitations may diminish its overall effectiveness. Thus, plant management deficiencies *per se* frequently abort pollination success. This feature has been examined in tetraploid red clover where higher seed yields were produced at a row spacing of 60 cm as compared to that at 15 cm (Clifford and Scott, 1989). As a consequence, of equal concern for all insect-pollinated crops is the managerial requirement of pollination enhancement through an effort/reward advantage, followed by adequate nutrient partitioning to all fertilized ovules to maximize fruition.

7.7 PRACTICAL MANAGEMENT OF INSECT POLLINATORS

7.7.1 Honey Bees (*Apis mellifera* L.)

Life history

Honey bees can be made available readily for the pollination of large crop areas, and are the most widely used pollinators for forage legumes. A honey bee colony consists of several thousand morphologically identical sterile females (the workers), one fertile female (the queen), a few hundred males (the drones), and a series of hexagonal cells containing brood, honey and pollen (for further details see Seely, 1985; Gould and Gould, 1988; Roubik, 1989). A hive (Fig. 7.1) consists of a wooden box containing a series of removable wooden frames in which the bees build their combs. A removable roof and floorboard encloses the hive. The floor board contains an aperture through which the bees enter and leave the hive. Eggs are laid singly, one at the bottom of each cell, and are hatched into larvae after about 3 days. The larvae are fed on a special diet rich in protein (from pollen), honey (from nectar) and water. After 5 days of feeding, each larva develops into a pupa, and the workers build a canopy of wax to completely enclose the cell. An adult worker bee emerges 13–15 days later.

Fig. 7.1. Honey bee hives in a red clover seed field.

There is a very distinct division of labour in a colony. The queen bee lays eggs, the drones fertilize virgin queens and the workers, as the name implies, have a number of duties (cleaning cells, building combs, hive repair, foraging . . .) depending on their age. There is an overlap in the ages at which each task is done and the number performing these tasks are adjusted according to the requirements of the colony.

Management for pollination

It is important that colonies used for pollination fulfil the minimum size requirements. Brood rearing is greatly diminished during the winter and gradually increases in early spring, reaching a peak in summer. Large colonies have a greater foraging population at all times and are generally preferred for pollination. However, small colonies are sometimes more useful because deterioration in foraging conditions discourages foraging relatively more with large colonies than with small ones (Free and Preece, 1969).

It is important to provide an adequate stocking rate of active colonies within a legume crop; the recommended stocking rates for different species (Crane and Walker, 1984) are given in Table 7.2. These stocking rates are guidelines and could vary with a number of factors such as the size of the field or the presence of native pollinators. Colonies must be removed soon after the crop being pollinated has ceased flowering, to avoid the depletion of pollen and nectar from nearby native flowering plants that provide sustenance to native pollinators nesting in the vicinity of legume seed fields (Slessor *et al.*, 1988; Williams *et al.*, 1991).

Lack of attention to a number of details often results in ineffective pollination of legume crops by honey bees. Many of these problems are related to a mismatch in the requirements of the crop and insect pollinator at different times during the

Table 7.2. Recommended stocking rate (honey bee colonies ha^{-1}) to pollinate forage legumes; only those species for which recommendation data are available are presented (Crane and Walker, 1984).

Crop	Stocking rate[1]
Clovers (*Trifolium* spp.)	
Alsike (*T. hybridium* L.)	Range: 2–8
Arrowleaf (*T. vesiculosum* Savi.)	2.5
Crimson (*T. incarnatum* L.)	2.5–12
Persian (*T. resupinatum* L.)	Range: 4–5
Red (*T. pratense* L.)	Range: 3–15
White (*T. repens* L.)	Range: 2–3
Lucerne (*Medicago sativa* L.)	Range: 5–10
Sainfoin (*Onobrychis viciifolia* Scop.)	Range: 2–10
Trefoil, birdsfoot (*Lotus corniculatus* L.)	2
Hairy vetch (*Vicia villosa* Roth)	Range: 1–2
Sweet vetch (*Hedysarum coronarium* L.)	Range: 5–8

[1]Range; number used depends on competing bloom and size of field.

growing season. Honey bees learn to recognize specific types of flowers by their colour, shape and odour (Free, 1970; Gould and Gould, 1988) and, when they are working one type of flower, are not likely to change to another quickly. Honey bees introduced into a legume crop before sufficient bloom is present will drift elsewhere to more attractive pasture. This occurs often with forage legumes because these crops usually flower relatively late in the season, after a number of surrounding native species are already in bloom. Free *et al.* (1960) demonstrated that when honey bees were introduced into birdsfoot trefoil, lucerne and red clover seed fields before and after these crops flowered, proportionally more bees (up to 12 times as many) introduced after legume bloom than the pre-bloom introduced bees visited the crop on the first day.

Control of parasites

Both the varroa mite (*Varroa jacobsoni* Oud.) and the tracheal mite (*Acarapis woodi* Rennie) are parasites of honey bees. The varroa mite, which is an external parasite, is considered the most serious pest affecting the honey bee (Fries *et al.*, 1994), reducing bee weight, longevity and foraging activity (DeJong *et al.*, 1982; DeJong and DeJong, 1983). Unless appropriate control measures are applied (Table 7.3), most honey bee colonies will not survive an infestation.

The tracheal mite lives in the trachea of adult bees, feeding on the bee's blood. These mites increase colony mortality, and reduce brood area and honey production (Eischen *et al.*, 1989). They can be controlled by the use of insecticides, chemicals such as formic acid and vegetable oils (Delaplane, 1996), although draining a colony of older bees reduces a large proportion of its tracheal mites because newly emerged bees are free of mites. This is achieved naturally by swarming (Royce *et al.*, 1993).

Table 7.3. Preparations in use to control varroa mites (Ritter, 1993).

Product name	Producer	Active ingredient	Type of product	Method of application
Api-Life-Var	LAIF	Essential oils	Plates	Evaporation
Apistan	Zoecon/Sandoz	Fluvalinate	Plastic strips	Hung in hive
Apitol	Ciba Geigy	Cymiazol	Aqueous solution	Trickled on
Bayvarol	Bayer	Flumethrin	Plastic strips	Hung in hive
Folbex-VA	Ciba Geigy	Bromopropylate	Fumigant strip	Fumigation
Formic acid	—	Formic acid	Aqueous solution on plates	Evaporation
Lactic acid	—	Lactic acid	Aqueous solution	Spray
Perizin	Bayer	Coumaphos	Aqueous solution	Trickled on

7.7.2 The Alfalfa Leafcutting Bee (*Megachile rotundata* (Fab.))

Leafcutting bees (*Megachile* spp.) are effective pollinators of forage legumes and in excess of 200 species are found in North America (Finnamore and Mitchner, 1993). However, these North American *Megachile* are not gregarious and to date have not been successfully increased to the large populations required for pollination on a commercial scale. The gregarious Eurasian species, *M. rotundata* (Fab.), often referred to as the alfalfa leafcutting bee, was unintentionally introduced into North America some time after the mid 1960s and is managed for lucerne pollination in Canada and the USA (Bohart, 1972). In the 1980s, the list of candidate crops was expanded to include diploid red clover (Fairey *et al.*, 1989), alsike clover (*T. hybridum* L.) (Fairey and Lefkovitch, 1993), birdsfoot trefoil (Pesson and Louveaux, 1984), cicer milkvetch (*Vicia astragalus rubyi* Greene and Morris) (Richards, 1987), sainfoin (*Onobrychis viciifolia* Scop.) (Richards, 1988) and oilseed rape (*Brassica campestris* L.) (Fairey and Lefkovitch, 1990). While the largest populations of the alfalfa leafcutting bee used in commercial seed fields currently exist in western Canada and the Pacific northwest in the USA, smaller populations are found in Europe and in New Zealand.

Life history

Large numbers of the alfalfa leafcutting bee can be made to nest in a given area with the provision of man-made nesting tunnels that are 5–7 mm in diameter (Stephen and Osgood, 1965). The female bees construct thimble-shaped cells (8–10 mm in length) in these tunnels from leaf cuttings, and provision them with nectar and pollen. An egg is then laid in the cell (Klostermeyer, 1982). A series of these cells are arranged linearly in each tunnel (Fig. 7.2d), with the first cell being placed at the back end of the tunnel. After all the cells in a tunnel have been constructed, the tunnel is capped with a number of leaf cuttings.

The egg hatches in its cell, the larva feeds on the nectar and pollen provisions, metamorphoses into the prepupal stage, and then spins a tough silken cocoon before becoming dormant for the winter. Adult bees die at the end of the growing season. The dormant pre-pupae spend the winter in the cocoon and can be conveniently stored at 5°C until the following spring when they are incubated at 30°C for 20–25 days before being released into legume seed fields.

Management for pollination

An annual schedule or 'calender' for the management of the alfalfa leafcutting bee in the northern hemisphere is presented in Table 7.4.

The management techniques for leafcutting bees in the northwestern USA, and in the southern and the northern growing areas in western Canada differ to some extent, mostly because of the differences in environmental conditions with increasing latitude. Most of the differences are in the types of materials provided for the bee to nest in, and the type of shelter that houses this material in the field. Nesting material is made from wood or polystyrene.

Solid nesting blocks are used predominantly in the USA while the 'loose-cell' system is used in Canada (Peterson *et al.*, 1992). In the solid-block system, the nesting tunnels do not go completely through the blocks and the progeny of the bees are allowed to overwinter in the tunnels. The blocks are taken to the field in the following growing season, and the adults that emerge from them re-use the nesting tunnels. However, this re-use of nesting blocks promotes a rapid increase in fungal diseases, particularly chalkbrood (*Ascosphaera* spp.), and parasites.

In the loose-cell system, grooved laminates that can be taken apart, or tunnels that go completely through the nesting material are used. The cells are removed from the nesting material at the end of the growing season. This system allows better control of diseases and parasites (Bohart, 1972).

Wood shelter panels are used in the northwestern USA and in the southern growing areas in western Canada (Hobbs, 1973; Stephen, 1981). The shelters in the northern growing areas are constructed with polyethylene or polyester (Pankiw and Lieverse, 1982; Fairey *et al.*, 1988). These 'northern' shelters

Table 7.4. A calender for the management of the leafcutting bee in the northern hemisphere. (Adapted from Fairey *et al.*, 1984.)

Months	Activity	Remarks
January to April	Obtain bee cells	A large proportion of the cells should be viable
May to June	Incubate cells	Commence incubation about 20–25 days before bees are required in field
June to July	Transfer bees to field	Bees are transferred to shelters in the field; nesting material is provided in the shelters
July to August	Bees in the field	Adult females gather pollen and nectar and in the process effect pollination
August to September	Nesting material containing bee cells is brought indoors to a temperature-controlled unit. Material gradually cooled down to 10°C before cells are removed	Adults die at the end of the growing season
October to December	Bee cells are removed from nesting material	Excess leaf material is removed from bee cells that are subsequently stored at 5°C

(Fig. 7.2a, b) utilize the 'greenhouse effect' to enhance temperatures and have been adapted by other countries such as Denmark (Holm, 1982), France (Tasei, 1982), New Zealand (Donovan *et al.*, 1982) and regions in the former USSR (Semin and Bourmistrov, 1982).

Control of diseases and parasites

Chalkbrood, a fungal disease of bee larva is caused by the fungus *Ascosphaera aggregata* Skou (Vandenberg and Stephen, 1982). In the USA, chalkbrood poses a

Fig. 7.2. (*and opposite*) Leafcutting bees in the field. a,b. 'Northern' shelters in the Peace River region, Canada. c. Leafcutting bee cells in nesting tunnels. d. Nesting tunnels in hive.

(c)

(d)

threat to the continued use of the alfalfa leafcutting bee (Peterson *et al.*, 1992), but the disease is less common in Canada. Several management practices, such as exposure of nesting material to sodium hypochlorite (Mayer *et al.*, 1988), high temperatures (Kish and Stephen, 1991), and paraformaldehyde or methyl bromide fumigants (Goerzen and Watts, 1991; Mayer *et al.*, 1991), can be used to reduce the incidence and spread of chalkbrood.

Parasites of leafcutting bees are predominantly from four genera, *Pteromalus*, *Monodontomerus*, *Tetrastichus* and *Melittobia*. An ultraviolet light placed over a container of water successfully attracts and drowns most parasites (Waters, 1970). Vapona (tradename) strips (2,2-dichlorovinyl dimethyl phosphate) (Hill *et al.*, 1984) are also effective.

7.7.3 Alkali Bees

The alkali bee (*Nomia melanderi* Ckll.) is a solitary, ground-nesting halictid that prefers to nest in beds of alkaline soil with little or no vegetation. Like the alfalfa leafcutting bee, the alkali bee is a native of the steppes of eastern Europe and central and southwest Asia and was accidentally introduced into the western states of the USA (O'Toole and Raw, 1991). Alkali bees are currently used for the pollination of lucerne in the intermountain states of the USA from central Utah to east-central Washington, and in the northern San Joaquin Valley of California (Ribble, 1968). The use of this pollinator has declined in most of the northwestern USA, partly as a result of difficulties associated with the maintenance of the bee beds, and also because of the widespread availability of alfalfa leafcutting bee cells. Unsuccessful attempts were made to introduce these bees into Montana and the Dakotas in the USA and into the Canadian province of Alberta; however, their importation and propagation into New Zealand has been successful (Donovan, 1979; Palmer and Donovan, 1980) and a small colony is currently maintained in the Marlborough flats in the South Island for lucerne pollination.

Life history

The alkali bee is a non-social, gregarious, ground-nesting species (Stephen, 1959). The adult bee emerges between late June to early July, about the time that lucerne blooms. The female mates after emergence and then begins to dig her main burrow. The next day, she prepares and provisions her first cell, and the following day she lays an egg on the pollen and seals the cell entrance. This pattern is repeated during the summer so that, on any given day, a female lays an egg in a cell provisioned the day before, provisions a new cell already excavated, and excavates a new cell. A complete nest then consists of a vertical shaft about 20 cm long leading from a mound of soil on the surface to a series of three or four branch burrows each about 5–8 cm long. The adults die at the end of the growing season. The pre-pupae overwinter in the burrow and complete their metamorphosis into pupae and adults in the spring of the following year.

Management for pollination of forage legumes

Alkali bees forage on several different species of crops and weeds (Packer, 1970), the primary legumes being lucerne and sweetclover (*Melilotus alba* Medik.;

M. officinalis Lam.). In lucerne seed fields, these bees initially forage on plants close to their nesting beds and progressively extend their foraging territory in search of unpollinated flowers, sometimes in excess of 3 km by the end of the growing season (Stephen, 1959).

Artificial nesting sites can be modelled after the natural nests (Johansen *et al.*, 1978). Natural bee beds can be moved to crop sites, or artificial beds can be 'seeded' with cores of soil prepupae obtained by driving steel cylinders into existing bee beds in the spring and autumn. The emerging bees nest in their own soil cores and in the surrounding soil.

Control of diseases and parasites

The bee fly (*Heterostylum robustum* (O.S.)) and the thick-headed fly (*Zodion obliquefasciatum* (Macq.)) are the predominant parasites of the alkali bee. Adult bee flies are often controlled by swatting them, stepping on teneral adults, or by insecticide stakes (Bohart *et al.*, 1960). No controls have been developed for the thick-headed fly which parasitizes adult bees. Other factors that cause a reduction in alkali bee populations are spoilage of the pollen by yeasts and fungi, flooding, excessive dryness, or a hard crust of sodium chloride on the soil surface.

7.7.4 Bumble Bees

Life history

Young, mated bumble bee queens hibernate during winter in small burrows in the soil. The rest of the colony, including the old queen, die in the late autumn (see Free and Butler, 1959; Heinrich, 1979).

The young queen emerges in the spring and forages on plants in bloom. She eats large quantities of pollen and nectar while her ovaries develop, and commences her search for a nest. She moulds the pollen into a mass, and on this pollen mass builds a cell of wax and lays about 8–10 eggs in it. The next stage of colony development finds the queen home bound to brood her eggs that hatch after 4–6 days. The larvae feed on pollen and nectar and remain together in their cell; after 10–12 days they separate from each other by spinning a silken cocoon around themselves. Metamorphosis into a pupa occurs inside the cocoon. The queen scrapes the pollen-wax mixture from the cocoons and uses it to make more egg chambers. Adult workers emerge after about 2 weeks as pupae and, after about 24 h, begin foraging and helping the queen tend the brood. More workers continue to be produced until there are enough of them to forage for the colony. After this occurs, the queen devotes herself to full-time egg laying and brooding.

After the colony is mature, some eggs develop into males and females instead of workers. The causes for this change are not well understood. Queens and males are the last brood to be reared, and are the colony's contribution to the next generation. The young queens may continue to return to the nest for a while even after mating; they eat large quantities of nectar and pollen and build up a fat body before each seeks a site to overwinter.

Management for pollination

The bumble bee is perhaps one of the most efficient pollinators of many legumes. However, they are usually too few to pollinate large areas of agricultural crops. Various suggestions to increase bumble bee populations have been made (e.g. Free and Butler, 1959; Holm, 1966), but the value of most of these suggestions in legume seed fields has yet to be tested.

Wild bumble bee populations have been built up around legume seed fields with the provision of artificial nesting boxes/domiciles for the young queens in the early spring. In Canada, Hobbs *et al.* (1962) used wood boxes containing upholsterer's cotton to increase nesting of four species of bumble bees, *B. borealis* Kirby, *B. fervidus* Fabricius, *B. huntii* Greene and *B. nevadensis* Cresson; these species are indigenous to the prairies and are useful pollinators of red clover. This procedure has been carried one-step further in New Zealand where the transfer of nesting boxes (Fig. 7.3) from various locations into red clover seed fields, in addition to the provision of empty nest boxes for young nesting queens in the early spring, has been recommended as a simple and effective way to increase bumble bee populations (Donovan and Wier, 1978).

In New Zealand, a three-fold increase in local bumble bee populations in red clover fields was obtained in three successive years by the introduction of 100 queens in the spring of each year (Clifford, 1973). However, there was no carry-over effect from one year to the next, and when no further queens were introduced, the density of the local bumble bee population decreased to pre-release levels. McFarlane *et al.* (1983) obtained a red clover seed yield of 210 kg ha^{-1} in a year when queens were released but it declined to 100 kg ha^{-1} in the two subsequent seasons. The decline in seed yield was attributed to low bumble bee populations because of insufficient nesting sites and early spring bloom prior to the commencement of flowering of red clover.

7.7.5 Mason Bees

Mason bees (*Osmia* spp.) belong to the family Megachilidae and are excellent pollinators of forage legumes (Tasei, 1972, 1976; Parker, 1981). However, management techniques for a commercial scale of operation have not been devised.

Life history

Mason bees nest in the wild in tubular burrows in wood or plant stems. They construct cells in these tubes with materials such as mud, resin or dung that they gather from the surrounding environment rather than from their own secretions. The female lays an egg in a cell that is provisioned with pollen and nectar for the developing larva. Adults die at the end of the growing season and their progeny overwinter in the dormant state and emerge the following spring.

Management for pollination of forage legumes

O. cornuta (Latr.) and *O. rufa* (L.) pollinate red clover and lucerne and utilize the pollen in nesting (Tasei, 1972, 1976; Parker, 1981). Successful pollination and seed set in lucerne in southern Nevada was obtained with another species,

Fig. 7.3. Nesting boxes for bumble bees that will be moved into legume seed fields in New Zealand.

O. sanrafaelae Parker, a native bee from the desert regions of southeastern Utah (Parker, 1986).

Osmia utilize artificial 'trap nests' that can be drinking straws, hollow bamboo or drilled wood boards, nest gregariously, overwinter in the dormant state, and their emergence can be timed to coincide with crop bloom. Parker (1986) has suggested that *O. sanrafaelae* appears to have many advantages over the alfalfa leafcutting bee as a pollinator of lucerne. *Osmia* populations overwinter as adults and require only a few days to emerge; the alfalfa leafcutting bee, on the other hand, requires about 3 weeks of incubation before emergence. Thus *Osmia* adult emergence could be easier to coordinate with the onset of flowering in the crop. Another important trait is that the number of generations produced in a growing season is only one for *Osmia*, whereas in the Pacific Northwest USA, there may be up to three generations per season for the alfalfa leafcutting bee. Consequently, large populations of the alfalfa leafcutting bee are lost or not effectively utilized because the later summer generation emerges after most of the lucerne seed crop has been pollinated. In addition, significant numbers of diapausing larvae are lost to emerging second-generation bees because cell series contain both diapausing and non-diapausing individuals (Tepedino and Frohlich, 1984).

7.7.6 Other Wild Bees

There are at least 20,000 described species of bees (O'Toole and Raw, 1991). However, the pollinating activity of just a few species, including honey bees, leafcutting bees, alkali bees, bumble bees, and some mason bees on temperate legumes, has been studied in some detail. There are a few reports on the contribution of wild bees that nest in the margins of seed fields; most of these reports are on either lucerne or red clover. Generally, the intensification of agriculture has had a large impact on feral pollinating insects and their distribution. The planting of large, solid tracts of cross-pollinated crops, and the frequent use of herbicides and insecticides on these same crops, has gradually reduced the habitats of wild pollinators, and domesticated or managed colonies of pollinators have been used to ensure pollination. There are few successful programmes that enhance native pollinator numbers for commercial application, such as apple pollination by mason bees in Japan (Maeta, 1978), despite the fact that enhancing native pollinator populations by habitat management could be a potentially cost-effective option (Corbet *et al.*, 1991). The successful use of wild bees depends on a number of factors such as the synchrony of peak flowering of the crop and wild bee numbers, the size of the field in relation to the proximity of natural vegetation used for nesting habitat, and the presence of competing flowering sources of nectar and pollen.

Parker *et al.* (1976) found that many genera of native leafcutting bees visit forage legume crops and are effective pollinators. Bohart (1957) listed about 75 species of wild bees that make a contribution to lucerne seed yields. Most of the wild bees that play a role in lucerne pollination in North America and Europe belong to *Andrena, Bombus, Eurera, Halictus, Megachile, Melitta* and *Xylocopa* spp. Only those with a body length greater than 6 mm can trip the lucerne flower. The small halictids and adrenids visit flowers that are already tripped and could help in cross-pollination (Linsley, 1946; Pengelly, 1953). The large impact that native pollinators can have on lucerne seed production is also evident from reports from Canada (Peck and Bolton, 1946; Stephen, 1955).

Wild bees do not appear to play a prominient part in the pollination of white clover (Green, 1957). However in New Zealand, Palmer-Jones *et al.* (1962) suggested that wild bumble bees play a role in the reseeding of white clover in permanent pastures, and *Anthidium punctatum* Latr. and *Melitta leporina* (Panzer) are very abundant in white clover fields in France (Lecomte and Tirgari, 1965).

7.8 CONCLUSIONS

Pollination and fertilization are essential steps in the formation of a mature seed. However, post-fertilization limitations often derail the success of the first two events. Greater attention should be paid to plant management because these deficiencies *per se* frequently influence seed production success. While the role of insect pollinators in legume seed production has been dealt with in some detail in the literature, scientific investigation on many potentially suitable bee species has been lacking. In some cases, consideration could be given to importing the natural pollinator with the introduced plant species; this occurred by accident with the alfalfa leafcutting bee in North America. Finally, measures to conserve native pollinators with proper attention to maintenance of their natural habitat could prove to be both a cost-effective and environmentally responsible strategy for enhancing pollination.

REFERENCES

Arcioni, S. and Maritti, D. (1983) Selfing and interspecific hybridization in *Lolium perenne* L. and *Lolium multiflorum* Lam. Evaluated by phosphoglucosisomerase as isozyme marker. *Euphytica* 32, 33–40.

Atwood, S.S. (1940) Genetics of cross-incompatibility among self incompatible plants of *Trifolium repens*. *Journal of the American Society of Agronomy* 32, 955–968.

Atwood, S.S. (1941) Controlled self- and cross-pollination of *Trifolium repens*. *Journal of the American Society of Agronomy* 33, 538–545.

Atwood, S.S. (1943) 'Natural crossing' of white clover by bees. *Journal of the American Society of Agronomy* 35, 862–870.

Baker, M.J. and Williams, W.M. (1987) *White Clover*. CAB International, Wallingford.

Bawa, K.S. and Buckley, D.P. (1989) Seed : ovule ratios, selective seed abortion, and mating in Leguminosae. In: Stirton, C.H. and Zarucchi, J.L. (eds) *Advances in Legume Biology. Monograph of Systematic Botany*, vol. 29. Missouri Botanical Gardens, pp. 243–262.

Beddows, A.R. (1931) *Seed Setting and Flowering in Various Grasses*. University of Wales, Series H, 12. Welsh Plant Breeding Station, Aberystwyth, pp. 5–99.

Berg, A.R. (1972) Grass reproduction. In: Younger, V.B. and McKell, C.M. (eds) *The Biology and Utilization of Grasses*. Academic Press, New York, pp. 334–347.

Bohart, G.E. (1957) Pollination of alfalfa and red clover. *Annual Review of Entomology* 2, 355–380.

Bohart, G.E. (1972) Management of wild bees for the pollination of crops. *Annual Review of Entomology* 17, 287–312.

Bohart, G.E., Stephen, W.P. and Eppley, R.K. (1960) The biology of *Heterostylum robustum* (Diptera: Bombyliidae), a parasite of the alkali bee. *Annals of the Entomological Society of America* 53, 425–435.

Brian, A.D. (1954) The foraging bumble bee. *Bee World* 35, 61–67.

Brink, R.A. and Cooper, D.C. (1938) Partial self-incompatibility in *Medicago sativa*. *Proceedings of the National Academy of Science, Washington DC* 24, 497–499.

Burbidge, A., Hebblethwaite, P.D. and Ivins, J.D. (1978) Lodging studies in *Lolium perenne* grown for seed. 2. Floret site utilization. *Journal of Agricultural Science* 90, 269–274.

Canadian Seed Growers Association (1986) *Regulations and Procedures for Pedigreed Seed Crop Production*, Circular 6–94. Canadian Seed Growers' Association, Ottawa, Ontario.

Chen, C-C. and Gibson, P.B. (1973) Effect of temperature on pollen-tube growth in *Trifolium repens* after cross- and self-pollination. *Crop Science* 13, 563–566.

Chubirko, M.M. (1965) Microsporogenesis and development of the male gametophyte in clover *Trifolium repens* L. *Botanicheski Zhurnal* 50, 1578–1584.

Clausen, J. (1954) Partial apomixis as an equilibrium system in evolution. *Proceedings of the 9th International Genetic Congress Carylogia* 6 (suppl.), 469–479.

Cleveland, R.W. (1985) Reproduction cycle and cytogentics. In: Taylor, N.L. (ed.) *Clover Science and Technology*. American Society of Agronomy, Madison, pp. 71–110.

Clifford, P.T.P. (1973) Increasing bumblebee densities in red clover seed production areas. *New Zealand Journal of Experimental Agriculture* 1, 377–379.

Clifford, P.T.P. and Scott, D. (1989) Inflorescence, bumble bee, and climate interactions in seed crops of a tetraploid red clover (*Trifolium pratense* L.). *Journal of Applied Seed Production* 7, 38–45.

Cooper, D.C. and Brink, R.A. (1940) Partial self-incompatibility and the collapse of fertile ovules as factors affecting seed formation in alfalfa. *Journal of Agricultural Research (Washington, DC)* 60, 453–472.

Corbet, S.A., Williams, I.H. and Osborne, J.L. (1991) Bees and the pollination of crops and wild flowers in the European community. *Bee World* 72, 47–59.

Cornish, M.A., Hayward, M.D. and Lawrence, M.J. (1980) Self-incompatibility in ryegrass. III. The joint segregation of S and PGI-2 *Lolium perenne* L. *Heredity* 44, 55–62.

Crane, E. and Walker, P. (1984) *Pollination Directory for World Crops*. International Bee Research Association, London.

DeJong, D. and DeJong, P.H. (1983) Longevity of Africanized honey bees (Hymenoptera : Apidae) infested by *Varroa jacobsoni* (Parasitiformes : Varroidae). *Journal of Economic Entomology* 76(4), 766–768.

DeJong, D., DeJong, P.H. and Goncalves, L.S. (1982) Weight loss and other damage to developing worker honeybees from infestation with *Varroa jacobsoni*. *Journal of Apicultural Research* 21(3), 165–167.

Delaplane, K.S. (1996) Practical science – research helping beekeepers. 1. Tracheal mites. *Bee World* 77(2), 71–81.

Dennis, B. (1980) Breeding for improved seed production in autotetraploid red clover. In: Hebblethwaite, P.D. (ed.) *Seed Production*. Butterworths, London, pp. 229–240.

Donovan, B.J. (1979) Importation, establishment and propagation of the alkali bee, *Nomia melanderi* Cockerell (Hymenoptera : Halictidae) in New Zealand. *Proceedings of the 4th International Symposium on Pollination. Maryland Agricultural Experimental Station Special Miscellaneous Publication* 1, 257–268.

Donovan, B.J. and Wier, S.S. (1978) Development of hives for field population increase, and studies on the life cycles of four species of introduced bumble bees in New Zealand. *New Zealand Journal of Agricultural Research* 21, 733–756.

Donovan, B.J., Read, P.E.C., Wier, S.S. and Griffin, R.P. (1982) Introduction and propagation of the leafcutting bee *Megachile rotundata* (F.) in New Zealand. In: Rank, G.H. (ed.) *Proceedings of the First International Symposium on Alfalfa Leafcutting Bee Management*, Saskatoon, Saskatchewan, August 16–18, 1982, University of Saskatchewan, Canada, pp. 212–222.

Eischen, F.A., Cardoso-Tamez, D., Williams, W.T. and Diet, A. (1989) Honey production of honey bee colonies infested with *Acarapis woodi* (Rennie). *Apidologie* 20(1), 1–8.

Elgersma, A. (1985) Floret site utilization in grasses: definitions, breeding perspectives and methodology. *Journal of Applied Seed Production* 3, 50–54.

Elgersma, A. and Sniezko, R. (1988) Cytology of seed development related to floret position in perennial ryegrass (*Lolium perenne* L.). *Euphytica* 59, 59–68.

Elgersma, A., Stephenson, A.G. and den Nijs, A.P.M. (1989) Effects of genotype and temperature on pollen tube growth in perennial ryegrass (*Lolium perenne* L.). *Sexual Plant Reproduction* 2, 225–230.

Ellis, W. (1974) The breeding system and variation in populations of *Poa annua* L. *Evolution* 27, 656–662.

Evans, L.T. (1964) Reproduction. In: Barnard, C. (ed.) *Grasses and Grasslands*. Macmillian, New York, pp. 126–153.

Faegri, K. and van der Pijl, L. (1980) *The Principles of Pollination Ecology*, 3rd edn. Pergamon Press, New York, pp. 34–40.

Fairey, D.T. (1993) Pollination and seed set in herbage species: a review of limiting factors. *Journal of Applied Seed Production* 11, 6–12.

Fairey, D.T. and Lefkovitch, L.P. (1990) Reproduction of *Megachile rotundata* Fab. foraging on *Trifolium* spp. and *Brassica campestris*. *Acta Horticulturae* 288, 185–189.

Fairey, D.T. and Lefkovitch, L.P. (1993) Pollination of *Trifolium hybridum* by *Megachile rotundata*. *Journal of Applied Seed Production* 11, 34–38.

Fairey, D.T., Lieverse, J.A.C. and Siemens, B. (1984) *Management of the Alfalfa Leafcutting Bee in Northwestern Canada*. Northern Research Group Publication 84–21, Agriculture Canada, Beaverlodge, Alberta.

Fairey, D.T., Lefkovitch, L.P., Lieverse, J.A.C. and Siemens, B. (1988) Materials for leafcutting bee (*Megachile rotundata* F.) shelters in north west Canada. *Journal of Applied Entomology* 106, 119–122.

Fairey, D.T., Lefkovitch, L.P. and Lieverse, J.A.C. (1989) The leafcutting bee, *Megachile rotundata* (Fab.): a potential pollinator of red clover. *Journal of Applied Entomology* 107, 52–57.

Fairey, D.T., Lefkovitch, L.P. and Owen, R.E. (1992) Resource partitioning: bumble bee (*Bombus*) species and corolla lengths in legume seed fields in the Peace River region. *Bee Science* 2(4), 170–174.

Finnamore, A.T. and Michener, C.D. (1993) Superfamily Apoidea. In: Goulet, H. and Huber, J.T. (eds) *Hymenoptera of the World: An Identification Guide to Families*. Agriculture Canada, Research Branch Publication, Ottawa, Ontario, pp. 279–357.

Foster, C.A. and Wright, C.E. (1970) Variation in the expression of self-fertility in *Lolium perenne*. *Euphytica* 19, 61–70.

Free, J.B. (1965) The ability of bumble bees and honey bees to pollinate red clover. *Journal of Applied Ecology* 2, 289–292.

Free, J.B. (1970) The effect of flower shapes and nectar guides on the behaviour of foraging honeybees. *Behaviour* 37, 269–285.

Free, J.B. (1993) *Insect Pollination of Crops*. Academic Press, London, pp. 247–297.

Free, J.B. and Butler, C.G. (1959) *Bumblebees*. Collins, London.

Free, J.B. and Preece, D.A. (1969) The effect of the size of a honeybee colony on its foraging activity. *Insects Society* 16, 73–78.

Free, J.B., Free, N.W. and Jay, S.C. (1960) The effect on foraging behaviour of moving honey bee colonies to crops before or after flowering has begun. *Journal of Economic Entomology* 53, 564–566.

Fries, I., Scott, C. and James, S. (1994) Population dynamics of *Varroa jacobsoni*: a model and review. *Bee World* 75(1), 5–28.

Gill, N.T. and Vear, K.C. (1966) *Agricultural Botany*. Gerald Duckworth & Company Ltd, London.

Goerzen, D.W. and Watts, T.C. (1991) Efficacy of the fumigant paraformaldehyde for control of microflora associated with the alfalfa leafcutting bee, *Megachile rotundata* (Fabricius) (Hymenoptera : Megachilidae). *Bee Science* 1, 212–218.

Gould, F.W. (1968) *Grass Systemics*. McGraw-Hill, New York.

Gould, J.L. and Gould, C.G. (1988) *The Honey Bee*. Scientific American Library, New York.

Green, H.B. (1957) White clover pollination with low honeybee populations. *Journal of Economic Entomology* 50, 318–320.

Gregor, J.W. (1928) Pollination and seed production in the ryegrasses (*Lolium perenne* and *Lolium italicum*). *Transcripts of the Royal Society of Edinburgh* 55, 773–794.

Hampton, J.G. and Hebblethwaite, P.D. (1983) The effects of the environment at anthesis on the seed yield and yield components of perennial ryegrass (*Lolium perenne* L.) cv. S24. *Journal of Applied Seed Production* 1, 21–22.

Harper, J.L. (1977) *Population Biology of Plants*. Academic Press, London.

Heinrich, B. (1979) *Bumble-bee Economics*. Harvard University Press, Cambridge, Massachusetts.

Hides, D.H. and Desroches, R. (1990) Role of seeds in forage production – factors limiting optimal utilization. *Proceedings of the XVI International Grassland Congress*, pp. 1777–1789.

Hill, M.J. (1980) Temperate pasture grass-seed crops: formative factors. In: Hebblethwaite, P.D. (ed.) *Seed Production*. Butterworths, London, pp. 137–149.

Hill, B.D., Richards, K.W. and Schaalje, G.B. (1984) Use of dichlorovos resin strips to reduce parasitism of alfalfa leafcutter bee (Hymenoptera : Megachilidae) cocoons during incubation. *Journal of Economic Entomology* 77, 1307–1312.

Hobbs, G.A. (1973) *Alfalfa Leafcutter Bees for Pollinating Alfalfa in Western Canada*. Agriculture Canada Publication 1495, Agriculture Canada, Ottawa, Ontario.

Hobbs, G.A., Nummi, W.O. and Virostek, J.F. (1962) Managing colonies of bumble bees (Hymenoptera : Apidae) for pollination purposes. *Canadian Entomologist* 94, 1121–1132.

Holm, S.N. (1966) The utilization and management of bumble bees for red clover and alfalfa seed production. *Annual Review of Entomology* 11, 155–182.

Holm, S.N. (1982) Management of *Megachile rotundata* for pollination of seed crops in Denmark. In: Rank, G.H. (ed.) *Proceedings of the First International Symposium on Alfalfa Leafcutting Bee Management*, Saskatoon, Saskatchewan, August 16–18, 1982, University of Saskatchewan, Saskatchewan, Canada, pp. 223–233.

Johansen, C.A., Mayer, D.F. and Eves, J.D. (1978) Biology and management of the alkali bee, *Nomia melanderi* Cockerell (Hymenoptera: Halictidae). *Melanderia* 28, 23–46.

Jones, M.D. and Brown, J.C. (1951) Pollination cycles of some grasses in Oklahoma. *Agronomy Journal* 43, 218–222.

Jones, R.N. and Jenabzadeh, P. (1981) Variation in self-fertility, flowering time and infloresence production in inbred *Lolium perenne* L. *Journal of Agricultural Science* 96, 521–537.

Julén, G. (1975) The current situation in the tetraploid clover. In: Nüesch, B. (ed.) *Ploidy in Fodder Plants*. Report of the meeting of the Fodder Crops Section of Eucarpia, 23–25 April, Zürich-Reckenholz, Switzerland, pp. 79–89.

Kish, L.P. and Stephen, W.P. (1991) Chalkbrood disease. In: *Alfalfa Seed Production and Pest Management*. Western Regional Extension Publication 12, University of Idaho, Moscow, Idaho, USA.

Klostermeyer, E.C. (1982) Biology of the alfalfa leafcutting bee. In: Rank, G.H. (ed.) *Proceedings of the First International Symposium on Alfalfa Leafcutting Bee Management*, Saskatoon, Saskatchewan, August 16–18, 1982, University of Saskatchewan, Saskatchewan, Canada, pp. 10–19.

Knowles, R.P. and Baenziger, H. (1962) Fertility indices in cross-pollinated grasses. *Canadian Journal of Plant Science* 42, 460–471.

Kreitner, G.L. and Sorensen, E.L. (1984) Stigma development and the stigmatic cuticle in alfalfa, *Medicago sativa* L. *Botanical Gazette* 145, 436–443.

Lecomte, J. and Tirgari, S. (1965) On some pollinators of the fodder legumes. *Annals Abeille* 8, 83–93.

Lindauer, M. (1971) The functional significance of the honeybee waggle dance. *American Naturalist* 105, 89–96.

Linsley, E.G. (1946) Insect pollinators of alfalfa in California. *Journal of Economic Entomology* 39, 18–28.

Lundqvist, A. (1961) A rapid method for the analysis of incompatibility in grasses. *Hereditas* 52, 705–707.

Maeta, Y. (1978) Comparative studies on the biology of the bees of the genus *Osmia* of Japan, with special reference to their management for pollination of crops (Hymenoptera : Megachilidae). *Bulletin, Tohoku National Experimental Station* 57, 1–221.

Marshall, A.H., Michaelson-Yeates, T.P.T. and Williams, I.H. (1995) The use of isoenzymes to quantify the efficiency of insect pollinators in mediating pollen flow in white clover (*Trifolium repens* L.) seed crops and consequences for seed yield. *Journal of Applied Seed Production* 13, 42.

Mayer, D.F., Lunden, J.D. and Kious, C.W. (1988) Effects of dipping alfalfa leaf-cutting bee nesting materials on chalkbrood disease. *Applied Agricultural Research* 3, 167–169.

Mayer, D.F., Lunden, J.D., Goerzen, D.W. and Simko, B. (1991) Fumigating alfalfa leaf-cutting bee [*Megachile rotundata* (Fabr.)] nesting materials for control of chalkbrood disease. *Bee Science* 1, 162–165.

McFarlane, R.P., Griffin, R.P. and Read, P.E.C. (1983) Bumble bee management options to improve 'Grasslands Pawera' red clover seed yield. *Proceedings of the New Zealand Grassland Association* 44, 47–53.

Morse, D.H. (1978) Size-related foraging differences of bumble bee workers. *Ecological Entomology* 3, 189–192.

Niklas, J. (1985) The aerodynamics of wind pollination. *Botanical Review* 51, 328–386.

Oertel, E. (1946) Effect of temperature and relative humidity on sugar concentration of nectar. *Journal of Economic Entomology* 39, 513–515.

O'Toole, C. and Raw, A. (1991) *Masons. Bees of the World, Facts on File, New York.* Blandford Publishing, London, pp. 61–76.

Owens, S.J. (1992) Pollination and fertilization in higher plants. In: Marshall, C. and Grace, J. (eds) *Fruit and Seed Production.* Cambridge University Press, Cambridge, UK.

Packer, J.S. (1970) The flight and foraging behaviour of the alkali bee (*Nomia melanderi* Ckll.) and the alfalfa leaf-cutter bee [(*Megachile rotundata* (F.)]. PhD Dissertation, Utah State University, Logan, Utah.

Palmer, T.P. and Donovan, B.J. (1980) Seed production of new cultivars of lucerne. In: Lancashire, J.A. (ed.) *Herbage Seed Production.* Grassland Research and Practice Series No. 1. New Zealand Grassland Association, Palmerston North, New Zealand, pp. 87–91.

Palmer-Jones, T., Forster, I.W. and Jeffery, G.L. (1962) Observations on the role of the honey bee and bumble bee as pollinators of white clover (*Trifolium repens* Linn.) in the Timaru district and Mackenzie country. *New Zealand Journal of Agricultural Research* 5, 318–325.

Pankiw, P. and Lieverse, J.A.C. (1982) Weights of adult leafcutter bees as affected by environment and source. In: Rank, G.H. (ed.) *Proceedings of the First International Symposium on Alfalfa Leafcutting Bee Management,* Saskatoon, Saskatchewan, August 16–18, 1982, University of Saskatchewan, Saskatchewan, Canada, pp. 234–238.

Parker, F.D. (1981) A candidate red clover pollinator, *Osmia coerulescens* (L.). *Journal of Apiculture Research* 20, 62–65.

Parker, F.D. (1986) Field studies with *Osmia sanrafaelae* Parker, a pollinator of alfalfa (Hymenoptera : Megachilidae). *Journal of Economic Entomology* 79(2), 384–386.

Parker, F.D., Torchio, P.F., Nye, W.P. and Pedersen, M. (1976) Utilization of additional species and populations of leafcutter bees for alfalfa pollination. *Journal of Apiculture Research* 15, 89–92.

Peck, O. and Bolton, J.L. (1946) Alfalfa seed production in northern Saskatchewan as affected by bees, with a report on means of increasing the populations of native bees. *Scientific Agriculture* 26, 338–418.

Pengelly, D.H. (1953) Alfalfa pollination in S. Ontario. *Report of the Entomological Society of Ontario* 84, 101–118.

Pesson, P. and Louveaux, J. (1984) *Pollinisation et productions vegetales*. Institut National de la Recherche Agronomique, Paris, p.663.

Peterson, S.S., Baird, C.R. and Bitner, R.M. (1992) Current status of the alfalfa leafcutting bee, *Megachile rotundata*, as a pollinator of alfalfa seed. *Bee Science* 2(3), 135–142.

Pohl, F. (1937) Die Pollenerzeugung der Windblutler. *Beihefte zur Botanisches Zentralblatt*, A 56, 365–470.

Ribble, D.W. (1968) A list of recent publications on the alkali bee, *Nomia melanderi*, with notes on related species of bees. Wyoming Agricultural Experimental Station Science Monograph 11.

Richards, K.W. (1987) Diversity, density and efficiency, and effectiveness of pollinators of cicer milkvetch, *Astragalus cicer* L. *Canadian Journal of Zoology* 65, 2168–2176.

Richards, K.W. (1988) Density, diversity, and efficiency of pollinators of sainfoin, *Onobrychis viciaefolia* Scop. *Canadian Entomology* 120, 1085–1100.

Ritter, W. (1993) Chemical control: options and problems. In: Matheson, A. (ed.) *Living with Varroa*. Proceedings of the IBRA Symposium, London, pp. 17–24.

Roubik, D.W. (1989) *Ecology and Natural History of Tropical Bees*. Cambridge University Press, Cambridge, UK.

Royce, L.A., Stringer, B.A., Kitprasert, C., Burgett, D.M. and Rossgnol, P.A. (1993) Infestation of tracheal mites (Acari : Tarsonemidae) in feral and managed colonies of honey bees (Hymenoptera : Apidae). *Journal of Economic Entomology* 86(3), 712–714.

Seely, T.D. (1985) *Honeybee Ecology*. Princeton University Press, Princeton, New Jersey.

Semin, A.S. and Bourmistrov, A.N. (1982) Leafcutter bees in USSR: experience of application, prospects of reproduction, problems. In: Rank, G.H. (ed.) *Proceedings of the First International Symposium on Alfalfa Leafcutting Bee Management*, Saskatoon, Saskatchewan, August 16–18, 1982, University of Saskatchewan, Saskatchewan, Canada, pp. 265–268.

Sherwood, R.T., Young, B.A. and Bashaw, E.C. (1980) Facultative apomixis in bufflegrass. *Crop Science* 20, 375–379.

Shuel, R.A. (1952) Some factors affecting nectar secretion in red clover. *Plant Physiology* 27, 95–110.

Shuel, R.A. and Pedersen, M.W. (1952) The effect of environmental factors on nectar secretion as related to seed production. *Proceedings of the 6th International Grasslands Congress* 1, 887–871.

Silverton, J.W. (1982) *Introduction to Plant Population Ecology*. Longman, London.

Slessor, K.N., Kaminski, L.A., King, G.G.S., Borden, J.H. and Winston, M.L. (1988) Semiochemical basis of the retinue response to queen honey bees. *Nature* 332, 354–356.

Spoor, W. (1976) Self-incompatibility in *Lolium perenne* L. *Heredity* 37, 417–421.

Stebbins, G.L. (1957) Self-fertilization and population varibility in the higher plants. *American Naturalist* 91, 337–354.

Stephen, W.P. (1955) Alfalfa pollination in Manitoba. *Journal of Economic Entomology* 48, 543–548.

Stephen, W.P. (1959) *Maintaining Alkali Bees for Alfalfa Seed Production*. Oregon Agricultural Experimental Station Bulletin 568. Oregon Experimental Station, Corvallis, Oregon.

Stephen, W.P. (1981) *The Design and Function of Field Domiciles and Incubators for Leafcutting Bee Management* (*Megachile rotundata* (Fabricus)). Oregon State University Agricultural Experiment Station Bulletin 654. Oregon Experimental Station, Corvallis, Oregon.

Stephen, W.P. and Osgood, C.E. (1965) Influence of tunnel size and nesting medium on sex ratio in a leaf-cutter bee, *Megachile rotundata. Journal of Economic Entomology* 58, 965–968.

Tasei, J.N. (1972) Observations preliminaires sur la biologie d'*Osmia* (Chalcosmia) *coerulescens* L. Pollinisatrice de la lucern. *Apidologie* 3, 149–165.

Tasei, J.N. (1976) Recolte des pollens et approvisionnement du nid chez *Osmia coerulescens* L. (Hymenoptera, Megachilidae). *Apidologie* 7(4), 277–300.

Tasei, J.N. (1982) Status of *Megachile rotundata* (F.) in France. In: Rank, G.H. (ed.) *Proceedings of the First International Symposium on Alfalfa Leafcutting Bee Management*, Saskatoon, Saskatchewan, August 16–18, 1982, University of Saskatchewan, Canada, pp. 239–246.

Tepedino, V.J. and Frohlich, D.R. (1984) Fratricide in *Megachile rotundata*, a non-social megachilid bee: impartial treatment of sibs and non-sibs. *Behavioural Ecology and Sociobiology* 15, 19–23.

Thomas, R.G. (1961) The influence of environment on seed production capacity in white clover (*Trifolium repens* L.). I. Controlled environment studies. *Australian Journal of Agricultural Research* 12, 227–238.

Thomas, R.G. (1981) Studies on inflorescence initiation in *Trifolium repens* L. The short-long day reaction. *New Zealand Journal of Botany* 19, 361–369.

Thomas, R.G. (1987) Reproductive development. In: Baker, M.J. and Williams, W.M. (eds.) *White Clover*. CAB International, Wallingford, pp. 63–109.

Vandenberg, J.D. and Stephen, W.P. (1982) Etiology and symptomatology of chalkbrood in the alfalfa leafcutting bee, *Megachile rotundata. Journal of Invertebrate Pathology* 39, 133–137.

Vines, A.E. and Reese, N. (1964) *Plant and Animal Biology*, vol. 1, Sr Isaac Pitman and Sons, London.

Walton, P.D. (1983) *Production and Management of Cultivated Forages*. Reston Publishing Company, Virginia.

Waters, N.D. (1970) *Lights and Water Traps for Alfalfa Leafcutter Bee Incubators*. University of Idaho Agricultural Extension Service No. 120. Parma, Idaho, USA.

Wiens, D. (1984) Ovule survivorship, brood size, life history, breeding systems, and reproductive success in plants. *Oecologia* 64, 47–53.

Wiens, D., Calvin, C.L., Wilson, C.A., Davern, C.I., Frank, D. and Seavey, S.R. (1987) Reproductive success, spontaneous embryo abortion and genetic load in flowering plants. *Oecologia* 71, 501–509.

Williams, I.H., Corbet, S.A. and Osborne, J.L. (1991) Beekeeping, wild bees and pollination in the European community. *Bee World* 72, 170–180.

Wilson, M.F. and Price, P.W. (1977) The evolution of inflorescence size in *Asclepis* (Asclepiadaceae). *Evolution* 31, 495–511.

Willmer, P.G. (1983) Thermal constraints on activity patterns in nectar-finding insects. *Ecological Entomology* 8, 455–469.

Harvest and Postharvest Management of Forage Seed Crops

8

U. Simon,[1] M.D. Hare,[2] B. Kjaersgaard,[3] P.T.P. Clifford,[4]
J.G. Hampton[5] and M.J. Hill[5]

[1]*Technische Universität München, 85350 Freising-Weihenstephan, Germany;*
[2]*Faculty of Agriculture, Ubon Ratchathani University, Ubon Ratchathani 34190, Thailand;* [3]*DLF-Trifolium A/S, 4000 Roskilde, Denmark;*
[4]*AgResearch, PO Box 60, Lincoln, New Zealand;* [5]*Seed Technology Centre, Department of Plant Science, Massey University, Palmerston North, New Zealand*

8.1 SEED HARVEST

8.1.1 Introduction

Not all the seed that is produced can be harvested (Lorenzetti, 1993), because of the range of seed ripeness, the effects of lodging, seed shattering/pod dehiscence, and losses during the harvest process. These losses can be substantial; e.g., in an on-farm survey, Clifford and McCartin (1985) measured harvesting losses in white clover (*Trifolium repens* L.) seed crops of 12–39%, the mean being 200 kg seed ha^{-1}. Similarly in grasses, losses ranging from 20 to 75% are often reported (Andersen and Andersen, 1980; Meijer, 1985; Hampton, 1991).

Hopkinson and Clifford (1993) suggested four stages of seed loss during the time from seed physiological maturity to collection of the seed from the combine harvester:

Stage 1: Environmental losses – the effects of wind, rain, lodging and the ability of a cultivar to retain its seed.
Stage 2: Cutting losses – seed shaken from the seed head during the cutting of the crop, or forced out of the head by wind or rain while in the swath.
Stage 3: Pickup losses – seed shaken from the seed heads during the lifting from the swath for presentation to the combine auger platform.
Stage 4: Separation losses – seed lost because of threshing inefficiency, excessive aspiration draught blowing seed directly over the straw walkers and out the back of the separation chamber, and seed entrapment in the offal deposited on the ground from the straw walkers.

8.1.2 Seed Shattering

Grass and legume cultivars are bred primarily for their forage production, and not their seed production ability. High seed retention is therefore not usually a plant breeding goal. Consequently, because of natural variation, there is a great difference in seed retention among species and cultivars (see Chapters 4 and 10).

Steen (1983) produced a seed shattering index for eight forage grasses, and concluded that Italian ryegrass (*Lolium multiflorum* L.), common meadow grass (*Poa trivialis* L.) and meadow fescue (*Festuca pratensis* Huds.) were species most prone to seed shattering, while Kentucky bluegrass (*Poa pratensis* L.) and red fescue (*Festuca rubra* L.) lost little seed through shattering. However, any forage species will shed seed through shattering if incorrect decisions as to the time of harvest are made (e.g. Table 8.1).

Differences in seed retention among cultivars within a species have been reported (McWilliam, 1980; Elgersma *et al.*, 1988; Simon, 1993). For example, Piccirilli and Falcinelli (1989) found large variation in seed shattering among different ecotypes of cocksfoot (*Dactylis glomerata* L.) and identified differences in genetic resistance to shattering in some of these ecotypes. In tall fescue (*Festuca arundinacea* Schreb.) Falcinelli (1993) found genotypes with delayed shattering which indicates that resistance to seed shattering is likely to exist (see Chapter 10).

Environmental conditions from flowering to maturity play a role in the ability of a crop to retain its seed (Steen, 1983; Elgersma *et al.*, 1988). For example, in perennial ryegrass, dry and warm conditions increase seed shattering; in tall fescue and Yorkshire fog (*Holcus lanatus* L.) strong winds through a non-lodged crop can result in 50–90% seed shattering within 24 h (J.G. Hampton, Palmerston North, 1996, personal communication).

Unfortunately, data on seed shattering of grass and legume species/cultivars are not well documented. This information is pivotal to minimizing harvest losses. The seed shattering potential of a crop influences the choice of harvest method, as well as the time of harvest.

8.1.3 Crop Growth and Conditions of Growth

Lodging

The risk of seed loss in a species which is prone to shattering is greatly decreased if the crop has lodged after anthesis (Ellegaard, 1971a, b; Jensen, 1976). This is

Table 8.1. Effect of date of harvest on seed yield of meadow fescue. (Adapted from Simon, 1993.)

Date	Seed moisture content (%)	Seed yield (kg ha^{-1})	Seed loss (%) from optimum[1]
29 June	51	1040	—
4 July	41	1221	—
8 July	31	1039	15
20 July	13	808	34

[1]Peak seed yield recorded on 4 July.

because the loss of fertile tiller fresh weight in the latter stages of seed development can result in the upward movement of the crop (i.e. a partial reversal of lodging) just before harvest, and this movement can increase shattering (Andersen and Andersen, 1980).

Alternatively when seed crops are grown in windy environments, some lodging after anthesis can offer protection from the seed shatter which results from wind action.

Vegetative growth

Once translocation of assimilates to the seed ceases (see Chapter 4), or when lodging allows light to penetrate to the base of the plant, new vegetative growth begins as long as water and nutrients are not limiting. This flush of vegetative growth can in severe cases virtually submerge the seed heads, making direct combining impossible. Even the presence of small amounts of new vegetative growth means that it becomes more difficult to thresh the seeds out of the straw, and threshing losses are increased. Vegetative regrowth in swathed crops may mean that the swath has to be lifted before feeding into the combine, or undercut by mowing. This process is likely to increase seed shattering (Clifford and McCartin, 1985).

Uneven ripening

The indeterminate growth habit of forage legumes means that the seed crop will not ripen uniformly, as flowering may be spread over several weeks. While forage legume seed crop management aims to produce a single peak of flowering (Hopkinson and Clifford, 1993; Chapter 6), a single harvest of a forage legume seed crop will include seed at all stages of development.

Ripening in grasses is usually more uniform, but factors such as varying soil quality, uneven fertilizer application and areas of lodging within a crop can all result in uneven seed ripening.

In a very unevenly ripened seed crop it may be economic to thresh more than once (Arnold and Lake, 1965, 1966).

Sprouting

Seed sprouting (germinating) in the seed head before harvest may occur in wet seasons, resulting in increased difficulties with harvesting, high yield losses and poor quality seed.

8.1.4 Determining the Stage of Development of the Crop

Delaying or advancing seed harvest by only a few days can result in substantial yield losses (Andersen and Andersen, 1975), and it is therefore important to know the stage of development of the crop during the period leading up to harvest. This can be determined in the following ways.

Days after anthesis

For most forage species, the number of days from peak anthesis to maximum seed dry weight (harvest ripeness) is around 30 days (Hyde *et al.*, 1959; Hill and Watkin,

1975a; Hare and Lucas, 1984; Hopkinson and Clifford, 1993; see Chapter 4), although this time period may vary within a cultivar depending upon climatic conditions (Hare and Lucas, 1984). For example, higher temperatures shorten the period from peak anthesis to harvest ripeness (Komatsu et al., 1979).

Harvest decisions based on a grower's local environmental knowledge and previous experience with the time it takes a crop to reach harvest ripeness can be extremely effective. For example, Marr (1990) stated 'as a general rule, when the cocksfoot crop flowers at Methven (Canterbury, New Zealand), there are 28 days until harvest'.

Crop colour

Crop colour has been proposed as a method for determining the stage of development of the crop (Hill and Watkin, 1975a; Andersen and Andersen, 1980), and Steen (1983) developed a scale of five ripeness groups, with a rating for the colour of the stem and spikelets from green to yellow. However, specific climatic conditions and cultivar differences influence the relationship between stem and seed dry matter, seed shattering and colour. This method is rarely used.

Endosperm consistency

Pegler (1976) suggested that endosperm consistency was a reasonably reliable indicator of seed maturity. For grass seeds, growers consider that when the endosperm has passed from the 'milk' to 'dough' stage, there are around 7 days until harvest (Marr, 1990).

Stripping ripeness

The 'stripping ripeness' of a crop was described by Kåhre (1964), i.e. a crop is considered ripe for harvesting when single seed heads can be stripped with a single sweep of the hand from the bottom to the top of the head. This is equivalent to the 'hat test', where a crop is judged to be ripe when seeds drop into a hat swept through the crop (Marr, 1990). This technique is considered especially useful in determining the time to harvest species prone to shattering (Jensen, 1976).

Seed moisture content

Seed moisture content (SMC) is the most reliable parameter for determining harvest timing (Hill and Watkin, 1975b; Klein and Harmond, 1971) provided that first the crop is sampled accurately, and second an accurate method is used to determine seed moisture. Because of differences in seed maturity within a crop, and the tendency to take a few seed heads from the nearest plants rather than a truly representative sample of seed heads from the crop, the SMC of the sample may differ significantly from that of the crop. SMC can be determined accurately by the use of an infrared lamp (Hill and Crosbie, 1966). Portable moisture meters such as those used for cereals cannot be reliably used for forage seed crops, as conductance meters become widely inaccurate at SMC > 20% (M.J. Hill, Palmerston North, 1996, personal communication).

8.1.5 Optimal Harvest Time

The optimal harvest time is that which provides the highest yield of quality seed, but this will depend on the species being grown and the method of harvest. Seed viability is not a factor in determining harvest date, as this reaches a maximum long before seed harvest is even considered. In perennial ryegrasses, for example, maximum viability is reached 14–17 days after anthesis (Hill and Watkin, 1975b; see Chapter 4). However threshing viable seed too early (i.e. when SMC is > 45%) can reduce germination because of physical damage to the seed (Nellist and Rees, 1963).

For forage grasses, the optimum time to begin pre-threshing operations is the point at which the yield of viable seed is maximal (i.e. before seed shattering begins). For perennial ryegrass this is usually at around 42–45% SMC, as seed shattering begins at around 40% SMC. For forage legumes such as white clover which can be managed to produce a single peak of flowering, optimum harvest date is usually from 5 to 6 weeks after peak anthesis (Hopkinson and Clifford, 1993), whereas in a species such as big trefoil (*Lotus uliginosus* Schkr.) where umbels mature over several weeks, the optimum harvest time is when the majority of pods are light-brown (around 30 days after anthesis – Hare and Lucas, 1984).

Recommendations for optimal harvest time vary from species to species and from country to country. A summary of recommendations is presented in Table 8.2 for grasses and Table 8.3 for legumes.

8.1.6 Harvest Methods

The harvesting method is determined by the growth habit of the crop, the climatic conditions, and the availability of machinery for both harvesting and drying.

Legumes

Pre-cutting. The decision to cut crops green is related to herbage moisture, crop bulk and weather pattern stability for drying (Hopkinson and Clifford, 1993). The advantage of additional seed maturation in the swath must be balanced against the likely chance of rain and wind damage in the 7–14 days the crop is on the ground, and the high losses associated with any 'undercutting' requirements (Clifford and McCartin, 1985).

Desiccation using diquat reduces vegetative bulk, but requires precise timing in relation to both crop maturity and stable weather patterns.

Cutting. Mowers are frequently used in all forage legume species (Hopkinson and Clifford, 1993), the main types being: (i) standard sicklebar (17 fingers per metre); (ii) double reciprocating knife (17 knife sections per metre); and (iii) a range of rotary types. In white clover, seed losses were 10%, 5% and 27% for (i)–(iii) respectively (Clifford and McCartin, 1985). Windrowers are used for large fields.

Pickup. Lifters used in association with the combine reel guide the crop over the combine knife and onto the auger platform. Murphy pickups are counter-rotating beaters which ensure centre feed to the auger platform. They are predominantly

Table 8.2. Optimal harvest time in grass seed crops.

Crop	Swathed before threshing	Direct combined
Cocksfoot	Morphological ripeness* – 3,4–3,6[1]; Moisture content – 44%[5], 35–40%[1]	26–30 days after peak anthesis[2]; Endosperm consistency – cream cheesy/cheesy[2]
Meadow fescue	Morphological ripeness – 3,5–4,0[3], 3,0[1]; Moisture content – 40–50%[3], approx. 43%[1]	17–29 days after peak anthesis[4], 18–26 days after peak anthesis[2]; Morphological ripeness – 4,0–4,8[3], 3,4–3,7[1]; Endosperm consistency – cream cheesy/cheesy/hard[2]; Moisture content – 25–35%[3], 27–50%[4], approximately 35–40%[1]
Red fescue	Morphological ripeness – 4,0–4,5[3], 4,0[1]; Moisture content – 30–45%[3], 25%[5], approx. 35%	24–38 days after peak anthesis[4]; about 30 days after first anthesis[6]; Morphological ripeness – 4,5–5,0[3]; Moisture content – 20–30%[3], 24–40%[4], about 35%[6]
Tall fescue	Moisture content – 35–41%[9]	29–30 days after peak anthesis[2]
Smooth meadowgrass	Morphological ripeness – approx. 3,3[1]; Moisture content – 28%[5], approx. 23%[1]	Endosperm consistency – cheesy[2]; About 23 days after first anthesis[2]; Moisture content – about 35%[6]
Hybrid ryegrass		Moisture content, approx. 40%[8]
Italian ryegrass	Morphological ripeness – 3,0[1]; Moisture content – 40–45%[3], approximately 45%[2]	28–32 days after anthesis[10], 28–30 days after peak anthesis[2]; Morphological ripeness – 3,3–3,4[1]; Endosperm consistency – doughy to solid[10], cheesy[2]; Moisture content – 30–40%[3], approximately 37–40%[1]
Perennial ryegrass	Morphological ripeness – 2,5–3,5[3], 2,9[1]; Moisture content – 40–50%[3], 40–47%[11], 35%[5], min. 37%[1]	Approx. 30 days after first anthesis[6], 28–30 days after peak anthesis[2], 23–34 days after peak anthesis[4]; Morphological ripeness – 3,0–4,5[3], 3,0[1]; Endosperm consistency – cheesy[2]; Moisture content – 25–35%[3], 25–46%[4], 20–25% in lodged crop[12], 30–40% in un-lodged crop[12], approx. 42%[6], 40–47%[11], approx. 30%[1]
Timothy	31–35 days after peak anthesis[12]; Morphological ripenesss – approx. 3,6[1]; Moisture content – 46–47%[11], approx. 40%[12], approx 37%[1]	33–38 days after peak anthesis[2]; Endosperm consistency – cheesy/hard[2]; Moisture content – 23–31%[11]; below 30%[12]

Morphological ripeness (Andersen and Andersen, 1974): completely green inflorescences, 1; not completely green or completely yellow, 3; completely yellow inflorescences, 5.

[1]Steen, 1983; [2]Pegler, 1976; [3]Andersen and Andersen, 1975; [4]Anonymous, 1972; [5]Klein and Harmond, 1971b; [6]Hebblethwaite and Ahmed, 1978; [7]Komatsu et al., 1979; [8]Williams, 1972, [9]Andrade et al., 1994; [10]Komatsu et al., 1971; [11]Hill and Watkin, 1975a; [12]Ellegaard, 1971; [13]Roberts, 1969.

Table 8.3. Optimal harvest time in legume seed crops.

Crop	Swathed before threshing	Direct combined
Red clover	Approximately 6 weeks after peak anthesis[1]	Approximately 6 weeks after peak anthesis, if desiccated[1]
White clover	Approximately 21–26 days after peak anthesis or optimal bee visits in the field[1]. At least 25% brown flowers on average per head[1]	
Lucerne	The crop should be swathed during periods of high humidity or when leaves are wet with dew after two-thirds to three-fourths of the seed pods have changed to dark brown. High proportion of green but fully formed, plump pods[2,3]	A chemical desiccant is usually used when all pods are nearly ripe. Combine within 3–10 days as soon as pods and leaves are 15–20% (wt/wt) moisture to avoid shattering[2,4]
Lotus	Swath when a majority of the pods are yellow-brown to brown. Do not delay until pods are black or swath early when a majority of pods are purple to green[5,6]	

[1]Borggaard *et al.*, 1991; [2]Jones, 1952; [3]Smith and Melton, 1967; [4]Bunnelle *et al.*, 1954; [5]Anderson, 1955; [6]Pieroni and Laverack, 1994.

used in white clover, with minor use in red clover and lucerne (Hopkinson and Clifford, 1993). Drapers are counter-rotating belts with bars and/or fingers attached to ease the crop from the ground to the auger platform. They are best used in bulky crops.

Direct-combining. Normally only desiccated upright crops such as red clover and lucerne are direct-combined.

Grasses

Cutting. For grass seed crops the two basic machines are top-hat rotary mowers and windrowers (Hopkinson and Clifford, 1993). Cutting the crop and laying it in a swath has the following advantages:

1. Seed ripeness is more likely to be uniform.
2. Seed dry down in a swath is relatively rapid.
3. Seed losses from wind are minimal (cf. a standing crop).
4. If required the crop may be lifted or turned to aid drying.
5. Drying costs are lower (cf. direct combining).

A crop such as perennial ryegrass would be cut at 43% SMC and either:

1. Left to dry for 2–3 days in the swath to 25% SMC, then threshed and artificially dried to 14% SMC or
2. Left to dry for 4–6 days in the swath to 14% SMC, then threshed.

For method (1) there is a necessity for artificial seed drying and some risk of seed loss from the swath as a result of inclement weather. Method (2) is the least expensive (no artificial drying) but the most risky, as seed may shatter and losses from environmental damage can be large.

Pickup. Counter-rotating tined belt pickups in a range of designs are used for grass seed crops (Hopkinson and Clifford, 1993). Major attachment concerns are: (i) the angle of incline to the auger platform and (ii) ensuring that there are no gaps where shattered seed can be returned to the field. Counter-rotation speed is adjusted to the combine forward speed to ensure that no windrow 'bunching' or 'stripping' occurs.

Direct-combining. Seed crops may be direct-combined at around 43% SMC, and then artificially dried to a storage moisture content of < 14%. This method minimizes the time a crop is in the field, and is most appropriate when the crop is prone to shattering, bird or insect damage is a problem, or when there is a risk of weathering damage induced by adverse weather. Disadvantages may include the high cost of artificial drying, and potential damage to seed; because the seed is still 'wet', a high drum speed and narrow concave setting need to be used to remove the seed from the seed heads, and this can bruise and/or crush the seed, reducing germination (physical damage) and vigour (physiological damage).

Threshing and separation mechanisms

Usually, large quantities of plant material have to pass through the combine harvester when seed crops are threshed. The moisture content of this material is often high, and in some years the material may be 'stringy'. As the seeds are small and light, heavy demands are made on both the knife and threshing equipment in combine harvesters primarily designed to thresh cereals and/or large seeded legumes, not forage seeds.

The threshing mechanism consists of a slotted concave, against which the plant material is beaten by a rotating barred drum. For forage legumes (e.g. white clover), plates are frequently attached to the concave wires to improve threshing efficiency (Hopkinson and Clifford, 1993). Threshing efficiency is the interrelationship between drum speed and distance from the concave in relation to crop bulk and moisture content. Too harsh a threshing damages seed, both physically and physiologically (Hampton, 1990), leading to lowered germination and vigour. It may also damage adhering structures on seeds of undesirable weeds, thereby reducing the ability to remove them during seed cleaning (Hartley, 1980; see Chapter 9). Incorrect air-blast settings can blow seed out of the combine.

While it is not possible to supply 100% clean seed directly from the combine, correct adjustments will provide a well-threshed raw product which has a high proportion of pure seed and an acceptable seed weight.

8.2 SEED DRYING

8.2.1 Introduction

The moisture content of seed is one of the most important factors affecting seed quality and longevity in storage. For most forage species, a seed moisture content (SMC) of between 8 and 12% is considered 'safe' for storage, and therefore for freshly harvested seed, SMC may have to be reduced from 40 to 45% (e.g. direct combined grasses) or 14 to 20% (desiccated red clover) to the desired storage SMC. Failure to reduce SMC promptly and adequately, or incorrect use of seed dryers, leads rapidly to heat damage which can ultimately result in seed death.

8.2.2 Heat Damage

Field heating in wet seed

When seed is harvested at SMC > 15% it is generally dry on the outside, but has a wetter interior. If this moisture is not removed, the moisture difference begins to 'even out', and the seed lot is referred to as 'going through a sweat' or 'wetting back' (Hill and Johnstone, 1985). At the same time, the humidity of the inter-seed space increases rapidly, creating an ideal environment for heat production (as a byproduct of metabolism) and increased activity of a range of microorganisms already present in or on the seed. The principal agents are *Aspergillus* and *Penicillium* fungi (see Chapter 9), and their activity can raise the temperature of the seed lot to around 55°C (Hill, 1975), at which point germination losses occur. The rate of deterioration in seed quality increases with increasing heat and with increasing time of exposure to high temperatures. A seed lot can be killed in 12–15 h (Hill, 1975).

Field heating in dry seed

Even when seed is not harvested until it is at a 'safe' SMC (e.g. 13–14%), heat damage can still occur. This is because when seed is threshed from the windrow on a hot, sunny day, it will invariably contain radiant heat, so that the seed temperature may be 10–12°C higher than ambient air temperature (Hill and Johnstone, 1985). This effect is greatest on hot, clear days, is reduced by cloud, and ceases at sunset (Hill and Crosbie, 1966).

If 'hot' seed is stored in bags, further heating does not occur because heat loss into the surrounding air occurs reasonably rapidly. However if such seed is stored in bulk, the insulation properties of seed result in heat retention in the mass, and subsequent seed damage through microbial activity (Matthews and Hill, 1967).

Artificial drying

Forage seeds are hygroscopic, meaning they can gain or lose water depending on the amount of moisture in the air around them. They also exhibit varying degrees of heat sensitivity, being more easily damaged by high temperatures and particularly when SMC is high. Drying conditions combining high temperature, high SMC and a fast drying rate are particularly injurious. High temperatures when drying moist seed rapidly remove free surface moisture, and this rapid evaporation removes heat from the seeds fast enough to keep the actual seed temperature depressed. As the SMC drops, however, the supply of moisture for evaporation is less readily available, and actual seed temperature will increase. Under these conditions, seed injury can occur (Hill and Johnstone, 1985).

8.2.3 The Drying Process

Drying occurs when there is moisture movement out of the seed and into the surrounding air. For this to happen, the relative humidity (RH) of the air must be reduced to a level which establishes the movement of moisture from the seed surface into the surrounding air. This evaporation process continues until the SMC and moisture in the air reach a balance. This point of equality is known as the equilibrium moisture content (EMC).

The SMC at EMC will differ for different species. For example, with drying air of 70% RH, the SMC at EMC is 8.2% for lucerne, 11.0% for cocksfoot, 9.3% for birdsfoot trefoil, and 12.8% for perennial ryegrass (Dexter, 1957). Such variations in EMC between species are mainly attributable to differences in composition, a range of forage seed examples being presented in Table 8.4.

Artificial drying is simply a way of accelerating the rate of natural movement (diffusion) of water from the inside of the seed to the surface where it is available for removal by evaporation. Diffusion rate depends on seed size and composition, speed of surface evaporation, temperature, initial moisture level, seed coat permeability and time (Hill, 1982).

Table 8.4. Equilibrium moisture content of seeds of a range of forage grasses and legumes at relative humidities from 15 to 85%. (Adapted from Harrington, 1968; Dexter, 1957; Justice and Bass, 1978.)

	Relative humidity (%)								
	15	30	45	55	65	70	75	80	85
Grasses									
Bromegrass	6.6	9.0	10.5	11.5	12.5	13.1	13.7	16.1	18.4
Browntop	6.2	8.6	9.6	10.0	10.7	11.0	12.5	13.5	15.0
Canary grass	6.6	8.8	9.8	11.4	12.0	12.5	13.5	14.7	15.7
Cocksfoot	6.0	8.0	9.1	9.8	10.5	11.0	12.0	13.4	14.9
Fescue	6.5	8.7	9.4	10.5	11.9	12.5	13.2	15.0	17.3
Ryegrass	6.5	8.6	10.7	11.0	12.1	12.8	13.4	14.9	16.6
Timothy	6.7	8.9	9.9	10.9	11.8	12.5	13.6	14.6	16.1
Legumes									
Lotus	4.8	5.6	6.2	7.1	8.8	9.3	11.7	15.9	18.9
Lucerne	4.5	5.2	6.0	6.5	7.1	8.2	9.3	14.5	18.3
Red clover	4.7	5.7	6.3	7.6	8.6	9.1	11.2	15.6	18.7
Sub. clover	4.3	5.1	6.0	6.9	7.8	8.7	11.0	15.0	17.8
White clover	4.6	5.4	6.5	7.2	8.4	9.0	10.9	15.4	18.0

Table 8.5. Maximum safe drying temperature without germination loss at different levels of seed moisture content. (Adapted from Hill and Crosbie, 1966.)

Seed moisture content (%)	Maximum drying temperature (°C)
> 20	32
18–20	34
14–17	37
11–13	40
9–10	42
< 10	43

Drying temperature

No single temperature can be quoted beyond which it is unsafe to heat seed during drying, because many factors are involved. These include the type of seed, maximum heat tolerance, SMC, duration of temperature rise and drying rate (Hill and Johnstone, 1985). However the drying temperatures provided in Table 8.5 are considered safe for forage seeds (Hill and Crosbie, 1966).

Drying systems

Low temperature dryers. The simplest artificial drying system is one which operates by increasing airflow through the seed bulk by means of a fan. Ambient air is used, and the seed gradually reaches a SMC in equilibrium with the RH of the air. Cool night air is often used, the aim being to cool the seed to within 1–2°C of ambient air temperature reasonably quickly (Crosbie, 1972).

If the air RH is around 70%, then seed can be dried to 'safe' SMC by this method. If air RH is > 70%, raising the air temperature will lower the RH (by approximately 4.5% RH for every 1°C increase in temperature).

Medium temperature dryers. Warmed air is blown through a bed of seed of controlled depth. The drying process stops before equilibrium SMC is reached. A temperature rise of up to 14°C above ambient will produce moisture extraction rates of up to 1% per hour. As each batch of seed dries, a moisture gradient develops from the bottom to the top layer. The heat is switched off and ambient air blown through to reduce the seed temperature and prevent possible subsequent condensation.

High temperature dryers. These subject the seed to a high temperature for a short period of time. The seed loses moisture, but does not come into equilibrium with the air RH which is very low. Wet seed passes through the dryer and emerges dry in 30–60 min. The process is continuous, the amount of moisture removed being determined by the air temperature and the time of exposure of the seed. In a continuous flow dryer with 30 min per pass, perennial ryegrass seed at 15–20% SMC can tolerate a dryer temperature of up to 46°C with no detrimental effects on germination, but an operating temperature of 43°C is recommended (Hill and Johnstone, 1985).

For a discussion of the advantages and disadvantages of various dryer types see Justice and Bass (1978).

It is important to understand the principles of seed drying to ensure a high quality product is obtained. The market acceptability of poorly dried seed is often low, resulting in downpricing due to its appearance and low germination. Such factors as scorching, heat damage, mould activity and cracking can all be important in affecting the quality of dried seed, and particularly the ability of seed to retain germination capacity even under the best storage conditions.

Perhaps surprisingly, seed drying damage is not usually reflected in a loss of germinability in seed lots tested immediately after drying. However, the subsequent and often rapid deterioration of damaged seed is shown in an often substantial loss of viability within a short time, even if seed is stored under supposedly 'ideal' conditions. This is a commonly observed but not generally understood situation; the cause being more commonly focused on the 'inadequacy of the storage system' than on the pre-storage drying history of the seed lot.

There is no doubting the importance of controlled drying systems in affecting seed quality and longevity. However, many seed processing systems do not focus on seed drying as an important seed quality control issue, suggesting it is one of the least understood and least appreciated aspects of seed production.

8.3 SEED PROCESSING

The success in meeting seed lot purity and any size standards for a particular harvested crop relies on a knowledge of the principles of seed separation in terms of size, weight, shape, texture and colour. Of equal importance is processor interpretation of these relationships as they affect the efficient use of the range of commercially available seed-cleaning equipment (Hartley, 1980). Where any contaminant species that share the same separation characteristics as the sown species are liable to occur, their removal prior to sowing the crop (see Chapters 5 and 6) is imperative.

8.3.1 Primary Equipment

Air screen cleaners

The combination of separation principles allowing sample differentiation in terms of weight and general sizing make these machines a basic component of any seed processing operation. Most air screen cleaners have three distinct sample separation operations; a coarse and a fine aspiration separated by screening.

Aspiration. The initial operation is a coarse aspiration or controlled airflow removal of particles based on differences in weight, specific gravity or density; akin to the age-old practice of winnowing. This initial step still remains a quick and inexpensive way of removing dust, small weed seeds and chaff (large unit area per unit of weight) from the sample. The removal of the chaff fraction helps increase screen efficiency. This coarse pre-screening operation does not require the air-column length of aspiration post-screening. For the latter aspiration, screen sizing has produced a more homogeneous remainder which needs greater column length for a much finer separation based on terminal velocity. Manipulation of the air-column velocity to gain the best separation is initially through fan speed adjustment. Further refinement beyond fan speed is by way of similar adjustment to the fan air intake.

Screens. Arranged between the coarse and fine aspiration columns are reciprocating inclined screens to promote seed-material separation on the basis of width or diameter. The main screen types are square or oblong holes woven in wire and slotted oblong or round holes punched in metal. Due to the roundness of wire this type has a greater ability to set grass seeds on edge for sizing. At the lowest end of the scale the gradient in hole size is very small to cater for fine differences. Sizing can be in metric (mm) or imperial (64ths) units. Graduation increase at the smallest sizings is by 0.1 mm from 0.5 mm upwards or by 0.25 64ths from 1.25 64ths. Air screen cleaner manufacturers present very comprehensive tables relating sown species/weeds to scalping and grading screen size requirements. Because of the wide cultivar variations that can occur, let alone those resulting from the seed production environment, these tables act as very useful guides only. Screens are normally used in pairs, fixed one above the other in cradles or boats. The size of the upper sieve diameter is such that it removes all material of greater width than the sown seed: the process is known as scalping. The lower sieve is chosen to just retain the sown seed on the screen surface with the smaller material falling through: this is called grading. Samples and rejection material are checked early on to ensure the desired separations are occurring. In air-screen cleaners used to produce a final product, a series of two boats is frequently used to ensure fineness of width separation. The depth of sample presented to the top of the screen is no more than that which will reduce to the height/diameter of the sown seed prior to leaving the sieve.

To ensure maximum area use, screens are continually cleaned by balls or brushes. Ball cleaning relies on a reasonable oscillation speed and is more effective on larger holed screens. Brushes work well for small screen sizes. Cleaning the brushes between cultivars and species is very important, to ensure contamination does not occur.

Indented cylinder separator

This piece of equipment may be needed where the sample has to be stratified in terms of length. Should this be the case, the cylinder follows in sequence after the air-screen cleaner. Because indented cylinder throughput is slower than that of the air-screen cleaner, several may have to be run in parallel if forming part of a continuous process.

The major principles involved are the revolving of a cylinder punched with pockets of distinctive shape and size, at a speed which is sufficient to retain the length/s of components until gravity, at a point above the collection trough, more than offsets the applied centrifugal force. At this point the component/s are removed. Depending on their length differential either species or contaminant may be indented. As for screens, indent size may be given either in metric or imperial units.

Examples of weed seed separation from grass seed using indented cylinders are *Avena fatua* L. (wild oats), awned goose grass (*Bromus mollis* L.) and hairgrass (*Vulpia* spp.), field madder (*Sherardia arvensis* L.) and docks (*Rumex* spp.) (Hartley, 1980). Sheep sorrel (*Rumex acetocella* L.), knotted clover (*Trifolium striatum* L.), clustered clover (T. *glomeratum* L.), plantain (*Plantago lanceolata* L.), Scotch thistle (*Cirsium vulgare* (Savi) Ten.), Californian thistle *(C. arvense* (L.) Scop.) and fathen (*Chenopodium album* R. Br.) can be separated from clovers.

Specific gravity table

This machine may be used to give very fine and often final separations as an adjunct to an air-screen cleaner or cylinders.

Separations are mainly by weight and texture differences in the seed sample being stratified as it travels over a vibrating shallowly inclined fabric surface which has an airblast applied to the underside. The relative fineness of fabric mesh size ensures uniformity of air throughput across the whole deck, while ensuring that the sample is retained on the fabric surface. Separations can be further enhanced through deck cloth choice to give the friction coefficient which promotes the greatest separation level. Lightest seed travels the shortest distance with largest separation levels being associated with maximizing the length of the deck used.

For legumes, impurities such as fathen, field madder, chickweed (*Stellaria media* (L.) Vill.), catchfly (*Silene gallica* L.), earth particles and cracked, damaged and broken seed can be removed (Hartley, 1980). Separations of ryegrass and cocksfoot can also be achieved.

Velvet roller mill

Roller or Dodder Mills are another alternative for sample separation based on texture. In this process the seed is fed into the groove created between two gently sloping counter-rotating velvet-covered rollers. The smooth seed gradually moves along this groove to be collected at the far end. In contrast, rough-coated seeds such as dodder (*Cuscuta* spp.), dock, sheep sorrel or even inert material get constrained by the downy surface of the velvet. This constraint lifts them through the sample mass finally to be deposited on the other side of the rollers.

Spiral separators

This technology is very important in separating components which have similar terminal velocities, widths and lengths. In this case shape, as it varies in reaction to centrifical force, is the separation element. For example, white clover is virtually impossible to separate from tetraploid *Lotus uliginosus* either by screens or aspiration. Their only exploitable dissimilarity is shape. Whereas lotus seed is predominantly spherical, that of white clover is flattish to oval and heart shaped. When this mixture is fed into the top of an inclined spiral, friction effects based on contact area will modify the terminal velocity of the two sample components. The greater friction coefficient of the white clover reduces the terminal velocity to contain these seeds close to the spiral axis. In contrast, the lack of seed surface contact due to the roundness of lotus enhances its terminal velocity. Travelling at higher speeds down

Colour sorters

This technology is expensive and as a consequence used only occasionally within the highest forage seed grade. Further adding to the cost is slowness of operation. Seed is channelled in as a single column for passing over a colour background similar to its own. Any major divergence away from the background colour activates the seed removal mechanism.

8.3.2 Secondary Equipment

Depending on harvest conditions, differences in species/cultivar, physical seed characteristics and end use, additional seed processing options may be needed.

Seed polisher

This machine performs a range of options on grasses and legumes. Seed is fed in at one end of the machine to be abraded by brushes rotating against a woven wire screen, with the seed plus screenings exiting at the opposite end. Intensity of abrasion can often be adjusted by alteration of brush speed, distance between brushes and screen, and throughput rate. For de-awning or de-hulling grass seed this machine is the initial processing step prior to the first aspiration of a pre-cleaner or air-screen separator. Where a species such as cocksfoot needs to be 'shelled' to improve uniformity of spread during aerial oversowing then this process is carried out immediately prior to sowing. For de-awning, brush-type polishers are not suitable for all grasses, as evidenced by the development of a flame-removal technology to de-awn prairie grass (*Bromus willdenowii* Kunth.) in order to stop seed bridging in the drill.

For legumes, polishers can perform two different functions which are carried out near or at the end of the cleaning process. Where a proportion of seed has not been cleanly threshed from its attendant floral parts, this sample is polished prior to a final aspiration. The major use of polishers for legumes, however, is for 'clean-seed' abrasion (scarification) to reduce hard seed and increase germination to a commercially acceptable level.

Dehullers

Some legume seeds such as sainfoin (*Onobrychis viciifolia* Scop.), sulla (*Hedysarum coronarium* L.) and serradella (*Ornithopus sativus* Brot.) are enclosed in a thick husk. At sowing, particularly where moisture limitations for germination occur, husk removal prior to sowing is advantageous. These machines use a much higher level of abrasion than polishers and vary in design. De-hulling is done prior to sowing, or in batches to service the species' seasonal sowing requirement, as long-term, germination levels can be impaired.

An alternative grass seed use in New Zealand has been with grazing brome (*Bromus stamineous* Desv.) cv. Grasslands Gala for seed coat smoothing to aid seed drill-flow characteristics.

8.3.3 Quality Verification

For certified seed, strict procedural requirements are imposed by the Certifying Authority to ensure that seed authenticity is maintained from the harvest field through to the final processed product (Anon., 1994). Field storage containers used for transport to the seed processing plant are field labelled and bulk inward weights of the seed lot taken. Final processed weight is recorded for the allocation of the exact tag numbers for baggings or containers. In New Zealand these figures must contain the kind and class of seed, registered number of the grower who produced the seed and a unique reference number. All certified seed must be labelled and tagged prior to sale.

8.3.4 Seed Treatment

The concepts and technologies of selected seed treatments as reviewed by Taylor and Harman (1990) comprise a very extensive subject. As only a brief discussion will be presented in this section, refer to Taylor and Harman (1990) for further information.

Simplistically, the seed is used as the most efficient vehicle for transport to the sowing site and intimate placement of a range of materials for its own betterment. This added cost of treatment, however, must be balanced by an increase in income. Forage seeds, in contrast to vegetable and flower seeds, have a much lower expectation of return. Therefore, apart from the more specialized amenity grass market, only relatively minor cost increases above that of bare seed can be supported. That forage seed treatment is a viable entity is evidenced in the amount of commercial investment into process and product development. As a consequence there is an exclusivity level through patent rights which, in some cases, detracts from exact knowledge of processing and product technology.

Fungicide and/or insecticide application

Machines originally developed for applying slurry or liquid cereal seed treatment to curtail seed-borne diseases are also used for forage seeds, either to control seed-borne or soil-borne pathogens, or insect pests; for example, captan formulations to protect grass and legume seedlings against damping-off fungi (e.g. *Pythium* spp.), and imidacropyrid and furathioicarb to control grass grub (*Costelytra zealandica* Wh.) larvae and Argentine stem weevil (*Listronotus bonariensis* L. Kuschel) respectively. The technology involved has to ensure a desired dosage of chemical is uniformly spread over a unit weight of seed.

Seed coating

This term as defined by Taylor and Harman (1990) covers a range of processes of which, for forage seeds, 'film coating' and 'pelleting' are the most important. Film coating is the application of material dissolved in a liquid adhesive to the seed surface. Seed weight increase from this process can be up to 10%. In contrast, seed pelleting usually involves the use of a solid inert material and may increase weight–size relationships well beyond that of film coating, depending on requirements. For example, pelleting can be used for seed sizing of highest grade seeds for precision sowing to ensure maximum seed yield per seed number sown. For very small seeds such as browntop (*Agrostis capillaris* L.) ballistic capacities to promote a greater uniformity of spread at sowing are considerably improved by pelleting. Even colour has been incorporated in coating materials to promote easy assessment of white clover sowing depth (Hopkinson and Clifford, 1993).

Materials appended to seed cover a broad spectrum. These include fungicides and insecticides for disease and pest control, trace elements and bird repellants. Probably the most important legume seed treatment had been inoculation with *Rhizobium* spp. to promote nitrogen fixation. To be effective the *Rhizobium* spp. must be compatible with the species to be treated. It is also of note that product acceptability and dosage rate of chemicals can vary from species to species.

Priming

Seed priming, matriconditioning or osmoconditioning describe technologies associated with a pre-sowing hydration treatment intended to improve seedling establishment. The aim of this technology is for an earlier more uniform emergence particularly at lower temperatures. These technologies relate mainly to the vegetable industry, and are not generally used for forage seeds.

8.4 SEED STORAGE

8.4.1 Introduction

Once forage seed has been harvested, it must be stored for a short, intermediate or extended period of time. Immediately after harvest, short-term storage may be necessary before seed drying, or between drying and processing. Forage seeds moving in international trade are often stored for various periods of time in warehouses and custom houses before overseas shipments. Seed supply in any one season may exceed demand, and seed lots may be stored for sale in the following season. Small seed lots may be kept in storage for many years as a means of maintaining cultivars or in order to preserve genetic resources. The common objective of all seed storage practices is to maintain the original seed quality, particularly viability and vigour, and to protect it from damage.

Literature on seed storage and various factors affecting seed viability and seed deterioration has been reviewed in general by Barton (1961), Roberts (1972a, 1979), Kozlowski (1972), Heydecker (1973), Doerfler (1976), Justice and Bass (1978), Bass (1979), Bewley and Black (1982) and Priestley (1986). Special reference to the storage of forage seeds has been made in reviews by Gunn (1972) and Bass *et al.* (1988).

8.4.2 Inherent Factors Affecting Seed Viability

Species and cultivar

The ability to maintain seed viability differs among species and cultivars. Justice and Bass (1978) devised a 'Relative Storability Index'. They classified those plant species for which 50% or more of the seeds could be expected to germinate after 1–2 years of storage under favourable ambient conditions (Category 1), after 3–5 years (Category 2) and after 5 or more years (Category 3). Many forage plants were in Category 1. Generally, legumes appear to retain germinability better than grasses (Hanson and Moore, 1959; Rincker, 1983).

Interactions between species and long-term storage conditions with regard to maintaining viability have been reported by Canode (1972b). In short-term storage experiments Luczynska (1980) found tall oat grass (*Arrhenatherum elatius* (L.) Presl.), Kentucky bluegrass and Italian ryegrass to be more sensitive to high RH than red top (*Agrostis alba* L.), timothy and perennial ryegrass.

Cultivar effects that have been observed in other species (Justice and Bass, 1978) have not yet been reported for forage crops, although they are likely to occur, considering the differences in seed volume between diploid and tetraploid cultivars of *Lolium* and *Trifolium* species, for example.

Seed maturity

Maximum longevity can only be expected when seeds are fully mature at harvest. Immature seeds have not only a higher initial SMC, but may also be less developed, lighter and, as a consequence, less viable than mature seeds. Gaspar *et al.* (1981) have shown that lighter weight seeds in a seed lot have lower germination and shorter longevity than heavier seeds.

The weather conditions during the pre-harvest period, such as extremely high temperature, drought or excessive rainfall, can also seriously affect seed quality and storability (Bass *et al.*, 1988).

Seed dormancy and hardseededness

Harvested seeds may remain dormant for a certain period of time. A phenomenon particularly in leguminous species is hardseededness, due to the impermeability of the seed coat. Dormancy of seeds has been treated very extensively by Bewley and Black (1982). Dormancy is greatest when the seed is physiologically ripe. It decreases with ageing at a rate that depends primarily upon the temperature and moisture conditions of the storage facility (Nakamura, 1962; Kendall and Stringer, 1985). In Kentucky bluegrass and many other grasses, dormancy usually vanishes within a few weeks after harvest (Delouche, 1958). Kentucky bluegrass seed grown in Norway, however, expressed significant dormancy even after 10 years' storage under warehouse conditions (Aamlid and Arntzen, 1993).

Hardseededness in relation to seed storage has been reviewed by Justice and Bass (1978). The temperature and RH of seed storage facilities have a pronounced effect on hardseededness of clover species (Kendall and Stringer, 1985). Low temperature and high humidity conditions during storage favour the softening of hard seeds, although the rate of softening differs among species and sources of seed.

Moisture content

The initial SMC of a seed lot at harvest may vary tremendously with maturity, atmospheric moisture, species and cultivar, and contamination with excessive fresh foreign plant material such as leaves, stems and seeds (see Section 8.1). Since

SMC during storage is the most influential factor affecting seed longevity, it is important to harvest mature, relatively dry seed or to reduce the SMC of high-moisture seeds soon after harvest (Justice and Bass, 1978; see Section 8.2). High SMC combined with high temperature causes most seeds to lose viability rapidly (see Section 8.2).

The effects of different initial SMC on germination of timothy and meadow fescue seed after 3 years' storage in a warehouse have been demonstrated by Roberts (1959). In timothy, when the initial SMC was 7.7%, germination decreased from 98 to 73%, but if the initial SMC was 13.5%, germination dropped to 33%. In meadow fescue the germination at 8.9% SMC fell from 96 to 77%, but to 3% at 16.6% initial SMC. As a rule of thumb the storage life of seed is doubled for each 1% decrease in SMC in the range from 5 to 14% (Harrington, 1960). Data for the most appropriate SMC for various kinds of seeds and storage conditions were reported by Nakamura (1975). Roberts (1972b) developed equations to predict the viability of seeds after any storage period under a wide range of conditions including initial seed quality, RH and temperature. These formulae were later revised by Ellis and Roberts (1980).

Endophytes

Seeds of *Festuca, Lolium, Poa, Agrostis* and other grass species may contain fungal endophytes of the genus *Acremonium* (see Chapter 9). These endophytes may cause both beneficial effects to the host plant and detrimental effects to grazing livestock and insect herbivores (Rolston *et al.*, 1993).

The endophytes can be transmitted only by seed. Therefore, from a livestock production point of view, it is usually desirable to establish a pasture with endophyte-free seed. Various procedures have been proposed to reduce viable endophyte in the seed without impairing germination. Seedborne infection can be reduced by treatment of the seed with fungicides or heat (Welty *et al.*, 1987). It has been shown that endophyte viability in perennial ryegrass decreased to zero just by storing the seed in calico bags under ambient warehouse conditions for 12 months, and that the endophyte declined more rapidly than germination (Rolston *et al.*, 1993). Welty *et al.* (1987) found the most rapid decrease of endophyte in tall fescue and perennial ryegrass at SMC between 14 and 24%. The optimum reduction of endophyte with least reduction of germination occurred at 10°C and 19% SMC in tall fescue. In perennial ryegrass, under these conditions, endophyte decreased by 50% while germination was maintained above 95%. Endophyte viability can be maintained for many years when seed is stored at SMC < 10% and temperature < 5°C (Rolston *et al.*, 1993).

8.4.3 Effects of Harvesting and Processing

Mechanical damage during harvesting, handling and processing may influence the viability of seed in storage. Improper adjustment of harvesting or cleaning equipment contributes to such damage (Bass *et al.*, 1988; see Sections 8.1 and 8.3). Damaged seeds do not store as well as intact seeds. According to Moore (1972) small and hidden injuries in seeds may not cause immediate loss in germination, but they can become increasingly critical with ageing of the seed. Scarified seed loses its ability to germinate at a significantly greater rate than unscarified seed (Brett, 1952).

Drying

The SMC of herbage seeds coming off the combine may vary from 13% to more than 40% (Simon, 1993; see Section 8.2). In contrast, seed is considered safe to remain in bulk after threshing only if the SMC is below 12%. As a general rule, seed especially of grasses must be dried soon after harvest in order to reduce the moisture content to the appropriate safe maximum for the species. If freshly harvested grass seed is kept in bulk at ambient temperature without drying, temperature rises sharply resulting in a dramatic decline of germinability (see Section 8.2).

Seed treatment

Fungicide and ultrasonic treatment, and coating and pelleting of seeds may affect longevity. The results of pertinent investigations are at variance depending on the type of treatment, species and environmental conditions. Seeds of lucerne, ladino types of white clover and red clover were not adversely affected when treated with various fungicides after storage for 30 months at 25°C (Kreitlow and Garber, 1946). Hafenrichter *et al.* (1965) reported that treatment with 2% ethyl mercury phosphate reduced the viability of grass seed. Ultrasonic treatment increased the germination percentage and reduced the level of hard seeds in a number of small-seeded forage legumes (Kövics-Tatár and Nagy, 1984).

Seed pellets of white clover and bermudagrass (*Cynodon dactylon* (L.) Pers.) developed with kaolin clay and polyvinyl alcohol retained excellent germination when stored for up to 6 months (Smith and Miller, 1987). The effects of various coating materials and minerals on the germination of seeds of five forage plant species were studied by Lee *et al.* (1987), who found differential responses among formulations and species.

Seed handling

Seed should be handled carefully in order to minimize mechanical damage. Harvesting, cleaning and transportation may injure seed to some extent. The degree of damage depends on how well harvesting, processing and transport equipment was adjusted. The effects of mechanical damage are minimized under favourable storage conditions, i.e. low temperature and RH (Bass *et al.*, 1988).

8.4.4 The Storage Environment

Temperature and RH are the most important environmental factors affecting the viability of seeds during storage. An intricate relationship exists between the RH of the surrounding air and the SMC, resulting after a certain period of time in a moisture equilibrium (see Section 8.2). Both RH and SMC are governed by temperature. Any increase in temperature decreases RH and vice versa. The effects of temperature and air/seed moisture on the viability of seeds have been discussed in detail by Roberts (1972a), Doerfler (1976) and Justice and Bass (1978).

Temperature

The effect of temperature on the viability of seeds has been studied by numerous authors. As a general rule, the longevity of seeds increases as temperature decreases. According to the rule of thumb proposed by Harrington (1960), the lifespan of seeds is halved for each 5°C increase in seed storage temperature within a temperature range from 0 to 50°C. The detrimental effect of increased temperature is more rapid at high RH and SMC. Rincker (1983) reported that 260 seed lots from 26 cultivars of nine forage species retained germinability for 20 years when stored in cotton bags at −15°C and 60% RH. Similar results were obtained by Canode (1972b) and Acikgöz and Knowles (1983) in grass seed, Bass (1978) in crimson clover, Hafenrichter *et al.* (1965) and Hanson and Moore (1959) in a number of forage species, and Nienhuis and Baltjes (1985) in different field crop and grass seeds.

Relative humidity of the atmosphere

The RH of the surrounding atmosphere affects seed quality in two ways (Doerfler, 1976):

1. The ambient RH is directly related to the SMC which has been shown to be the most important factor for the preservation of viability of seeds.
2. Infestation, growth and reproduction of storage moulds and insects increases with increasing RH and SMC.

RH is particularly important when seeds are stored in open or porous containers (Bass *et al.*, 1988; see Section 8.2)

Dexter (1957) and Harrington (1968) published data on equilibrium moisture contents of seeds of grasses and small-seeded legumes at 23/25°C (see Table 8.4). Relative humidity values of 30, 45, 65 and 85% correspond with approximately 8–9, 10–11, 11–12, 16–18% SMC in grass seed, respectively. In comparison with grasses the moisture equilibrium in small-seeded legumes is generally 1–3% less. Although RH is influenced by temperature, a change of temperature has little effect on SMC and RH in practice (Justice and Bass, 1978).

8.4.5 Effects of Pests

Stored seed is endangered by the activity of many harmful organisms. Certain fungi, insects and rodents may cause seed deterioration.

> Saprophytic and parasitic seedborne fungi remain dormant during seed storage unless SMC increases greatly. Storage fungi are principally *Aspergillus* and *Penicillium* spp. (see Chapter 9). They invade and destroy seeds at 4–45°C and 65–100% RH. Their activity is largely determined by the physical condition, vitality and moisture content of the seed and the ambient temperature and RH of the storage area. Under favourable conditions deterioration of seeds can occur in a few days.
> (Kulik, 1978).

Microflora and seed deterioration have been reviewed by Christensen (1972). Mills (1986) described insect, mite and mould characteristics associated with stored grains. Howe (1972) reviewed insects attacking seeds during storage.

8.4.6 Packaging of Seeds

A detailed review of types, material and problems of seed packages has been published by Justice and Bass (1978).

A large variety of packages and packaging materials are available. The type to be used depends on the kind and amount of seed to be stored, the storage time, the storage temperature, the relative humidity of the atmosphere, the geographical area, transportation facilities and transportation distance (local, overseas), and whether the seed is moved from the grower to the processing plant, or destined for wholesale, retail, or for special purposes such as storage for plant breeding programmes, seed testing, plant variety control agencies, or gene banks.

Bulk seed, as it moves from grower to the processing plant (Fig. 8.1), is usually stored in wood or steel pallet boxes (Fig. 8.2). The capacity of these boxes is approximately 500–1500 kg. Subsequent storage may be in such boxes, in holding bins or in sacks. Forage crops and turf grass seed for wholesale is usually marketed in bags of 25 kg capacity (Fig. 8.3). Turf grass seed for retail is often sold in smaller units packaged to appeal to the purchaser.

Packages for processed seed may be bags from burlap, cotton, paper or plastic/foil materials, fibreboard boxes, metal cans, glass jars or containers made of various combinations of materials. Seed stored under cool and dry conditions for a short time will retain good viability in porous fabric or paper containers, but seeds stored under humid tropical conditions will lose viability rapidly without moisture protection. Justice and Bass (1978) differentiate between moisture-proof materials such as metal and glass containers, and moisture-resistant materials such as polyethylene and polyvinyl films, cellophane, pliofilm, aluminium foil, and laminations of many of these and other materials. The construction and source of 18 flexible packaging materials tested for suitability as moisture barrier seed packages is listed by Justice and Bass (1978). These authors summarized the results of

Fig. 8.1.   Seed processing plant, USA.

Harvest and Postharvest Management 203

Fig. 8.2. Steel pallet boxes used for storage of forage seeds after processing.

Fig. 8.3. Processed seed stored in 25-kg multiwall paper bags.

experiments with grass and forage legume seeds under porous and moisture barrier conditions at different levels of temperature, SMC and RH. It can be concluded from this and other information (Bean *et al.*, 1984; Bekendam and van Pijlen, 1987) that simply sealing seeds in moisture-proof containers does not necessarily guarantee safe long-term storage: 'Adequate drying before sealing is absolutely essential for safe storage of seeds in airtight moisture-proof containers' (Justice and Bass, 1978). In general, a maximum of 6% is considered to be safe for the long-term storage of seed in airtight containers although for both grasses and forage legumes, up to 10% SMC is safe. Under optimal conditions seed can retain germinability for a very long time. Evidence is provided by the report by Aufhammer and Simon (1957) on barley and oats seeds which had lain in the foundation stone of the Nuremberg city theatre and retained germinability for 123 years. The seeds were enclosed in sealed glass tubes at 7.3% SMC in barley and 8.0% SMC in oats. The average storage temperature was calculated by Roberts (1972b) to have been 10.6°C. For special-purpose long-term storage in sealed containers, the use of desiccants has been advocated; a certain amount of a drying agent like calcium oxide, calcium chloride or silica gel is included in the container. For adequately dried seeds a desiccant may not be necessary (Justice and Bass, 1978).

8.4.7 Field Performance of Stored Seed

A loss of seed vigour during storage may result in a reduction in the yield of the plants grown from such seed. The relationship between age or deterioration of seed and yield has been reviewed by Roberts (1972a) and Justice and Bass (1978). In this context it must be emphasized that chronological age of seed within certain limits is not usually correlated with viability or vigour. Deterioration of seed may affect the yield of the crop derived from it in two ways: the crop may be reduced due to a decreased plant population, and a reduced vigour of the seedlings may result in weaker plants.

Several investigations have shown that normal forage crop yields have been obtained from seed that has been stored for a prolonged period of time, provided that the initial germination and vigour of the seeds and the storage conditions were good. Plants of perennial ryegrass derived from seed lots stored for 20 years did not differ in growth habit from plants derived from 1-year-old seed (Griffiths and Pegler, 1964). However, the seed setting capacity of the plants from old seeds was markedly reduced. Rincker (1981) reported the results of experiments with lucerne, red clover, alsike clover and birdsfoot trefoil grown from 14- to 18-year-old seed stored at −15°C and 60% RH. He concluded that forage yields would not be affected by long-term subfreezing of seed if the seed had maintained good germinability during storage.

8.5 POSTHARVEST SEED CROP MANAGEMENT

8.5.1 Introduction

Many perennial forage grass and legume seed crops are harvested more than once in their lifetime. To ensure continued seed production, appropriate management of the newly harvested crop is important. Following seed harvest in late summer, preparations must begin immediately to remove crop residue and then stimulate new vegetative growth before the winter.

In perennial grasses, particularly those that require vernalization, reproductive tillers for the next season's harvest are produced in the autumn and early winter (see Chapter 2) and accumulated straw, debris or stubble from the previous harvest can seriously impair new tiller development, thereby lowering seed yields. With many perennial legumes, the stand density may increase, because of volunteer plants growing from fallen seed and regrowth of old plants, to a level where too much interplant/interstem competition will reduce seed yields (Hare, 1992). Cultural practices aimed at reducing crop density may be required in the postharvest period (Hare, 1992).

In grasses, shading can seriously reduce new tiller development. In timothy (*Phleum pratense* L.) and cocksfoot, tillers may either fail to initiate, or fail to become reproductive (Ryle 1961, 1966). Ryle (1967) similarly reported that shading inhibited tiller production in meadow fescue and perennial ryegrass. Langer (1972) showed that perennial ryegrass tiller numbers per plant declined continuously as natural light was reduced from 100 to 5%, but once shade was removed they resumed tillering at the same rate as plants that were not shaded. However, in tall fescue Hare and de Ruiter (1993) found that plants that had been shaded, tillered eight times more than the unshaded plant rate once shading was removed. Also, if sunlight penetrates to the plant base after harvest, more tillers emerge before the autumn (Ensign *et al.*, 1983). Where animals form an important part of the farming system, quick regrowth of plants after harvest is necessary for autumn forage production (Hare, 1993).

Removal of harvest stubble and straw also helps to control pests and diseases which can seriously reduce seed yield in grasses such as perennial ryegrass, Kentucky bluegrass and creeping red fescue (*F. rubra* L.) (Chilcote *et al.*, 1980) and in legumes such as lucerne (Pedersen *et al.*, 1972). Some methods of postharvest management also control weeds (Mueller-Warrant *et al.*, 1994a) and prevent thatch build-up (Chilcote *et al.*, 1980).

8.5.2 Postharvest Management Practices

Burning

Burning of grass seed crop residues began nearly 40 years ago in the USA (Hardison, 1976) for the control of the fungus *Gloeotinia granigena* (Quelet) Schumacher which causes blind seed disease in perennial ryegrass. Burning then became a common practice in many grass seed crops to dispose of residue, control diseases and weeds and, in particular, to enhance seed yields (Youngberg, 1980). In the USA open-field burning of perennial ryegrass seed crop residues nearly always produces better seed yields than non-burn methods (Chilcote, 1969; Canode and Law, 1975, 1979; Chilcote *et al.*, 1980; Ensign *et al.*, 1983; Hickey and Ensign, 1983; Young *et al.*, 1984a).

A more open canopy after postharvest burning allows more vigorous tillering in perennial grasses, and better flower induction and higher panicle production in the spring (Chilcote *et al.*, 1980). In the USA, burning stubble and straw as soon as possible after harvest increased tall fescue seed yields by more than 20% compared with baling and removing the straw (Youngberg, 1980). However, by contrast in New Zealand, burning tall fescue straw and stubble immediately after harvest produced similar seed yields to cutting or grazing (Hare, 1993). These contrasting results may be explained by the different climates. In Oregon, where most of the published work on burning has been carried out, the dry summer results in little growth of perennial grasses after harvest. The grasses are almost dormant (Youngberg, 1980). It is not until the autumn rains that tillering commences in Oregon, whereas in the North Island of New Zealand, the moist summer and autumn enables tillering to commence immediately after harvest.

In creeping red fescue seed crops in New Zealand, Hare et al. (1990) found that burning produced similar seed yields to cutting. In the USA, however, autumn tillering in burned plots of creeping red fescue began earlier and at a greater rate than tillering in unburned plots (Chilcote et al., 1980). The tillers on burned plots were exposed to a longer period of low-temperature induction, resulting in more reproductive tillers. Furthermore, tillers on burned plots received more light during the winter because of reduced canopy cover, and thus were more receptive to vernalization. Subsequently, they initiated spikelets and florets earlier which lead to a longer period of differentiation. This generally results in a larger number of florets per spikelet, more spikelets per tiller and more seeds per tiller, although the response has not been consistent in all trials involving burning (Chilcote et al., 1980; Young et al., 1984a, b; Hare et al., 1990).

It seems that burning of perennial grasses is an advantage over other postharvest practices in places with a dry summer and autumn, but not in places which have a moist summer and autumn. If burning is practised it must be completed as soon as possible after harvest and before autumn regrowth commences. Late burning has resulted in seed yields being reduced by 12–35% compared with mid-summer burning (Youngberg, 1980). Similarly, autumn burning of creeping red fescue seed fields in New Zealand has lowered seed yields by 14% in one cultivar and 66% in another (Hare et al., 1990).

Burning of legume seed crop residues has not been well documented. Generally, legumes will not tolerate as much burning as grasses, and so burning should be quick. Large windrows which burn slowly will cause excessive plant death.

Burning will control some diseases and pests in the fields. Stubble and weeds should be removed and burned from around seed crop edges, as these are potential sites for overwintering pests such as aphids.

Field burning as a practice in many seed production areas in the USA and Europe is being phased out gradually for environmental reasons, i.e. concerns over air pollution, public safety and human health from smoke inhalation (Mueller-Warrant et al., 1994a). In addition, many areas that have very dry summers and autumns have very strict laws for fear of fires getting out of control. Permits to burn must be obtained from local councils and wide firebreaks must be ploughed around the fields to be burned. In some areas the fire risk may be so high as to impose a total ban on open fire burning for several weeks. For these reasons, systems of removing crop residues by mechanical means or by grazing animals, combined with herbicides to control weeds, have been developed.

Cutting

Cutting grass seed crop stubbles to approximately 25 mm can give seed yields nearly equal to those obtained following burning (Chilcote et al., 1980; Young et al., 1984a; Coats et al., 1990). However, some species respond better to cutting than others species (Table 8.6).

In creeping red fescue fields, close-clipping (early or late) maintained seed yields (Table 8.6), but more volunteer seedlings established which could cause

Table 8.6. Effect of differing postharvest management on seed yield of three perennial grass species. (Adapted from Young et al., 1984a.)

Postharvest	Seed yield (kg ha^{-1})		
	Perennial ryegrass[1]	Smooth meadowgrass[2]	Red fescue[3]
Close-clip[4]	579	828	298
Burn[5]	713	798	305
Flail-chop[6]	588	692	261
Late close-clip[7]	595	806	328
Late flail-chop[7]	603	709	278
LSD ($P < 0.05$)	56	51	44

[1]Mean yield 1979–1982.
[2]Mean yield 1980–1982.
[3]Mean yield 1980–1982.
[4]Cut to 25 mm and straw removed.
[5]Mobile field-sanitizing machine.
[6]Cut to 75 mm and straw removed.
[7]After autumn regrowth.

rejection of fields from certification (Young et al., 1984a). In Kentucky bluegrass fields, close-clipping also gave yields equal to burning, but in perennial ryegrass, burning was the only treatment capable of maintaining seed yield over several years. In these trials (Table 8.6), cutting resulted in increased weed seed content in the grass seed crops compared with burning (Young et al., 1984b).

In the North Island of New Zealand, where regrowth commences within days of harvest, cutting creeping red fescue stubble to 10 mm after harvest and removing the straw, and cutting tall fescue stubble to 100 mm after harvest and removing the straw, produced seed yields similar to burning the stubble (Hare et al., 1990; Hare, 1993). In contrast, cutting creeping red fescue seed fields to 10 mm in early autumn significantly reduced seed yield (Hare et al., 1990), but cutting tall fescue seed fields three times in the autumn to 100 mm and leaving the cut forage to decompose on the field did not affect seed yield (Hare, 1993).

No published research comparing cutting and burning has been conducted on legume seed crops. Good quality hay can be made from legume seed-crop straw if it is baled immediately after combine harvesting, when the vegetation is still green and before too much leaf has been lost. Very stemmy vegetation can still be baled, as it is useful as bedding for animals or as compost and mulch. Some stemmy hay may not be palatable to sheep and dairy cows, but may be ideal for goats and deer. The residue from seed crops that have been chemically desiccated usually cannot be used for hay following the seed harvest.

Where livestock are an important part of the farming system, crop residues can be conserved as hay or straw or grazed 'in situ'. Even if baling or grazing crop residues does not give a seed yield equal to burning, the increased animal output and performance, combined with a lower seed yield, may still give a better economic farm return than a higher seed yield and lower animal performance.

Grazing

There has been very little work on the effectiveness of immediate postharvest grazing compared with burning or cutting. Hare (1993) found that using sheep (1000 ha^{-1}) to graze tall fescue stubble to near ground level (30–50 mm) after harvest produced seed yields similar to burning or cutting. Hare et al. (1990) also found that immediate postharvest grazing was just as effective as cutting or burning in creeping red fescue seed crops. In contrast, Coats et al. (1990) found that ungrazed stubble areas of Kentucky bluegrass outyielded grazed stubble areas. These authors did comment though that farmers generally found grazing to be better than not grazing. However, proper precautions should be taken with grasses infected with the endophyte fungi (*Acremonium* spp.) because of their association with animal health problems that include staggers and heat stress, resulting in low animal growth rates (see Chapter 5; Hoveland et al., 1983).

Legume seed crop stubbles often contain high quality residues which provide good livestock feed. However, grazing livestock should be closely monitored in legumes such as lucerne because of the anti-quality factors associated with ruminant bloat (see Howarth, 1988). On the positive side, grazing lucerne stands after seed harvest can remove a high portion of overwintering aphids (*Acrythosiphon* spp.) (Penman et al., 1979) and reduce the number of sitona weevil adults (*Sitona* spp.) laying eggs (Trought, 1981).

Cultivation

With age, grass seed crops can become overpopulated with tillers and an excessive tiller population can lead to a drop in seed yield, as intertiller competition reduces seed yield per tiller. With legumes, seed yields can decrease due to interplant and interstem competition.

Edes (1968) stated that after seed harvest, severe harrowing or gapping of tall fescue effectively reduced the density of the stand. Bean (1978) also considered that tall fescue appears to benefit from gapping in second and subsequent harvest years. Some grass crops drilled in rows less than 60 cm apart have benefited more from gapping than crops planted in wider rows. Cocksfoot seed yields, for example, have been increased 33% when 30 cm of grass was removed every 30 cm of drill row, in crops originally drilled in rows 30 or 60 cm apart (Lambert, 1963). However, when cocksfoot was grown in 91-cm rows, removing 30 cm of grass from the row reduced seed yield by 29% (Canode, 1972a).

Gapping does not benefit all grass species. Large 'clump-like' grasses such as tall fescue and cocksfoot appear to derive some benefit from it in their second and subsequent harvest years, but seed yields of timothy and meadow fescue have been reduced by gapping (Lambert, 1964). Gapping has increased seed yield of Kentucky bluegrass at low rates of nitrogen but decreased it at high nitrogen rates (Evans and Canode, 1971).

Gapping is usually done by rotary cultivators. However, herbicide spraying can also be used. Some tall fescue seed crops in New Zealand have been sown in 15-cm rows and once established have been hand sprayed with glyphosate at right angles (spray 10 cm, leave 15 cm), with no loss of seed yield (Hare et al., 1990).

In legume seed crops, inter-row cultivation or gapping has improved seed yields of white clover, lucerne and big trefoil (Table 8.7). Timing of inter-row

Table 8.7. The effect of inter-row or gapping cultivation on seed yields (kg ha^{-1}) of white clover, lucerne and big trefoil.

Crop	Inter-row or gapping	Cultivation method	
		Cultivated	Non-cultivated
White clover cv. Kent[1]	8 cm cultivated and 30 cm of plant left Cut both ways	338	275
White clover cv. Nesta[1]	8 cm cultivated and 30 cm of plant left Cut one way	339	316
Lucerne[2]	60 cm rows. Every 75 cm of row left, 45 cm of row removed	1271	1166
Lucerne[2]	91 cm rows. Every 15 cm of row left, 30 cm of row removed	1305	1166
Big trefoil cv. Grasslands Maku[3]	45 cm rows. 47 cm strips cultivated across rows, 13 cm of plant left	177	64

Adapted from [1]Lewis *et al.*, 1984; [2]Jones and Pomeroy, 1962; [3]Hare, 1992.

cultivation is important. Hare (1992) found that the most successful method was to inter-row cultivate in both spring and early summer (Table 8.7). If cultivation was only done once in mid-winter, the rows usually closed up again with the vigorous spring growth of stolons or rhizomes. Hare (1992) did find that both inter- and cross-row cultivation of the same field reduced seed yields 33% below that of uncultivated fields.

In lucerne inter-row cultivation or thinning has produced seed crops that are shorter, lodge less, are less susceptible to frost injury, flower earlier, have more upright growth allowing bees greater access to flowers, and have increased nectar secretion and concentration resulting in better pollination (Pedersen *et al.*, 1959; Pedersen and Nye, 1962; Pedersen *et al.*, 1972). Thinned plants of lucerne also have higher root carbohydrate reserves and produce more seed, more pods per stem and seeds per pod than non-thinned plants which have less carbohydrate (Dobrenz and Massengale, 1966).

Using herbicides in strips across a field or using an inter-row precision sprayer (de Lacey, 1986) will reduce white clover density and remove volunteer plants. Band spraying white clover seed crops in Wales has given a mixed response. Creosote and glyphosate sprayed in strips increased flower production by 70% (Lewis *et al.*, 1993) but further work with ethofumesate and TDA with dicamba, and MCPA and mecoprop, decreased seed yield (Marshall and James, 1986). Herbicides used in strip spraying must be chosen with care, as any systemic herbicide will lower seed yields in the unsprayed strip unless a cutting disc is used at spraying (de Lacey, 1986).

Herbicides

Herbicides sprayed in strips to reduce crop density have been discussed in the previous section. In this section the application of herbicides where open field burning is not practised will be discussed. Preliminary efforts in the USA to control weeds without field burning met with disappointing results (Mueller-Warrant *et al.*, 1994a). Some of the agressive chemicals, such as atrazine, often destroyed stands or reduced yields, while gentler treatments failed to control volunteer crop seedlings and weeds. In New Zealand though, broadcast spraying of atrazine (2.0 kg a.i. ha^{-1}) in mid winter increased seed yields of big trefoil 119% above non-sprayed treatments (Hare, 1992) and in tall fescue an atrazine rate of 3 kg a.i. ha^{-1} in the autumn initially reduced tiller numbers, but subsequently had no effect on seed yield (Hare, 1993). In the USA, simazine (0.5 kg a.i. ha^{-1}) has been used to reduce stand density of mature lucerne seed crops (Peters and Peters, 1972).

However, in recent years, countries with strict herbicide registration requirements have withdrawn many herbicides, such as atrazine, simazine and chlorprophan, from use on forage seed crops (Mueller-Warrant *et al.*, 1994a). Recent studies (Mueller-Warrant *et al.*, 1994a) have shown that in perennial ryegrass seed crops where stubble was cut after seed harvest, an application of metochlor before weeds or seedlings emerged, followed by oxyfluorfen postemergence, gave the best weed control. When both weed control and seed yield were considered in non-burned treatments, while pendimenthalin was the best pre-emergence herbicide, it still required applications of postemergence herbicides, such as diuron or oxyfluorfen, to reduce competition from volunteer perennial ryegrass seedlings and maintain seed yields (Mueller-Warrant *et al.*, 1994a). If appropriate herbicide combinations are used, the seed yield and seed quality of perennial ryegrass from non-burned residue systems are usually similar to those from burned systems in first-year stands, and can be slightly greater in second-year stands (Mueller-Warrant *et al.*, 1994b).

Nevertheless, the effect of herbicides varies greatly from site to site, depending on species, crop age, soil type, weeds present and rainfall. For example, many of the herbicides used by Mueller-Warrant *et al.* (1994 a,b) have not proven very effective in controlling seedling perennial ryegrass in second-year stands in New Zealand (R. Maxwell, Palmerston North, 1996, personal communication). Herbicides are discussed in further detail in Chapters 5 and 6.

Irrigation

In areas which have very dry summers, irrigation applied after crop residues have been removed is often an advantage. Irrigation will help to maintain plant vigour, rot away any debris left and germinate fallen seeds which can then be eradicated by inter-row cultivations or herbicide application. Irrigation enables seed crops to regrow quickly and provide autumn grazing where livestock are a part of the farming system.

REFERENCES

Aamlid, T.S. and Arntzen, D. (1993) Long-term seed dormancy in *Poa pratensis* L. in northern Norway. *Proceedings of the XVII International Grassland Congress*, pp. 1869–1871.

Acikgöz, E. and Knowles, R.P. (1983) Long-term storage of grass seeds. *Canadian Journal of Plant Science* 63, 669–674.

Anderson, S.R. (1955) Development of pods and seeds of birdsfoot trefoil, *Lotus corniculatus* L., as related to maturity and to seed yield. *Agronomy Journal* 47, 483–487.

Andersen, S. and Andersen, K. (1980) The relationship between seed maturation and seed yield in grasses. In: Hebblethwaite, P.D. (ed.) *Seed Production*. Butterworths, London, pp. 151–172.

Andersen, S. and Andersen, S. (1975) Høsttidsforsøg med Frøgraes. *Tidsskrift for Frøavl* 63, 176–184.

Andrade, R.P., Grabe, D.F. and Ehrensing, D. (1994) Seed maturation and harvest timing in turf type tall fescue. *Journal of Applied Seed Production* 12, 34–46.

Anon. (1972) Direkte mejetaerskning af graesfrø på forskelligt tidspunkt 1966–71. *Statens Redskabsprøver Medd.* no. 1089. Bygholm, Denmark.

Anon. (1994) *Seed Certification 1994–1995, Field and Laboratory Standards*. MAF Quality Management, Christchurch.

Arnold, R.E. and Lake, J.R. (1965) Direct, indirect and double threshing in herbage seed production. I: S.48 Timothy. *Journal of Agricultural Engineering Research* 10, 204–211.

Arnold, R.E. and Lake, J.R. (1966) Direct, indirect and double threshing in herbage seed production. II: S.143 Cocksfoot. *Journal of Agricultural Engineering Research* 11, 276–281.

Aufhammer, G. and Simon, U. (1957) Die Samen landwirtschaftlicher Kulturpflanzen im Grundstein des ehemaligen Nürnberger Stadttheaters und ihre Keimfähigkeit. *Zeitschrift für Acker – und Pflanzenbau* 103, 454–472.

Barton, L.V. (1961) *Seed Preservation and Longevity*. Interscience Publishers, New York.

Bass, L.N. (1978) Sealed storage of crimson clover. *Seed Science and Technology* 6, 1017–1024.

Bass, L.N. (1979) Physiological and other aspects of seed preservation. In: Rubinstein, I., Phillips, R.L., Green, C.E., and Gengenbach, B.G. (eds) *The Plant Seed: Development, Preservation, and Germination*. Academic Press, New York, pp. 145–170.

Bass, L.N., Gunn, C.R., Hesterman, O.B. and Roos, E.E. (1988) Seed physiology, seedling performance and seed sprouting. In: Hanson, A.A. (ed.) *Alfalfa and Alfalfa Improvement*. American Society of Agronomy, Madison, Wisconsin, pp. 961–983.

Bean, E.W. (1978) *Principles of Herbage Seed Production*, 2nd edn. Welsh Plant Breeding Station, Plas Gogerddan, Aberystwyth, Wales.

Bean, W.W., Sengul, S. and Tyler, B.F. (1984) The germination of grass seeds after storage at different temperatures in aluminium foil and manilla paper packets. *Annals of Applied Biology* 105, 399–403.

Bekendam, J. and van Pijlen, J.G. (1987) Doorlatendheid van enige zaadverpakkingsmaterialen voor waterdamp en vluchtige giftige stoffen. *Prophyta* 41(5), 96–99.

Bewley, J.D. and Black, M. (1982) *Physiology and Biology of Seeds*, vol. 2. Springer-Verlag, Berlin, Heidelberg, New York.

Borggaard, T., Christoffersen, E., Møller-Jensen, T. and Ubbesen, H. (1991) *Vaerd at vide om frøtaerskning*, DLF-Trifolium, Roskilde, Denmark, 56 pp.

Brett, C.C. (1952) Factors affecting the viability of grass and legume seed in storage and during shipment. *Proceedings of the VI International Grassland Congress*, pp. 878–884.

Bunnelle, P.R., Jones, L.G. and Goss, J.R. (1954) Combine performance in small legume seed harvesting. *Agricultural Engineering* 35, 554–558.

Canode, C.L. (1972a) Grass seed production as influenced by cultivation gapping and post harvest residue management. *Agronomy Journal* 64, 148–151.

Canode, C.L. (1972b) Germination of grass seed as influenced by storage conditions. *Crop Science* 12, 79–80.

Canode, C.L. and Law, A.G. (1975) Seed production of Kentucky bluegrass associated with age of stand. *Agronomy Journal* 67, 790–794.

Canode, C.L. and Law, A.G. (1979) Thatch and tiller size as influenced by residue management in Kentucky bluegrass seed production. *Agronomy Journal* 71, 289–291.

Chilcote, D.O. (1969) Burning fields boosts grass seed yield. *Crops and Soil* 21, 18.

Chilcote, D.O., Youngberg, H.W., Stanwood, P.C. and Kim, S. (1980) Post-harvest residue burning effects on perennial grass development and seed yield. In: Hebblethwaite, P.D. (ed.) *Seed Production*. Butterworths, London, pp. 91–103.

Christensen, C.M. (1972) Microflora and seed deterioration. In: Roberts, E.H. (ed.) *Viability of Seeds*. Chapman and Hall, London, pp. 59–93.

Clifford, P.T.P. and McCartin, S.J.M. (1985) Effects of pre-harvest treatment and mower and header types on seed loss and hard seed content at mowing, recovery and separation when harvesting a white clover seed crop. *New Zealand Journal of Experimental Agriculture* 13, 307–316.

Coats, D.D., Young, W.C. and Crowe, F.J. (1990) *Effects of Post-harvest Residue Management on Kentucky Bluegrass Seed Yield and Seed Quality in Central Oregon*. Seed Production Research at Oregon State University. USDA-ARS Cooperating 4–5.

Crosbie, C.J. (1972) *Heat Damage in Small Seeds*. Advisory Services Division, Ministry of Agriculture and Fisheries, Christchurch.

de Lacey, H. (1986) Precision sprayer will make our clover seed the cleanest. *New Zealand Farmer* May 22, 22.

Delouche, J.C. (1958) Germination of Kentucky bluegrass harvested at different stages of maturity. *Proceedings of the Association of Official Seed Analysts* 48, 81–84.

Dexter, S.T. (1957) Moisture equilibrium values in relation to mold formation of seeds of several grass and small-seeded legumes. *Agronomy Journal* 49, 485–488.

Dobrenz, A.K. and Massengale, M.A. (1966) Change in carbohydrates in alfalfa (*Medicago sativa* L.) roots during the period of floral initiation and seed development. *Crop Science* 6, 604–607.

Doerfler, T.H. (1976) *Seed Production Guide for the Tropics*. Germany Agricultural Team in Sri Lanka. Colombo, Sri Lanka.

Edes, R. (1968) *Grass and Clover Crops for Seed*. Bulletin No. 204. Ministry of Agriculture, Fisheries and Food, London.

Elgersma, A., Leeuwangh, J.E. and Wilms, H.J. (1988) Abscission and seed shattering in perennial ryegrass (*Lolium perenne* L.). *Euphytica* S. 51–57.

Ellegaard, H.C. (1971a) Høsttidsforsøg med almindlig rajgraes til frø samt undersøgelse over frøspildet i frømarker. *Dansk Frøavl* 54, 237–246.

Ellegaard, H.C. (1971b) Høsttidsforsøg med sildig almindlig rajgraews udført på Hofsmansgave i årene 1967–70. *Dansk Frøavl* 55, 323–331.

Ellis, R.H. and Roberts, E.H. (1980) Improved equations for the prediction of seed longevity. *Annals of Botany* 45, 13–30.

Ensign, R.D., Hickey, V.G. and Bernards, M.D. (1983) Effects of sunlight reduction and post-harvest residue accumulations on seed yields of Kentucky bluegrass. *Journal of Applied Seed Production* 1, 19–20.

Evans, D.W. and Canode, C.L. (1971) Influence of nitrogen fertilization, gapping and burning on seed production of Newport Kentucky bluegrass. *Agronomy Journal* 63, 575–580.

Falcinelli, M. (1993) Seed shattering in tall fescue. *Proceedings of the XVII International Grassland Congress*, pp. 1666–1667.

Gaspar, S., Bus, A. and Banyai, J. (1981) Relationship between 1000-seed weight and germination capacity and seed longevity in small seeded fabaceae. *Seed Science and Technology* 9, 457–467.

Griffiths, D.J. and Pegler, R.A.D. (1964) The effect of long-term storage on the viability of S.23 perennial ryegrass seed and on subsequent plant development. *Journal of the British Grassland Society* 19, 183–190.

Gunn, C.R. (1972) Seed characteristics. In: Hanson, C.H. (ed.) *Alfalfa Science and Technology*. American Society of Agronomy, Madison, Wisconsin, pp. 667–687.

Hafenrichter, A.L. Foster, R.B. and Schwendiman, J.L. (1965) Effect of storage at four locations in the west on longevity of forage seeds. *Agronomy Journal* 57, 143–147.

Hampton, J.G. (1990) Herbage seed lot vigour – do problems start with seed production? *Journal of Applied Seed Production* 9, 87–93.

Hampton, J.G. (1991) Temperate herbage seed production: an overview. *Journal of Applied Seed Production* 9 (suppl.), 2–13.

Hanson, C.H. and Moore, R.P. (1959) Viability of seeds of eight forage crop plants stored under subfreezing conditions. *Agronomy Journal* 51, 627–628.

Hardison, J.R. (1976) Fire and flame for plant disease control. *Annual Review of Phytopathology* 14, 355–379.

Hare, M.D. (1992) Inter- and cross row cultivation, atrazine application and band spraying effects on 'Grassland Maku' lotus (*Lotus uliginosus* Schk.) seed production. *Journal of Applied Seed Production* 10, 78–83.

Hare, M.D. (1993) Post-harvest and autumn management of tall fescue seed fields. *New Zealand Journal of Agricultural Research* 36, 407–418.

Hare, M.D. and de Ruiter, J.M. (1993) Seed production of tall fescue (*Festuca arundinacea* Schreb.) established under a barley (*Hordeum vulgare* L.) cover crop. *New Zealand Journal of Agricultural Research* 36, 419–428.

Hare, M.D. and Lucas, R.J. (1984) 'Grasslands Maku' lotus seed production I. Development of Maku lotus seed and the determination of time of harvest for maximum seed yields. *Journal of Applied Seed Production* 2, 58–62.

Hare, M.D., Rolston, M.P., Archie, W.J. and McKenzie, J. (1990) Grasslands Roa tall fescue seed production: research and practice. *Proceedings of the New Zealand Grassland Association* 52, 77–80.

Harrington, J.F. (1960) Thumb rules of drying seed. *Crops and Soil* 13, 16–17.

Harrington, J.F. (1968) Moisture equilibrium values for several grass and legume seeds. *Agronomy Journal* 60, 594–597.

Harrington, J.F. (1972) Seed storage and longevity. In: Kozlowski, T.T. (ed.) *Seed Biology*, vol. 3. Academic Press, New York, pp. 145–150.

Hartley, J.R. (1980) Handling herbage seed from grower to consumer. In: Lancashire, J.A. (ed.) *Herbage Seed Production*. Grassland Research and Practice Series No. 1, New Zealand Grassland Association, Palmerston North, pp. 96–98.

Hebblethwaite, P.D. and Ahmed, M. el H. (1978) Optimum time of combine harvesting for amentiy grasses grown for seed. *Journal of the British Grassland Society* 23, 35–40.

Heydecker, W. (1973) *Seed Ecology*. Pennsylvania State University Press, University Park, Pennsylvania, USA.

Hickey, V.G. and Ensign, R.D. (1983) Kentucky bluegrass seed production characteristics as affected by residue management. *Agronomy Journal* 75, 107–110.

Hill, M.J. (1975) Heat damage in freshly harvested ryegrass seed with emphasis on thermophilic fungal action during seed storage. *Australian Seed Science Newsletter* 1, 43–52.

Hill, M.J. (1982) Drying of cereals and legumes in the tropics. In: Hor Y.L. (ed.) *Proceedings of the 3rd Regional Seed Technology Workshop for ASEAN and the Pacific*, Serdang, Malaysia, pp. 110–139.

Hill, M.J. and Crosbie, C.J. (1966) Bulk handling of cereal and herbage seed crops can affect quality. *New Zealand Journal of Agriculture* 112(1), 48–53.

Hill, M.J. and Johnstone, C.R. (1985) Heat damage and drying effects on seed quality. In: Hare, M.D. and Brock, J.L. (eds) *Producing Herbage Seeds*, Grassland Research and Practice Series No. 2. New Zealand Grassland Association, Palmerston North, pp. 53–57.

Hill, M.J. and Watkin, B.R. (1975a) Seed production studies on perennial ryegrass, timothy and prairie grass. 2. Changes in physiological components during seed development and time and method of harvesting for maximum seed yield. *Journal of the British Grassland Society* 30, 131–140.

Hill, M.J. and Watkin, B.R. (1975b) Seed production studies on perennial ryegrass, timothy and prairie grass. 1. Effect of tiller age on tiller survival, ear emergence and seedhead components. *Journal of the British Grassland Society* 30, 63–71.

Hopkinson, J.M. and Clifford, P.T.P. (1993) Mechanical harvesting and processing of temperate zone and tropical pasture seed. *Proceedings of the XVII International Grassland Congress*, pp. 1815–1822.

Hoveland, C.S., Schmidt, S.P., King, C.C. Jr, Odom, J.W., Clark, E.M., McGuire, J.A., Smith, L.A., Grimes, H.W. and Holliman, J.L. (1983) Steer performance and association of *Acremonium coenophialum* fungal endophyte on tall fescue pasture. *Agronomy Journal* 75, 821–824.

Howarth, R.E. (1988) Antiquality factors and nonnutritive chemical components. In: Hanson, C.H. (ed.) *Alfalfa Science and Technology*. American Society of Agronomy Monograph 29, Madison, Wisconsin, pp. 493–510.

Howe, R.W. (1972) Insects attacking seeds during storage. In: Kozlowski T.T. (ed.) *Seed Biology*, vol. 3. Academic Press, New York, pp. 247–300.

Hyde, E.O.C., McLeavey, M.A. and Harris, G.S. (1959) Seed development in ryegrass and white clover. *New Zealand Journal of Agricultural Research* 2, 947–952.

Jensen, H.A. (1976) Investigation and anthesis, length of caryopses, moisture content, seed weight, seed shedding and stripping-ripeness during development and ripening of *Festuca pratensis* seed crop. *Acta Agriculturae Scandinavica* 26, 264–268.

Jones, L.G. (1952) Preharvest spraying to condition small-seeded legume crops for threshing. *Down Earth* 8, 2–4.

Jones, L.G. and Pomeroy, C.R. (1962) Effect of fertilizer, row spacing and clipping on alfalfa seed. *California Agriculture* February, 8–10.

Justice, O.L. and Bass, L.N. (1978) *Principles and Practices of Seed Storage*. USDA Agriculture Handbook 506. Washington, DC.

Kåhre, L. (1964) Frömognad hos vallväxter. *Växtodling* 20.

Kendall, W.A. and Stringer, W.C. (1985) Physiological aspects of clover. In: Taylor, N.L. (ed.) *Clover Science and Technology*. American Society of Agronomy, Madison, Wisconsin, pp. 111–159.

Klein, L.M. and Harmond, J.E. (1971) Seed moisture – a harvest timing index for maximum yield. *Transactions ASAE* 14, 124–126.

Komatsu, T., Shimizu, N. and Suzuki, S. (1979) Studies on development and ripening in temperate grasses. *Bulletin of the National Grasslands Research Institute* 15, 59–69.

Kövics-Tatár, M. and Nagy, J. (1984) Effect of ultrasonic treatment on the germination of the seeds of certain species of legume fodder plants. *Növénytermelis* 33, 125–138.

Kozlowski, T.T. (ed.) (1972) *Seed Biology*. Academic Press, New York.

Kreitlow, K.W. and Garber, R.J. (1946) Viability of stored seeds of forage crops treated with different fungicides. *Phytopathology* 36, 403 (abstract).

Kulik, M.M. (1978) Effects of pests and chemicals on seed deterioration in storage. In: Justice O.L. and Bass, L.N. (eds) *Principles and Practices of Seed Storage*. USDA Agriculture Handbook 506, Washington, DC, pp. 81–91.

Lambert, D.A. (1963) The influence of density and nitrogen in seed production stands of S37 cocksfoot (*Dactylis glomerata* L.). *Journal of Agricultural Science (Cambridge)* 61, 361–373.

Lambert, D.A. (1964) The influence of density and nitrogen in seed production stands of S48 timothy and S215 meadow fescue (*Festuca pratensis* L.). *Journal of Agricultural Science (Cambridge)* 63, 35–42.

Langer, R.H.M. (1972) *How Grasses Grow.* The Institute of Biology's studies in biology, No. 34. Arnold, London.

Lee, H.W., Jung, B.Y. and Kim, H.K. (1987) Studies on establishment of oversown pasture seed. I. Effects of coating materials and minerals on germination. *Journal of the Korean Society of Grassland Science* 7, 113–119.

Lewis, J., James, I. and Marshall, A.H. (1983) *Mechanical Gapping of Second Year Crops.* Report for 1982. Welsh Plant Breeding Station, Aberystwyth, pp. 122–123.

Lewis, J., James, I. and Marshall, A.H. (1984) *Mechanical Gapping of Second Year Seed Crops.* Report for 1983. Welsh Plant Breeding Station, Aberystwyth, pp. 106–109.

Lorenzetti, F. (1993) Achieving potential herbage seed yields in species of temperate regions. *Proceedings of the XVII International Grassland Congress*, pp. 1621–1628.

Luczynska, J. (1980) The rate of deterioration in germinating capacity of seed of some agricultural plants after storage at high humidity. *Biuletyn Instytutu Hodowli i Aklimatyzacji Roslin* 141, 133–143.

Marshall, A. and James, I. (1986) Evaluation of chemicals for spring gapping white clover seed crops. *Annals of Applied Biology* 108 (Suppl.), 110–111.

Marr, G. (1990) Harvest – the week before and on the day. In: Rowarth, J.S. (ed.) *Management of Grass Seed Crops.* Grassland Research and Practice Series No. 5, New Zealand Grassland Association, Palmerston North, 28–29.

Mathews, B.D. and Hill, M.J. (1967) Keeping seeds healthy in storage. *New Zealand Journal of Agriculture* 114, 25–27.

Meijer, W.J.M. (1985) The effect of uneven ripening on floret site utilization in perennial ryegrass seed crops. *Journal of Applied Seed Production* 3, 55–57.

McWilliam, J.R. (1980) The development and significance of seed retention in grasses. In: Hebblethwaite, P.D. (ed.) *Seed Production.* Butterworths, London, pp. 51–60.

Mills, J.T. (1986) Postharvest insect–fungus associations affecting seed deterioration. In: West, S.H. (ed.) *Physiological–Pathological Interactions Affecting Seed Deterioration.* Special Publication Number 12. Crop Science Society of America, Madison, Wisconsin, pp. 39–51.

Mueller-Warrant, G.W., Young, W.C. III. and Mellbye, M.E. (1994a) Influence of residue removal method and herbicides on perennial ryegrass seed production. I. Weed control. *Agronomy Journal* 86, 677–684.

Mueller-Warrant, G.W., Young, W.C. III. and Mellbye, M.E. (1994b) Influence of residue removal method and herbicides on perennial ryegrass seed production. II. Crop tolerance. *Agronomy Journal* 86, 684–690.

Nakamura, S. (1962) Germination of legume seeds. *Proceedings of the International Seed Testing Association* 27, 694–709.

Nakamura, S. (1975) The most appropriate moisture content of seeds for their long life span. *Seed Science and Technology* 3, 747–759.

Nellist, M.E. and Rees, D.V.H. (1963) A comparison of two methods of harvesting cocksfoot seed. *Journal of Agricultural Engineering Research* 8, 136–146.

Nienhuis, K.H. and Baltjes, H.J. (1985) Seed storage and germination in testing varieties for distinctness, uniformity and stability. *Seed Science and Technology* 13, 19–25.

Pedersen, M.W. and Nye, W.P. (1962) *Alfalfa Seed Production Studies.* Utah State University Agricultural Experiment Station, Utah, Bulletin 436.

Pedersen, M.W., Bohart, G.E., Levin, M.D., Nye, W.P., Taylor, S.A. and Haddock, J.L. (1959) *Cultural Practices for Alfalfa Seed Production*. Utah State University Agricultural Experiment Station, Bulletin 408.

Pedersen, M.W., Bohart, G.E., Marble, V.L. and Klostermeyer, E.C. (1972) Seed production practices. In: Hanson, C.H. (ed.) *Alfalfa Science and Technology*. American Society of Agronomy Monograph 15, Madison, Wisconsin, pp. 689–720.

Pegler, R.A.D. (1976) Harvest ripeness in grass seed crops. *Journal of the British Grassland Society* 31, 7–13.

Penman, D.R., Rohitha, B.H., White, J.G.H. and Smallfield, B.M. (1979) Control of blue-green lucerne aphids by grazing management. *Proceedings of the 32nd New Zealand Weed and Pest Control Conference* 186–191.

Peters, E.J. and Peters, R.A. (1972) Weeds and weed control. In: Hanson, C.H. (ed.) *Alfalfa Science and Technology*. American Society of Agronomy Monograph 15, Madison, Wisconsin, pp. 555–573.

Piccirilli, M. and Falcinelli, M. (1989) Anatomy of seed dispersal mechanisms in high and low seed shattering cultivars of orchardgrass. *Crop Science* 29, 972–976.

Pieroni, S.J. and Laverack, G.K. (1994) Determination of harvest date in *Lotus corniculatus* L. by pod colour. *Journal of Applied Seed Production* 12, 62–65.

Priestley, D.A. (1986) *Seed Ageing*. Comstock, Ithaca and London.

Rincker, C.M. (1981) Long-term subfreezing storage of forage seed crops. *Crop Science* 21, 424–427.

Rincker, C.M. (1983) Germination of forage crop seeds after 20 years of subfreezing storage. *Crop Science* 23, 229–231.

Roberts, E.H. (1972a) *Viability of Seeds*. Chapman and Hall, London.

Roberts, E.H. (1972b) Storage environment and the control of viability. In: Roberts, E.H. (ed.) *Viability of Seeds*. Chapman and Hall, London, pp. 14–58.

Roberts, E.H. (1979) Seed deterioration and loss of viability. *Advances in Research and Technology of Seed* 4, 25–42.

Roberts, H.M. (1959) The effect of storage conditions on the viability of grass seeds. *Proceedings of the International Seed Testing Association* 24, 184–213.

Roberts, H.M. (1969) Harvesting S352 timothy for seed. *Journal of the British Grassland Society* 24, 14–16.

Rolston, M.P., Crush, J.R., Hare, M.D. and Moore, K.K. (1993) *Lolium* endophyte viability: effect of seed storage. *Proceedings of the XVII International Grassland Congress*, pp. 1876–1877.

Ryle, G.J.A. (1961) Effects of light intensity on reproduction in S48 timothy (*Phleum pratense* L.) *Nature* 191, 196–197.

Ryle, G.J.A. (1966) Physiological aspects of seed yield in grasses. In: Milthorpe, F.L. and Ivins, J.D. (eds) *The Growth of Cereals and Grasses*. Butterworths, London, pp. 106–120.

Ryle, G.J.A. (1967) Effects of shading on inflorescence size and development in temperate perennial grasses. *Annals of Applied Biology* 59, 297–308.

Simon, U. (1993) Effects of date of harvest on yield and quality of meadow fescue seed. *Proceedings of the XVII International Grassland Congress*, pp. 1829–1830.

Smith, A.E. and Miller, R. (1987) Seed pellets for improved seed distribution of small seeded forage crops. *Journal of Seed Technology* 11, 42–51.

Smith, G. and Melton, B.V. (1967) Effects of pod maturity on yield and quality of alfalfa seed. *New Mexico Agricultural Experimental Station Bulletin* 516.

Steen, P. (1983) *Modningsforløbet i frøafgrøder af graesarter*, 154 pp.

Taylor, A.G. and Harman, G.E. (1990) Concepts and technologies of selected seed treatments. *Annual Review of Phytopathology* 28, 321–339.

Trought, T.E.T. (1981) *Sitona Weevil in Lucerne; Biology and Control.* Aglink FPP 548. New Zealand Ministry of Agriculture and Fisheries, Wellington.

Welty, R.E., Azevedo, M.D. and Cooper, T.M. (1987) Influence of moisture content, temperature, and length of storage on seed germination and survival of endophyte fungi in seeds of tall fescue and perennial ryegrass. *Phytopathology* 77, 893–900.

Williams, S. (1972) The effects of harvest date on yield and quality of seed of tetraploid hybrid ryegrass. *Journal of the British Grassland Society* 27, 221–227.

Young, W.C. III, Youngberg, H.W. and Chilcote, D.O. (1984a) Post-harvest residue management effects on seed yield in perennial grass seed production. I. The long term effect from non-burning techniques of grass seed residue removal. *Journal of Applied Seed Production* 2, 36–40.

Young, W.C. III, Youngberg, H.W. and Chilcote, D.O. (1984b) Post-harvest residue management effects on seed yield in perennial grass seed production. II. The effect of less than annual burning when alternated with mechanical residue removal. *Journal of Applied Seed Production* 2, 41–44.

Youngberg, H.W. (1980) Techniques of seed production in Oregon. In: Hebblethwaite, P.D. (ed.) *Seed Production.* Butterworths, London, pp. 203–213.

Youngberg, H.W. and Wheaton, H.N. (1979) Seed production. In: Buckner, R.C. and Bush, L.P. (ed.) *Tall Fescue*, Agronomy Series 20, ASA-CSSA-SSSA, Madison, USA, pp. 141–153.

Seed Quality of Grasses and Legumes

9

M.J. Hill, J.G. Hampton and K.A. Hill
Seed Technology Centre, Department of Plant Science, Massey University, Palmerston North, New Zealand

9.1 INTRODUCTION

The recent upsurge in worldwide interest in seed quality has created increasing awareness of its importance in crop production and has stimulated research interest in what was previously a 'second class' aspect of seed technology. Voluminous literature exists on a wide variety of research into factors affecting production and yield of seed crops. We seem to have been conditioned by the fact that seed is a 'volume' commodity which is traded on the basis of weight. For this reason seed price per unit weight is given more emphasis than its quality – the latter being often only required to be 'acceptable' rather than 'unacceptable'. As a result many seed traders fail to price seed at markedly differential prices depending on the quality level of individual seed lots. The suggestion that purchasers will often not pay a premium price for high quality but will 'overpay' for low quality is as true for seed as it is for other commodities.

The traditional concept of seed quality has been the germination ability of pure seed of the species concerned. However there are other facets involved, and the increased interest in the term 'quality' in seed has led to a greater interest in the development of relatively quick and reliable seed quality testing methods which can be used to expose seed weaknesses not detected in normal laboratory tests for purity and germination (Hill and Johnstone, 1984). Various tests for assessing seed quality have been developed to allow distinction to be made between seed lots of high potential storability and vigour, and seed lots which have already begun to deteriorate. Such tests can also be used to detect possible causes of deterioration and whether this has occurred as a result of mechanical injury during threshing, poor drying technology, damage during processing or seed treatment, or to deterioration due to respiration heating, fungal or insect activity, or poor storage conditions.

This chapter considers the components of seed quality, the use of post-harvest seed quality assessment techniques and suggests areas of technological development in seed quality control. It is our intention to develop a more expansive outlook on the term 'quality' than the simple reliance on purity and germination analysis results to describe seed planting value.

9.2 DEFINITIONS AND DIMENSIONS OF SEED QUALITY

Quality can be defined as those features and characteristics of a product that bear on its ability to satisfy a given need. Esbo (1980) defined seed quality as 'a collection of seed properties which are considered to be of importance for sowing purposes', and Thomson (1979) suggested that these properties included analytical purity, species purity, freedom from weeds, cultivar purity, germination capacity, vigour, size, uniformity, health and moisture content. Coolbear and Hill (1988) preferred to use three categories for the components of seed quality – accurate description; hygiene; viability and performance (Table 9.1).

To control quality, seed should ideally be monitored effectively at all stages involved in the growing, processing and distribution of the crop. In other words, seed quality control concerns quantifying the extent of seed care prior to sowing (Coolbear and Hill, 1988). At the simplest level, quality control systems should make it possible to screen against poor or misrepresented seed lots. At the most sophisticated level, quality control techniques should be capable of predicting seed performance in the field, or of diagnosis when things go wrong in the seed production or distribution network. Many countries have some form of seed quality control, or seed quality assurance scheme in operation, most by legislation, but others (e.g. New Zealand) by means of a voluntary system (Scott, 1980).

Procedures for assessing most of the quality components (either directly or indirectly) are published periodically by the International Seed Testing Association (e.g. International Seed Testing Association, 1996). Such procedures are internationally acceptable and widely used.

Table 9.1. Components of seed quality. (Modified from Coolbear and Hill, 1988.)

1. Description
   Species identification
   Cultivar purity
   Analytical purity
   Seed size
2. Hygiene
   Pathogens
   Storage fungi
   Noxious or prohibited weed seed contamination
   Insect and mite infestation
3. Viability and performance
   Germination potential
   Vigour and actual field performance
   Potential storability

9.3 COMPONENTS OF SEED QUALITY

9.3.1 Seed Lot Description and Uniformity

Description of the seed lot is attained primarily by means of seed certification (see Sections 9.3.2 and 9.3.3) which operates to maintain the identity of cultivars through successive generations of multiplication and involves species and cultivar verification by crop inspection, tagging and sealing of seed containers after harvest and cleaning, analytical purity analysis, and postharvest plot testing. However, before a seed lot can be accurately described, it must be as uniform as possible. Every seed lot is to some extent a mixture of pure seed, inert matter, crop seeds, weed seeds, and of live or dead seeds (Thomson, 1979). It is desirable that within a lot, the contents of each bag or bin should be exactly the same, but in practice complete uniformity is rarely achieved. Reasons for this include: differences within the crop from which the seed was harvested (e.g. maturity, lodging, disease); timing and duration of harvesting operations; lack of uniformity in threshing and subsequent processing and storage of seed from the same crop; poor blending of different seed lots; and the segregation of light and heavy seed fractions within the bulk or bag (Scott and Hampton, 1985).

The amount of variation within a seed lot can be measured by testing any seed lot attribute (e.g. number of weed seeds) for uniformity or heterogeneity (ISTA, 1996). For example, Tattersfield and Johnston (1970) showed that the variation found in perennial ryegrass (*Lolium perenne* L.) and cocksfoot (*Dactylis glomerata* L.) seed lots differed significantly from that expected from random variation. In perennial ryegrass this variation was caused by the presence of weed seeds (*Vulpia* spp. and *Bromus mollis* L.) and in cocksfoot by the percentage of multiple seed units which differed from bag to bag within the lots tested. However, white clover (*Trifolium repens* L.) and red clover (*T. pratense* L.) seed lots were generally more uniform for both purity and germination. Nevertheless, when seeds of *Plantago lanceolata* L. were present in red clover seed lots the variation within lots often exceeded that expected from random variation (Tattersfield, 1977).

Because seed analysis results and pre- and postharvest cultivar verification results are determined from a sample taken from a seed lot, they can only be applied to the seed lot as a whole if the sample truly represents that seed lot. Seed stores must be operated in such a way as to ensure the highest possible levels of seed lot uniformity (Scott and Hampton, 1985). Such variables as seed sampling intensity, sampling position in containers, and sampling instrument and technique, can all contribute to the inaccurate drawing of seed samples, particularly in species which have a reputation for heterogeneity (e.g. cocksfoot).

9.3.2. Analytical Purity

Analytical purity indicates the proportion of a seed lot which is pure seed of the species concerned. The laboratory analysis also identifies and quantifies impurities (seeds of other crops and weeds, inert matter) which may occur in the seed lot. These components are always expressed as a percentage by weight of the seed sample analysed. However, as well as the percentage figure, the nature of the

Table 9.2. Percentage of New Zealand seed lots downgraded or rejected from seed certification through failure to meet analytical purity standards. (From Rowarth et al., 1993.)

Reason	Species			
	White clover	Perennial ryegrass	Cocksfoot	Tall fescue
Presence of undesirable weed seeds	0.4	5.7	0.0	3.4
Excess other weed seeds	8.8	2.7	2.1	4.6
Total due to weeds	9.2	8.4	2.1	8.0
Total due to inert matter and seed of other crops	2.9	3.7	2.6	19.8
Total downgraded or rejected	12.1	12.1	4.7	27.8

impurities has to be considered, e.g. whether they are innocuous empty florets or potentially harmful weed seeds (Thomson, 1979).

Seed certification schemes have analytical purity standards which must be met (see Section 9.3.3) and failure to do so means downgrading or rejection from certification (Table 9.2). Although the proportion of such seed lots is generally low, reasons for the failure to meet the required purity standards may differ.

Seeds of a large range of weed species occur regularly in forage grass and legume seed crops. Some are of little consequence because they are readily removed during seed cleaning or are easily controllable in the field. Others, however, such as *Avena fatua* L., *Bromus mollis* and *Vulpia* species in grass seed crops such as ryegrass and *Trifolium dubium* Sibth., *Chenopodium album* L., *Stellaria media* (L.) Vill. and *Rumex acetosella* L. in white clover are difficult to remove, particularly when seeds of these species have been damaged during threshing by the removal of awns, lemma tips or sides in grass seed contaminants or the removal of crown awns or rough perianth coverings in legume seed contaminants. Such shape or size alteration of weed seeds during threshing of grass or legume seed crops may modify them to the extent that they are similar in size to the pure seed. Examples include *Bromus mollis* in ryegrass (*Lolium* spp.) and *Stellaria media* in white clover (Scott and Hampton, 1985). High content of 'other crop' seeds can also be a common reason for downgrading or rejecting seed lots (Table 9.2), although the problem species will depend on the forage seed crop. One example is *Trifolium dubium* in white clover seed lots.

The presence of weed seed contaminants in seed lots can create problems for seed exporters, since many countries have restrictions or prohibitions on the importation of seed lots contaminated with specific weed seeds. *Amsinckia calycina* (Moris) Chater, *Carduus nutans* L., *Cirsium arvense* (L.) Scop and *Conium maculatum* L., all of which can occur as contaminants in New Zealand forage seed lots (Scott and Hampton, 1985) are all prohibited entry into Australia; while Japan allows a maximum of only 0.03% earth particles in imported seed lots. These examples stress the importance of clean seed production and processing to ensure forage seed quality, and hence saleability, is not jeopardized on account of undesirable or excessive quantities of contaminants.

9.3.3 Cultivar Purity

For temperate forage species, a wide range of cultivars are often available for use. For example, 16 'early', 13 'intermediate', 19 'late' and 15 amenity cultivars of perennial ryegrass were inspected and approved for certification in the UK in 1986 (Anon., 1987). Each of these cultivars has been bred or selected to incorporate desirable production features, some for general use and others for specific environments.

As a rule, cultivar purity differs from analytical purity in that it cannot usually be easily determined by visual examination of seeds in the laboratory. To ensure that seed of a given cultivar is not replaced by, or contaminated with, seed of other cultivars, seed certification schemes have been set up to control seed multiplication and seed production procedures. For recent reviews of seed certification in the UK and New Zealand see Kelly and Bowring (1990) and Hampton and Scott (1990) respectively. However, with the worldwide intensive effort being made in local cultivar selection, the greater acceptance of Plant Variety Rights, and the consequent increase in specialized breeding programmes, there has been a rapid increase in new and often closely related cultivars. Cultivar verification is proving to be one of the most difficult areas of seed quality control, particularly because of the lack of standardization of even simple methods (Coolbear and Hill, 1988).

Under a seed certification scheme such as that of the Organisation for Economic Cooperation and Development (OECD), the first and last assessment of cultivar purity is carried out by means of plot testing, whereby seed lots are checked to determine whether the plants produced conform to the morphological description of the cultivar and published standards for cultivar purity (OECD, 1971). Scott and Hampton (1985) reported that 466 New Zealand forage grass seed lots, and 424 white clover and 52 red clover seed lots were tested in 1984 under this system, and only one (a red clover cultivar) was contaminated and no substitutions were detected. However, there are a number of problems:

1. Published cultivar descriptions are often very brief and inadequate to allow similar cultivars to be distinguished (Miller and Hampton, 1988).
2. In cultivars of cross-pollinated species, greater genetic mixing occurs with every generation, and not all individual plants will show the same cultivar characteristics; therefore the mean value or frequency of the trait in the cultivar will be characteristic (Young, 1987). To establish differences between cultivars and to detect contaminants, it is therefore necessary to measure plant-to-plant variation within each cultivar (Hawkins *et al.*, 1964).
3. Field plot testing is costly, labour intensive and time consuming (Scott and Hampton, 1985), and results may not be available for 8–12 months.

As the number of cultivars within a species increases, the need to maintain cultivar purity becomes even more important. Recognition of the problems with existing techniques has led to a search for more rapid and efficient cultivar identification techniques, based on laboratory or glasshouse methods (Young, 1987; Steiner, 1993). For example, Miller and Hampton (1988) and Hampton and Musukwa (1993) found that morphological data for white clover obtained from glasshouse evaluation showed no more variation than that recorded from many more plants in field plot tests, and suggested that glasshouse evaluation of Breeders and Basic seed lots may provide better cultivar quality assurance than plot testing.

Although yet to be accepted as seed certification procedures, modern methods for cultivar identification are all laboratory based. They include the use of computerized systems to capture and process morphological information (machine vision), and the use of biochemical methods to analyse various components of seeds (chemotaxonomy). These techniques have been most recently reviewed by Cooke (1995).

The use of machine vision offers considerable potential for cultivar identification (Cooke, 1995), although most work has been with self-pollinating cereals (e.g. Keefe and Draper, 1986). However, Dehghan-Shoar (1996) has recently demonstrated that even with a cross-pollinating species such as lucerne (*Medicago sativa* L.), machine vision can be used to discriminate among cultivars using a combination of individual seed characters.

As noted by Cooke (1995), it is the various techniques of protein electrophoresis which have so far had the largest impact on cultivar identification, although the approach taken is governed by whether a species is self- or cross-pollinating (see Cooke, 1988). Banding patterns of seed storage proteins separated on the basis of their molecular weight are characteristic of a cultivar and independent of environmental effects during seed production. Forde and Gardiner (1988) and Gardiner and Forde (1988) have reported that grass cultivars of perennial and Italian ryegrass, cocksfoot, *Festuca* spp., *Agrostis* spp., and *Bromus* spp., and legume cultivars of white and red clover, subterranean clover (*T. subterraneum* L.), lotus (*Lotus uliginosus* Schkuhr.) and birdsfoot trefoil (*L. corniculatus* L.), could all be separated using sodium dodecyl sulphate (SDS–PAGE) gel electrophoresis of proteins extracted from ground seed. They also noted that although individual seeds of outbreeding species such as perennial ryegrass and red and white clover produce different banding patterns, the combined population representing the cultivar remains constant unless there had been genetic shift during seed multiplication. The technique thus offers a useful and rapidly performed adjunct to the more traditional methods of cultivar identification, particularly where either complete substitution or gross contamination has occurred. The International Rules for Seed Testing (International Seed Testing Association, 1996) now contain a standardized version of SDS–PAGE for distinguishing between and identifying commercial seed lots of ryegrass cultivars and species. Several molecular methods are now also available to distinguish and differentiate cultivars, using the variation which exists in DNA. The methods include the use of restriction fragment length polymorphisms (RFLPs) using cloned probes derived from complementary DNA libraries, or random amplified polymorphic DNA (RAPDs) derived from oligonucleotide primers of arbitrarily chosen sequence using the polymerase chain reaction (Steiner, 1993). Dehghan-Shoar (1996) has reviewed the limited literature for the use of these techniques in forage species, and also demonstrated that both RFLP and RAPD could discriminate among lucerne cultivars. Whether these techniques will have a role in maintaining cultivar purity in the future is yet to be determined.

Cultivar verification ascertains that seed certification schemes are operating satisfactorily (OECD, 1982). Although field plot testing has problems, it will continue to be an important part of seed certification either in association with various laboratory cultivar verification techniques, or until it is replaced by a technique such as image analysis or some form of biochemical identification (Cooke, 1995).

As part of seed certification purity requirements, analytical purity standards must be met. One of the requirements is for species purity, e.g. in the New Zealand scheme for Basic seed of tall fescue (*Festuca arundinacea* Schreb.) no more than 0.1% of *Lolium* spp. is allowed in the purity analysis, while for Basic and first-generation seed of lucerne no more than 0.5% of red clover is allowable (Anon., 1995). When it is especially desirable to avoid contamination of one crop species by another of similar type, a larger sample may be examined, and the number of seeds of the other species counted (Thomson, 1979).

9.3.4 Seed Size

Any seed lot when harvested includes seed of different sizes. This variation is due to differences between the seeds harvested from different plants, and to differences between seeds borne on the same plant. Seed size, usually measured as thousand seed weight (TSW), may be an indicator of quality, as increasing TSW can result in improved seedling growth (Evans, 1973; Hampton, 1986). Significant positive correlations between seed size and seedling dry weight have been demonstrated for many temperate grasses and legumes (Evans, 1973; Hampton, 1986). For example, increasing seed weight increased seedling performance in field sowings in Italian ryegrass and shoot and root dry weight and root length in red clover (Table 9.3).

Despite this, some studies on the effect of seed size on germinability, plant growth and final crop performance have often been quite contradictory, and the superiority of smaller seeds to bigger ones has been reported in small-seeded

Table 9.3. Effect of seed weight on seedling performance and size in Italian ryegrass (Hampton, 1986) and red clover (Evans, 1973).

Italian ryegrass

	Thousand seed weight (g)[1]	% Field emergence	Tillers (no.m^{-2})[2]	Dry matter (g m^{-2})[2]
	3.2	68	1535	266
	4.0	82	1922	330
	5.0	85	2184	352

Red clover

Seed weight (mg)	Shoot weight (mg)	Root weight (mg)	Shoot/root ratio	Root length (cm)
1.62	5.5	1.8	3.2	35
2.20	8.0	2.8	2.9	54
2.59	9.1	3.1	3.0	60

[1] Equal number of seeds sown.
[2] At 151 days after sowing.

legumes and grasses (Moore, 1943; Wang and Hampton, 1989, 1991), the reason being that small seed may be of higher vigour than large seed (see 9.3.8).

Thousand seed weight can also be used as a cultivar descriptor where a standard is set for certified seed production. The first forage grass cultivar to have such a standard set in New Zealand was Grasslands Moata tetraploid Italian ryegrass for which a minimum TSW of 4.0 g is required. In the first 3 years of certification 19%, 42% and 9% of seed crops were rejected through failure to meet this TSW value (Scott and Hampton, 1985). Similarly, seed of Grasslands Pawera tetraploid red clover must meet a minimum TSW standard of 2.5 g to remain within the certification scheme, and any seed lot with a TSW of 2.5–3.0 g must undergo a ploidy test to check for diploid cultivar contamination (Anon., 1995).

9.3.5 Seed Hygiene

Seed hygiene is a highly significant aspect of agricultural production, with both national and international implications. Nationally, seed hygiene is an important factor in the reproduction, utilization and storage of seed crops. Internationally, seed hygiene has quarantine and quality assurance implications for both the importing and exporting of seed lots. Seed hygiene may be considered from two points of view: either qualitatively (whether seed is infected with pathogens, infested with pests or contaminated by noxious or prohibited weed seeds) or quantitatively (the level of infection, infestation or contamination).

Pathogens

A pathogen can be defined as 'any entity capable of causing a disease' and in plants, the major organisms causing disease are fungi, bacteria, viruses and nematodes. Those organisms carried in, on, or with seed are termed seed-borne pathogens, and since most crops are propagated by seed, they are subject to infection by a number of seed-borne diseases unless appropriate strategies are implemented. Some of the more important seed-borne diseases of herbage grasses and legumes are listed in Table 9.4.

Control of seed borne diseases should be directed against those parts of the life cycle of the pathogen that are most sensitive (Neergaard, 1979) through:

1. Avoidance or elimination of the pathogen, by means of appropriate quarantine precautions for imported seed, or eliminating pathogens in domestic seed lots through health testing, seed certification, seed treatment, etc. For example, New Zealand quarantine regulations require imported seed of various grasses to be tested for the presence of dwarf bunt (*Tilletia contraversa* Kuhn), a pathogen not established in New Zealand; and seed certification regulations for *Bromus* spp. require that seed crops be rejected from certification at field inspection if standards for the presence of head smut (*Ustilago bullata* Berk.) are exceeded, and recommend that all *B. willdenowii* Kunth and *B. sitchensis* Trin. seed be appropriately fungicide treated before sowing. Similarly, seed certification regulations for lucerne cultivar Wairau state that the crop will be rejected if *Corynebacterium insidiosum* (McCulloch) Jensen (bacterial wilt) is found at field inspection or following a serological health test (Anon., 1995).

Table 9.4. Some important[1] seedborne pathogens in common forage grasses and legumes. (Modified from Richardson, 1979.)

Pathogen	Disease	Hosts
Grasses		
Claviceps purpurea	Ergot	*Agropyron, Bromus, Lolium, Dactylis, Festuca, Paspalum, Phleum, Poa*
Gloeotinia granigena	Blind seed disease	*Agrostis, Bromus, Cynosurus, Dactylis, Festuca, Holcus, Lolium, Phleum, Poa*
Drechslera spp.	Leaf spots	*Agrostis, Bromus, Dactylis, Festuca, Lolium, Phleum, Poa*
Ustilago bullata	Head smut	*Agropyron, Agrostis, Bromus, Festuca.*
Legumes		
Alfalfa mosaic virus	AMV	*M. sativa*
Aureobasidium caulivorum	Scorch	*T. pratense*
Botrytis anthophila	Anther mould	*T. pratense*
Cercospora zebrina	Black stem/leaf spot	*M. sativa, T. pratense, T. repens, T. subterraneum*
Colletotrichum trifolii	Anthracnose	*M. sativa, T. pratense*
Corynebacterium insidiosum	Bacterial wilt	*M. sativa*
Ditylenchus dipsaci	Stem eelworm	*M. sativa, T. pratense, T. repens*
Peronospora trifoliorum	Downy mildew	*M. sativa, T. pratense, T. repens*
Phoma medicaginis	Spring black stem	*M. sativa*
Pleospora herbarum	Leaf/ring spot	*M. sativa, T. pratense, T. repens*
Rhizoctonia leguminicola	Black patch	*T. pratense, L. uliginosus*
Sclerotinia spp.	Sclerotinia rot	*M. sativa, T. pratense, T. repens, T. subterraneum*
Verticillium alboatrum	Wilt	*M. sativa, T. repens*

[1]Economic importance may vary between hosts, countries and seasons.

2. Reducing established inoculum in the soil, in alternate hosts and plant residues and in seed. For example, burning or deep ploughing of crop residues (including sclerotia and blind seed) after harvest is one method of reducing inoculum levels of *Claviceps* spp. (ergot) and *Gloeotinia granigena* (Quel.) T. Schumacher (blind seed disease). However, for environmental and agronomic reasons this is not always possible, and both these diseases continue to adversely affect quality in grass seed crops. Cagas (1987) reported that ergot levels had increased in Czechoslovakia, affecting 13% of Italian ryegrass, 42% of red fescue (*Festuca rubra* L.) and 77% of Kentucky bluegrass (*Poa pratensis* L.) seed crops in 1985 and resulting in downgrading or rejection from certification and important losses to grass seed growers. In New Zealand, poor germination in perennial ryegrass seed lots is still often explained by the presence of blind seed disease, as although nationally the disease incidence has declined (Hampton and Scott, 1980), localized outbreaks may still occur (Hampton *et al.*, 1995). Although sclerotia of clover rot (*Sclerotinia trifoliorum* (Lib.) deBary) are often removed during

seed cleaning, control of the disease in the field is difficult because of the longevity of sclerotia in soil and the pathogen's wide host range (Latch and Skipp, 1987).

3. Slowing down development and spread of inoculum by utilizing climatic differences, influencing the microclimate, or impeding the spread of inoculum through isolation, vector control and hygiene. For example, Hampton (1984) recommended that seed crops of paspalum (*Paspalum dilatatum* Poir) be grown in an area with a reliable dry spell in early to mid summer in a position preferably isolated from volunteer plants, to reduce the risk of ergot infection caused by *Claviceps paspali* Stev. and Hall. Also, resistant cultivars of lucerne are used to control bacterial wilt and nematodes (*Ditylenchus dipsaci* Kuhn) Filipjev (Close et al., 1982).

4. Improving conditions for plant development through good agronomic management. For example, perennial ryegrass seed crops require around 130 kg of available nitrogen per hectare and Hampton (1987) has demonstrated that providing this requirement can improve seed quality by reducing blind seed disease incidence (Table 9.5).

In legumes such as red clover Fulton and Hanson (1960) have concluded that control of *Fusarium* spp. (root rot) involves encouraging vigorous plant growth through suitable lime and fertilizer use, proper crop rotation and avoidance of plant stress.

Storage fungi

Storage fungi can invade and destroy seeds over a wide range of temperatures (4–45°C) and relative humidities (65–100%) (Christensen and Kaufmann, 1969). Their activity is largely determined by the physical condition, vitality and moisture content of the seed and the ambient temperature and relative humidity of the storage area. Consequently, the population of storage fungi reflects the kind of postharvest handling, conditioning and storage environment of a given seed lot (Kulik, 1995).

Much of the information we have on seed storage fungi has come from studies on stored cereals, but their activity is also an important dimension of quality in forage grasses and legumes (Hill and Crosbie, 1966). The strong insulative properties of small seeds rapidly result in high seed temperatures as a result of fungal growth. Fungally heated 'hot spots' in seed commonly reach temperatures in

Table 9.5. Effect of nitrogen application rate on seed quality in perennial ryegrass. (After Hampton, 1987.)

Applied-N (kg ha^{-1})[1]	Total-N (kg ha^{-1})[2]	% Blind seed disease[3]	% Germination[3]
0	30	44	13
50	80	41	21
100	130	24	62
150	180	20	64

[1]As urea at spikelet initiation.
[2]Applied plus soil residual.
[3]Of harvested seed.

excess of 50°C, and may rise higher if the seed lot contains damaged seed. Even in seeds stored at a low moisture content, pockets of moist seeds can arise due to roof leaks, insect activity and moisture translocation when temperature gradients within the seed mass are allowed to occur (Kulik, 1995). As Christensen (1957) has stated 'heating is likely to be the final and violent effect of mould invasion of the seed, and not an indication of the beginning of deterioration'.

In grasses and legumes the predominant storage fungi which reduce seed quality are members of the *Aspergillus glaucus* group (*A. glaucus*, Link: Fr., *A. hollandicus* R.A. Samson and W. Gams, *A. chevalieri* (L. Mangin) Thom and Church, *A. reptans* R.A. Samson and W. Gams, *A. rubrobrunneus* R.A. Samson and W. Gams, *A. restrictus* G. Sm., *A. ochraceus* K. Wilh., *A. niger* Tiegh, *A. flavus* Link: Fr.), *Penicillium* species and *Rhizopus stolonifer* (Ehreub: Fr.) Vuill. Excellent backgrounds to the effects of fungi on stored seeds have been published by Christensen (1982) and Kulik (1995).

Storage fungi contaminate seed lots after they have entered a seed store, and little if any infection is of field origin (Christensen and Kaufmann, 1974). Their principal deteriorative effects involve invasion and killing of the embryo, reducing germination, discoloration of seed, increased free fatty acid levels, decreased non-reducing sugars, production of mycotoxins, rapid increases in seed lot temperature through the heat produced as a by-product of respiration, and eventually caking and decay (Hill and Johnstone, 1985). Market acceptability of such seed is low, and seed lots may deteriorate quickly to a stage where they are unsaleable.

Seed lots free of pathogens and with high germination may rapidly deteriorate in storage because of the effects of storage fungi (Table 9.6).

Detection of 'heating damage' by aseptic plating of surface sterilized seed onto 7.5% salt (NaCl)-enhanced malt agar or potato-dextrose agar is becoming a more widely used and recognized technique for detecting storage-fungal heating injury. The level of storage fungi and the species present can be used to determine whether microbiological heat, and hence seed deterioration, have occurred in individual seed lots (Matthews and Hill, 1967; Hill and Johnstone, 1985).

Insect and mite infestation

During the analytical purity analysis, larvae and/or adults of various insects may be detected. These may be species which contaminate the seed lot during seed production e.g. *Bruchophagus* spp. (legume seed chalcid) in *Trifolium*, *Medicago* and *Lotus* spp. (Berlage *et al.*, 1986) or storage pests such as the grain weevil *Sitophilus*

Table 9.6. Relationship between time of storage, incidence of *Aspergillus glaucus* and percentage germination in seed of hybrid ryegrass *(Lolium × boucheanum* L.). (After Matthews and Hill, 1967.)

Time in storage (h)	% A. glaucus	% Germination
16	6	91
32	89	72
56	100	20
96	100	15

granarius L. There are many major seed storage insect pests (Neergaard, 1979), but they are usually of greater significance in cereals than temperate grass seeds. However, poor seed store hygiene can allow insect and mite infestation of grass and legume seed lots resulting in seed heating, moisture migration and lowered germination.

The presence of live insects in seed lots can create problems for seed exporters (Berlage *et al.*, 1986), as quarantine regulations of importing countries may either prohibit the importation of such lots (e.g. Japan) or require proof of fumigation of such lots before importation (e.g. Australia).

Endophyte

Endophytic fungi such as the *Acremonium* spp. have a wide distribution in many grasses. They are seed borne but not pathogenic, their relationship with host grasses being described as mutualistic symbiosis (Siegal *et al.*, 1987). In perennial ryegrass, *A. lolii* Latch, Christensen and Samuels acts as a biological control agent against at least four New Zealand pasture pests (Prestidge *et al.*, 1994), but can also cause severe animal health problems (Rowan, 1993). The discovery of low- or zero-alkaloid producing endophytes has led to the production of 'novel' endophytes which do not affect animal health. These fungi are protected under Plant Variety Rights (PVR) legislation, and ryegrasses with novel endophytes are treated as a new cultivar under the New Zealand Seed Certification Scheme (Anon., 1995). Seed production systems need to ensure that either endophyte viability is protected, or that endophyte-free status is maintained (see Chapter 6).

9.3.6 Seed Damage and Deterioration

Simple visual examination of seed may not enable the detection of mechanical damage unless such injuries are very extensive. Moore (1972) has listed the mechanical injuries detected in standard growth tests as detached seed structures, breaks within structures, abnormally shaped structures, scar tissues, infections, restricted growth, uneven placement of cotyledons, unnatural shrinkage of cotyledons, abnormally twisted hypocotyls and primary roots, and dwarfed and twisted roots with blunt tips of dull appearance. Although tetrazolium tests have proved very useful in detecting and assessing the extent of mechanical injuries (Moore, 1969), this test suffers from the fact that 2-3-5 triphenyl tetrazolium chloride/bromide is toxic to seed, preventing verification of the effect of seed damage in subsequent germination or seedling growth tests. The increasing interest in the use of X-ray radiography has overcome this problem. The X-ray method allows visual expression of internal structural damage or deformity but is non-destructive, allowing the examination of seedlings from X-rayed seed in positional germination tests (Hill and Hill, 1994).

While major seed damage is immediately reflected in reduced germination, more minor damage may not show up in germination tests carried out immediately after harvesting or processing. The adverse effects of even minor damage, including non-visual bruising are, however, magnified in storage, resulting in significant and often rapid reduction in the viability and vigour of stored seeds.

The extent of seed damage depends on the type of seed, its shape and size, seed-coat thickness, embryo structure and position, and seed moisture level. Large-seeded dicotyledonous seeds such as peas are remarkably more susceptible to mechanical damage than small-seeded grasses. However, the hardseededness of many legumes does provide some protection against mechanical damage, as do the lemma and palea coverings in grass seeds (Basu, 1995).

It has been general experience that very dry seeds suffer fractures or mechanical impact damage while, in the same conditions, high-moisture seeds are bruised but intact. Different species have different optimum moisture levels for minimizing mechanical damage and subsequent poor storability following rough manual or machine handling operations (Hill and Crosbie, 1966).

Because there is often a need to remove moisture from harvested seeds prior to storage, seed drying technology has become a major field of study for seed technologists and engineers. Commercial seed drying is a highly specialized job, the success of which, in terms of retaining high seed quality, depends on not only the final seed moisture content, but the initial seed moisture, drying temperature and drying rate (Nellist, 1981).

Over the past 25 years there has been considerable research into the mechanisms of seed deterioration. Despite this, the literature in some areas is, at best, confusing. It is clear there is still a great deal to learn about the relative importance of the range of events which occur as seeds age, how they interact with each other, and how they are affected by the prestorage history of seeds and different ambient storage conditions (Coolbear, 1995).

Two fallacies exist about seed deterioration, as Coolbear (1995) has clearly shown. The first is 'that seed deterioration is always irreversible'. This simply overlooks the fact that seeds not only have inbuilt systems to counter the impact of deteriorative events, but also, given the right conditions, have active mechanisms for self-repair. The second fallacy is 'that seed deterioration is a sequence of events', a suggestion which goes back to the early 1970s (Heydecker, 1972; Delouche and Baskin, 1973). Sequential models as proposed imply that all aspects of loss of seed quality (decreased germination rate, decreased storability, decreased stress tolerance, etc.) are in the same continuum which ultimately leads to loss of seed viability. There is increasing evidence this is not the case. Coolbear *et al.* (1984) and pioneering work on self-repair in seeds by Villiers (1973) suggest that one of the reasons partially deteriorated seeds germinate more slowly is that they take additional time to undergo the necessary processes of self-repair.

9.3.7 Germination Potential

Seed testing has been developed to minimize the risk of sowing seed that does not have the capacity to produce an abundant crop of the required cultivar. Seed quality is therefore assessed to ultimately determine the value of seed for planting (International Seed Testing Association, 1996). To be of any value for planting purposes, pure seeds of the cultivar concerned must be viable (capable of growing), be able to germinate (capable of producing the structures essential to develop a strong healthy seedling) and be able to establish (capable of producing a plant in the field).

Historically, seed quality has been synonymous with germination, i.e. the measurement of the percentage of seeds growing normally under standardized, controlled, optimum laboratory conditions, set so that the seed is given every chance to germinate to its full potential. Germination in a laboratory test is defined as the emergence and development of the seedling to a stage where the aspect of its essential structures indicates whether or not it is able to develop further into a satisfactory plant under favourable conditions in soil (International Seed Testing Association, 1996). Germination data are therefore used to provide information regarding the field planting value of the seed lot, and to provide results that can be used to compare the value of different seed lots.

The germination of forage grasses and legumes may vary among species, among cultivars of the same species and among seed lots of the same cultivar. In legumes this situation is further complicated by the presence of hard seeds; that is, seeds which fail to imbibe water within the prescribed test period. In hand-harvested legumes, seed is often more than 90% hard, but mechanical harvesting and cleaning (including scarification), usually results in sufficient physical stress on the seed to reduce the hard seed content to acceptable levels. This is particularly so for white clover where hard- seed levels are rarely greater than 10%. However, hard seed can be a problem in other *Trifolium* spp. and *Medicago* and *Lotus* spp. (Scott and Hampton, 1985; Charlton *et al.*, 1986).

In the absence of dormancy and excluding hard seed, low germination is an indication of seed lot deterioration, because of the production of abnormal seedlings and presence of dead seed. Abnormal seedlings can be a result of harvesting immature seed, harvesting seed at high-moisture content, wetting and drying of a swathed crop prior to harvest and heating or insect damage (e.g. *Coleophora* spp. – clover case bearer moth), (Hill and Johnstone, 1984). In forage grasses low germination is more often associated with dead seeds than abnormal seedlings (Table 9.7). In *Lolium* spp., dead seeds usually result from one or more of three factors – blind seed disease, heating damage or immature seed (Hill and Crosbie, 1966; Hill, 1975; Hampton and Young, 1985).

Little published information exists as to reasons for low germination in species other than *Lolium*, but Scott and Hampton (1985) suggested that seed lots of *Cynosurus*, *Dactylis* and *Bromus* often contain immature seed, while seed lots of *Phleum* and *Holcus* contain hulled seed, the embryo having been damaged during threshing. Blind seed disease may also be a factor affecting germination in species other than ryegrass (Richardson, 1979).

Table 9.7. Mean percentage of germinated seed, abnormal seedlings and remainder (dead seeds) within two germination categories for Italian ryegrass cultivar Grasslands Moata seed lots. (Hampton and Young, 1985.)

Germination category	Total number of seed lots	Mean percentage		
		Germination	Abnormal seedlings	Remainder
< 70%	29	58	2	40
> 90%	38	93	1	6

9.3.8 Vigour and Field Performance

Information about how seed vigour is acquired during seed development and subsequently lost through deterioration is important in planning seed production and predicting likely crop performance. Seed vigour is seldom a priority objective in plant breeding and its development in the field is not understood. Despite this we know that quantitative vigour differences occur between seed lots grown in the same environment and harvested in the same year (Hampton, 1991). Continued research to explain such differences and the development and acceptance of consistent vigour testing methods in seed testing laboratories which will provide accurate, reliable and meaningful information to the crop producer remain a high priority. Vigour tests have been described and appraised by Hampton (1995).

Forage grass and legume seed lots with high germinability often differ in vigour, reducing establishment, dry matter production, and performance in storage (Table 9.8). Seed vigour is not a single measurable property like germination, but a concept describing seed performance associated characteristics (Perry, 1981) and which encompasses potential seed performance both in the field and in storage (Hampton and Coolbear, 1990). Loss of vigour precedes loss of germination (Hampton and Hill, 1990), so although two seed lots may have similarly high laboratory germination values, one can be physiologically older than the other (more deteriorated), so its vigour is low and it cannot perform as well as the high vigour lot (Table 9.8).

Seed lot vigour becomes important when environmental stresses occur at sowing (e.g. low autumn soil temperatures). This is particularly important in species which are more temperature sensitive, and also when species are sown in mixtures and seedling competition becomes a factor (Hampton and Hill, 1990). If, as a seed buyer, there is a choice between several seed lots with similar high germination values, it would be advantageous to know the vigour status of each seed lot before selecting the one to buy.

Germination results often correlate well with field emergence under favourable field conditions, although it is usual to expect a lower value in the field than is indicated by a germination test result. Correlations of 0.66 and 0.89 have

Table 9.8. Field and storage performance of herbage seed lots which germination data indicate are of similar quality. (Adapted from Hampton, 1991.)

	Red clover[1]				Italian ryegrass[1]			
Seed lot	1	2	3	4	1	2	3	4
% Germination	90	90	90	90	96	95	92	94
% field emergence	76	56	78	80	90	67	79	87
	Red clover[2]				Tall fescue[2]			
Seed lot	1	2	3	4	1	2	3	4
Prestorage % germination	90	90	90	90	90	91	90	88
Poststorage % germination	90	66	71	91	90	73	58	24

[1]Sown in replicated adjacent rows.
[2]Ambient storage for 12 months.

been reported for perennial and Italian ryegrass respectively (Naylor and Hutcheson, 1986; Naylor and Syversen, 1988; Table 9.8). However, in both these experiments 'field conditions' were similar to laboratory germination conditions. More usually, field conditions at sowing are somewhat more removed from laboratory conditions, and seed lots which have similar laboratory germination values may differ in emergence when sown in the field (Naylor and Syversen, 1988). Such differences in field performance are termed differences in seed vigour and may be explained in terms of the physiological age of the seeds. Maximum germination and vigour is achieved at physiological maturity (prior to harvest). From this point on, seed begins to die, the rate and extent of this physiological deterioration depending upon many factors (Scott and Hampton, 1985). Physiologically older seeds (which are not necessarily chronologically older) may be successful at producing high germination test results and seedlings in good field conditions, but are less able to perform under conditions of stress (e.g. wet or cool soils) in the field, producing poorer emergence and establishment. For example, Naylor (1981) demonstrated that vigour differences existed among Italian ryegrass seed lots. Also, Wang and Hampton (1989, 1991) showed that vigour differences existed among red clover seed lots. These authors have suggested that selection of seed lots with a high vigour rating could result in improved emergence and establishment.

Compared with other species, the vigour status of forage seed lots has received little attention. Various vigour tests have been developed to identify differences in planting value, e.g. the use of techniques which accelerate the ageing of seed lots has led to improved estimates of field emergence in a range of species (Hampton and TeKrony, 1995). In such tests, the greater the reduction in germination after ageing, the lower the vigour. Clark (1982) and Hampton and Bell (1989) have used such tests to investigate seed vigour in *Cynosurus cristatus* and *Bromus willdenowii* respectively, while Helmer et al. (1962) and Wang and Hampton (1993) have used such tests to investigate seed vigour in *Trifolium* spp. and *Medicago sativa*. The conductivity test (Hampton and TeKrony, 1995) has also been used successfully to detect vigour differences among forage seed lots. Standardized methodology for the conductivity test is now available for perennial and Italian ryegrass, birdsfoot trefoil, lotus and red clover. Similarly, standard procedures for accelerated ageing have been developed for smooth brome (*Bromus inermis* Leyss.), prairie grass, crested dogstail, tall fescue, perennial ryegrass, lucerne, red clover and crimson clover (*Trifolium incarnatum* L.) (Hampton and TeKrony, 1995).

9.3.9 Potential Storability

The interaction between storage temperature, seed moisture content and seed longevity is well established and there have been many reports of the deleterious effects of unsuitable storage conditions on the viability of forage seeds (Hampton and Bell, 1989). However, differences between seed lots in their ability to retain germination during storage may also be attributed to differences in seed vigour (Table 9.8; Delouche and Baskin, 1973).

Ellis and Roberts (1981) have demonstrated that seed survival in storage can be predicted by accelerated ageing techniques, but further work is required to confirm their usefulness with forage grasses and legumes.

9.4 AREAS OF TECHNOLOGICAL DEVELOPMENT IN SEED QUALITY CONTROL

The general appreciation that dry seed will gain moisture by absorbing water from humid environments in storage has led to an increasing interest in the use of moisture barrier storage systems, whether these involve controlled storage environments or the use of moisture-proof/resistant packaging materials. Perhaps on the grounds of cost alone in many of these systems, future technology may investigate the commercial feasibility of coating seeds themselves with moisture-resistant barriers which effectively resist the adverse effect of humid climate storage. An interesting possibility is the coating of seeds with polyvinyl alcohol which has been shown to effectively resist the adverse effect of accelerated ageing in soybean (West *et al.*, 1985), or, as Basra (1995) suggests, the paradoxical situation where high storability of dormant seeds is maintained in a fully imbibed state. The physiological basis of this improved storability has been discussed by Villiers and Edgecumbe (1975) in terms of activity of an endogenous repair system in the hydrated state. Also, despite well-documented evidence that the storability of many seed species significantly increases with reduction in oxygen levels, and the success of sealed storage of seeds at low moisture contents, variable results are often obtained commercially. The likely involvement of lipid peroxidation in seed ageing implicates a profound role of oxygen (Basra, 1995). This suggests that the beneficial effect of antioxidants on seed longevity has considerable physiological relevance, and potential commercial interest. Similarly, the future may see increased development of the role of inert-gas sealed storage, e.g. nitrogen, to maintain long-term seed viability in storage.

Considerable research interest, particularly over the past 20 years, has been generated in the area of physiological seed treatments, whether these involve dry permeation techniques employing solvents such as acetone for introducing anti-ageing chemicals into seed or so called 'wet' treatments in which seeds are hydrated and then dehydrated to safe levels of moisture for storage (Francis and Coolbear, 1987). The mechanism of action of the seed vigour and viability maintenance these seed treatments confer is not fully understood, but may be related to reduced lipid peroxidation during storage (Saha *et al.*, 1990). This is an exciting research area for the future development of simple methods for maintaining seed quality in storage.

Separate hydration has also been used to improve seed performance prior to sowing. Techniques include pregermination, osmotic priming and matric priming, and the topic has been recently reviewed by Pill (1995). While success has been achieved for many vegetable species (Pill, 1995), there are as yet few reports of the use of these techniques for forage species. This also is a potential area for further research.

Many attributes of seed quality are affected by husbandry and harvesting procedures, and are also subject to varying degrees of postharvest modification. Good knowledge of plant morphology and physiology promotes a better understanding of the management required to consistently produce quality seed crops. Differences in seed quality may occur between seed lots of the same cultivar because of variation in climate and management between farms. As one example, between-lot variation in certified seed of S23 perennial ryegrass in different UK counties or regions has been demonstrated by Bean (1980). Management factors which may affect seed quality during crop growth include the quality of the seed sown, mineral nutrition (particularly nitrogen), weathering damage in the field, moisture stress, pest and disease infestation, e.g. blind seed disease in ryegrass; interpollination contamination between grass species (Hampton, 1994); poor weed control, etc. These are all important research targets for future work on seed quality improvement.

Environmental stresses occurring during grass and legume seed-crop development have the potential to greatly reduce yield, but their effects on the ability of seed to acquire maximum germination ability and vigour have received very little attention (Dornbos, 1995). For example, it is unclear if stress during flowering and pollination predisposes the ability of ovules to develop into viable and vigorous seeds in a positive or negative way. Similarly, little is understood about the interaction between the production environment and subsequent seed quality during the period between pollination and pod elongation in legumes or early caryopsis development in grasses. This is despite considerable information on the high levels of abortion (lack of seed set) which occur irrespective of the presence or absence of stress in both grasses (Hill, 1975) and legumes such as birdsfoot trefoil (Hill and Supanjani, 1993) and white clover (Pasumarty et al., 1993). We do, however, have considerable information on the contribution environmental stress (particularly for moisture) can have on reduction in seed quality (both germination and vigour) during the period from physiological- to harvest-maturity (Green et al., 1966; Delouche, 1980).

9.5 CONCLUSION

The most obvious conclusion about much of the research on seed quality must be that it is generally inconclusive; many data are highly variable, contradictory and/or simply not sound. Major differences in the susceptibility of seeds to loss of quality occur with different species, cultivars and seed lots. There is well-documented evidence of species and cultivar differences in their tolerance to the postharvest/prestorage treatment of seeds, and to genotypic variation in vigour and storability. Quite a lot is known about seed quality – we trade on it, we research it and we pay reverence to it. What we really need, however, is a clearer and more precise understanding of why and how seeds lose quality on the plant, at harvest, during processing and during storage, and to take into account the interactions between different storage conditions and the seed's prestorage history. Such an approach to the term 'seed quality' would give us clearer ideas on how to more fairly assess the planting value of seeds prior to sowing.

Many of the basic techniques of seed certification and germination testing are relatively straightforward, with well-established and standardized procedures. However, although many agricultural economies have new or specialized cultivars available, the supply of reliable quality improved seed is often a limiting factor.

So, too, is an appropriate extension programme designed to motivate and maintain the confidence of farmers (Hill, 1977).

In many regions of the world, seed quality control schemes have been established but have lost impetus. Such lack of progress is not necessarily due to lack of available technology, but may often occur through a failure to apply existing procedures properly in the right circumstances or through inefficiency in detecting malpractice. In such situations it is common to find a loss of confidence by the very farmers for whom the programmes were designed (Hill, 1977).

While continued improvements in the technologies of regulation will unquestionably make the task of quality control easier, failures in this area often stem from the lack of personnel adequately trained in the application of existing techniques or the lack of a fully independent, properly equipped, impartial control agency with adequate legal authority. Although there are sufficient market incentives for countries such as New Zealand to maintain voluntary certification systems, it is presently difficult to see how such voluntary systems can be maintained efficiently in developing agricultural economies. The importance of a Seed Act in the maintenance and development of a seed programme cannot be overemphasized. Governments must be prepared to support such legislation with adequate finance. This support must be channelled in two directions: first, to provide for adequate training, facilities, salaries and status to enable employees of regulating bodies to carry out their tasks efficiently and impartially; and second, to develop and maintain the necessary extension package for farmers, including an incentive price for the purchase and use of high-quality seed.

Ultimately, the success or failure of a new cultivar depends on its acceptance by local farmers. Often the criteria on which they make their decisions are unclear, but experience has repeatedly shown that high seed quality is one of the most important factors contributing to the acceptance of improved cultivars over home-saved seed. It would be a tragedy if a great deal of the effort put into plant breeding and its ancillary sciences continued to be wasted because of inadequate seed quality control.

By now we have probably confused, for ever, the definition of what 'seed quality' really is. It is more than just 'good seed'. It is a complicated physiological, morphological and environmental interactive property which we are only slowly beginning to unravel.

9.6 REFERENCES

Anon. (1987) *Bulletin of Crop Varieties and Seeds 1986/87*. National Institute of Agricultural Botany, Cambridge, UK.

Anon. (1995) *Seed Certification 1995–1996. Field and Laboratory Standards*. MAF Quality Management, Lincoln, New Zealand.

Basra, A.S. (ed.) (1995) *Seed Quality – Basic Mechanisms and Agricultural Implications*. Food Products Press, Binghamton, New York.

Basu, R.N. (1995) Seed viability. In: Basra, A.S. (ed.) *Seed Quality – Basic Mechanisms and Agricultural Implications*. Food Products Press, Binghamton, New York, pp. 2–42.

Bean, E.W. (1980) Factors affecting the quality of herbage seeds. In: Hebblethwaite, P.D. (ed.) *Seed Production*. Butterworths, London, pp. 593–604.

Berlage, A.G., Kamm, J.A. and Bilsland, D.M. (1986) Separation of viable from chalcid-infested alfalfa seed. *Journal of Applied Seed Production* 4, 26–29.

Cagas, B. (1987) Important problems of plant protection in grass seed production in Czechoslovakia. In: *Preprint of Lectures, International Seed Conference*, Tune Landboskole, Denmark.

Charlton, J.F.L., Hampton, J.G. and Scott, D.J. (1986) Temperature effects on germination of New Zealand herbage grasses. *Proceedings of the New Zealand Grassland Association* 47, 165–172.

Christensen, C.M. (1957) Deterioration of stored grains by fungi. *Botanical Review* 23, 108–134.

Christensen, C.M. (ed.) (1982) *Storage of Cereal Grains and their Products*, 3rd edn. American Association of Cereal Chemists, Minnesota.

Christensen, C.M. and Kaufmann, H.H. (1969) *Grain Storage: The Role of Fungi in Quality Loss*. University of Minnesota Press, Minnesota.

Christensen, C.M. and Kaufmann, H.H. (1974) Microflora. In: *Storage of Cereal Grains and their Products*. American Association of Cereal Chemists, Minnesota.

Clark, S.M. (1982) An evaluation of seed quality in crested dogstail in relation to storability. *Seed Science and Technology* 10, 517–526.

Close, R.C., Harvey, I.C. and Sanderson, F.R. (1982) Lucerne diseases in New Zealand and their control. In: Wyn-Williams, R.B. (ed.) *Lucerne for the 80s*. Agronomy Society of New Zealand, Lincoln.

Cooke, R.J. (1988) Electrophoresis in plant testing and breeding. *Advances in Electrophoresis* 2, 3–13.

Cooke, R.J. (1995) Variety identification: modern techniques and applications. In: Basra, A.S. (ed.) *Seed Quality – Basic Mechanisms and Agricultural Implications*. Food Products Press, Binghamton, New York, pp. 279–318.

Coolbear, P. (1995) Mechanisms of seed deterioration. In: Basra, A.S. (ed.) *Seed Quality – Basic Mechanisms and Agricultural Implications*. Food Products Press, Binghamton, New York, pp. 223–277.

Coolbear, P. and Hill, M.J. (1988) Seed quality control. In: *Rice Seed Health*. International Rice Research Institute, Los Banos, Philippines, pp. 331–342.

Coolbear, P., Francis, A. and Grierson, D. (1984) The effect of low temperature pre-sowing treatment on the germination performance and membrane integrity of artificially aged tomato seeds. *Journal of Experimental Botany* 35, 1609–1617.

Dehghan-Shoar, M. (1996) The evaluation of morphological and molecular techniques for discrimination among and verification of lucerne (*Medicago sativa*)cultivars. PhD Thesis, Massey University, Palmerston North, New Zealand.

Delouche, J.C. (1980) Environmental effects on seed development and seed quality. *Horticultural Science* 15, 775–779.

Delouche, J.C. and Baskin, C.C. (1973) Accelerated ageing techniques for predicting the relative storability of seed lots. *Seed Science and Technology* 1, 427–452.

Dornbos, D.L. (1995) Production environment and seed quality. In: Basra, A.S. (ed.) *Seed Quality – Basic Mechanisms and Agricultural Implications*. Food Products Press, Binghamton, New York, pp. 45–80.

Ellis, R.H. and Roberts, E.H. (1981) The quantification of ageing and survival in orthodox seeds. *Seed Science and Technology* 9, 373–409.

Esbo, H. (1980) Seed quality. *Advances in Research and Technology of Seeds* 5, 9–24.

Evans, P.S. (1973) Effect of seed size and defoliation at three development stages on root and shoot growth of seedlings of some common pasture species. *New Zealand Journal of Agricultural Research* 16, 389–394.

Forde, M.B. and Gardiner, S.E. (1988) The use of seed protein banding patterns for identification of pasture cultivars. *Proceedings of the New Zealand Grassland Association* 49, 87–91.

Francis, A. and Coolbear, P. (1987) A comparison of changes in the germination responses and phospholipid composition of naturally and artificially occurring aged tomato seeds. *Annals of Botany* 59, 167–172.

Fulton, N.D. and Hanson, E.W. (1960) Studies on root rots of red clover in Wisconsin. *Phytopathology* 50, 541–550.

Gardiner, S.E. and Forde, M.B. (1988) Identification of cultivars and species of pasture legumes by sodium dodecylsulphate polyacrylamide gel electrophoresis of seed proteins. *Plant Varieties and Seeds* 1, 13–26.

Green, D.E., Pinnell, E.L. and Cavanaugh, L.E. (1966) Effect of seed moisture content, field weathering, and combine cylinder seed on soybean quality. *Crop Science* 6, 7–10.

Hampton, J.G. (1984) Control measures for ergot in the paspalum (*Paspalum dilatatum*) seed crop. *Journal of Applied Seed Production* 2, 32–35.

Hampton, J.G. (1986) Effect of seed and seed lot 1000 seed weight on vegetative and reproductive yields of 'Grasslands Moata' tetraploid Italian ryegrass (*Lolium multiflorum*). *New Zealand Journal of Experimental Agriculture* 14, 13–18.

Hampton, J.G. (1987) Effect of nitrogen rate and time of application on seed yield in perennial ryegrass cv. Grasslands Nui. *New Zealand Journal of Experimental Agriculture* 15, 9–16.

Hampton, J.G. (1991) Herbage seed lot vigour – do problems start with seed production? *Journal of Applied Seed Production* 9, 87–93.

Hampton, J.G. (1994) Quality and seed production in New Zealand. In: Coolbear, P., Cornford, C.A. and Pollock, K.M. (eds) *Seed Symposium, Seed Development and Germination*. Agronomy Society of New Zealand special publication No. 9, Christchurch, pp. 87–95.

Hampton, J.G. (1995) Methods of viability and vigour testing: a critical appraisal. In: Basra, A.S. (ed.) *Seed Quality – Basic Mechanisms and Agricultural Implications*. Food Products Press, Binghamton, New York, pp. 81–118.

Hampton, J.G. and Bell, D.D. (1989) Seed quality and storage performance of prairie grass (*Bromus willdenowii* Kunth cv. Grasslands Matua). *New Zealand Journal of Agricultural Research* 17, 139–143.

Hampton, J.G. and Coolbear, P. (1990) Potential versus actual field performance – can vigour testing provide an answer? *Seed Science and Technology* 18, 215–225.

Hampton, J.G. and Hill, M.J. (1990) Herbage seed lots: are germination data sufficient? *Proceedings of the New Zealand Grassland Association* 51, 59–64.

Hampton, J.G. and Musukwa, E.J. (1993) Growing methods for the morphological assessment of white clover cultivar purity. *Proceedings of the XVII International Grassland Congress*, pp. 1867–1868.

Hampton, J.G. and Scott, D.J. (1980) Blind seed disease of ryegrass in New Zealand II. Nitrogen fertiliser: effect on incidence, and possible mode of action. *New Zealand Journal of Agricultural Research* 23, 149–153.

Hampton, J.G. and Scott, D.J. (1990) New Zealand seed certification. *Plant Varieties and Seeds* 3, 173–180.

Hampton, J.G. and TeKrony, D.M. (eds) (1995) *Handbook of Vigour Test Methods*, 3rd edn. International Seed Testing Association, Zurich, Switzerland.

Hampton, J.G. and Young, K.A. (1985) Quality herbage seed – germination and contamination. *Proceedings of the New Zealand Grassland Association* 46, 191–194.

Hampton, J.G., Rolston, M.P., Grbavac, N., Hill, K.A., Hill, M.J. and Finnerty, A.M. (1995) Germination in perennial ryegrass seed lots from the 1993 New Zealand harvest. *Journal of Applied Seed Production* 13, 60 (abstract).

Hawkins, R.P., Horne, F.R. and Kelly, A.F. (1964) Identifying cultivars of grass and clover. *Proceedings of the International Seed Testing Association* 29, 837–850.

Helmer, J.D., Delouche, J.C. and Lienhard, M. (1962) Some indices of vigour and deterioration in seeds of Crimson clover. *Proceedings of the Association of Official Seed Analysts* 52, 154–161.

Heydecker, W. (1972) Vigour. In: Roberts, E.H. (ed.) *Viability of Seeds*. Chapman and Hall, London, pp. 209–252.

Hill, M.J. (1975) Heat damage in freshly harvested ryegrass seed with emphasis on thermophilic fungal action during seed storage. *Australian Seed Science Newsletter* 1, 43–52.

Hill, M.J. (1977) The role of seed certification in agricultural development. In: Chin, H.F., Enoch, I.C. and Raja Harun, R.M. (eds) *Seed Technology in the Tropics*. Faculty of Agriculture, Universiti Pertanian, Malaysia, pp. 219–222.

Hill, M.J. and Crosbie, C.J. (1966) Bulk handling of cereal and seed crops can affect quality. *New Zealand Journal of Agriculture* 112, 48–53.

Hill, M.J. and Hill, K.A. (1994) X-ray radiography for rapid assessment of seed quality. *International Herbage Seed Production Research Group Newsletter* 18, 9–12.

Hill, M.J. and Johnstone, C.R. (1984) Postharvest seed quality assessment. *Proceedings of the New Zealand Grassland Association* 45, 243–246.

Hill, M.J. and Johnstone, C.R. (1985) Heat damage and drying effects on seed quality. In: Hare, M.D. and Brock, J.L. (eds) *Producing Herbage Seeds*. New Zealand Grassland Association, Palmerston North, pp. 53–57.

Hill, M.J. and Supanjani (1993) Reproductive abortion in birdsfoot trefoil (*Lotus corniculatus* L.). *Proceedings of the XVII International Grassland Congress*, pp. 1645–1646.

International Seed Testing Association (1996) International Rules for Seed Testing. *Seed Science and Technology* 24 (suppl.), 335 pp.

Keefe, P.D. and Draper, S.R. (1986) The measurement of new characters for cultivar identification in wheat using machine vision. *Seed Science and Technology* 14, 715–724.

Kelly, A.F. and Bowring, J.D.C. (1990) The development of seed certification in England and Wales. *Plant Varieties and Seeds* 3, 139–150.

Kulik, M.M. (1995) Seed quality and microorganisms. In: Basra, A.S. (ed.) *Seed Quality – Basic Mechanisms and Agricultural Implications*. Food Products Press, Binghamton, New York, pp. 153–171.

Latch, G.C.M. and Skipp, R.A. (1987) Diseases. In: Baker, M.J. and Williams, W.M. (eds) *White Clover*. CAB International, Wallingford.

Matthews, B.D. and Hill, M.J. (1967) Keeping seeds healthy in storage. *New Zealand Journal of Agriculture* 114, 25–27.

Miller, J.E. and Hampton, J.G. (1988) Evaluation of cultivar purity in white clover (*Trifolium repens* L.) seed lots. *Proceedings of the Agronomy Society of New Zealand* 18, 59–63.

Moore, R.P. (1943) Seedling emergence of small seeded legumes and grasses. *American Society of Agronomy* 35, 370–381.

Moore, R.P. (1969) History supporting tetrazolium seed testing. *Proceedings of the International Seed Testing Association* 34, 233–242.

Moore, R.P. (1972) Effects of mechanical injuries on viability. In: Roberts, E.H. (ed.) *Viability of Seeds*. Chapman and Hall, London, pp. 94–133.

Naylor, R.E.L. (1981) An evaluation of various germination indices for predicting differences in seed vigour in Italian ryegrass. *Seed Science and Technology* 9, 593–600.

Naylor, R.E.L. and Hutcheson, H.J.A. (1986) The germination behaviour in soil and compost of different seed lots of perennial ryegrass. *Crop Research* 25, 123–132.

Naylor, R.E.L. and Syversen, M.K. (1988) Assessment of seed vigour in Italian ryegrass. *Seed Science and Technology* 16, 419–426.

Neergaard, P. (1979) *Seed Pathology*. Macmillan, London, 839 pp.

Nellist, M.E. (1981) Predicting the viability of seeds dried with heated air. *Seed Science and Technology* 9, 439–455.

OECD (1971) OECD scheme for the varietal certification of herbage seed moving in international trade. *Proceedings of the International Seed Testing Association* 36, 421–469.

OECD (1982) *OECD Seed Scheme Methods for Plot Tests and Field Inspections*. Organisation for Economic Cooperation and Development, Paris.

Pasumarty, S.V., Matsumura, T., Higuchi, S. and Yamada, T. (1993) Cultivar variation for seed development in white clover (*Trifolium repens* L.). *Euphytica* 65, 211–217.

Perry, D.A. (ed.) (1981) *Handbook of Vigour Test Methods*. International Seed Testing Association, Zurich, Switzerland, 72 pp.

Pill, W.G. (1995) Low water potential and presowing germination treatments to improve seed quality. In: Basra, A.S. (ed.) *Seed Quality – Basic Mechanisms and Agricultural Implications*. Food Products Press, Binghamton, New York, pp. 319–359.

Prestidge, R.A., Popay, A.J. and Ball, O.J.P. (1994) Biological control of pastoral pests using *Acremonium* spp. endophytes. *Proceedings of the New Zealand Grassland Association* 56, 33–38.

Richardson, M.J. (1979) *An Annotated List of Seed-borne Diseases*. Commonwealth Agricultural Bureau, Surrey, UK.

Rowarth, J.S., Rolston, M.P. and Johnson, A.A. (1993) Weed seed contamination: use of a seed testing database to identify processing limitations. *Proceedings of the XVII International Grassland Congress*, pp. 1872–1874.

Rowan, D.D. (1993) Lolitrems, paxalline and peramine: mycotoxins of the ryegrass/endophyte interaction. *Agricultural Ecosystems and Environment* 44, 103–122.

Saha, R., Mandal, A.K. and Basu, R.N. (1990) Physiology of seed invigoration treatment in soybean (*Glycine max.* L.). *Seed Science and Technology* 18, 269–276.

Scott, D.J. (1980) The role of seed testing. In: Lancashire, J.A. (ed.) *Herbage Seed Production*. New Zealand Grassland Association, Palmerston North, pp. 103–109.

Scott, D.J. and Hampton, J.G. (1985) Aspects of seed quality. In: Hare, M.D. and Brock, J.L. (eds) *Producing Herbage Seeds*. Grasslands Research and Practice Series No. 2. New Zealand Grassland Association, Palmerston North, pp. 43–52.

Siegal, M.R., Latch, G.C.M. and Johnson, M.C. (1987) Fungal endophytes of grasses. *Annual Review of Phytopathology* 25, 293–315.

Steiner, J.J. (1993) The benefits of seed quality in intensive temperate forage systems. *Proceedings of the XVII International Grassland Congress*, pp. 1857–1861.

Tattersfield, J.G. (1977) Further estimates of heterogeneity in seed lots. *Seed Science and Technology* 5, 443–450.

Tattersfield, J.G. and Johnston, M.E.H. (1970) The H-value heterogeneity test: New Zealand experience. *Proceedings of the International Seed Testing Association* 35, 719–734.

Thomson, J.R. (1979) *An Introduction to Seed Technology*. Leonard Hill, Glasgow.

Villiers, T.A. (1973) Ageing and longevity of seeds in field conditions. In: Heydecker, W. (ed.) *Seed Ecology*. Butterworths, London, pp. 265–288.

Villiers, T.A. and Edgecumbe, D.J. (1975) On the cause of seed deterioration in dry storage. *Seed Science and Technology* 3, 761–774.

Wang, Y.R. and Hampton, J.G. (1989) Red clover (*Trifolium pratense* L.) seed quality. *Proceedings of the Agronomy Society of New Zealand* 19, 63–69

Wang, Y.R. and Hampton, J.G. (1991) Seed vigour and storage in 'Grasslands Pawera' redclover. *Plant Varieties and Seeds* 4, 61–66.

Wang, Y.R. and Hampton, J.G. (1993) Seed vigour in lucerne and white clover. *Proceedings of the XVII International Grassland Congress*, pp. 1868–1869.

West, S.H., Loftkin, S.K., Wahl, M. and Batich, C.D. (1985) Polymers as moisture-barriers to maintain seed quality. *Crop Science* 25, 941–944.

Young, K.A. (1987) Keeping track of cultivars In: Hampton, J.G. (ed.) *Official Seed Testing Station Report 1986*. Ministry of Agriculture and Fisheries, Palmerston North, New Zealand.

10 Breeding for Higher Seed Yields in Grasses and Forage Legumes

A. Elgersma[1] and A.J.P van Wijk[2]
[1]Department of Agronomy, Wageningen Agricultural University, Haarweg 333, 6709 RZ Wageningen, The Netherlands; [2]VanderHave Grasses BV, PO Box 127, 5250 AC Vlijmen, The Netherlands

10.1 INTRODUCTION

The potential seed yield of forage species is high, whereas realized seed yields are generally low and unpredictable. Fluctuations in seed yield are due to environmental factors, management practices and genetic factors. This chapter focuses on breeding for improvement of seed yield of the most important temperate grasses and forage legumes.

According to Hides and Desroches (1990), forage breeders have not attached importance to improving seed yield potential, due to the supposed negative relationship between herbage and seed characters, and to commercial competition for a limited market that requires breeding for agronomic performance. Moreover, in cultivar evaluation systems the criteria for acceptance are herbage characters. Seed yield is not a discriminating factor for inclusion of a cultivar in a national cultivar list. The seed yielding capacity of a cultivar is considered to be the responsibility of the breeder, and no official evaluation is required. However, the ability of a cultivar to produce seed in commercial quantities ultimately determines its success. Seed yield is a decisive character for a cultivar (Lewis, 1966) and the breeder is therefore challenged to combine vegetative qualities and high stable seed yield in one and the same cultivar.

In cereals the objectives of breeding and seed production interests are similar. Grain yields have been improved by increasing harvest index, but this is not an option in forage crops. There is no reason to assume, however, that improvements in herbage characteristics necessarily reduce the seed yielding capacity. Relationships between vegetative and generative traits depend on genotypes. Late-flowering, leafy, persistent, densely tillering plants generally produce less seed than less persistent, early-flowering, low-tillering plants (Griffiths et al., 1980; Bugge, 1987), but among cultivars of comparable maturity, such associations may be absent or positive. In 'wild' populations a much wider range in seed yield is present than among cultivars, because for cultivars low-yielding populations are not economically viable. Large seed yield differences between existing cultivars are therefore not usual.

Nevertheless, there are significant and consistent cultivar differences for seed production (Rincker *et al.*, 1988; Nordestgaard and Andersen, 1991), indicating that high seed yielding capacity and good agronomic performance are not mutually exclusive. However, selection criteria for high seed production are lacking due to the absence of clear relationships between seed yield and its components.

10.2 GENETIC VARIATION AND HERITABILITY OF FACTORS AND PROCESSES DETERMINING SEED YIELD

10.2.1 Genetic Variation, Heritability and Breeding

The genetic structure of a population determines its capacity to be changed by selection. Most temperate perennial forage species are cross-fertilizing natural

Table 10.1. Grasses and forage legumes and their ploidy levels. Induced polyploids and artificial hybrids are printed in **bold**. Autoploids and alloploids are indicated where known.

Species	Common name	2×	4×	6×	8×	Auto/alloploid
Agrostis capillaris	Common bent (brown top)			28		
Agrostis stolonifera	Creeping bent		28	42		
Alopecurus pratensis	Meadow foxtail		28			
Bromus catharticus	Rescue grass			42		
Bromus sitchensis	Alaska bromegrass				56	
Dactylis glomerata	Cocksfoot		28			Auto
Festuca arundinacea	Tall fescue			42		Allo
Festuca pratensis	Meadow fescue	14				
Festuca rubra	Red fescue					
	Chewings fescue			42		
	Slender creeping fescue			42		
	Creeping fescue				56	
Lolium perenne	Perennial ryegrass	14	**28**			Auto
Lolium multiflorum (Lm)	Italian ryegrass	14	**28**			Auto
Lm var. *westerwoldicum*	Westerwold ryegrass	14	**28**			Auto
Lolium × boucheanum	Hybrid ryegrass	**14**	**28**			Allo
Phalaris arundinacea	Reed canarygrass		28			
Phleum pratense	Timothy			42		Auto
Poa pratensis[1]	Smooth stalked meadowgrass[2]					
Lotus corniculatus	Birdsfoot trefoil					
Medicago sativa	Lucerne		24			
Trifolium hybridium	Alsike clover	16	32			Auto
Trifolium pratense	Red clover	14	**28**			Auto
Trifolium subterraneum	Subterranean clover	12 or 16				
Trifolium repens	White clover		32			Allo

[1]Apomict; 2n = 28 to > 100.
[2]Syn. Kentucky bluegrass

polyploids, either auto- or allopolyploids (Table 10.1). Perennial and Italian ryegrass (*Lolium perenne* L. and *L. multiflorum* Lam.), red and alsike clover (*Trifolium pratense* L. and *T. hybridum* L.) are natural diploids, but induced autotetraploids have been developed and bred into cultivars.

Most traits are complex, being under the control of a number of genes as well as being considerably influenced by the environment. Yield, forage quality and maturity date are typical examples. These characters do not show clear discontinuity between genotypes. It is assumed that several genes are responsible for the variation in these traits, such genes being referred to as polygenes or quantitative trait loci. Diploids and allopolyploids show disomic inheritance patterns, but in autopolyploids inheritance is complex, in part due to the nature of meiosis.

Plant breeding starts with the collection of plants to create a variable population, followed by an efficient selection of the best plants for the composition of a potential cultivar. Variation may arise from ecotypes, mutations, hybridization or polyploidization. Selection is only possible if genetic variation exists. Heritability (h^2) relates to the genetic control of the variation. The proportion of the observed variation between individuals in a population arising from genetic segregation is known as the 'broad-sense heritability' (h^2_b) of the trait. Usually this is assessed in clonal material of spaced plants. Progeny testing provides estimates of narrow-sense heritability (h^2_n), which is the proportion of observed variation due to the additive effects of genes. In synthetic cultivars h^2_n is more meaningful because most dominance effects which are included in h^2_b are lost during multiplication. Estimates for h^2_n are lower than those for h^2_b. The size of h^2 depends on the population and on the environment(s) in which the population is studied. The previous degree of selection within a population reduces the heritability of the selected character. Heritability estimates provide an indication of the expected response to selection in a segregating population, and are useful in designing an effective breeding programme. Correlated responses in other traits should be investigated to prevent undesirable effects in such traits which could over-rule the progress in the target trait. The progress made in a breeding programme depends on the ability to identify desirable genotypes. Furthermore, heritability of a trait should be high to facilitate selection for that trait.

In breeding, spaced plants are used to select the best phenotypes prior to crossing (Fig. 10.1). Selected plants can be combined into a candidate cultivar (candivar) or can be pair-crossed, topcrossed or polycrossed, followed by progeny testing, to select the best genotypes. The candivars are then tested in agronomic trials in comparison with standard cultivars, whereafter the best cultivar is selected for further use.

The seed yielding capacity of the candivar is often only assessed in the later stages of the breeding process. Selection for seed yielding ability in spaced plants at an early breeding stage might be advantageous, provided that identified spaced plant characters have a high heritability, and a high correlation with seed yield of their progenies under the specific agronomic conditions of a seed crop. Therefore, phenotypic relationships between parents and their offspring are influenced by genetic (heritability) as well as agronomic factors (plant spacing, crop husbandry).

Fig. 10.1. A spaced-plant collection of perennial grasses.

Even if considerable genetic variation exists and heritability estimates are high, environmental factors and management practices can have large effects on the expression of the selected characters, and over-rule genetic aspects. The highest seed yields can only be obtained when cultivars with a high seed yield potential are grown under optimal conditions with proper management.

Seed production has traditionally been analysed through the components of seed yield, but this is primarily a static concept. Dynamic processes, such as synchronization of flowering and ripening, and rate of seed or pod shedding also determine the amount of seed harvested. Measurements on the ontogeny and duration of various stages are required to understand dynamic aspects of seed yield (Hill and Loch, 1993).

10.2.2 Transition to the Reproductive Stage

Is there genetic variation for the timing of reproductive development? Three distinct phases can be distinguished:

1. Floral induction – a chemically or hormonally induced differentiation in vegetative primordia.
2. Floral initiation – the morphological transformation of an induced growing point from a vegetative to a generative primordium. In most temperate grasses, both induction and initiation often result from exposure to cool temperatures and/or short-day photoperiods.
3. Floral development – the phase of spikelet differentiation and inflorescence elongation, often resulting from exposure to warm temperatures and long-day photoperiods.

Rhoads *et al.* (1992) found reproductive variation among five smooth-stalked meadowgrass (*Poa pratensis* L.) cultivars, although the transition from a vegetative to a reproductive condition was also influenced by year and management. The stage of plant development, the length of the vernalization period and the cultivar, significantly affected the mean panicle number per plant, the time from vernalization to panicle emergence, and the probability of floral induction (Carlson *et al.*, 1995). In Italian ryegrass Nelson *et al.* (1996) found significant effects of genotype, vernalization time and their interaction for days to heading and tillering.

Some grass species (e.g. *Dactylis glomerata* L., *Festuca* spp., *Phleum pratense* L.) have a juvenile stage during which the plant is insensitive to factors which subsequently induce flowering; only after this stage is the plant able to respond. Havstad (1996) reported a close relationship between plant age before primary induction and subsequent panicle development in meadow fescue (*Festuca pratensis* L.), although there were cultivar differences. Especially in non-adapted material, there is scope for breeding. However, if multiplication of an established cultivar is done outside the area where breeding was done, there will be a risk of genetic shift (Davies, 1954) (see also Section 10.4.3).

The precise control which daylength and temperature have over flowering in grasses is not found in legumes, but daylength has an effect on these characters in most temperate legumes. In lucerne (*Medicago sativa* L.), increasing daylength reduces the number of days to flowering; at a constant daylength high light intensity increases the percentage of stems which develop into reproductive buds. For most legumes temperature does not influence flowering (Walton, 1983).

The reproductive development of white clover (*Trifolium repens* L.) was reviewed by Thomas (1987). The critical daylength for the long-day reaction varies greatly with genotype and the variation in physiological responses is related to the different origin of populations. The self-fertilizing subterranean clover (*Trifolium subterraneum* L.) is day-neutral (Walton, 1983).

10.2.3 Size of the Reproductive System

The reproductive potential of a seed crop is determined by the number of inflorescences per unit area and the number of potential seed sites (ovules) per inflorescence. In grasses, the latter results from the number of spikelets per inflorescence and the number of florets per spikelet, whereas in legumes it results from the number of florets per inflorescence and the number of ovules per floret. Summarizing, the reproductive potential of a seed crop is defined as ovule number per unit area, usually assessed at anthesis (Elgersma, 1985). There is a large difference between potential and actual seed production (Table 10.2).

The size of the reproductive system is determined by genetic, environmental and management factors. Knowledge of the basic patterns of vegetative and reproductive development common to a species and its cultivars is essential to understand the responses of those cultivars to the environment and to cultural practices (see also Section 10.2.5).

In white clover, inflorescences are produced in the leaf axils of actively growing stolons. The number of inflorescences per unit area is the major component of seed yield (Clifford and Baird, 1993) and maximizing inflorescence production should therefore increase seed yield. There are also interactions between genetic and

Table 10.2. Estimates of potential and actual seed yield and harvest index of some forage crops (Lorenzetti, 1993).

	Infl (no. m^{-2})	Fl/infl.	Ov/fl	TSW (g)	OSU (%)	PSY (t ha^{-1})	RSP (t ha^{-1})	% PSY	HI
Grasses									
Lolium perenne	2000	200	1	2.0	40	8.0	1.0	13	10
Dactylis glomerata	600	760	1	1.0	40	4.6	0.8	17	10
Festuca arundinacea	660	680	1	2.0	50	9.0	1.0	11	6
Legumes									
Medicago sativa	3750	16	10	2.0	8	12.0	0.5	4	12
Trifolium repens	600	100	6	0.5	50	1.8	0.4	22	10
Trifolium pratense	750	110	2	1.6	25	2.6	0.6	23	12
Lotus corniculatus	400	6	40	1.2	40	1.2	0.2	17	10

Infl, inflorescences; Fl/infl, flowers per inflorescence; Ov/fl, ovules per flower; TSW, thousand seed weight; OSU, ovule site utilization; PSY, potential seed yield; RSP, realized seed potential in agriculture; %PSY, the percentage of the potential seed yield realized in agriculture; HI, harvest index.

management factors (see Chapter 6). Clearly cultivar and seed crop management have to match.

Realized seed yields are low, while the potential seed yield is high (Table 10.2). Forage species have been relatively recently domesticated and many 'wild' seed characters are still present, such as a high potential number of seed sites, small seed size, dormancy, indeterminate growth habit, seed shattering and uneven ripening. Is there scope for improving these traits by selection and breeding? From an evolutionary viewpoint, selection for an even higher number of potential seed sites does not seem logical. Breeding for higher inflorescence numbers is not advocated in grasses because of the reduction in pasture quality that is caused by increased stem production. It is therefore not desirable to increase the size of the reproductive system, and instead, the efficiency of the reproductive system should be increased (Bean, 1972).

10.2.4 Efficiency of the Reproductive System

The efficiency of the reproductive system and the realization of actual yield are largely determined by processes from anthesis until seed cleaning. Losses occur during pollination, fertilization, and seed development (Hill, 1980). Furthermore, in an economic sense, floret (ovule) site utilization (FSU/OSU) is also strongly decreased by shattering and losses during harvest and processing (Elgersma, 1985).

Pollination, fertilization and early stages of seed development are critical periods for the realization of the yield potential. Is there genetic variation for tolerance against stress during pollination, fertilization and seed development? Rapid and reproducible methods are needed for screening a large number of genotypes in various environments (Fig. 10.2).

Fig. 10.2. Pollination of detached perennial ryegrass ovaries in a Petri dish.

Anthesis

Flowering is affected by several environmental factors. In white clover Davies (1962) showed that low light intensities generally diminished flowering, but certain genotypes were more seriously affected than others. There is variation among perennial ryegrass genotypes for the daily onset of flowering; it is not clear if these early starters are less sensitive to low temperatures or low light intensity. Selection for uniformity of flowering would allow a more closely synchronized crop and higher harvested seed yield.

In wind-pollinated grasses, pollen release and transport are affected by lodging and by weather conditions (reviewed by Hill, 1980). Genetic variation for lodging resistance before and during anthesis might offer possibilities. If retarded pollen tube growth at low temperatures is a major bottleneck for obtaining high seed yields, tolerance to adverse temperatures at anthesis might be a useful breeding objective. Under controlled temperature conditions both the genotype of the mother and temperature affected pollen grain performance in perennial ryegrass (Elgersma *et al.*, 1989) (Fig. 10.3). In lucerne pollen grain germination and pollen tube growth are influenced by both environment and genetic factors (Viands *et al.*, 1988). Improved retention of stigma receptivity might be another option. Unpollinated stigmas of perennial ryegrass kept their receptivity for 1 or 2 days after the day of flowering (A. Elgersma and J. van Hateren, 1991, unpublished). In legumes,

Fig. 10.3. Germinating perennial ryegrass pollen grains (fluorescence microscopy).

the stigma remains receptive for up to 2 weeks if the flower is not visited by a bee (Walton, 1983). In most temperate forage legumes, pollination is possible only if a suitable insect pollinator is available (see Chapter 7). Pollination is not a major problem in white clover but it can be in red clover, especially in tetraploid cultivars (Dennis, 1980). Pollinating insects such as honeybees discriminate against the tetraploids because of their long style which means the bees cannot reach the nectaries with their mouthparts. Consequently they do not revisit the tetraploid flowers. Spiss and Góral (1989) studied two red clover populations with floral modifications, but unfortunately corolla tube shortness was associated with male sterility. Breeding studies to improve the seed setting capacity in a mutant with a deeply split corolla tube are continuing. In alsike clover there are few pollination problems, because the corolla tubes are much shorter than in red clover (Julén, 1975).

In tetraploid red clover, preferential pollination of diploids by bees can cause a shift towards diploidy if more than 4% diploids are present in the tetraploid crop (reviewed by Elgersma, 1992). This is important for cultivar maintenance (see also Section 10.4.3).

Fertilization

Fertilization is a fusion of gametes: one male gamete fuses with the egg cell to form the zygote (embryo) and another male gamete fuses with the central cell to form the endosperm. Most cross-fertilized forage grasses and legumes are self-incompatible; i.e. there is a genetically controlled physiological hindrance to self-fertility. The mechanism is either retarded or inhibited pollen grain germination or pollen tube growth, or abortion of zygotes following self-fertilization. If the embryo develops successfully, the plant growing from the selfed seed often shows inbreeding depression.

Grasses have a gametophytic two-locus system, each with multiple alleles, which remains intact after polyploidization. Pollen tube growth is inhibited on the stigma. Legumes have a gametophytic one-locus system and here polyploidization leads to compatibility. Dennis (1975) found considerably higher self-fertility in tetraploid red clovers than in diploid ones and reported a high correlation between (cross) fertility and self-fertility. Thus selection for higher seed yield (higher fertility) resulted in increased self-fertility, which might lead to reduced productivity. In lucerne, simple inheritance patterns have not been established for self-incompatibility and self-sterility, but the loss of higher order allelic interactions through inbreeding appears to be of major importance (Viands *et al.*, 1988).

Seed development

The percentage of florets (ovules) successfully fertilized determines biological FSU or OSU, together with successful seed development. The major reduction in OSU occurs during the first week after flowering (Elgersma, 1990b). Genetic or cytological factors are probably responsible for the failure of seed development, rather than physiological stress due to competition between sinks. The genetic load of the embryos and developmental selection may be more important for FSU than pollination failure or resource limitation (Marshall and Ludlam, 1989). Elgersma and van Hateren (1991) detached spikelets of perennial ryegrass prior to flowering and cultured them on water. Viable seeds were produced, indicating that spikelets can be self-supporting and seed development was not resource limited (but see Chapter 3).

Many authors have reported FSU in grasses to be a fairly stable and highly heritable character, while others noted large environmental effects on FSU (reviewed by Elgersma, 1985). In spaced plants of three perennial ryegrass cultivars, h^2_b estimates ranged from 0.46 to 0.73 over 3 years, but in the fourth year estimates in the same genotypes were not significantly different from zero. Estimates of h^2_n did not differ significantly from zero during these 4 years (Elgersma, 1990d).

White clover florets mostly have five or six ovules, but normally only 50% develop into a seed. Pasumarty *et al.* (1993) found no significant difference in ovule number of four cultivars, but the high yielding cultivar Grasslands Huia from New Zealand had more fertile ovules and more seeds per floret than three Japanese cultivars. The major loss of seed per floret was due to ovule sterility (30%) and seed abortion (33%) during the first week after pollination. Van Bogaert (1977) reported no difference in florets per head for white clover in a wet year (75.3) versus a dry, sunny year (77.7); average numbers of seeds per pod were 2.5 and 3.9, the percentages of seed-bearing florets were 82.2 and 97.6, and fertilization values were 42 and 65%, respectively.

The rate and duration of seed fill determine the seed's final size. In research trials, samples are often taken in the final stages of seed development to determine yield components like FSU or OSU. Losses due to shattering and mechanical harvesting are then often lower than in practice.

Seed retention

Seed shattering occurs before and during harvest. It is an important source of yield loss in forage seed production (see Chapter 4), although seed shattering and pod dehiscence can be desirable in annual self-reseeding species (*Medicago* spp., *Trifolium* spp., *Lolium rigidum* Gaud.) to ensure natural regeneration of the pasture (Lorenzetti, 1993). Shattering resistance (seed retention) does not affect other seed yield components or forage quality, and therefore seems a very desirable trait, provided that the recovery of seeds during threshing is not impaired (Elgersma *et al.*, 1988).

In cereals many non-shattering mutants with either a partly developed or no abscission layer have been identified (McWilliam, 1980). Unlike most traits (see Section 10.2.1), elimination of the abscission layer is usually controlled by a few major genes. Is there scope for gene transfer? Variation for seed retention has been observed in perennial ryegrass (Elgersma *et al.*, 1988), Italian ryegrass (Harun and Bean, 1979), cocksfoot (Falcinelli *et al.*, 1984) and tall fescue (Falcinelli, 1993), but abscission layers were still present. Successful seed retention has been achieved in cocksfoot. Seed yield was increased by selecting for seed setting and seed retention, and seed quality was increased after several cycles of phenotypic selection for seed weight (Falcinelli *et al.*, 1996).

Successful breeding for resistance to seed shattering in grasses was reported for meadow foxtail (*Alopecurus pratensis* L.) where mutation breeding resulted in 38 shattering-resistant genotypes out of 6300 inbred progenies. From this material the cultivar Alko was bred. It outyielded the standard cultivar Lipex by 50–100% when seed harvest was delayed for 1–3 weeks. A similar programme in meadow fescue produced cultivar Fesco, for which yield reduction after delayed harvest was only 39% of that of the standard cultivar Cosmos 11 (Simon, 1996).

During maturation the moisture content in the seed declines. The moisture level is an indicator of physiological ripeness and of the likely date of abscission or seed shedding. In Italian ryegrass Hides *et al.* (1993) found a greater variation in the moisture content at which shedding began than in the rate of shedding thereafter. Selection for improved seed retention in ecotypes has been possible, although limited. In the improved material, shedding began at a lower moisture content than in the ecotype from which it was selected, whereas the pattern and rate of shedding were similar.

In birdsfoot trefoil (*Lotus corniculatus* L.) seed set is abundant, but losses are high due to pod dehiscence upon ripening as indeterminate flowering and hence varied maturity cause problems in selecting the optimal harvest time. Selection for uniformity of flowering and ripening would give a more closely synchronized crop which facilitates the choice of the optimal harvest time. Better seed retention leads to a higher harvested seed yield.

Harvesting and processing

Seed losses during processing can be considerable (see Chapter 8), and are often larger (in terms of numbers of seeds) than losses due to low biological FSU. Elgersma (1991) estimated that in perennial ryegrass out of 100 florets counted at anthesis, 65 florets produced a seed that could be counted prior to harvest, but after processing only 25 seeds were left. In grass species such as timothy the threshing technique may greatly affect the number of recovered seeds (Bean, 1972). Smooth-stalked meadowgrass has a hairy lemma, requiring special threshing techniques for seed cleaning. *Poa longifolia* Trin. has non-hairy-lemmas. By crossing both species, progeny with non-hairy lemmas were obtained (Van Dijk and Winkelhorst, 1982) which were further developed in commercial breeding programmes. However, turf performance and seed production of the resulting candivars were not acceptable for commercialization.

Resistance to biotic stresses

Breeding for resistance to biotic stress factors is carried out where it affects vegetative performance. However, some diseases are important only in seed crops.

If ryegrass comes into flower under cool, very moist conditions, infection by the fungus *Gloeotinia granigena* (Quel.) T. Schumacher can cause blind seed disease, resulting in a loss of germination of the harvested seed. By backcrossing the perennial ryegrass cultivar S24 with resistant plants originating from the local cultivar Irish Commercial, the resistant cultivars Callan and Lagan were bred (Wright and Faulkner, 1982). The genetic control of the resistance was found to be polygenic with a relatively uncomplicated gene action which was mainly additive, but with a degree of incomplete dominance. Stem rust (*Puccinia graminis* Pers. subsp. *graminicola*) is a serious seed production disease of perennial ryegrass in western Oregon, USA. In most years, stem rust is controlled by one to three applications of a fungicide (Welty and Barker, 1994), but breeding for stem rust resistance is an alternative. Rose-Fricker *et al.* (1986) studied the inheritance of stem rust resistance and concluded that this was predominately quantitatively inherited with minor and possibly some major genes. No relationship was found between the seedling reactions and the adult plant response to the disease. It was suggested that selection could be most effectively carried out in the booting stage. Cultivars have been developed with increased resistance to stem rust. Stem rust also affects seed production of turf-type tall fescue in Oregon, USA. Welty and Barker (1993) developed a seedling screening procedure to identify resistant sources of stem rust for incorporation in a breeding programme.

Rust (*Uromyces trifolii-repentis* Liro.) in subterranean clover can cause herbage and seed yield losses; Barbetti and Nichols (1991) reported reductions in seed yield ranging from 40 to 89% in five cultivars, but cultivar Larisa was highly resistant and the disease caused no significant reduction in herbage or seed yield.

Herbicide tolerance

Breeding grasses for herbicide tolerance creates a means of manipulating the composition of the grass sward. By using herbicide-tolerant cultivars the sward of both forage and turf grasses can be kept free of indigenous weed grasses. Herbicide tolerance is also a useful character in the production of herbage seed. Tolerant cultivars can be sprayed for the removal of weed grasses and rogues of the same species, which contributes to the purity of the seed crop (see Chapter 9). Genetic variation for herbicide tolerance between and within plant species is present, and particular herbicides are more active against some grass species than others (Johnston and Faulkner, 1991). Herbicide-tolerant cultivars of perennial ryegrass have been developed and are being commercialized. In white clover breeding, very little has been reported on this aspect, but cultivar differences were demonstrated in tolerance to paraquat at the seedling stage (Williams, 1987).

Annual species such as subterranean clover rely on replenishment of seed reserves each year to maintain a large bank of germinable seed for the following autumn. Herbicides may be used to manipulate the proportion of grasses and broadleaf weeds, but the sensitivity of the legume to the herbicide is critical, in terms of both herbage yield and seed set. Subterranean clover cultivars exhibit substantial variation in tolerance to a number of broadleaf herbicides (Evans *et al.*, 1989), and there may be scope for selection for improved herbicide tolerance.

10.2.5 Realized Seed Yield and Relationships Among Yield Components

Seed weight and seed number

There is a wide range in individual seed weight within a seed lot. Very small and light seeds may be cleaned out and thus the minimum seed weight in a cleaned seed lot depends on the cleaning intensity. The thousand seed weight (TSW) of a crop is assessed after seed processing, and tends to be relatively constant over a range of conditions (Marshall, 1985).

In perennial ryegrass seed crops, significant cultivar differences were found for TSW, but the ranking was different from that for seed yield. TSW was less affected by the environment than seed yield (Elgersma, 1990a,b). Yield is the product of TSW and seed number. As yield fluctuates more than TSW, there must be a close relationship between the number of pure seeds and seed yield (Hebblethwaite *et al.*, 1980; Elgersma, 1990b).

Yield is a function of total seed growth rate and seed fill duration. The total seed growth rate is the product of seed number per unit area and individual seed growth rate (ISGR). Are there genotypic differences for ISGR and duration of seed growth? The number of seeds per unit area is biologically determined by seed yield components that are very variable. There is also an important yield component compensation. Negative correlations between TSW and seed set or seed number have been reported. Lewis (1970) studied six genotypes of perennial ryegrass cultivar S23 in a diallele crossing scheme and found a negative correlation between spikelet number and TSW, suggesting a physiological barrier to full seed development associated with high floret numbers. In smooth-stalked meadow-grass Van Wijk (1985) found positive effects of high numbers of inflorescences and spikelets on seed yield in two breeding populations, but these effects were reduced by a smaller seed weight.

In white clover seed size varies considerably, but is not considered an important agronomic character; its inheritance has apparently not been studied (Williams, 1987). Large seed does not necessarily establish better than small seed. In white clover any gain in unit seed weight was at the expense of yield per inflorescence (Clifford and Baird, 1993).

Seed yield

Many studies have been carried out on spaced plants. However, seed yield per plant may not be an indicator of seed yield in seed crops, and especially in tufted (non-spreading) species, relationships between yield components found within single plants may not be identical to relationships within seed crops!

Spaced plants. In plant breeding, individual genotypes are screened as spaced plants because in dense stands they cannot be easily distinguished. Many genetic studies have been carried out on single plants, with seed yield per plant as the target characteristic (Figs 10.1 and 10.4). The genotypic variation within cultivars between spaced plants is often large, and highly significant genetic differences between genotypes have been reported for tiller number and seed yield per plant in perennial ryegrass (Bugge, 1987; Elgersma, 1990b), tall fescue (Bean, 1972), timothy (Bean, 1972) and cocksfoot (Stratton and Ohm, 1989).

Bugge (1987) found h^2_b values in perennial ryegrass for ear length, numbers of spikelets per ear and florets per spikelet, seed yield per plant and fertile tiller number of 0.92, 0.83, 0.71, 0.64 and 0.49, respectively. In perennial ryegrass, earliness, ear length, the number of spikelets per ear and flag leaf width, had a high narrow-sense heritability in spaced plants and therefore showed promise for selection (Elgersma, 1990d). Most other seed yield components (fertile tillers, seed yield per tiller, numbers of florets and seeds and FSU) and seed yield per plant were not promising for selection in spaced plants in the material studied.

Fig. 10.4. Spaced perennial ryegrass plants grown for seed.

Fig. 10.5. Single plants and small plots of smooth-stalked meadowgrass.

Smooth-stalked meadowgrass forms rhizomes. Evaluation of genotypes therefore takes place in a closed sward, approaching tiller densities of a seed crop (Fig. 10.5). Seed yield assessment in single plants can therefore be done more reliably in stoloniferous or rhizomatous species than in tufted grasses (see Section 10.2.3). Van Wijk (1985) determined seed yield and its components in F1 hybrids of smooth-stalked meadowgrass. Ample genetic variation was present. It was concluded that the seed yield per plant was the best predictor for seed yield, irrespective of how the components contributed to the final yield. A higher number of inflorescences and spikelets, for example, had a positive effect on seed yield, but this effect was reduced by a smaller seed weight.

Negative correlations between seed yield components are often found. In a red clover genotype with an elongated flower head, floret, and seed numbers per head were higher, but the number of flower heads and seed setting were lower, whilst the seed yield was not affected (Góral and Spiss, 1996).

In 63 agronomically superior spaced plants of white clover a high harvest dry matter was generally associated with high numbers of florets, seeds per inflorescence, inflorescence density and total seed yield (Clifford and Baird, 1993). Five flowering types were defined, ranging from mainly crown-flowering to mainly stolon tip-flowering. Dominantly crown-flowering types gave the highest seed yield. White clover seed yields are closely related to numbers of seed heads produced, but plants have the ability to compensate for a reduced inflorescence number. Binek (1983) found in cultivar Podkowa that removal of up to 48% of the inflorescences did not reduce seed yield. A compensatory increase of 58–78% occurred in both the number and weight of seeds per head, although the TSW altered less than 8%. From a plant breeding viewpoint, reduction to a standard number of heads per plant improved the selection efficiency (Williams, 1987). Large differences among plants for seed set per flower have been found. Dessureaux

Table 10.3. Distribution of white clover ovule number per floret. (From Williams, 1987.)

Source	Number of ovules per floret (total of 100 florets)					
	8	7	6	5	4	< 4
van Bogaert (1977)	2	13	62	21	2	0
Dessureaux (1951)	0.1	0.6	21.1	26.1	47.7	4.3

(1951) reported significant differences among Ladino clover clones for the number of ovules per floret, seeds set per floret and percentage of seed-bearing florets. Selection was successful in changing the numbers of ovules and seeds. A high number of ovules and a high degree of pollination were the main factors favouring high seed setting. Selection based on seed set often increased ovule number and generally improved efficiency of pollination (Dessureaux, 1951; Cebrat *et al.*, 1982).

In a study of 18 white clover clones of five origins, van Bogaert (1977) found that the number of florets per head was highly variable and highly heritable (h^2_b was 0.75). The number of seeds per pod was also highly heritable. The number of ovules per floret can differ widely (Table 10.3).

Seed crops. In seed crops, seed-yield components are variable and show interactions (Lorenzetti, 1993). Elgersma (1990a) studied 12 seed crops of each of nine perennial ryegrass cultivars and found significant seed-yield differences among the cultivars. The highest-yielding cultivar was superior over a wide range of environments and the seed yield of the poorest cultivar was on average only 64% of that of the best cultivar. This implies that vegetative quality and high stable seed yield can be combined in one cultivar. An interesting question is how a high, stable seed yield is realized in a crop. Which yield components are most important for high seed yield? Are they suitable for indirect selection, and how can a breeder select for high stable seed yield?

Significant cultivar differences also occurred for TSW, but the rankings of the cultivars for seed yield and for TSW were different. Therefore, TSW cannot be used for indirect selection for seed yield. The calculated number of seeds per unit area was strongly correlated with seed yield. The number of seeds is determined by seed-yield components such as fertile tiller number, numbers of spikelets per ear, seeds per spikelet and FSU. However, these components were very variable within plots, and often cultivar differences were not significant within a trial and not consistent in other trials. The number of fertile tillers did not differ among the cultivars and was not associated with seed yield. There were clear cultivar differences for spikelet number per ear, but these were not associated with seed yield. No consistent cultivar differences were found in the seed crops for the number of florets and seeds per spikelet, or for FSU, and these traits were not associated with seed yield. Cultivar differences for seed yield could thus neither be predicted from seed-yield components assessed in seed crops, nor from crop physiological traits such as growth pattern or dry matter accumulation and partitioning (Elgersma, 1990b). No independent characters were identified in seed crops that correlate with total seed yield per plot.

In white clover Evans *et al.* (1986) observed differences in seed-yielding ability among 15 cultivars grown as seed crops. The high potential seed yield of cultivars Menna and Olwen was related to vigorous inflorescence production over a narrow time interval, resulting in a high proportion of ripe inflorescences at harvest. The lower seed set of cultivar Olwen when compared with cultivar Menna was compensated for by the higher percentage of ripe inflorescences of Olwen at harvest. The candivar Anna had a low seed yield owing to a prolonged flowering period, resulting in a low percentage of ripe inflorescences at harvest. Another low-yielding candivar AC 20 flowered late and produced few inflorescences. There was an overriding effect of weather conditions on several of the components (see also Chapter 6).

In lucerne, seed production is not generally considered to be a pervasive problem. Cultivars, however, clearly differ in their genetic potential for seed yield. Occasionally cultivars have been developed with the potential to produce high yields of quality forage in humid regions of the USA without the genetic capacity to achieve economic levels of seed production in the western, drier states, where most seeds are produced. Although several studies have been conducted to develop criteria for use by forage breeders in predicting seed yield, it has not been possible to do so. Thus, seed-yield evaluation of promising genetic material remains an important step in lucerne improvement (Rincker *et al.*, 1988).

Spaced plants versus seed crops. Much of the published data have been concerned with the elucidation of component factors that together make up seed yield, with only a few examples of studies carried through a selection phase (Davies, 1981). For successful selection, spaced-plant traits should have a high correlation with crop seed yield in addition to a high heritability (Fig. 10.6). Bean (1972) studied clonally replicated spaced plants from both a tall fescue and a timothy

Fig. 10.6. Harvest of selected red fescue (*Festuca rubra* L.) plants.

cultivar. The genotypes with the highest seed yield were interpollinated and progeny seed yield was significantly higher than that of the cultivars. Elgersma (1990c) calculated correlations for many traits between the average of 50 spaced plants of a cultivar and seed yield in a seed crop for nine perennial ryegrass cultivars. Correlations were generally low; for example, the two highest-yielding cultivars differed for most spaced-plant traits, and spaced-plant averages of the lowest-yielding cultivar could not be distinguished from those of most other cultivars (Fig. 10.7). A small plant girth at ear emergence and a high seed yield per spike showed the best correlation with a high seed yield per plot, but unfortunately the narrow-sense heritabilities of these traits were very low. The traits with a high heritability were not correlated with seed yield per plot. Therefore, no useful selection criteria for seed yield per plot could be identified in spaced plants.

A parent–offspring analysis was also made (Elgersma et al., 1994). Spaced-plant traits of 31 genotypes in each of two perennial ryegrass cultivars were related to seed production characteristics in seed crops of open-pollinated progenies of these plants. The number of spikelets per ear in the maternal plants was negatively correlated with the seed yield of the progenies, explaining 17% of the variation for seed yield in cultivar Barenza and 14% in cultivar Wendy. No (combinations of) other plant traits consistently explained a major portion of the variation for seed yield of the progenies. The results indicate that in these cultivars, spaced-plant data

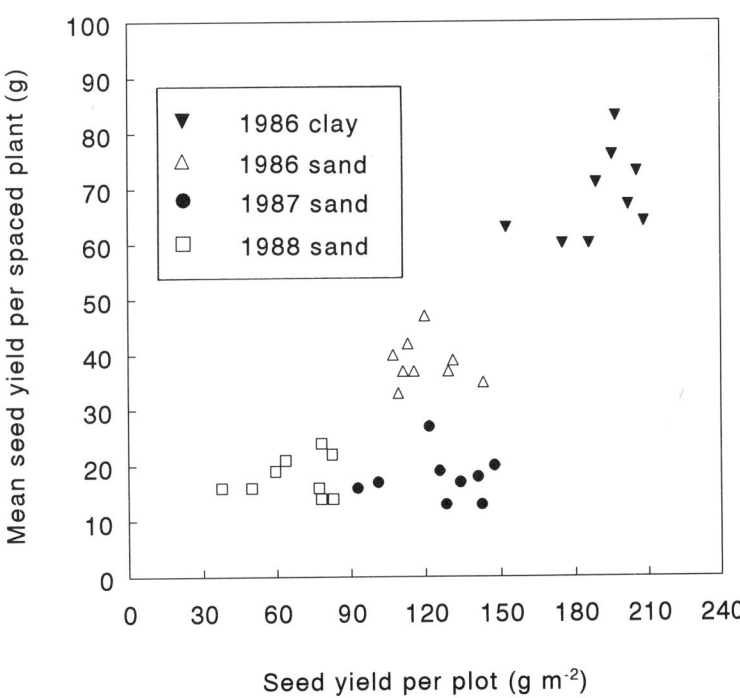

Fig. 10.7. The relationship between seed yield per spaced plant and seed yield per plot for nine perennial ryegrass cultivars in four environments. Symbols refer to environments (see Elgersma, 1990c).

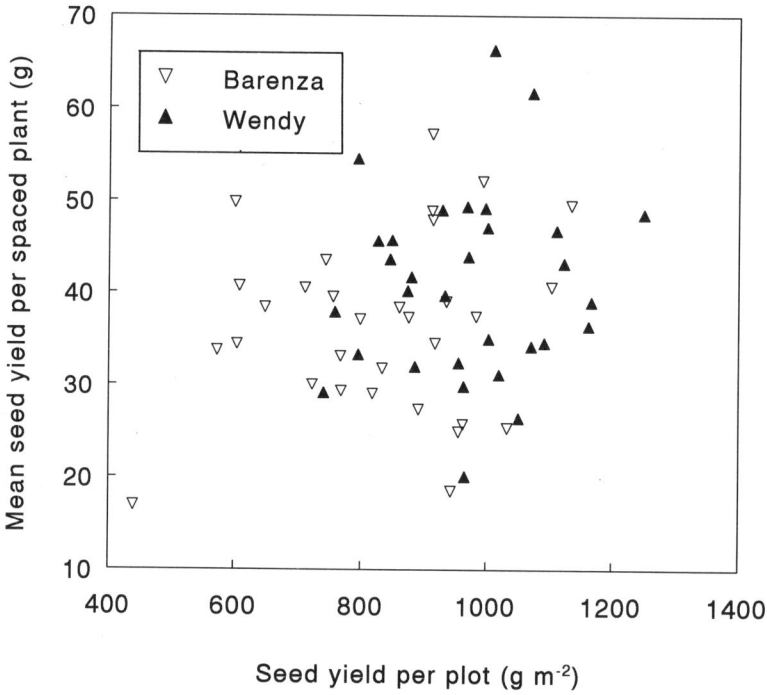

Fig. 10.8. The relationship between maternal seed yield per spaced plant and progeny seed yield per plot for two perennial ryegrass cultivars. Symbols refer to cultivars (see Elgersma et al., 1994).

are of limited value in predicting seed production (Fig. 10.8). Similarly, Bugge (1987) found in perennial ryegrass a low correlation ($r = 0.21**$) between the seed yield of 100 single plants and that of their topcross offsprings in drilled plots, although seed yield per plot increased by 6% when the best 10 plants were selected (see also Section 10.3.1).

In tetraploid red clover, selections for single characters such as pollen grain fertility, meiosis regularity and corolla tube length, were not very promising, and better results were later obtained by simply selecting for high number of seeds per head. The cultivar Sally was improved in seed set by this technique and was the first Swedish tetraploid red clover cultivar that, under practical field conditions, gave an acceptable seed yield (Sjödin, 1981).

10.3 BREEDING METHODS AND ACHIEVEMENTS

Most forage grasses and legumes are naturally cross-fertilized because of self-incompatibility. With the exception of subterranean clover, there are no temperate self-fertilizing grass and forage legume species of major agronomic importance. Commercial F1 hybrids are presently not available. In lucerne, hybrid production has been suggested to improve the yield of synthetic varieties. Two types of male-sterility systems, genetic and cytoplasmic, have been identified. Genetic male sterility is controlled by recessive genes, and is therefore not very useful in field production of hybrids. Cytoplasmic male sterility (CMS) is conditioned by single recessive genes interacting with cytoplasmic factors, and offers the best mechanism for the production of hybrid seed. Attempts were made using pairs of two self-sterile clones. However, unanticipated self-pollination problems were encountered, probably due to individual clonal preferences of bees (see also Section 10.4.3). Commercial hybrids using CMS were marketed in 1968, but seed yield on male sterile plants has remained a serious economic problem and is presently not economic. The random distribution of male-sterile and pollen-donor plants in the field, the increased ratio of pollen donor to male-sterile plants, and the use of high seed-setting or high pollen-producing pollen-donor plants are important considerations in increasing yield of male-sterile plants. An alternative method of producing hybrid seed is to use female-sterile plants in pollinating male-sterile plants. Large-scale application of this system is limited because female steriles, controlled by a single recessive gene, must be maintained either by outcrossing to female-fertile plants or by vegatative propagules (Viands *et al.*, 1988).

The identification of easily measured characteristics with a predictive value for seed yield is important because of the large numbers of plants that have to be screened. A relatively easy character to assess before flowering on single plants is inflorescence number. The determination of yield components such as spikelet and floret number is more time consuming. Components like seed weight and seed yield can only be assessed after harvest. In practical plant breeding, plants with an insufficient number of inflorescences and with a low seed yield are discarded in an early stage of the selection programme.

When a breeder wants to assess the seed-yielding capacity of a population in plots, border effects should be minimized and each plot should be harvested individually according to its harvest ripeness. In a seed-yield trial with various cultivars a pollen grain mixture is present, in contrast to commercial multiplication where only pollen from the cultivar itself is present. When new populations with insufficient compatibility are tested, this might result in overestimation of the seed production, because the presence of compatible pollen grains is not limited in a comparative trial, but could be limiting in large-scale seed production.

10.3.1 Cross-fertilizing Species

Most forage species are relatively easy to multiply vegetatively. Clonal evaluation is often practised to study the (total) genetic variation in the base population. To separate additive and non-additive variation, progeny testing is applied. Possible systems are half-sib mating, including polycross, topcross and open-pollination, and full-sib mating (pair crosses). Open-pollination and polycross are commonly used to assess the general combining ability of the parental clones for synthetic cultivar development, to recombine selected entries in recurrent selection programmes, and to obtain quantitative genetic information.

In a study on the breeding value of clones, Thomas and Frakes (1967) compared clonal evaluation with five methods of progeny testing for increased seed yield in tall fescue and found no advantage from using progeny tests. This suggests that clonal evaluation for seed yield is an effective method for selecting parents of higher seed yielding cultivars. Ross and Adams (1955) studied 15 inbred and 15 open-pollinated clones of smooth bromegrass (*Bromus inermis* Leyss.) and their progenies. Seed-yielding capacity was reliably estimated by measuring FSU and differences were highly heritable. They recommended mass selection for FSU based on clonal means.

Spaced-plant data were of limited value in predicting seed production of perennial ryegrass (Bugge, 1987; Elgersma *et al.*, 1994) (see 'Seed weight and seed number' above). In this case, direct selection in seed crops of progenies in later stages of the breeding programme remains the best method for obtaining cultivars with sufficient seed production. Breeders should test progenies for their seed-yielding capacity as soon as possible in drilled plots. Limitations of this method are the availability of seed, and the space and labour requirement of drilled plots. Since in seed crops no indirect traits were identified, seed yield itself must be determined, irrespective of how the components contribute to the resulting yield.

In order to register a cultivar for plant breeders' rights, the cultivar has to comply with DUS (Distinctness, Uniformity and Stability) requirements. Vegetative and generative characters are measured and observed on 60 spaced plants, and then analysed to determine the novelty and uniformity of the cultivar. Correlating these measurements with the time of submission for cultivar registration allowed the determination of whether recent cultivars differed from older cultivars for the characters measured. Data are presented for the most intensively bred forage grass species, diploid perennial ryegrass (Table 10.4). Newer cultivars of diploid perennial ryegrass have longer ears and more spikelets per ear. Vegetative characters have changed over the years as well. Recent cultivars are more robust and vigorous as expressed by their larger spring height and height at ear emergence. The increased length and width of the flag leaf as observed in the Netherlands could not be confirmed by the Irish data, while a longer stem length of new cultivars was only significant in Northern Ireland.

Table 10.4. Simple correlation coefficients between time of submission for DUS testing and plant characters for cultivars of diploid, forage perennial ryegrass in the Netherlands and Northern Ireland.

	Netherlands (Scale 1–9)[1]	Northern Ireland[2]
Number of cultivars	145	273
Growth habit	−0.120	−0.077
Spring height	0.482**	0.149*
Height at ear emergence	0.542**	0.147*
Length of flag leaf	0.214**	0.056
Width of flag leaf	0.373**	−0.117
Stem length	−0.033	0.216**
Ear length	0.215**	0.213**
Number of spikelets	0.358**	0.202**

[1]Scale: 1 = erect, short, narrow, few; 9 = prostrate, long, wide, many (from CPRO-DLO DUS tests).
[2]From actual morphological data recorded in 1995 (M.S. Camlin, Northern Ireland, 1995, personal communication).
*$P < 0.05$, **$P < 0.01$.

Table 10.5. Simple correlation coefficients between ear length and spikelet number and other plant characters of cultivars of diploid, forage perennial ryegrass based on DUS observations in the Netherlands and Northern Ireland.

	Ear length		Number of spikelets	
	Netherlands[1]	Northern Ireland[2]	Netherlands[1]	Northern Ireland[2]
Number of cultivars	145	273	145	273
Growth habit	−0.185	0.076	−0.165*	0.104
Spring height	0.274	0.073	0.151	0.148
Date of ear emergence	0.349**	0.280**	0.658**	0.164**
Height at ear emergence	0.366**	0.327**	0.519**	0.269**
Length of flag leaf	0.384**	0.281**	0.225**	0.021
Width of flag leaf	0.371**	0.050	0.163*	−0.132**
Stem length	0.103	0.742**	−0.046	0.546**
Ear length	1.000	1.000	0.599**	0.582**
Number of spikelets	0.559**	0.582**	1.000	1.000

[1] Data from CPRO-DLO Dus tests.
[2] M.S. Camlin, Northern Ireland, 1995, personal communication.
*$P < 0.05$, **$P < 0.01$.

The main breeding objective for the development of these cultivars has been herbage yield. Selection for herbage yield has been correlated with an increase in the size of the reproductive system. Data were not available to determine if the seed yield of these newer cultivars has been simultaneously increased. However, a steady increase in seed yield has been achieved in forage crops over the last 50 years (Nordestgaard and Andersen, 1991). In legumes, seed yield was positively correlated with forage yield in lucerne (Veronesi and Falcinelli, 1987) and birdsfoot trefoil (Onokpise et al., 1987).

Cultivars with longer ears and a greater spikelet number were strongly correlated with vigorous growth, late heading and longer leaves (Table 10.5). A significant correlation with stem length was only present in the data from Northern Ireland. From these data it can be concluded that selection for a higher dry matter will lead to a positively correlated response in the size of the reproductive system (see Section 10.2.3).

10.3.2 Apomictic Species

Most plants reproduce sexually through seeds. After the fusion of a female gamete (egg cell) and a male gamete (sperm cell) a zygote is formed, which develops into an embryo (amphimixis). In apomicts the chromosome number in the female gamete is not reduced and the embryo is formed without fertilization. The process resembles normal sexual reproduction, but is a method of vegetative reproduction through seed. Apomixis occurs in many polyploid tropical grass species and also in smooth-stalked meadowgrass. If apomixis is the only mode of reproduction, the apomixis is obligate. Facultative apomixis occurs if as well as the apomictic mode of reproduction, embryos are also formed sexually.

An advantage of obligate apomixis is that a single outstanding plant composes a cultivar, whereby the genetic constitution is fixed during seed multiplication, remaining genetically stable and uniform. Cultivars are therefore uniform, stable and easy to describe. Heterosis can be fixed. For seed production the isolation distances between cultivars can be reduced. However, apomixis is regarded as a barrier for the breeder, as the apomictic nature of reproduction makes recombination by crossing very difficult or even impossible. Breeding apomictic species is therefore hampered by the lack of variation. In smooth-stalked meadowgrass ecotype breeding is used extensively. However, the genotypic variability for the combination of good turf quality and a high seed yield is restricted. Ecotypes with such a rare combination can often not be distinguished from existing cultivars. Breeding methods for apomictic species have been reviewed by den Nijs and van Dijk (1993).

In smooth-stalked meadowgrass, a facultative apomict, an aposporous embryo sac is formed from a nucellus or integument cell, and the embryo develops parthenogenetically with basically the same genetic constitution as the mother plant. Pollination is required for fertilization of the central nucleus and development of the endosperm. The degree of facultative apomixis (sexuality) in smooth-stalked meadowgrass can be increased by high temperatures during flowering. Hybridization under high temperatures shows a higher degree of off-types in the offspring (an advantage for breeders), but also seed multiplication in unusually hot summers often reveals more heterogeneity in the offspring of commercial cultivars (a disadvantage for cultivar maintenance). Selection for sexuality leads to a lower degree of apomixis and more heterogeneity, which makes the candivar ineligble for registration (Hintzen and van Wijk, 1985). If for F1 hybrids the conditions of good turf performance and the requirements for cultivar registration have been met, poor seed productivity can stop the candivar from being commercialized. Often poor seed setting and proneness to diseases during flowering (e.g. ergot – *Claviceps purpurea* (Fr. ex Fr.) Tul.) lead to low seed yields.

10.4 CULTIVAR MAINTENANCE

Once a cultivar has been released, a suitable cultivar maintenance system is essential to ensure its genetic stability. 'Cultivar maintenance' should not be confused with 'maintenance breeding', which implies the continued development of a cultivar. A cultivar has to remain genetically stable over successive generations and under the OECD seed certification scheme (OECD, 1988) the first submitted sample seed of the cultivar in question serves as a source of reference against which all later generation seed samples will be checked.

10.4.1 Breeders Seed

Most bred cultivars of grasses and legumes are synthetic cultivars. A synthetic cultivar is a cultivar produced by crossing *inter se* a number of genotypes selected for good combining ability in all possible hybrid combinations, with subsequent maintenance of the cultivar by open pollination.

As outlined in Section 10.3.1, a cultivar may be composed from selected clones or from remnant seed of selected progenies. The first generation of the seed harvested from this material is called Syn (Syntheticum) 1. The Syn 1 generation is not in genetic equilibrium. For diploid material a further multiplication without selection will bring the cultivar in the Hardy–Weinberg equilibrium (Syn 2). Successive generations thereafter are supposed to remain genetically stable. The Syn 2 generation is called breeders seed out of which the subsequent generations are grown (OECD, 1988). For tetraploid species this equilibrium is reached in the Syn 3 or Syn 4. The genetic stability depends on the number and genetic similarity of the parental clones that build up the cultivar.

Through its breeder's seed the cultivar can be maintained if a large enough stock has been produced (250 kg) and stored at low temperature (3°C) and low humidity (30%). Another approach is to maintain the selected clones *in situ* or through tissue culture and build up the Syn 1 and Syn 2 generations according to requirement. However, vegetative maintenance *in situ* is prone to virus diseases, or clones can be lost due to biotic factors, while tissue culture requires sophisticated laboratory facilities. Cultivar maintenance by storing a reasonably large quantity of breeders seed under controlled conditions therefore is the most economical and feasible method of maintaining a cultivar.

10.4.2 Prebasic, Basic and Certified Seed

The breeder's seed is used for the production of prebasic seed (in the USA, prebasic seed is called foundation seed). A production area of 2–3 ha of prebasic seed, divided over two locations to spread the risks of production, will suffice for several years for the production of basic seed (in the USA, basic seed is called registered seed). Certification of a seed lot of a cultivar (prebasic, basic or certified seed) is done in accordance with the rules of the OECD Seed Scheme (OECD, 1988) by the nationally designated authorities. These rules are applied by the participating countries, but each country may have its own additions and requirements.

10.4.3 Cultivar Stability

The genetic stability of a cultivar is to a large extent determined by the careful maintenance of breeder's, prebasic and basic seed. Between the selection of a population and its commercial use, several generations of multiplication are necessary. The essential feature in cultivar maintenance is that no genetic changes occur during multiplication. In seed-propagated species, however, during each multiplication cycle natural selection takes place at the gamete, zygote, seed and plant level, owing to competition among pollen grains and/or egg cells, incompatibility, seed abortion, poor germination, seedling and plant competition and reduced fertility (Elgersma, 1992). This may cause genetic shift and affect cultivar stability.

By careful field inspection and pre- and postcontrols of the prebasic and basic seed by the certifying agency, the chances that the cultivar is not genetically stable are small. However, the breeder's seed that is used to build up the various generations should be in genetic equilibrium (see Section 10.4.1) and the production of breeders, prebasic and basic seed should be carried out under comparable climatic conditions (temperature) and at similar latitudes (photoperiod).

As the time of harvesting differs more from the mean ripeness date, the risk of genetic shift increases (Davies, 1954). Buss and Barnes (1976) reported genetic shift in seed lots from a second seed harvest late in the season in lucerne. Steiner *et al.* (1992) found interactions in lucerne between the location of seed production, bee preference and the level of inbreeding, resulting in genetic shifts in plant introductions, germplasms, genetic stocks and cultivars. However, when Japanese red clover was multiplied in the western USA (Maki *et al.*, 1974) and in a preliminary study when North European white clover cultivars were multiplied in the mediterranean zone (Rijckaert and Vandepitte, 1989), there was no evidence of genetic shift. Cases where genetic shift in white clover multiplication did occur were reviewed by Williams (1987). The choice of locations with ideal ripening conditions and the proper choice of harvest date can help to prevent genetic shift.

10.5 CONCLUSIONS

There is ample genetic variation both in spaced plants and in seed crops for seed yield and its components. The broad-sense heritability is often high, but the narrow-sense heritability is not.

Seed yield and relations between its components in spaced plants are often different from those in seed crops. Because there are negative correlations between components, selection for one component only is not advisable. Selection for seed retention is promising, because this is not correlated with other yield components. Selection for higher seed yield *per se* is an option, irrespective of which components contribute most to it. During selection for seed yield, vegetative yield and quality must not be impaired. However, it is possible to select for a combination of good agronomic performance and high seed-yielding potential.

ACKNOWLEDGEMENTS

We thank Professor Dr P.C. Struik, Dr A.C. Zeven and Professor Dr L. 't Mannetje for their valuable comments and suggestions.

REFERENCES

Barbetti, M.J. and Nichols, P.G.H. (1991) Herbage and seed yield losses in six varieties of subterranean clover from rust (*Uromyces trifolii-repentis*). *Australian Journal of Experimental Agriculture* 31, 225–227.

Bean, E.W. (1972) Clonal evaluation for increased seed production in two species of forage grasses, *Festuca arundinacea* Schreb. and *Phleum pratense* L. *Euphytica* 21, 377–383.

Binek, A. (1983) [The structure of seed yield in clones of white clover (*Trifolium repens* L.) after the reduction of inflorescences to a standard number per plant.] *Acta Agraria et sylvestria, Agraria* 22, 21–29. (In Polish.)

Bugge, G. (1987) Selection for seed yield in *Lolium perenne* L. *Plant Breeding* 98, 149–155.
Buss, G.R. and Barnes, K.D. (1976) Evidence for seasonal genetic shift in alfalfa seed production. In: *Report of the Twenty-Fifth Alfalfa Improvement Conference*, July 13–15, Cornell, USA, ARS-NC-52, p. 32.
Carlson, J.M., Ehlke, J.N. and Wyse, D.L. (1995) Environmental control of floral induction and development in Kentucky bluegrass. *Crop Science* 35, 1127–1132.
Cebrat, J., Kobierzynska-Golab, Z. and Ramenda, S. (1982) [The variability of quantitative characters which affect fertility in five varieties of white clover (*Trifolium repens* L.).] *Hodowla Foslin, Aklimatyzacjai Nasiennictwo* 26, 11–34. (In Polish.)
Clifford, P.T.P. and Baird, I.J. (1993) Seed yield potential of white clover: characteristics, components and compromise. *Proceedings of the XVII International Grassland Congress*, pp. 1678–1679.
Davies, W.E. (1954) 'Shift' in a late-flowering strain of perennial ryegrass (*Lolium perenne*). In: *European Grassland Conference*, OEEC, Paris, pp. 102–106.
Davies, W.E. (1962) The flowering of white clovers under glasshouse conditions. In: *Welsh Plant Breeding Station Report for 1961*, Welsh Plant Breeding Station, Aberystwyth, p. 44.
Davies, W.E. (1981) Selection for high seed production in leguminous forage plants – a review. In: *Breeding High Yielding Forage Varieties Combined with High Seed Yield*. Report of the Meeting of the Fodder Crops Section of Eucarpia, Merelbeke, Belgium, pp. 139–151.
den Nijs, A.P.M. and van Dijk, G.E. (1993) Apomixis. In: Hayward, M.D., Bosemark, N.O. and Romagosa, I. (eds) *Plant Breeding. Principles and Prospects*. Chapman & Hall, London, pp. 229–245.
Dennis, B. (1975) Self-compatibility and inbreeding in autotetraploid red clover. In: Nüesch, B. (ed.) *Ploidy in Fodder Plants*. Report of the Meeting of the Fodder Crops Section of Eucarpia, Zürich-Reckenholz, Switzerland, pp. 43–45.
Dennis, B. (1980) Breeding for improved seed production in autotetraploid red clover. In: Hebblethwaite, P.D. (ed.) *Seed Production*. Butterworths, London, pp. 229–240.
Dessureaux, L. (1951) Ovule formation as a factor influencing seed setting of Ladino white clover. *Science in Agriculture* 31, 373–382.
Elgersma, A. (1985) Floret site utilization in grasses: definitions, breeding perspectives and methodology. *Journal of Applied Seed Production* 3, 40–54.
Elgersma, A. (1990a) Genetic variation for seed yield in perennial ryegrass (*Lolium perenne* L.). *Plant Breeding* 105, 117–125.
Elgersma, A. (1990b) Seed yield related to crop growth and to yield components in nine cultivars of perennial ryegrass (*Lolium perenne* L.). *Euphytica* 49, 141–154.
Elgersma, A. (1990c) Spaced-plant traits related to seed yield in plots of perennial ryegrass (*Lolium perenne* L.). *Euphytica* 51, 151–161.
Elgersma, A. (1990d) Heritability estimates of spaced-plant traits in three perennial ryegrass (*Lolium perenne* L.) cultivars. *Euphytica* 51, 163–171.
Elgersma, A. (1991) Floret site utilization in perennial ryegrass (*Lolium perenne* L.). *Journal of Applied Seed Production* 9 (suppl.), 38–43.
Elgersma, A. (1992) Seed yield and seed yield selection in polyploid forage crops. In: *Ploidy and Chromosome Manipulation in Forage Breeding. Proceedings of the 17th Meeting of the Fodder Crops Section of Eucarpia*, Alghero, Italy, 1991, pp. 124–131.
Elgersma, A. and van Hateren, J. (1991) Development of *in vitro* methods for fertilization studies in perennial ryegrass. In: Nijs, A.P.M. den and Elgersma, A. (eds) *Fodder Crops Breeding: Achievements, Novel Strategies and Biotechnology. Proceedings of the 16th Meeting of the Eucarpia Fodder Crops Section*, Wageningen, the Netherlands, 1990, Pudoc, Wageningen, pp. 179–180.

Elgersma, A., Leeuwangh, J.E. and Wilms, H.J. (1988) Abscission and seed shattering in perennial ryegrass (*Lolium perenne* L.). *Euphytica* S, 51–57.

Elgersma, A., Stephenson, A.G. and den Nijs, A.P.M. (1989) Effects of genotype and temperature on pollen tube growth in perennial ryegrass (*Lolium perenne* L.). *Sexual Plant Reproduction* 2, 225–230.

Elgersma, A., Winkelhorst, G.D. and den Nijs, A.P.M. (1994) The relationship between progeny seed yield in drilled plots and maternal spaced-plant traits in perennial ryegrass (*Lolium perenne* L.). *Plant Breeding* 112/3, 209–214.

Evans, D.R., Williams, T.A. and Davies, W.E. (1986) Potential seed yield of white clover varieties. *Grass and Forage Science* 41, 221–227.

Evans, P.M., Smith, R.S., Carpenter, J.A. and Koen, T.B. (1989) Tolerance of subterranean clover cultivars and balansa clover to selective herbicides in Tasmania. *Australian Journal of Experimental Agriculture* 29, 785–789.

Falcinelli, M. (1993) Seed shattering in tall fescue. *Proceedings of the XVII International Grassland Congress*, pp. 1666–1667.

Falcinelli, M., Veronesi, F. and Negri, V. (1984) Seed dispersal of Italian ecotypes of cocksfoot (*Dactylis glomerata* L.). *Journal of Applied Seed Production* 2, 13–17.

Falcinelli, M., Russi, L. and Lorenzetti, F. (1996) Breeding strategies for yield and quality of forage grass seeds in a mediterranean environment. In: *Proceedings of the 3rd International Herbage Seed Conference*, Halle, Germany, pp. 114–118.

Góral, H. and Spiss, L. (1996) Long inflorescence in red clover (*Trifolium pratense* L.) and its significance for seed yield. In: *Proceedings of the 3rd International Herbage Seed Conference*, Halle, Germany, pp. 139–142.

Griffiths, D.J., Lewis, J. and Bean, E.W. (1980) Problems of breeding for seed production in grasses. In: Hebblethwaite, P.D. (ed.) *Seed Production*. Butterworths, London, pp. 37–49.

Harun, R.M.R. and Bean, E.W. (1979) Seed development and seed shedding in North Italian ecotypes of *Lolium multiflorum*. *Grass and Forage Science* 34, 215–220.

Havstad, L.T. (1996) The effect of plant age on the receptiveness for primary induction in *Festuca pratensis* Huds. In: *Proceedings of the 3rd International Herbage Seed Conference*, Halle, Germany, pp. 74–78.

Hebblethwaite, P.D., Wright, D. and Noble, A. (1980) Some physiological aspects of seed yield in *Lolium perenne* L. (perennial ryegrass). In: Hebblethwaite, P.D. (ed.) *Seed Production*. Butterworths, London, pp. 71–90.

Hides, D.H. and R. Desroches (1990) Role of seeds in forage production – factors limiting optimal utilization. *Proceedings of the XVI International Grassland Congress*, pp. 1777–1789.

Hides, D.H., Kute, C.A. and Marshall, A.H. (1993) Seed development and seed yield potential of Italian ryegrass (*Lolium multiflorum* Lam.) populations. *Grass and Forage Science* 48, 181–188.

Hill, M.J. (1980) Temperate pasture grass-seed crops: formative factors. In: Hebblethwaite, P.D. (ed.) *Seed Production*. Butterworths, London, pp. 137–149.

Hill, M.J. and Loch, D. (1993) Achieving potential herbage seed yields in tropical regions. *Proceedings of the XVII International Grassland Congress*, pp. 1629–1635.

Hintzen, J.J. and A.J.P. van Wijk (1985) Ecotype breeding and hybridization in Kentucky bluegrass (*Poa pratensis* L.). *Proceedings of the Vth International Turfgrass Research Conference*, pp. 213–219.

Johnston, D.T. and Faulkner, J.S. (1991) Herbicide resistance in the Graminaceae – a plant breeders' view. In: *Herbicide Resistance in Weeds and Crops, 11th Long Ashton International Symposium*, 1989, Bristol, UK, pp. 319–330.

Julén, G. (1975) The current situation in the tetraploid clover. In: Nüsch, B. (ed.) *Ploidy in Fodder Plants*. Report of the meeting of the Fodder Crops Section of Eucarpia, Zürich-Reckenholz, Switzerland, pp. 79–89.

Lewis, J. (1966) The relationship between seed yield and associated characters in meadow fescue (*Festuca pratensis*). *Journal of Agricultural Science, Cambridge* 67, 243–248.

Lewis, J. (1970) Reproductive growth in *Lolium*. 1. Evaluation of genetic differences within an established variety by means of a diallele cross. *Euphytica* 19, 470–479.

Lorenzetti, F. (1993) Achieving potential herbage seed yields in species of temperate regions. *Proceedings of the XVII International Grassland Congress*, pp. 1621–1628.

Maki, Y., Matsu-ura, M., Suginobu, K., Miyashita, Y., Hayakawa, R., Sato, H., Murakami, K., Kaneko, K. and Matsuura, M. (1974) Genetic shift in agronomic characteristics of the Japanese red clover cultivar 'Sapporo' grown from advanced generation seed multiplied at diverse latitudes in the United States. Plant introduction, breeding and seed production. *Proceedings of the XII International Grassland Congress*, pp. 236–239.

Marshall, C. (1985) Developmental and physiological aspects of seed production in herbage grasses. *Journal of Applied Seed Production* 3, 43–49.

Marshall, C. and Ludlam, D. (1989) The pattern of abortion of developing seeds in *Lolium perenne* L. *Annals of Botany* 63, 19–27.

McWilliam, J.R. (1980) The development and significance of seed retention in grasses. In: Hebblethwaite, P.D. (ed.) *Seed Production*. Butterworths, London, pp. 51–60.

Nelson, L.R., Ward, S.L. and Crowder, J. (1996) Effect of vernalization on heading date, tillering and plant biomass of *Lolium multiflorum*. In: *Proceedings of the 3rd International Herbage Seed Conference*, Halle, Germany, pp. 35–38.

Nordestgaard, A. and Andersen, S. (1991) Stability of high production efficiency in perennial herbage seed crops. *Journal of Applied Seed Production* 9 (suppl.), 27–32.

OECD (1988) *Scheme for the Varietal Certifiction of Herbage Seed Moving in International Trade*. Organisation for Economic Cooperation and Development, Paris.

Onokpise, O.U., Bowley, S.R., Tomes, D.T. and Twamley, B.E. (1987) Evaluation of self and polycross progeny testing in birdsfoot trefoil (*Lotus corniculatus* L.) for forage and seed yield. *Plant Breeding* 98, 141–148.

Pasumarty, S.V., Matsumura, T., Higuchi, S. and Yamada, T. (1993) Ovule fertility – a tool for selecting high-fertility populations of white clover. *Proceedings of the XVII International Grassland Congress*, pp. 1648–1649.

Rhoads, J.L., Dunn, J.H., Minner, D.D. and Hunt, K.L. (1992) Reproductive morphology of five Kentucky bluegrass cultivars. *Agronomy Journal* 84, 144–147.

Rijckaert, G. and Vandepitte, H. (1989) Preliminary study of the effect of the Mediterranean climate on the stability of N-European cultivars of white clover. *Revue de l'Agriculture* 42, 1067–1075.

Rincker, C.M., Marble, V.L., Brown, D.E., and Johansen, C.A. (1988) Seed production practices. In: Hanson, A.A., Barnes, D.K. and Hill, R.R. (eds) *Alfalfa and Alfalfa Improvement*. Agronomy Monograph no. 29. ASA-CSSA-SSSA, Madison, USA, pp. 985–1021.

Rose-Fricker, C.A., W.A. Meyer and W.E. Kronstad (1986) Inheritance of resistance to stem rust (*Puccinia graminis* subsp. *graminicola*) in six perennial ryegrasses (*Lolium perenne*) crosses. *Plant Disease* 70, 678–681.

Ross, J.G. and Adams, M.W. (1955) The influence of heredity on seed and forage production in smooth bromegrass. *Proceedings of the South Dakota Academy of Science* 34, 16–20.

Simon, U. (1996) Breeding for resistance to seed shattering in forage grasses. In: *Proceedings of the 3rd International Herbage Seed Conference*, Halle, Germany, pp. 119–123.

Sjödin, J. (1981) Selection for seed setting capacity in tetraploids of clover and grasses. In: *Breeding High Yielding Forage Varieties Combined with High Seed Yield*. Report of the Meeting of the Fodder Crops Section of Eucarpia, Merelbeke, Belgium, pp. 163–168.

Spiss, L. and Goral, H. (1989) A new mutant with open corolla tube in red clover. *Proceedings of the XVI International Grassland Congress*, pp. 419–420.

Steiner, J.J., Beuselinck, P.R., Peaden, R.N., Kojijs, W.P. and Bingham, E.T. (1992) Pollinator effects on crossing and genetic shift in a three-flower-color alfalfa population. *Crop Science* 32, 73–77.

Stratton, S.D. and Ohm, H.W. (1989) Relations between orchardgrass seed production in Indiana and Oregon. *Crop Science* 29, 908–913.

Thomas, J.R. and Frakes, R.V. (1967) Clonal and progeny evaluations in two populations of tall fescue (*Festuca arundinacea* Schreb.). *Crop Science* 7, 55–58.

Thomas, R.G. (1987) Reproductive development. In: Baker, M.J. and Williams, W.M. (eds) *White Clover*. CAB International, Wallingford, UK, pp. 63–123.

van Bogaert, G. (1977) Factors affecting seed yield in white clover. *Euphytica* 26, 233–239.

van Dijk, G.E. and G. Winkelhorst (1982) Interspecific crosses as a tool in breeding *Poa pratensis* L. 1. *Poa longifolia* Trin × *P. pratensis* L. *Euphytica* 31, 215–223.

Veronesi, F. and Falcinelli, M. (1987) Seed yield selection in *Medicago sativa* L. and correlated responses affecting dry matter yield. *Plant Breeding* 99, 77–79.

Viands, D.R., Sun, P. and Barnes, D.K. (1988) Pollination control: mechanical and sterility. In: Hanson, A.A., Barnes, D.K. and Hill, R.R. (eds) *Alfalfa and Alfalfa Improvement*. Agronomy Monograph no. 29. ASA-CSSA-SSSA, Madison, USA, pp. 931–960.

Walton, P.D. (1983) *Production and Management of Cultivated Forages*. Reston Publishing Company, Virginia, 336 pp.

Welty, R.E. and Barker, R.E. (1993) Reaction of twenty cultivars of tall fescue to stem rust in controlled and field environments. *Crop Science* 33, 963–967.

Welty, R.E. and Barker, R.E. (1994) Management of stem rust (*Puccinia graminis* subsp. *graminicola*) in perennial ryegrass (*Lolium perenne* L.) grown for seed. *IOBC/WPRS Bulletin* 17, 241–246.

van Wijk, A.J.P. (1985) Factors affecting seed yield in breeding material of Kentucky bluegrass (*Poa pratensis* L.). *Journal of Applied Seed Production* 3, 59–66.

Williams, W.M. (1987) Genetics and breeding. In: Baker, M.J. and Williams, W.M. (eds) *White Clover*. CAB International, Wallingford, UK, pp. 343–419.

Wright, C.E. and Faulkner, J.S. (1982) A backcross programme introducing resistance to blind seed disease (*Gloeotinia temulenta*) into the cultivar S24 of the cross pollinated species perennial ryegrass (*Lolium perenne*). *Record of Agricutural Research (Department of Agriculture, Northern Ireland)* 30, 45–52.

The Forage Seed Trade 11

A. Burgon,[1] O.B. Bondesen,[2] W.H. Verburgt,[3] A.G. Hall,[4]
N.S. Bark,[1] M. Robinson[5] and G. Timm[6]

[1]British Seed Houses Ltd, Portview Road, Avonmouth, Bristol BS11 9JH, UK; [2]Danish Seed Council, WA Vestrbrogade, DK-1020 Copenhagen, Denmark; [3]Mommersteeg International BV, Postbus 135250 AA, Vlijmen, The Netherlands; [4]New Zealand Grain and Seed Trade Association Ltd, PO Box 1208, Wellington, New Zealand; [5]Seed Research of Oregon Inc., PO Box 1416, Corvallis, Oregon 97339, USA; [6]Research Seeds Inc., St Joseph, Missouri, USA

11.1 INTRODUCTION

From the Willamette Valley in Oregon to the Polder region of the Netherlands, and from the Canterbury Plains of New Zealand to Jutland in Denmark, forage seed production forms an important part of agribusiness in many countries. While many factors such as climate, soil types and livestock systems affect the production and use of forage seeds around the world, the possibilities for trade and marketing both within individual domestic markets and through export opportunities have an important role to play.

Domestic marketing is widely variable, as illustrated by the fact that uncertified seed may be traded between certain states within the USA, but only certified seed may be sold within member countries of the European Union (EU). Even within the EU, some member states operate additional quality standards over and above those laid down by the European Economic Commission (EEC) Rules and Standards, e.g. the High Voluntary Standards (HVS) being operated in the UK.

Where international multiplication contracts are agreed between EU member states and producers outside of the EU, such produce must usually be accompanied by a statement that the seed 'meets EEC Rules and Standards', e.g. white clover (*Trifolium repens* L.) from New Zealand or perennial ryegrass (*Lolium perenne* L.) from the USA. While all countries may have satisfactory certification schemes in operation, protection for the importing country is offered by the Organization for Economic Cooperation and Development (OECD)* scheme for the varietal certification of seed moving in international trade. The objective of this scheme is to encourage the use of seed of consistently high quality in participating countries, and it is open on a voluntary basis to all members of the OECD, as well as to other

---

*OECD, 2 Rue André Pascal, 75775 Paris, Cedex 16, France.

countries who are members of the United Nations (UN) or its specialized agencies desiring to participate in the OECD scheme in accordance with the procedures for such participation as set out in the rules.

Most international trade is carried out under OECD certification, which includes only those cultivars which are officially recognized as distinct and having an acceptable value for cultivation and use (VCU) in at least one country. In addition, the progeny of these cultivars must have sufficiently uniform and stable characters. Only results from official trials, including comparative field tests, may decide the distinctness, uniformity, and stability (DUS) and VCU of a cultivar.

It should be emphasized at this point that OECD certification relates only to the cultivar characteristics of the seed, and not the analyses of the product. The analytical testing of the seed must be carried out under the methods as designated by the 'International Rules for Seed Testing' of the International Seed Testing Association (ISTA)*, these tests to be carried out in the Official Seed Testing Station of the Designated Authority. The results of these tests are usually reported on an Orange International Certificate (OIC), issued by the Authority, which with the OECD Certificate will guarantee to the buyer that the cultivar characters and seed quality are within acceptable standards. Universal acceptance of the OIC is vital to the continued international passage of seed, and as such the testing methods are continually being updated to ensure worldwide acceptance of the quality stated.

Whilst all seed traded must meet specified purity and germination standards, many countries have additional requirements relating to the importing of seed containing certain noxious weeds such as wild oats (*Avena fatua* L.) or black grass (*Alopecurus myosuroides* Huds.) and because of this, weed search requirements are often carried out by the Official Seed Testing Stations, and are also reported on the OIC. However, protection from the importation of harmful organisms, such as seed-borne diseases, is offered by means of a phytosanitary certificate issued by the respective Government Health Laboratories, without which import permits will not be issued by the receiving country or state. As with most industries, the seed trade has adopted guidelines covering global trading, and these are laid out in the 'Rules and Usages for the International Trade in Herbage Seeds' issued by the Federation International du Commerce des Semences (FIS)†. Such rules apply in full when the letters FIS have been embodied in a bid/offer/contract pertaining to seeds for sowing purposes. Protection of both parties is given covering quantity, quality and analyses, and arbitration under the FIS Arbitration Procedure Rules applies where disputes cannot be settled amicably.

This introduction has been concerned with explaining the parameters within which the seed trade operates on a worldwide basis, but in order to understand the different marketing strategies of individual countries, both domestically and internationally, representatives of a few seed-producing nations have been invited to prepare a short paper explaining their own country's trading philosophies.

---

*ISTA Secretariat, Reckenholz Strasse 191, PO Box 412, 8046 Zurich, Switzerland.
†FIS, Chemin du Reposoir 5–7, 1260 Nyon, Switzerland.

11.2 DENMARK

11.2.1 Introduction

Denmark is a small country covering an area of 43,000 km² of which 2.7 million ha are devoted to agriculture and around 55,000 ha or 2% of the agricultural land to the cultivation of forage seed. As most of the forage seed area is planted with grasses, Denmark is one of the world's most intensive grass seed-producing countries.

The Danish climate and the geographical position of the country favour the seed production of most grasses and white clover. The climate is temperate with a yearly average temperature of 7.9°C. July is the warmest month at 16.6°C, and February the coldest at −0.4°C. Average annual rainfall is 664 mm. Denmark is situated at the 56th parallel.

Danish forage seed production has developed to its present state from a modest start at the end of the nineteenth century and the beginning of this century. Although Danish production was initially directed to the domestic market, it rapidly developed into an export commodity, and for many years now Denmark has been the world's largest exporter of seed of cold season grasses.

Since 1973 Danish seed production has operated within the framework of the Common Agricultural Policy (CAP), which as a main objective has to ensure equal conditions for farming including seed production in Europe. The CAP has provided the opportunity for an optimal localization of the production among crops based on economic factors.

For several decades Denmark has been a leading European producer of grass seed. Since 1973 it has had 40–50% of the EU production, and the trend has been for an increase of this share. This is because within the EU, grass seed production of many species has comparative advantages in Denmark. First and foremost of these advantages is the relatively high seed yield of most grasses compared to those of alternative crops like cereals and oil seed rape. The same circumstances are considered to be the reason why Denmark is also by far the largest European producer of white clover seed. However, the optimal production region for white clover seed is found only in the southeastern part of Denmark.

11.2.2 Production

In the last decade grass and clover seed production in Denmark has occupied between 45,000 and 72,000 ha. The average area in Denmark is around 54,000 ha of grass seed and 3000 ha of clover seed. In approximate figures the grass seed area can be divided as follows: perennial ryegrass 24,000 ha; red fescue (*Festuca rubra* L.) 12,000 ha; Kentucky bluegrass (*Poa pratensis* L.) 6000 ha; Italian ryegrass (*Lolium multiflorum* Lam.) 4000 ha; cocksfoot (*Dactylis glomerata* L.) 2000 ha; meadow fescue (*Festuca pratensis* Huds.) 2000 ha; hybrid ryegrass (*L. perenne* × *L. multiflorum*) 1000 ha; timothy (*Phleum pratense* L.) 1000 ha; other grasses 2000 ha. The clover seed area consists of around 2500 ha with white clover and 500 ha with red clover (*Trifolium pratense* L.) The long-term tendency has been an increasing area of perennial ryegrass and red fescue and a declining area of white clover.

On average the annual production is 57,000 t of grass seed. The seed yield per hectare of grasses is relatively high in Denmark compared to the production per hectare of the main alternative, cereals. The 10-year average is 5.3 t ha^{-1} for cereals, while the 10-year average is around 1200 kg ha^{-1} for perennial ryegrass, 1050 kg ha^{-1} for red fescue and 950 kg ha^{-1} for Kentucky bluegrass.

The good production conditions for grass seed have resulted in the large-scale multiplication of foreign cultivars in Denmark. Through the last decade the area devoted to production of foreign cultivars has fluctuated between one-third and one-half of the total grass seed area. In the EU, Denmark is the only large-scale multiplier of foreign cultivars, and Danish production is, in that respect, in direct competition with multiplication outside the EU.

Grass seed production in Denmark is carried out in rotation with other crops like cereals, oil seed rape and protein plants. Thus a typical grass seed grower will have no more than 10–20% of his agricultural land planted with grass seed crops. Grass seed is produced all over the country, although among species the production intensity varies from one region to another, as the individual species have their optimums in different environments with regard to soil types, subclimates and types of farming. In 1993, 4741 out of a total number of 72,887 farms (6.5%) produced forage seed. The average forage seed area per farm was 12 ha. In general, the intensity of seed growing is increasing with increasing farm size; thus, one-third of Danish farms with an area of more than 100 ha are involved in forage seed production.

11.2.3 Marketing

All Danish seed growing is carried out on the basis of a contract between seed growers and seed companies. Each contract specifies the obligations of the seed growers as well as the company.

The seed companies constitute a very important and integrating element in the chain between the breeder and the user of seed. Among the obligations of the company are to contract areas for seed growing, to supply basic seed, and to process and sell the seed. Since 1995 the following seven herbage seed companies have been operating in Denmark: DLF-TRIFOLIUM A/S, Hunsballe Frø A/S, A/S Chr. Kehlets Frøforretning, A/S Morsø Frøkontor, A/S Anton Neilsens Frøavl og Frøeksport, Wilboltt Frø A/S, Østergaards Frøavl A/S.

These companies are of varying size with DLF-TRIFOLIUM, the largest, having around 70% of the Danish grass and clover seed production. Furthermore, the only Danish breeder of fodder plant species in the public sector, Danish Plant Breeding, now belongs to DLF-TRIFOLIUM. Østergaards Frøavl, Wiboltt Frø and Hunsballe Frø are middle size companies. The first two mentioned are affiliates to the Dutch seed companies Van der Have and CEBECO, respectively, while Hunsballe and the three smallest companies are Danish owned. The seven forage seed companies have a total of more than 20 departments for seed handling spread out over Denmark.

The Danish seed companies take care of all operations from the stage of delivery of the field-dressed seed to the company, and onwards until the distribution of the final certified seed to the customers at home or abroad. Generally, these companies use advanced techniques and large-scale operations which make high-volume Danish seed production possible.

11.2.4 Export and Domestic Consumption

Danish seed production is first and foremost for export. On average 90% of the grass and clover seed produced in Denmark is exported; an annual quantity of around 51,000 t of grass seed and 700 t of clover seed. The European Communities have always been by far the largest market for Danish seed. On a 10-year average, 83% of the total export went to the EU, 7% to the Nordic countries, 5% to other European countries, 4% to North America and 1% to South America. In the enlarged EU operating from the beginning of 1995 the Danish export share to other member states will increase to more than 90%. Denmark is therefore the world's dominant exporter of grass seed of cold season species.

Although domestic consumption only accounts for 10% of the Danish production, it is nevertheless a valuable market. On average the Danish market represents around 6100 t of grass and clover seed annually. Denmark imports negligible amounts of grass and clover seed.

11.3 THE NETHERLANDS

11.3.1 Introduction

The trade in seed of a large number of grass species for agricultural and horticultural use developed during the nineteenth and twentieth centuries. Since the largest area of cultivated land is under pasture, the trade in seed to renew those grazing grounds has been particularly popular. Initially, commercial 'wild' seed was in demand, until more cultivated species were developed. At the beginning of this century properly selected cultivars became the trend. In 1927 the first grass cultivars were entered in the descriptive list of cultivars of field crops.

11.3.2 History

The trade of forage seed in Holland goes back to the early part of the nineteenth century when a demand developed for commercial seed of popular forage species. Before that period, farmers used natural re-seeding or took seed from the hay-barn. Commercial seed was mostly imported from Canada, New Zealand and Ireland and was not always suited to the climatic conditions in Holland. Therefore a demand for locally grown seed developed and, as is usual, supplies followed demand in the form of seed gathered by 'stem cutters'; people who cut ripe ears from roadsides and other large public areas and were careful to keep the ears of the different species apart. Surprisingly large quantities were collected and traded in this manner, not only for the local market but also for export. After 1930 this practice became less popular and seed from old established pastures was gathered and provided a better quality seed for farmers. At the same time some specialists began to select plants within the well-known land races of the annual Westerwolds ryegrass (*L. multiflorum* var. Westerwoldicum) and perennial ryegrass, which in Holland, presumably due to the early imports from Northern Ireland, is popularly called 'English ryegrass'.

From the beginning of this century certain traders began specializing in fodder crops and the competition was lively. So when selected or bred cultivars became popular, those traders engaged graduates from the Wageningen Agricultural University to carry out the work of grass breeding, thus enabling these traders to become specialists and have a comprehensive range of cultivars of all popular species available. Remarkably those 'early starters' from the beginning of this century, who specialized in forage seed, are still the leading seed houses and plant breeders today. Some of these companies are under different ownership or working in alliances. Another interesting fact is that while a large number of smaller traders have left the field, hardly any newcomers are noticeable. The so called 'barrier to entry' has become quite high. A separate chapter could be written on the various organizations which represented the trade, often on a regional basis. Approximately 10 years ago, after various mergers, the seed trade was concentrated in a Seed Traders Association and Plant Breeders Association, both for agricultural crops. They merged into one association for 'Breeders and Trade', and were later joined by the Association for Horticulture Seed. In this connection it should be mentioned that the trade in agricultural seed has developed separately from the horticultural sector. Only some regional distributors are active in the distribution of both forage seed and horticultural seed.

11.3.3 Current Situation

Since the introduction of the Variety List in 1924 and the first description of grass cultivars in 1927, the trade in forage seed has been confined to the seed of recommended cultivars only. Those cultivars are bred and owned by the major seed companies whose mandate consists of an integrated package of activities: breeding, seed production, processing and wholesale distribution. These companies do not confine their activities to the domestic market only; their trading activities are worldwide. The forage seed trade in Holland is quite large considering the geographical area of the country. This can be attributed to the intensive way the dairy farmers use their resources, as is evident from the quantities of forage seed mixtures used in agriculture (Table 11.1).

Table 11.1. Quantities of forage and amenity seed used in the Netherlands, 1989–1994.

	1989	1990	1991	1992	1993	1994
Forage seed (t)[1]	6704	6475	6145	5428	4515	6137
Amenity seed (t)[2]	1776	1904	1926	1910	1967	2113

[1] For agricultural use.
[2] Grass seed only.

Virtually all grass seed in Holland is sold as mixtures. The dominant part of these mixtures consists of officially recommended cultivars. The descriptive cultivar list not only provides comprehensive information concerning the recommended cultivars (of which there were more than 80 in 1995), but also makes a recommendation to the farmer as to the composition of the mixtures for his/her particular use according to the kind of husbandry preferred. Farmers have always shown great faith in the recommended cultivar list, and it is difficult to sell any quantity of forage seed of a cultivar which is not on the list. In the last couple of years, however, some companies have sold their own grass seed mixtures in addition to the official ones. All seed sold in Holland must be certified according to the rules issued by the OECD and comply with quality standards which are established by the local Certification Agency, the 'Nederlandse Algemene Keuringsdienst (NAKJ)'. This requirement applies equally to locally or foreign produced seed.

The physical distribution of seed for the domestic market is relatively simple. Distances are short and a near perfect infrastructure is in place. Because of this, and also because of the competitive nature of the market, the end user can be assured of very efficient supply at short notice. Most major seed companies use a form of dealer network for their distribution, including regional cooperatives which are quite powerful in some parts of the country. The regional dealer will also provide technical information to the farmer, sourced from the major seed companies as well as the independent extension services, which operate on a fee-for-service basis.

During the last 30 years or so, an important market for specialist grass seed cultivars for non-agricultural purposes has developed. Previously agricultural cultivars were used for parks, sporting facilities and other amenity uses. However, this market demands certain characteristics, and to cater for this market the main commercial breeders bred specialized cultivars of several species. Quite a number of those cultivars are suitable over a large geographical area, and Holland is now a popular supplier of many species and cultivars of grasses for non-agricultural or amenity purposes (Table 11.1). The domestic marketing and distribution patterns of amenity grass seed are rather different from the trade in agricultural forage seed. Because of the very specialized nature of this market, the distribution chain is even shorter, and often the breeder supplies the seed directly to the 'consumer'. The marketing of this seed is backed by comprehensive technical and practical information.

11.3.4 The Future

Having briefly described the history and current way of marketing forage seed in Holland, it would appear to be reasonable to look into the future. Many changes will need to take place simultaneously. The total demand for forage is not increasing and will be steady at best. To increase their turnover and thereby spread their breeding costs over a larger volume, seed companies must compete fiercely. This kind of competition cuts into the profit margin, and ultimately will result in a reduction of the number of participants.

Government-subsidized support for research and other supplementary services like VCU trials is dwindling, thereby making the seed industry and the consumers more self-reliant in providing those services. This will cost money and the industry will be searching for different, more cost-effective ways to provide these services. The number of distributors will be reduced. The major seed companies will therefore have to look for different ways of distribution; they cannot become too reliant on just a few large distributors and thereby lose their options to be flexible towards the end user. In Europe, national borders will disappear or be reduced. This will result in easier movement of products, people and also practices, thereby reducing the significance of the 'domestic market'.

It will be possible for the seed industry in Holland to cope with these changes and other future developments. Traditionally the industry has been transnationally orientated, and therefore seed trade and marketing in this country will survive and the seed industry will accept the challenges of the future.

11.4 NEW ZEALAND

11.4.1 Introduction

New Zealand is an agriculturally based country which has relied on the quality of its pastoral lands to produce a varied range of agricultural products such as wool, meat and dairy products for both internal consumption and to service export markets. New Zealand pastures and the quality of its agricultural products are internationally recognized. Pastoral agriculture makes up 43% of the country's exports, accounts for 10% of its Gross Domestic Product and directly employs approximately 13% of the workforce.

The extensive use of temperate pastures to drive agricultural productivity has encouraged the development of a dynamic seed industry in New Zealand; an industry which has gained respect throughout the world. Initially New Zealand relied heavily on the use of imported seed for the development of its pastoral regions. However the need to provide more productive and adaptable cultivars became apparent, and this resulted in the establishment of plant breeding infrastructures to service the country's needs. New Zealand plant breeders, both public and private, play a major role in providing the improved quality forage cultivars that have enhanced New Zealand's pastoral regions and established the base for a seed industry that serves not only its internal customers, but ranks as the third largest exporter of forage seeds internationally. As an island nation, well removed from other land masses, New Zealand is free from many pests and diseases found elsewhere. Strict quarantine rules seek to keep it that way.

11.4.2 Production

The majority of seed production in New Zealand is undertaken on the Canterbury Plains situated in the South Island. Small areas of seed production are situated on the lower east coast of the North Island for ryegrass and in the southern areas of the South Island for timothy and dogstail (*Cynosurus cristatis* L.). Approximately

35,000 ha are committed to forage seed production annually. The temperate maritime climate, 500 mm average rainfall, and warm summers, are an ideal environment for the production of most temperate forage species, especially white clover and ryegrass. A significant portion of the seed production regions is under irrigation. New Zealand is recognized as the most efficient and reliable producer of white clover seed, supplying over 60% of the world crop.

Over the past decade forage seed growers have become very specialized and skilled and have the ability to produce a varied range of species. Specialization in production has tended to bring stability into seed supply, with the major variances being due to weather and offshore demand for re-export multiplications, such as the turf-types of perennial ryegrasses or tall fescue (*Festuca arundinacea* Schreb.) (Table 11.2). Another trend for the New Zealand industry has been northern hemisphere recognition of New Zealand's out-of-season seed production capability. Significant tonnages of white clover, forage and amenity seed are produced under contract for re-export. Northern Hemisphere seed producers recognize New Zealand as either a regular supplier of seed to their markets or as a country that can provide production alternatives where there is a shortfall in a given season.

New Zealand growers are very willing to produce for export markets, but look towards crops that can offer long-term consistency in production. Areas suitable for seed production are coming under increased pressure in New Zealand due to the expansion of vegetables grown for processing, increased areas being used for life style farming operations, and the continual development of the dairy industry.

11.4.3 Marketing

New Zealand is a free market economy. Members of the seed trade are all within the private sector, with no government trading corporation or similar body. Sales are by private treaty, frequently through brokers. There is no seed business conducted through commodity exchanges. There is no Seed Act or similar legislation, but the country has fair trading legislation in place. Plant Variety Rights (PVR) were introduced in New Zealand in 1972. Their introduction provided the catalyst for investment in plant breeding by private enterprise. Most cultivars now being released for marketing are covered by PVR. New Zealand is a member of the

Table 11.2. Seed production (tonnes) for some forage and amenity species in New Zealand.

	1989	1990	1991	1992	1993
White clover					
New Zealand cultivars	2790	3750	4050	4970	3860
Grown for re-export	170	410	450	1130	590
Total certified production	2960	4160	4500	6100	4450
Perennial ryegrass					
New Zealand cultivars	6180	7770	5400	4730	6810
Grown for re-export	110	1170	1050	680	450
Total certified production	6290	8940	6450	5410	7260
Tall fescue					
Total certified production	57	815	1796	1990	398

International Union for the Protection of New Varieties of Plants/Union Internationale pour la Protection des Obtentions Végétales (UPOV).

While the trading in common species such as white clover, cultivar Grasslands Huia and ryegrass cultivar Grasslands Nui is an important aspect of the free trade environment, the disciplines imposed on the release of proprietary cultivars covered by PVR have made a significant impact on the marketing systems now operating in New Zealand. This has led to greater involvement of companies in the promotion of their products, improved technology transfer to customers and consumers, and a greater awareness of the industry. There has been a steady move to increase areas of proprietary production in all species categories, with a decline in the production of public cultivars.

New Zealand forage seeds are vigorously promoted by individual companies both internally and on international markets. In addition, the New Zealand Seeds Promotion Council, which includes representatives from the public and private plant breeders, growers and members of the trade, seeks to promote New Zealand seeds generically both within New Zealand and towards target export markets.

11.4.4 Organizations

The seed trade organization in New Zealand is the New Zealand Grain and Seed Trade Association (NZGSTA) which is the official affiliate of FIS and recognized as representing private industry's views on matters relating to the seed industry. The members of NZGSTA comprise those organizations primarily involved in the contracting of production, distribution, marketing, exporting and general servicing of the seed industry. The Herbage Seed Subsection of Federated Farmers (HSSS) provides the growers' views on industry matters relating to forage seed, and represents growers' concerns generally.

AgResearch Grasslands remains the major public breeding institution in New Zealand, with a focus on improved forage species. Releases from this research institute are made mainly to members of the NZGSTA for both internal and international marketing. Private breeding companies also play a significant role in the introduction of new forage cultivars as a result of their own breeding endeavours or through their representation of overseas breeders material in the domestic market. There are four privately owned research centres operating in New Zealand with a focus on forage seeds. The New Zealand Plant Breeding and Research Association represents private industry's views on plant breeding and matters relating to PVR legislation. It is affiliated to the International Association of Plant Breeders for the Protection of Varieties (ASSINSEL) and organizes a voluntary merit testing service for forage species.

Seed certification has operated in New Zealand since the 1920s and provides the cornerstone of the industry. It is administered by the Ministry of Agriculture and Fisheries Regulatory Authority (MAFRA) to international requirements. Internally the Seed Quality Management Authority (SQMA) provides an industry forum to advise MAFRA on aspects of certification, and to convey to SQMA members changes in the rules and regulations of certification. Phytosanitary assurance is administered by MAFRA. The Government-administered and University seed testing stations at Palmerston North are the designated authorities recognized by ISTA to issue OIC and official certificates. New Zealand is also a member of the Association of Official Seed Certifying Agencies (AOSCA). Several private seed testing laboratories complement the role of the official testing stations, providing industry with ancillary seed testing services.

11.5 UNITED KINGDOM

11.5.1 Introduction

Forage seed markets and production are fluid situations which change, not only in the long term, but also within a short span of years, as illustrated at particular times within this chapter. Movements not only take place due to market forces, but also more recently due to political decisions. The current situation within the UK is as follows.

11.5.2 Production

Very few personnel remaining in the forage seed trade are aware that between 1950 and 1960, Northern Ireland was one of the major producers of ryegrass in the world, with the peak level of production being in the early 1950s at about 30,000 t. Between 1960 and 1965 this production disappeared with the advent of new agronomically improved cultivars appearing on the Recommended Lists.

Ryegrass dominates the UK market, and this applies to UK production. Throughout the 1980s the level of production of perennial ryegrass was consistently 10,000–12,000 ha, producing approximately 1 t ha^{-1}. Over the last 2–3 years this level of production has fallen to about 7000 ha, with an increasing percentage being cultivars bred for amenity purposes. Italian-hybrid ryegrass production has fallen in a similar period from approximately 2200 to 1200 ha.

There have been several reasons for this decline. Undoubtedly under the EU CAP policy, it has been more attractive for UK farmers to grow arable crops such as cereals and oilseed rape rather than forage seed. As a result of this, there has recently been a very significant increase in the prices paid to forage seed growers. A further reason for the fall within the UK has been the domination of the respective Recommended Lists by overseas cultivars and, therefore, lack of the availability of Basic seed to the UK grower. There are indications that this situation is changing as cultivars from the Institute of Grassland and Environmental Research (IGER) at Aberystwyth and the Northern Irish Breeding Station at Loughgall appear on the lists.

Over the past 5–10 years there has been an effort to produce a larger percentage of the fescues used in the amenity sector within the UK, with the area of production increasing from little more than 100 ha to about 1000 ha in 1992/1993. However, difficulties have been experienced in obtaining the very high quality standards required, and unless clearance can be obtained from the authorities for the use of chemicals to control the coarse grasses, there is a possibility that this production will decline.

There are several methods of contracting with farmers; some companies operate on a participation basis, i.e. paying the grower a certain percentage of the price obtained for the seed, others contract on the basis of a totally fixed price or a combination of a fixed price and a price to be negotiated after harvest.

11.5.3 Marketing

Agriculture

The UK agricultural market has reduced substantially in size over the past 30 years, and has become much more ryegrass dominated due to the present heavy emphasis on silage making as compared with hay making in the 1960s, and the very significant movement from short-term leys to long-term ones. It is impossible to produce precise figures on the annual sowing of leys within the UK. However, a reasonable figure to use for this is 400,000 ha. Approximately 70% of this area is in England and Wales, and approximately 30% in Scotland and Northern Ireland, although with the higher seeding rates in the latter areas the consumption of seed is nearer to 65% England and Wales, and 35% Scotland and Northern Ireland.

The approximate annual consumption of the species within these mixtures is as follows: perennial ryegrass 1200 t (25% tetraploid); Italian ryegrass 1800 t (10% tetraploid); hybrid ryegrass 600 t (almost 100% tetraploid); cocksfoot 150 t; timothy 750 t; meadow fescue 80 t; other grasses (Westerwold, tall fescue, etc.) 120 t; white clover 600 t, other clovers 100 t; for a total of 16,200 t. Ryegrasses in one form or another account for in excess of 80% of agricultural forage seed usage.

The UK agricultural market is extremely cultivar conscious, principally because of the independent trials conducted by the National Institute of Agricultural Botany (NIAB) for England and Wales, the Scottish Agricultural Colleges (SAC) for Scotland and the Department of Agriculture in Northern Ireland (DANI) for Northern Ireland. Each of these organizations produce Recommended Lists for the principal agricultural species and unless a cultivar appears on these lists in a favourable position, it will not achieve a significant market share.

Whilst two-thirds of the national retail companies/organizations sell ley mixtures to farmers, the bulk is sold by relatively small local merchants who obtain their supplies from six-sevenths of the wholesale companies, all of whom are either breeders in their own right or have close connections with a breeding organization.

Amenity

Although the agricultural market has reduced by approximately one-third during the past 30 years, the amenity market has shown an increase, perhaps as much as 50%, with the consumption of species as follows: perennial ryegrass 5500 t; creeping red fescue 2500 t; fine fescue (chewings, hard, slender) 1100 t; *Agrostis* spp. 450 t; Kentucky bluegrass 350 t; other species 100 t, for a total of 10,000 t.

Approximately two-thirds of this seed is sold to the professional market (sports grounds/golf courses/parks, etc.) and one-third as lawn seed to the domestic market. As with the agricultural market, the professional area of the amenity market is very cultivar conscious; in this case it is necessary for cultivars to appear prominently on the Annual Turfgrass Seed Guide produced by the Sports Turf Research Institute (STRI) at Bingley. The domestic area is not as cultivar conscious, although it is anticipated the position could change over the next few years.

National brands are more evident in the marketing of amenity mixtures to the professional sector. Five-sixths of the wholesalers supply the seed, in two instances selling direct to the consumer, with others supplying through local distributors.

During the past 10 years there has been a significant movement in the supply of lawn seed to the home owner by large superstores, who currently have in the region of 50% of this sector and sell seed in cartons. The balance is sold through garden centres in the form of a combination of cartons and bulk seed, which is weighed at the time of purchase, although the trend is distinctly towards cartons.

11.5.4 Organizations

The important organizations from a marketing point of view are NIAB, SAC, DANI and the STRI. In addition to these, other organizations that play a role are the British Society of Plant Breeders (BSPB) who negotiate with the respective government bodies on behalf of breeders, United Kingdom Agricultural Supply Trade Association (UKASTA) who carry out a similar function on behalf of seed multipliers, and the Amenity Grass Marketing Association (AGMA) who obviously concentrate on the amenity area. Whilst the National Farmers Union (NFU) does much of the negotiations with the government and the trade associations on behalf of the growers, there are two local organizations which encourage seed production and distributing, and the marketing of technical information to growers. They are the Wilts, Hants, Dorset Seed Growers Association (WHD) and the Lincolnshire Seed Growers Association (LSGA).

11.6 UNITED STATES OF AMERICA

11.6.1 Introduction

The United States has a diverse range of climatic zones, extending from the Canadian border to the Mexican border, and from the Pacific Ocean to the Atlantic Ocean. Within this area there are cold arid, cold humid, warm humid, warm dry, hot dry and hot humid zones. These different climatic conditions require a diverse range of forage species. The only native grasses existed on the prairies of the middle part of the continent. Basically, the majority of the seeded forage crops used today were introduced by the early European settlers. A few have been introduced over the last 100 years from collection trips to other continents. Most of these have successfully adapted to the wide range of climatic conditions that exist throughout the USA.

11.6.2 Tall Fescue

Tall fescue is the main range grass grown for forage in the humid lower, mid-western and south-eastern states. Since its introduction from Europe, it is now grown on over 10 million ha. It is the primary source of feed both as pasture and hay for beef cattle in this region. Missouri is the main seed production area for the most commonly used cultivar, Kentucky-31.

Since 1985 Oregon, with its excellent climatic conditions and experienced seed growers, has also become a major producing state for both forge and amenity tall fescue seed. Most of the breeding work on improved tall fescue cultivars has been done by public breeders. The University of Missouri and Auburn University have led the industry with the most recent releases. There are a few private programmes that also are making advances in forage quality.

One of the main efforts in recent years has been in dealing with the endophytic fungus present in tall fescue that causes serious health problems in cattle. Seed is marketed by a wide range of private seed dealers and farmers' cooperatives. Whilst production of tall fescue stands at around 114,000 t at present, the trend is increasing, with the major markets being France, Japan, South America and Australia.

11.6.3 Annual Ryegrass

Annual ryegrass is grown extensively across the hot humid regions of the southern USA as a source of winter forage for cattle. The main forage grasses in this region are bahiagrass (*Paspalum notatum* Flugge) and bermudagrass (*Cynodon dactylon* (L.) Pers.). These are warm season grasses however, and are dormant during the winter months. Annual ryegrass is used to overseed these dormant pastures to provide winter feed. It provides an excellent source of high quality forage.

A limited number of private and public university programmes are breeding improved annual ryegrasses. Texas A & M and Mississippi State have made a major effort to develop improved cultivars and have a few in production. Several European cultivars have been marketed on a limited basis. The seed is produced in Oregon and is marketed by a wide range of private seed dealers and farmers' cooperatives. Production areas are steady, and around 100,000 t are traded, the major exports going to Japan, South America, Europe and Korea.

11.6.4 Cocksfoot/Orchardgrass

Cocksfoot is used throughout the northern half of the USA. It has good winter hardiness and provides an excellent pasture and hay. It is usually planted in a mixture with other grasses and/or legumes. There is very limited breeding work being done on cocksfoot. The seed is produced by experienced growers in Oregon and marketed throughout its area of adaptation by private seed dealers and farmers' cooperatives.

Production of cocksfoot is declining, and now stands at just over 7000 t, with Japan and Korea being the major overseas markets.

11.6.5 Perennial Ryegrass

Perennial ryegrass is used on a very limited basis, primarily in the milder north-western USA. It has difficulty surviving the extremely cold temperatures of the upper mid-west and northeastern states. It also cannot survive the hot humid summer conditions of the southeastern states. Several of the northern European cultivars have had limited success in the northeastern states. They have been able to survive most winters but are subject to winter damage.

There are no active breeding programmes on forage perennial ryegrasses. The seed is produced in Oregon and marketed through private seed dealers and farmers' cooperatives. As with tall fescue, there is an increasing market in perennial ryegrass, with Japan being a large importer, but also Western Europe and the Middle East are important international customers.

11.6.6 Lucerne/Alfalfa

Lucerne (*Medicago sativa* L.) is the most important forage crop in the USA today. In total forage production and feed value per hectare, lucerne leads all other grasses and legumes. It is grown in every state of the USA. The seeded area of lucerne in the USA has increased to over 12 million ha.

Private and public research programmes have developed lucerne cultivars that are much higher in forage yield, pest resistance, persistence and forage quality. Improved strains of lucerne are now available for every area and forage use, adding greatly to farm income, providing more flexibility in crop rotations, and higher yields of better quality lucerne. A consistent supply of seed of these cultivars is obtained through carefully planned lucerne seed production by experienced growers in the western states. This supports a highly developed US marketing system which includes most national and regional farm and seed marketing companies.

Much of the 36,000 t of lucerne produced in the USA is exported to Asia, South America, Mexico and Saudi Arabia, and marketing of this species is steady at present.

11.6.7 Timothy

One of the earliest grasses to be introduced into the USA, timothy is the easiest to grow and obtain satisfactory pure stands and grass legume mixtures. It is one of the easiest to handle and cure for hay, and for many years was the best liked hay for horses. Timothy is grown for both forage and seed over a wide area of the northcentral and northeastern states and in Canada.

A very limited number of private and public breeding programmes for timothy exist in the USA today. However, new proprietary cultivars with improved quality and forage yields have been released during the last 10 years. Additionally, timothy cultivars from international sources are being produced and marketed in the USA. The northern US seed production by experienced grass seed producers supports a well-developed regional farm seed marketing system. Production remains steady at around 1000 t, with most being for the domestic market.

11.6.8 Smooth Bromegrass

Smooth bromegrass (*Bromus inermis* Leyss.) is the most widely used cool season grass in North America. It is grown extensively in Canada, northcentral and northeast USA. Smooth bromegrass provides excellent quality forage and pastures. It is used in many grass and legume mixtures. Smooth brome also provides excellent permanent cover for sites such as waterways, eroded areas, rocky areas and farm lanes.

A very limited number of private and public breeding programmes have released an adequate number of improved cultivars during the past 10 years. These cultivars are higher in forage yield, forage quality and pest resistance. Seed is produced by experienced growers in the northcentral and northwestern USA. A well-developed regional farm seed marketing system exists in the cool season usage areas throughout the USA. Production and marketing of this species is declining, with about 2000 t being sold annually, the main export opportunities being Europe.

11.7 CONCLUSION

It is clear that it would be impossible to amalgamate all countries' production and marketing philosophies into a single thesis. Even within localized production areas one can see differences in trading; for example, Denmark exports 90% of its forage seed, and the UK imports approximately 50%. Also, the European Union Common Agricultural Policy can have considerable influence over decision-making for crop production within its member states, with any changes in crop support, quota systems or livestock reductions having a significant effect within a relatively short period of time.

Changes in political climate may also have an effect in the future, particularly in areas such as Eastern Europe, where countries are now able to compete in a global capacity for the first time in over 40 years. There, growers and seed merchants are anxious to promote their enthusiasm and potentially good production land into much-needed monetary benefit, and it is therefore important that organizations such as FIS, OECD and ISTA maintain their international control and goodwill for the benefit of all involved in the seed trade.

Festuca arundinacea Schreb. (Tall Fescue) in the USA

W.C. Young III
Department of Crop and Soil Science, Oregon State University, Corvallis, Oregon 97331, USA

12.1 INTRODUCTION

An adequate seed supply of adapted and productive cultivars of tall fescue (*Festuca arundinaceae* Schreb.) is essential to any pasture or turfgrass improvement programme where this is the species of choice. Historically, the United States Department of Agriculture (USDA) and state universities were the major source of new cultivars. More recently, however, there has been a transition from public to private plant breeding programmes leading to a proliferation of new cultivars. Today, forage producers, home owners and other purchasers of seed can select from numerous cultivars specifically adapted to their farm and livestock production system, or amenity grass needs (Fig. 12.1). The development of new tall fescue cultivars is an especially interesting example of this.

As demand for high-quality seed increased, certain areas of the USA and other parts of the world were found to be better adapted to seed production than others. Thus, seed production has changed from a by-product of forage production that is grown and marketed in the local area to one that is international in scope. Improved cultivars bred in one area may be sent to areas adapted to seed multiplication, with the progeny returned to areas that are adapted for their intended use, and this has been the case with tall fescue seed produced in Oregon.

12.2 HISTORY

Tall fescue has a long history as a valuable forage grass in the USA and is the dominant cool-season perennial grass in the country. This plant is adapted to a wide range of climatic conditions, but grows best in the large transition zone that separates the northern and southern regions of the USA. The species is a versatile perennial grass used for livestock feed, turf purposes and for erosion control.

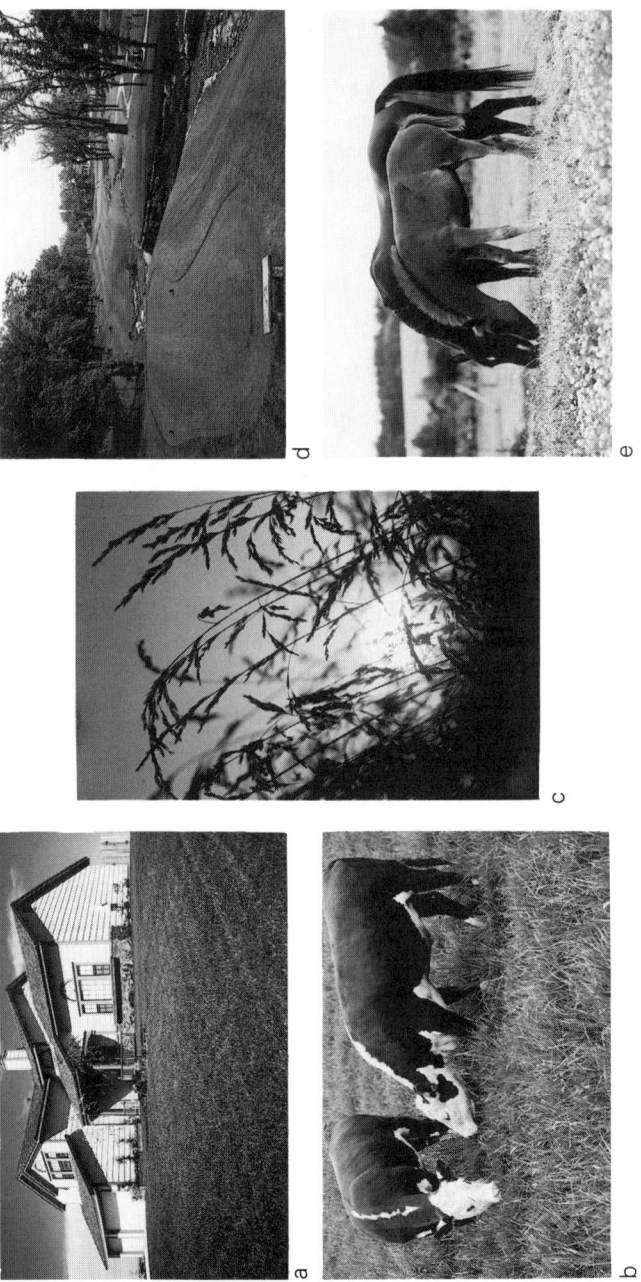

Fig. 12.1. Some of the many uses of tall fescue. a. A residential lawn. b. Pasture for horses. (Photograph credit: Larry Kassell, Kassell Concepts, Silverton, Oregon, USA. Sponsor: Oregon Tall Fescue Commission, Salem, Oregon, USA.)

Tall fescue is native to Europe and was probably introduced into the USA by the early settlers, although the exact date of introduction is unknown (Terrell, 1979). Historically, the plant was known as a relatively coarse textured, deep-rooted, bunch-type grass. Although not widely planted in the USA before 1940, it now occupies some 14.2 million ha (Ball et al., 1993).

Tall fescue did not become widely used in the USA until the development and release of the cultivars Alta and Kentucky 31 in 1940 and 1943, respectively (Cowan, 1956). Cultivar Alta, developed over a number of years from an ecotype selection by H.A. Schoth in Corvallis, Oregon, gained rapid acceptance by farmers throughout the Pacific Northwest, due to its wide adaptation and the ability to produce high yields of nutritious forage. As an indication of its rapid acceptance, in 1940, the value of this cultivar as a seed crop in Oregon was about $31,000, but by 1951, it had reached a peak of $2.5 million.

Kentucky 31 is a tall fescue ecotype that was found in 1931 and after being evaluated at the University of Kentucky was released as cultivar Kentucky 31 in 1943 (Fergus and Buckner, 1972). Quick acceptance of this cultivar by Kentucky farmers created a rapid increase in seed production in the state, similar to that in Oregon. Thus, the release of the two cultivars, Alta and Kentucky 31, moved tall fescue into a position of an important grass crop in the agriculture of the USA. During a 15-year period from 1940 to 1955 this crop, which was grown on 16,200 ha in 1940, had increased to more than 1.6 million ha by 1955 (Cowan, 1956).

12.3 SEED PRODUCTION AREAS

Prior to the mid-1940s, most seed available for commerce was harvested from pasture or meadows when a promising seed crop was present. Although harvesting of seed in this manner continues today for some older 'common' cultivars that have acceptance in the market as uncertified production, all proprietary cultivars are produced by specialist seed producers. The emergence of the specialist seed grower caused a shift in cool-season production from the Midwest to the Pacific Northwest, particularly the Willamette Valley of Oregon. Missouri, however, came to dominate the production of tall fescue seed crops, particularly uncertified Kentucky 31.

The cool-season grasses are well adapted to the mild winters and dry harvest season found in the valleys of the Pacific Northwest. Oregon has established a reputation as a dependable supplier of high-quality forage and turf grass seed, and provides that seed for much of the forage and turf markets in the USA. Factors favouring the state include: (i) climate and soils adapted to high yields of quality seed; (ii) experienced growers with modern equipment and conditioning facilities; and (iii) an established processing and marketing system.

During the years 1950–1977, tall fescue seed production in the USA rose from 8.7 to 45.5 million kg, while the area harvested increased from 35,000 to 182,000 ha (Youngberg and Wheaton, 1979). Missouri harvested 134,865 ha in 1977. Annual production statistics historically collected by the USDA Crop Reporting Board were discontinued in 1981, making a continuation of statistics on area and production more difficult to estimate nationally. Currently, the United States Department of Commerce Census of Agriculture (published every 5 years) is the most comprehensive source for national agricultural production, although state-produced statistics are often available.

Missouri has continued to harvest tall fescue seed from 80,000 to 130,000 ha of pasture annually; thus, seed production tends to be incidental to forage operations. Seed yield from pastures closed to grazing in advance of seed harvest are much lower than that of the specialized seed producer. Industry estimates of seed entering the market from Missouri production range from 25 to 45 million kg annually based on seed yields of 200–300 kg ha^{-1}. Recent figures from the United States Department of Commerce *1992 Census of Agriculture* (1994) reported 70,000 ha of tall fescue seed harvested in Missouri, yielding 13.4 million kg (190 kg ha^{-1}). Data collected 5 years earlier, in 1987, reported over 117,000 ha harvested in Missouri, which produced almost 30 million kg, averaging 250 kg ha^{-1}.

Tall fescue seed production in Oregon has exceeded 20 million kg per year since 1988. Production statistics for Oregon tall fescue seed crops are shown in Table 12.1. Note that for 20 years (1963–1982) the ha and yield were stable, with an average of 5525 ha and 780 kg ha^{-1}, respectively. Prior to 1983, almost all production was of forage cultivars. The introduction of turf-type cultivars led to the rapid increase in production shown after 1983, reaching a peak of over 55 million kg in 1990.

12.4 PLANT BREEDING DEVELOPMENTS

Tall fescue was not considered a good turf grass for many years. Although some testing of forage cultivars for turf potential was done, most conclusions were that the species acceptance as turf would be limited to play areas, air field and road sides, where its relatively coarse texture was not objectionable (Asay et al., 1979). However in the early 1970s, breeding programmes were initiated to develop cultivars of tall fescue more suitable for turf use. Today, many new tall fescue cultivars have been released with finer leaf texture; these new cultivars are known as 'turf-types'.

Most of the new turf-type tall fescues cultivars have been derived from germplasm selected from old turf areas in the northeastern and southeastern USA (Meyer and Funk, 1989). Plants with a lower growth habit, finer leaves, greener colour and a reduced rate of vertical growth have been selected from naturalized stands in these regions. The first turf-type cultivar was Rebel, released by the New Jersey Agricultural Experimental Station (Funk *et al.*, 1981). Other turf-type cultivars soon followed, all possessing leaves approximately 30–40% finer than Kentucky 31 and tiller densities almost twice that of Kentucky 31 (Meyer and Funk, 1989). Turf-type tall fescue is marketed where it is valued for greater drought tolerance, lower fertility needs and better persistence than traditional cool-season turf grasses.

The availability and usage of improved tall fescues for turf has increased dramatically in the past 15 years. At present, the USA production of turf-type tall

Table 12.1. Oregon tall fescue seed crop statistics, 1963–1995. (Source: Oregon State University, Extension Economic Information Office, Corvallis, Oregon, USA.)

Harvest year	Area harvested (ha)	Yield (kg ha^{-1})	Production (kg × 10^3)	Value of production ($US × 10^3)
1963	4860	678	3296	1343
1964	5468	835	4566	1006
1965	6075	751	4563	1005
1966	5873	813	4773	894
1967	5670	835	4735	1033
1968	5670	717	4068	1210
1969	6075	807	4903	1998
1970	6278	684	4293	1154
1971	6683	930	6218	1452
1972	7290	650	4740	1879
1973	7493	807	6047	3330
1974	6480	706	4576	1512
1975	5063	852	4313	1235
1976	4050	673	2724	1410
1977	3848	729	2803	1976
1978	3848	785	3019	1862
1979	4050	807	3269	1872
1980	4455	953	4245	2618
1981	4657	785	3655	2657
1982	6602	782	5169	4782
1983	7703	963	7419	6880
1984	10,238	1033	10,576	8363
1985	13,171	1221	16,082	13,248
1986	18,488	1132	20,895	26,004
1987	23,587	1110	26,061	29,734
1988	27,743	1267	35,097	49,167
1989	33,818	1065	36,171	42,156
1990	37,382	1345	50,454	55,558
1991	40,905	1435	58,605	55,300
1992	36,450	1087	39,627	35,245
1993	32,242	1457	46,916	37,161
1994	27,734	1081	33,506	28,048
1995	30,181	1255	38,015	41,857

fescue seed is estimated at 35–40 million kg annually. Records provided by the Oregon Tall Fescue Commission show that assessment was collected on 38.8 million kg in the 1994–1995 market year, of which 27.7 million kg (71%) was from turf-type cultivars. Estimates vary as to how much production from forage cultivars enters turf markets, but turf industry specialists continue to report cultivar Kentucky 31 on retail shelves where consumers would purchase seed for amenity use.

Oregon seed growers have had almost exclusive control over the production of turf-type tall fescue cultivars since the early 1980s. Much of their success can be attributed to Oregon's Seed Certification program and its ability to meet the challenge of maintaining cultivars as developed and described by plant breeders during the seed increases necessary to meet market demand. In 1994, total USA, Canada and New Zealand production of certified tall fescue was 21,559 ha (AOSCA, 1994); Oregon growers produced 21,436 ha, 99% of the production. As Oregon growers produced a total of 27,734 ha of tall fescue seed in 1994, 77% of the crop was in the certification programme.

Investment by the private sector in cultivar development increased greatly with the passage of the Plant Variety Protection Act in 1970. In Oregon, 38 grass cultivars were certified in 1970; in 1995 there were 655 grass cultivars in production. There were 162 tall fescue cultivars with certified production in Oregon in 1995, the abundance of these being turf types. Between 1984 and 1991, Oregon's tall fescue seed production area quadrupled, primarily due to the demand for high quality turf-type cultivars.

12.5 THE ROLE OF ENDOPHYTES

12.5.1 Forages

Another important development impacting on seed producers was the elucidation of the biology of the fungal endophyte, *Acremonium coenophialum* Morgan-Jones & Gams, and its association with numerous livestock disorders attributed to tall fescue forage. Since the late 1970s our understanding of the characteristics of the endophyte have yielded two practical points of interest. First, the fungus lives within fescue plants and does not affect the appearance of the grass. A seed or plant microscopic analysis is required to detect the presence of the fungus. Second, it is transmitted only by the seed, so that once an endophyte-free stand of tall fescue is established it will remain non-infected.

Since discovering the role of endophytes in reducing performance of animals grazing on infected tall fescue pastures, emphasis has been given to developing endophyte-free forage tall fescue cultivars. The principal use for tall fescue in the USA is in cow-calf enterprises, and low calf weaning weights have often been reported due to the effects of endophyte (Schmidt and Osborn, 1993). The Oregon seed industry's dedication to providing the highest quality seed to forage markets resulted in the development of another test for quality assurance. The Oregon Tall Fescue Commission petitioned the Oregon Department of Agriculture (ODA) to research the feasibility of establishing standards for endophyte content in seed lots.

In September 1983, the ODA became the first state regulatory agency to establish a testing and labelling programme issuing an Oregon Forage Grass Seed Endophyte Test and tag for forage grass seed. Labelling seed offered for sale for the level of endophyte is now required on tall fescue marketed in certain states. The cost of the test is $40 per sample whether or not the seed lot qualifies for tags.

Samples for the Oregon Forage Grass Seed Endophyte Test tag must be drawn as an official sample by the ODA. If the lot is eligible for certification under the Oregon Seed Certification programme, a portion of the sample that was drawn by a seed certification sampler may be forwarded to the ODA for examination. If a seed lot is not eligible for Oregon certification, or the sample to be taken is for endophyte tests only, the ODA must be contacted for sampling. Fifty seeds are selected, stained and examined under the microscope for the presence of endophyte mycelium. A test report is issued stating the percentage of endophyte in the lot.

Results from the ODA (G.M. Milbrath, ODA, Salem, Oregon, USA, 1995, personal communication) on endophyte testing of tall fescue seed from 1983 to 1995 report over 90% of Oregon's forage tall fescue tested contained 5% or less endophyte. Over 4200 tall fescue seed lots were tested representing 73 million kg of tall fescue seed produced in Oregon. Those lots with 5% or less infection received a state endophyte tag for seed bags. The programme is structured so the Department would not receive or test many turf-type varieties, where an enhanced level of infection is desired.

12.5.2 Turfgrass

The same fungal endophyte associated with livestock disorders has, however, imparted many desired characteristics in turfgrass. Endophyte-infected turfgrasses have frequently shown dramatic enhancement or resistance to many foliage-feeding insect pests, plus in tall fescue modest improvements in resistance to root-feeding white grubs and nematodes. In addition, improvements in stress tolerance and modification in water relations have also been observed under certain conditions (Funk *et al.*, 1990).

These beneficial characteristics of endophytes have resulted in many plant breeders releasing most of their genetically improved turf-type tall fescue cultivars with endophyte infection. Other turfgrass breeders have not emphasized endophytes in their turf-type cultivars. Reasons for avoiding endophyte-infection on turf types include concern for off-grade seed being used in forage mixes, the problems for growers of utilizing high endophyte straw, and fewer superior turf-type germplasm containing endophytes.

A recent survey of tall fescue cultivars entered in the Oregon Seed Certification programme recorded the level of endophyte infection reported by the contractor or owner (Table 12.2). This list, which includes forage cultivars, reports 55 cultivars (41%) with an endophyte level of 5% or less. Thirteen per cent of the cultivars report a level of endophyte infection greater than 75%.

12.6 THE FUTURE

In the future, breeders are likely to introduce novel endophytes into tall fescue. These novel endophytes should continue to improve stand persistence and resistance to pests, but would not decrease animal performance (Sleper and Buckner, 1995). This is based on the premise that deleting livestock toxins from the tall fescue-endophyte association will not reduce plant persistence or pest resistance. Forage breeders will continue working on improving other forage quality components, such as digestibility.

The demand for turf-type tall fescue cultivars is expected to grow. However, consumers will require that new releases provide improved turf quality and reduced energy inputs (Meyer and Funk, 1989). Lower-growing turfgrasses have

Table 12.2. Tall fescue cultivars and their endophyte level: 1995. (Source: Oregon State University, Seed Certification Service, Corvallis, Oregon, USA.)

Cultivar	Endophyte level (%)	Cultivar	Endophyte level (%)	Cultivar	Endophyte level (%)
Adobe	> 80	Finelawn 5GL	2	Penngrazer	0
Adventure	10	Finelawn 88	30	Phoenix	35
Adventure II	84	Finelawn Petite	12	Phyter	0
AFA	47	Forager	0	Pixie	90
Alamo	90	Galway	0	Poly 6–38	0
Alta	0	Georgia 5	70	Rebel 3D	6
Amigo	16	Grande	85	Rebel II	28
Anthem	0	Guardian	60	Rebel III	6
Apache	18	Hokuryo	0	Rebel Jr	45
Apache II	64	Homestead	4	Renegade	22
Arid	48	Houndog	0	Richmond	0
Au Triumph	0	Houndog 5	80	SBD	2
Austin	5	Jaguar	0	SR 8200	90
Avanti	0	Jaguar II	0	SR 8210	88
Aztec	0	Jaguar III	24	SR 8300	90
Barcel	0	Jessup	0	SRX 8350	90
Barnone	0	Jessup E +	70	Safari	34
Bonanza	12	Johnstone	0	Shenandoah	75
Bonsai	62	Jubilee	8	Shortstop	29
Cajun	0	Kenhy	0	Sibilla	0
Carefree	0	Kentucky 31	0	Silverado	0
Chesapeake	90	Kitty Hawk	8	Southern Choice	6
Chieftain	6	Lancer	20	Southern Cross	0
Chieftain II	3	Leprechaun	6	Stagecoach	42
Cimarron	0	Lexus	45	Starlet	12
Cochise	16	Marksman	16	Taurus	25
Coronado	99	Martin	0	Tempo	0
Conestoga	0	Maverick	29	Thoroughbred	16
Crewcut	< 20	Maverick II	0	Titan	98
Crossfire	12	Mesa	80	Titan 2	90
Crossfire II	98	Mic 18	0	Tomahawk	0
Debutant	55	Micro	56	Tradition	60
DMV	0	Mirage	< 20	Trailblazer	0
Dovey	0	Missouri 96	0	Trailblazer II	0
Duke	34	Mojave	10	Tribute	58
Duster	18	Monarch	0	Trident	15
Earthsave	5	Morgan	10	Thunderbird	24
Eldorado	0	Murietta	14	Twilight	30
Emperor	0	Mustang	11	Vegas	60
Era	20	N25	0	Veranda	30
Falcon	17	ND	10	Virtue	4
Falcon II	16	Nanryo	0	Willamette	0
Fawn	0	Oasis	85	Winchester 2	24
Festorina	0	Olympic	0	Wrangler	0
Finelawn 1	16	Olympic II	20	Wranger II	> 85

great appeal in efforts to decrease mowing costs and improve performance. The germplasm of turf-type tall fescue needs to be further expanded for it to be used in the deep south. Stress tolerance to a wider range of soil types is also anticipated as expanded markets are developed. This will enable the market to grow and allow turf-type improvements to be passed on to the homeowner who continues to use cultivar Kentucky 31.

For new turf-type cultivars the greatest challenge is to have high levels of resistance to brown patch disease (*Rhizoctonia solani* Kuhn), as this disease causes severe devastation to tall fescue lawns each summer in the transition zone of the USA (Meyer and Funk, 1989). Also, higher levels of resistance to net blotch (*Drechslera* spp.) and grey leaf spot (*Pyricularia* spp.) are needed to aid in the establishment of tall fescue (W.A. Meyer, Turf-Seeds, Hubbard, Oregon, USA, 1995, personal communication).

ACKNOWLEDGEMENTS

The author wishes to express his appreciation to the following persons for their helpful reviews of this publication: Mr Mike Baker, Pennington Seed, Inc., Lebanon, Oregon, USA; Dr Leah Brilman, Seed Research of Oregon, Inc., Corvallis, Oregon, USA; Mr Mark Mellbye, Oregon State University Extension, Linn County, Albany, Oregon, USA; Dr Bill Meyer, Turf-Seed, Inc., Hubbard, Oregon, USA; and Mr Dave Nelson, Oregon Tall Fescue Commission, Salem, Oregon, USA.

REFERENCES

Asay, K.H., Frakes, R.V. and Buckner, R.C. (1979) Breeding and cultivars. In: Buckner, R.C. and Bush, L.P. (eds) *Tall Fescue*. Agronomy No. 20. American Society of Agronomy, Madison, Wisconsin, pp. 111–139.

AOSCA (Association of Official Seed Certifying Agencies) (1994) *Acres Applied For Certification in 1994 by Seed Certifying Agencies of the AOSCA*. Production Publication No. 48, Association of Official Seed Certifying Agencies, Mississippi State, Mississippi.

Ball, D.M., Schmidt, S.P., Lacefield, G.D., Hoveland, C.S. and Young III, W.C. (1993) *Tall fescue/Endophyte/Animal Relationships*. Oregon Tall Fescue Commission, Salem, Oregon.

Cowan, J.R. (1956) Tall fescue. *Advanced Agronomy* 8, 283–320.

Fergus, E.N. and Buckner, R.C. (1972) Registration of Kentucky 31 tall fescue. *Crop Science* 12, 714.

Funk, C.R., Engel, R.E., Dickson, W.K. and Hurley, R.H. (1981) Registration of Rebel tall fescue. *Crop Science* 21, 632.

Funk C.R., White, R.H. and Breen, J.P. (1990) Importance of endophytes in turfgrass breeding and management. In: Joost, R.E. and Quisenberry, S.S. (eds) *Program and Abstracts, International Symposium on Acremonium/Grass Interactions*. Louisiana Agricultural Experiment Station, Louisiana State University, Baton Rouge, Louisiana.

Meyer, W.A and Funk, C.R. (1989) Progress and benefit to humanity from breeding cool-season grasses for turf. In: Sleper, D.A., Asay, K.H. and Pedersen, J.F. (eds) *Contribution* from *Breeding Forage and Turf Grasses*. CSSA No. 15. Crop Science Society of America, Madison, Wisconsin, pp. 31–48.

Schmidt, S.P. and Osborn, T.G. (1993) Effects of endophyte-infected tall fescue on animal performance. *Agricultural Ecosystems Environment* 44, 233–262.

Sleper, D.A. and Buckner, R.C. (1995) The fescues. In: Barnes, R.F., Miller, D.A. and Nelson, C.J. (eds) *Forages Volume I: An Introduction to Grassland Agriculture*. Iowa State University Press, Iowa, pp. 345–356.

Terrell, E.E. (1979) Breeding and cultivars. In: Buckner, R.C. and Bush, L.P. (eds) *Tall Fescue*. Agronomy No. 20. American Society of Agronomy, Madison, Wisconsin, pp. 31–39.

United States Department of Commerce (1994) *1992 Census of Agriculture*. Vol. 1: Geographic Area Series, Part 51: United States Summary and State Data. United States Department of Commerce, Washington, DC.

Youngberg, H.W. and Wheaton, H.N. (1979) Seed production. In: Buckner, R.C. and Bush, L.P. (eds) *Tall Fescue*. Agronomy No. 20. American Society of Agronomy, Madison, Wisconsin, pp. 141–153.

Festuca rubra L. (Creeping Red Fescue) in Canada

13

N.A. Fairey
*Northern Agriculture Research Centre, Agriculture and Agri-Food Canada,
Beaverlodge, Alberta T0H 0C0 Canada*

13.1 EVOLUTION OF RED FESCUE AS A SEED CROP

According to Stacey (1974), seed production of red fescue (*Festuca rubra* L.) was first undertaken in Canada in the 1930s by Andrew Anderson and John Olson, two growers from central Alberta (Fig. 13.1). Subsequently, James Murray of the nearby Alberta School of Agriculture and Home Economics at Olds, Alberta, recognized the value of the species for pasture. He developed the creeping-rooted

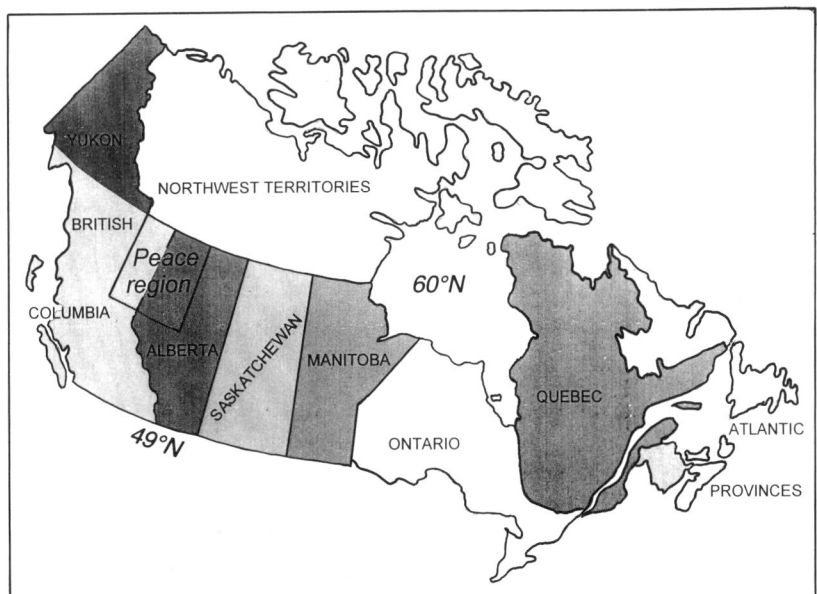

Fig. 13.1. The Provinces and the Peace region of Canada.

cultivar Olds using mass selection from the genetic stock that had been introduced in 1932 from Czechoslovakia (Elliott and Bolton, 1970). These authors reported that cultivar Olds was hardy throughout western Canada, moderately drought tolerant, and not subject to disease in Alberta. The suggested uses for creeping red fescue cultivar Olds were for lawns, fairways, pastures and soil conservation.

In 1935 Spence Morrison, an Advisory Officer with the Alberta Department of Agriculture, brought a small quantity of seed of cultivar Olds to Beaverlodge, a small community in the Peace region of northwestern Alberta. The Peace region of Canada encompasses a massive area of land drained to the Arctic Ocean via the Peace River (Fig. 13.1). Over the next few years, the combined efforts of two local farmers, Leslie and Rowe Harris, and the Canada Department of Agriculture at Beaverlodge, developed the potential of creeping red fescue as a new, dual-purpose crop for the region; its herbage remained green and extended the grazing season into the long, harsh winters, and its ability to set seed gave farmers an alternative cash-crop. According to Stacey (1974), seed yields of creeping red fescue were particularly good following 'open autumns and under conditions of favourable nutrition and moisture'.

These pioneering developments were the foundation for the herbage seed research conducted by Dr C.R. (Bob) Elliott of the Canada Department of Agriculture at Beaverlodge during his tenure from 1952 to 1982. His efforts nurtured an industry that supplied seed of creeping red fescue for home lawns, sports fields, golf courses, parks, school playgrounds and cemeteries (Elliott et al., 1961). When reflecting back on the rationale for his research on red fescue, he stated:

> Canada's lawn seed crop (creeping red fescue) was a war baby. It was first grown extensively during the early years of World War II, when a need arose for a turf-forming grass to seed on airfields and around military bases.
>
> (Elliott, 1975)

Elliott's research in the Peace region determined that floral induction of creeping red fescue occurs in well-developed tillers during the late autumn, when the temperature is cool and the photoperiod is short, and that floral initiation is promoted in the spring when the temperature is cool and the photoperiod is long (Elliott, 1966). These reproductive processes in red fescue are favoured by the environmental conditions that prevail in the Peace region and may be enhanced by proper management, particularly the use of nitrogen fertilizer in the autumn or very early spring prior to floral initiation. In an article highlighting the Peace region's northern agriculture, Dr Elliott expanded on why grass seed crops do so well:

> Short days in the cool, moist autumn coupled with long, cool days in the prolonged spring break-up period favour grass seed production. The profitable crop is ready for harvest during the warm, dry periods of late July and early August.
>
> (Elliott, 1974)

By 1966, Elliott had developed an improved cultivar of creeping red fescue, namely Boreal (Elliott, 1968a,b; Elliott and Bolton, 1970). It was a selection out of rejuvenated commercial seed fields originally sown to cultivar Olds; its improved uniformity and seed yield made it attractive in the Peace region and, as foreseen

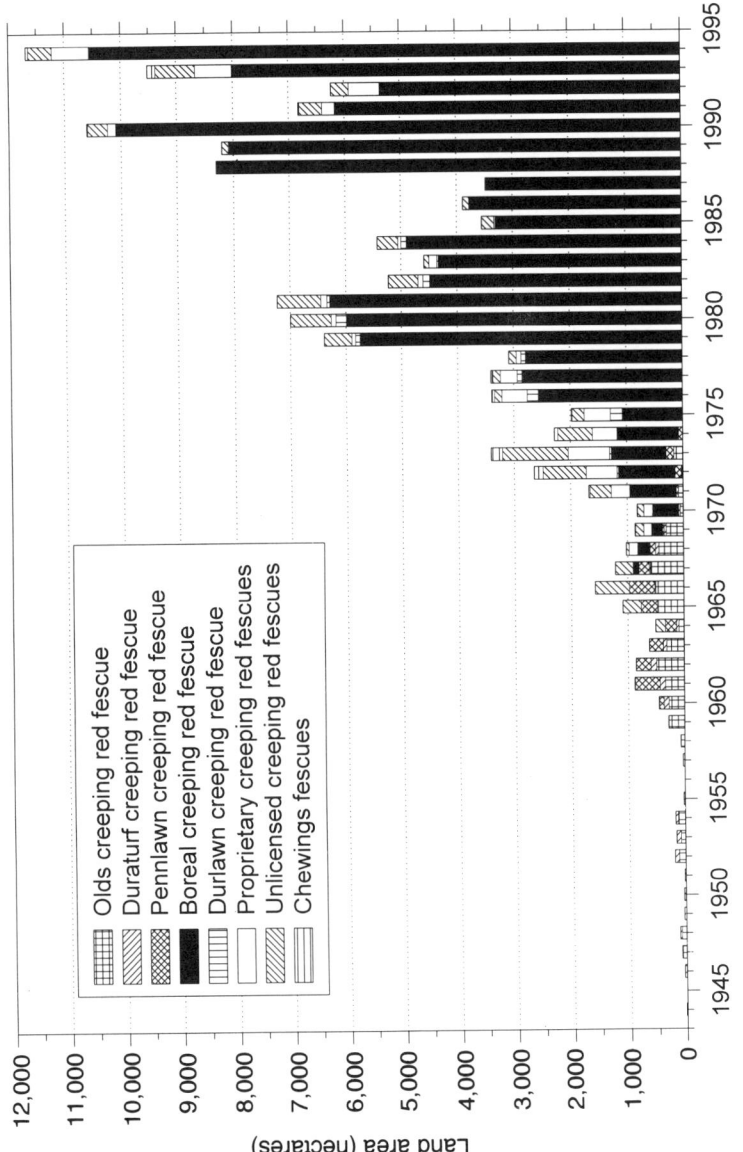

Fig. 13.2. The land area of red fescue cultivars inspected for certified seed production (1944–1994). (Prepared from statistics provided by the Canadian Seed Growers' Association.)

by its developer (Elliott, 1968a), its broad genetic base ensured its acceptance for end-use in foreign countries and in other regions of Canada, particularly for turf because of its strongly creeping growth habit. In 1995, almost 30 years after its introduction, cultivar Boreal creeping red fescue remains the mainstay of the grass seed industry in Canada (Fig. 13.2). Some of the early days of red fescue seed production in the Peace region are illustrated in the photographs shown in Fig. 13.3.

13.2 CROP STATISTICS

Although creeping red fescue was grown in Canada by the mid- to late-1930s, and commercial production of seed occurred from about 1940, the Agricultural Census

(a)

(b)

(c)

(d)

Fig. 13.3. Photographs of seed production of red fescue in Canada. (Reproduced from the historical collection of the Northern Agriculture Research Centre, Agriculture & Agri-Food Canada, Beaverlodge, Alberta.) a. Seed harvest in a 115-ha field of certified creeping red fescue at the farm of the Harris Brothers and E.C. Stacey near Beaverlodge, Alberta, in 1942. The field produced an average of 400 kg ha^{-1} of clean seed with some areas yielding as high as 900 kg ha^{-1}. b. Small strips of foundation 'Duraturf' creeping red fescue being grown in rows interseeded in an orchard at the Dominion Experimental Farm at Beaverlodge, Alberta, in 1944. The seed yield was 800 kg ha^{-1}. (Photograph taken by E.C. Stacey.) c. Four hundred tonnes of uncleaned seed of creeping red fescue stockpiled in jute bags near Beaverlodge, Alberta, during the winter of 1960–1961. Each individual pile contained more than 20 t of seed, most of which was destined for export. d. A small patch of creeping red fescue being harvested for seed (in the late 1960s), just within sight of Beaverlodge, Alberta. (Photograph taken by Dr C.R. Elliott.)

of 1961 was the first to gather specific information on the distribution and extent of seed production of the species. In 1961, almost 1600 farms reported growing in excess of 8000 t of seed. By 1971 the number of growers had increased somewhat but, in the two more recent assessments of 1981 and 1991, the numbers were considerably lower. In spite of the decrease in the number of growers over the past two decades, the area of land in seed production of red fescue in 1991 was similar to that in 1971, approximately 50,000 ha. Throughout the expansion in importance of this crop, the principal area of production was the Peace region of Alberta and British Columbia (Fig. 13.1, Table 13.1), with the remainder of the crop emanating from the Olds/Innisfail area of central Alberta (Elliott *et al.*, 1961; Elliott and Baenziger, 1977).

The annual volume of seed production exceeded 2000 t by 1953 and increased steadily until the early-1970s. From the mid-1970s until the early-1980s, it decreased steadily. Since then, the volume of production has increased somewhat, although output has fluctuated markedly from year to year; during the last decade, production was only 5500 t in 1985 but rebounded to an all-time high of 20,000 t in 1990 (Fig. 13.4). In 1948, the province of Alberta and the Peace region of British Columbia produced Canada's entire crop of creeping red fescue seed, over 700 t, and by 1960 a record crop of over 7500 t was produced from a growing area of over 20,000 ha (Elliott *et al.*, 1961). By 1973, production had increased to 16,000 t from a growing area of 34,000 ha and was worth in excess of Canadian $14 million (Elliott and Baenziger, 1977).

Canada started to export seed of creeping red fescue in the mid-1940s. The majority of this seed has been, and still is, not certified and the predominant market has been the USA. With the exception of the early- to mid-1980s, the magnitude of the exports increased steadily over the years, both in quantity (Fig. 13.5) and value (Fig. 13.6).

Table 13.1. The historical distribution and extent of seed production of creeping red fescue in Canada. (Source: The Agricultural Census of Statistics Canada.)

Census year	Canada	Manitoba	Saskatchewan	Alberta[1]	British Columbia[1]	Other provinces
Number of farms producing creeping red fescue for seed						
1961	1594	3	11	1283 (96%)	290 (100%)	7
1971	1718	13	6	1357 (99%)	330 (99%)	12
1981	531	3	1	385 (98%)	142 (99%)	0
1991	748	4	0	559 (99%)	185 (99%)	0
Land area in production of creeping red fescue for seed (ha)						
1971	49,910	173	78	39,597 (99%)	9992 (99%)	70
1981	25,975	49	16	16,007 (99%)	9903 (99%)	0
1991	50,765	54	0	35,169 (98%)	15,541 (100%)	0
Production of seed of creeping red fescue (t)						
1961	8027	9	27	6373 (96%)	1617 (100%)	1

[1]Percentage in Peace region shown in parentheses.

Festuca rubra L. in Canada 303

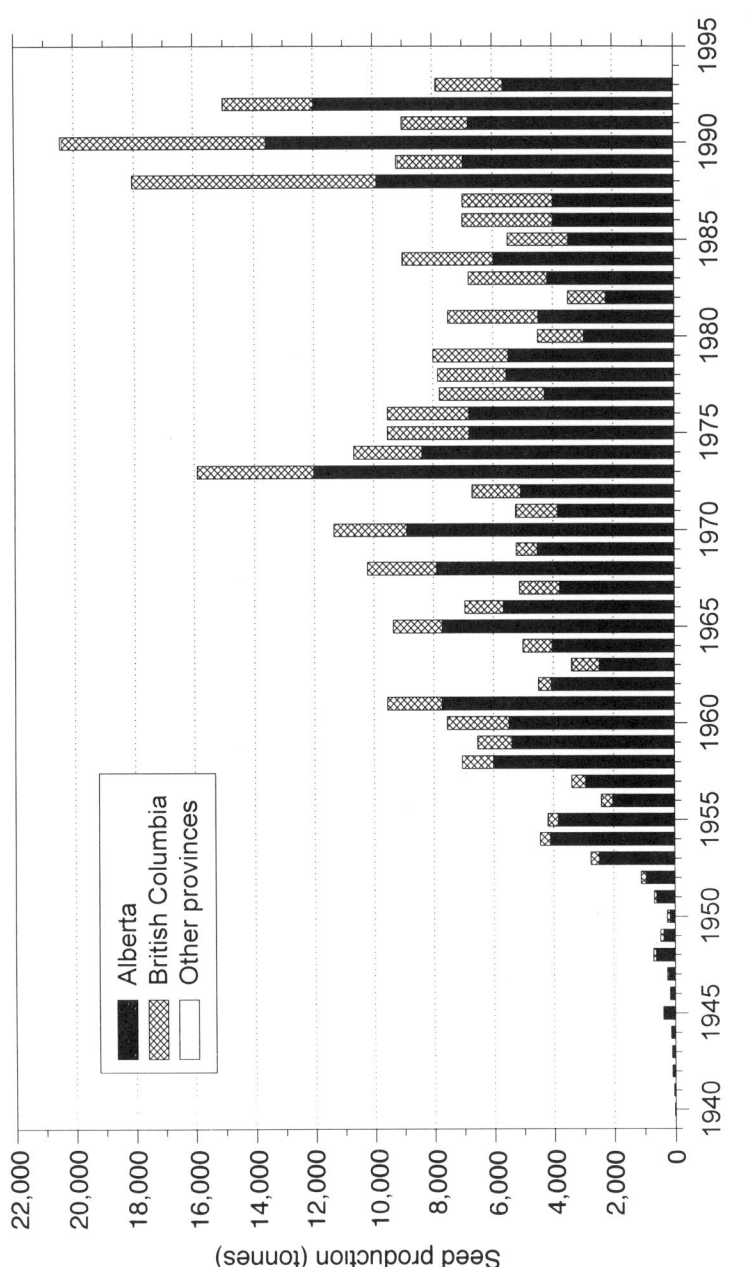

Fig. 13.4. The amount and distribution of production of seed of creeping red fescue in Canada (1940–1993). (Prepared from statistics provided to the Food Production and Inspection Division of Agriculture & Agri-Food Canada and the Canada Grains Council by Statistics Canada.)

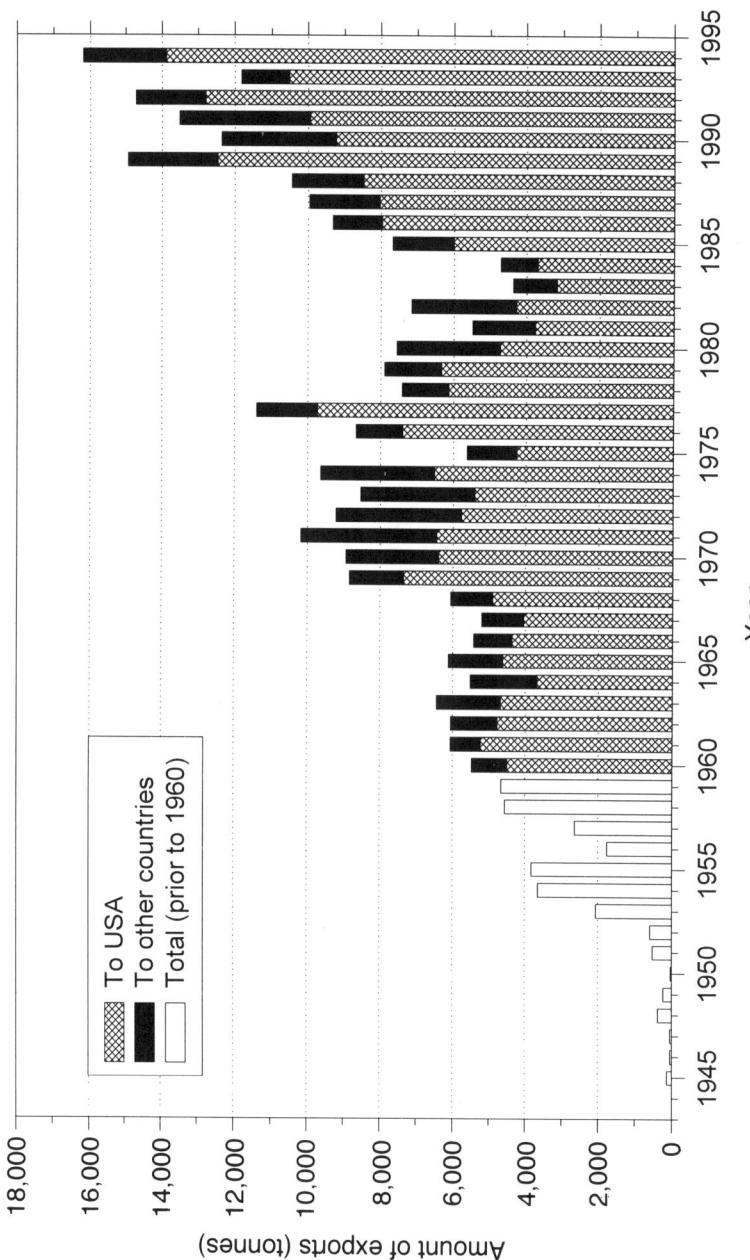

Fig. 13.5. The amount and destination of seed of creeping red fescue exported from Canada (1944–1994). Values for 1988 to 1994 include a small amount of seed of other fescues. (Prepared from statistics published by Statistics Canada.)

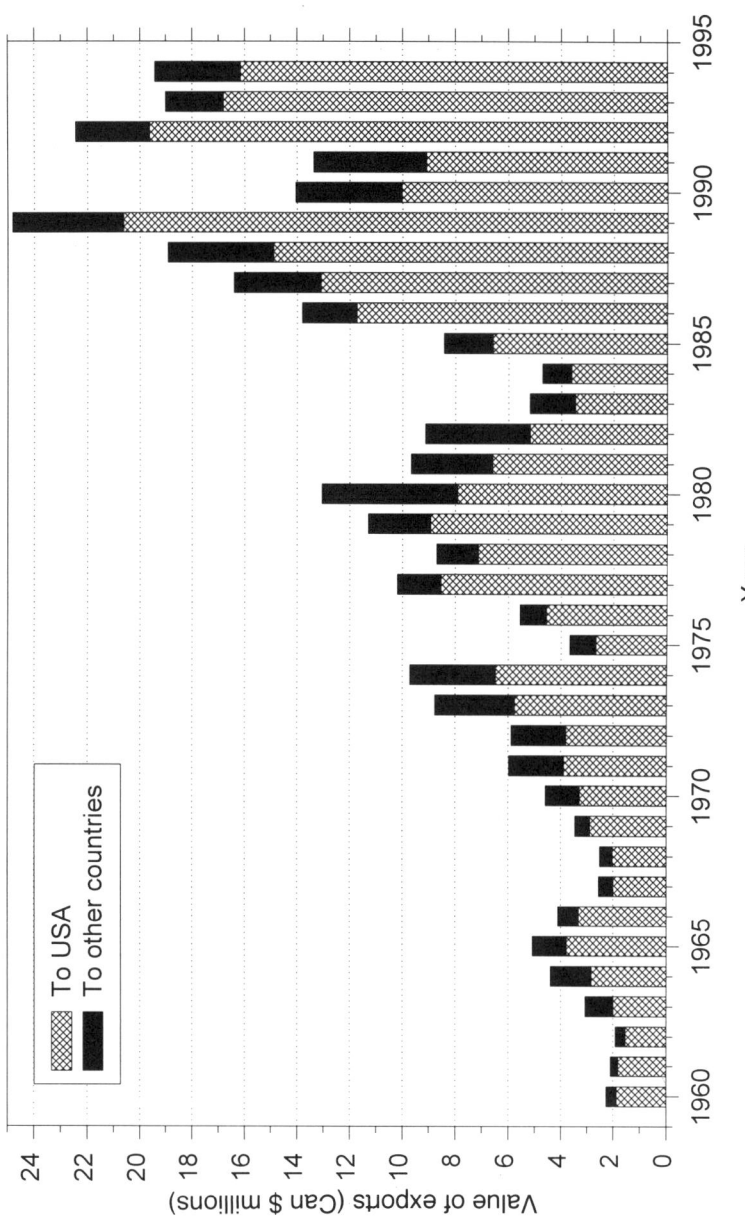

Fig. 13.6. The value of seed of creeping red fescue exported from Canada to the USA and other countries (1960–1994). Values not recorded prior to 1960, and values for 1988 to 1994 include a small amount of seed of other fescues. (Prepared from statistics published by Statistics Canada.)

In a report on the production and marketing of creeping red fescue seed, Maisonneuve *et al.* (1975) indicated that Canada was the world's largest producer (50% of the total, mostly from the Peace region). They pointed out that Denmark and the USA (primarily the state of Oregon) were the other two major growing areas but, whereas their seed was principally certified, 95% of that produced in Canada was uncertified. Furthermore, the expected seed yield in Canada was only about 800 kg ha^{-1} over two production years, at least 150 and 300 kg ha^{-1} below that from the USA and Denmark, respectively. The seed grown in Canada, however, was reported to be of higher physical purity and greater viability than that grown in Europe, but weed infestation from quack grass (*Agropyron repens* [L.] Beauv.) and wild oats (*Avena fatua* L.) was threatening the quality of the Peace region's seed.

The land area of red fescue inspected annually for certified seed production has increased from a range of 1000 to 3500 ha in the early 1970s to one of 6000 to 11,500 ha in the early-1990s. However, since the early days of the crop, the land area in certified seed production has generally not exceeded 10% of the total; this production has been the seed source for the majority of Canada's red fescue crop and, since the mid-1970s, it has been predominantly of one cultivar – Boreal (Fig. 13.2).

13.3 SPECIALIZED OCCURRENCES AND PRACTICES IN THE PEACE REGION

13.3.1 Stem Eyespot Disease

In the late-1960s, the red fescue seed crop in the Peace region, particularly in the Beaverlodge area, became infected with a stem eyespot disease; in some instances, infection levels were in excess of 90% (Smith *et al.*, 1968) and seed yield losses from a moderate to severe infection were about 30% (Smith and Elliott, 1970). The severity of the infection varied with the year of production; it was light in 1970 but severe in 1969 and 1971, and with a severe infection some yields were only 20–25% of the expected levels. The effects of the disease were relieved by the application of nitrogen fertilizer but removal of organic litter had little effect (Smith and Elliott, 1972). These same authors reported that the causal organism of the stem eyespot disease was *Didymella festucae* (Weg.) Holm, the imperfect state of *Phleospora idahoensis* Sprague.

In the 1980s a combination of crop protection chemicals was identified that controlled the disease but, unfortunately, the treatment also suppressed the seed-producing ability of the plants; genetic resistance is now being sought to counteract this disease. Although the impact of this disease in some growing seasons has been quite devastating, its seriousness over the past two decades has been contained somewhat by the practice of stand rejuvenation (discussed below). This practice is generally required after harvesting two consecutive seed crops in order to reduce the tiller density of the crop to a range suitable for another 2 years of seed production.

13.3.2 Establishment and Rejuvenation of Stands

The recommended practice for establishment of red fescue for seed production is to sow alone (i.e. without a companion crop) before mid-June, so that the seedlings are well developed prior to the long, harsh winter. Subsequently, two successive seed crops are harvested before the stand requires mechanical rejuvenation. Many growers, however, prefer to establish their fescue with a companion grain crop to help control weeds and to supplement the annual farm revenue. When a companion crop is sown during initial crop establishment or in the spring of the rejuvenation year, crop competition may restrict the growth of the fescue and suppress its seed yield in the subsequent year. In some cases, seed production is too low to justify harvesting, and the stand is 'clipped and shredded' with a rotary mower to control weeds and stimulate tillering for the next seed crop (Elliott and Baenziger, 1977), the so-called 'clip year'.

In the Pacific Northwest USA, creeping red fescue stands have been maintained in a healthy physiological condition for seed production, for many consecutive years, by postharvest burning of crop residues (Canode, 1968; Chilcote *et al.*, 1980; Youngberg, 1980). In the more northerly, shorter-season environment of the red fescue seed-growing zone of Canada (i.e. the Peace region), however, postharvest burning has not been adopted widely for two reasons. First, the volume of crop residue is relatively low and cannot be distributed evenly enough to ensure a complete, uniform burn. Second, the recovery period between burning and the cessation of growth for winter is too short. Hence, a physical method of stand rejuvenation, that conserves organic residues, has been adopted (Elliott and Baenziger, 1977), and is usually applied after harvesting the second seed crop.

The reported procedure for stand rejuvenation involves breaking the fescue sod in the fall with a mouldboard or disc plough, developing a seedbed in the spring, allowing the fescue to regenerate from rhizome buds and volunteer seedlings during that growing season (either with or without a companion grain crop sown at a lower-than-normal seeding rate), and applying nitrogen fertilizer in the autumn to stimulate tillering and subsequent seed production. Some growers prefer to delay the initial ploughing/discing operation until spring or early summer.

13.4 MARKETING STRATEGY AND SEED STORAGE

Since the early days of production, red fescue seed has been traded in an open market with no price, quota or delivery controls, and the vast majority of the seed has been uncertified (also known as commercial or common seed). This seed is purchased from growers on a cleaned basis after estimates of dockage are made by the processing companies. Small quantities of certified seed are also grown and each lot is purchased on a cleaned, bagged and graded basis.

Most of the uncertified seed of red fescue grown in Canada is exported to the USA (Fig. 13.5). The smaller lots of certified seed are normally grown on contract for direct export, primarily to countries in Europe. These markets have evolved because the prevailing rules of international trade have precluded the export of uncertified seed to countries in the European Union (and the former European Economic Community), and because the seed-producing potential of many European cultivars, when grown in the Peace region, is far below that of the locally adapted cultivar used for uncertified production, namely cultivar Boreal. This has meant that, to date, profit margins for Peace region growers have not been consistently high enough to foster the large-scale production and export of high-quality, certified seed of the red fescue cultivars recommended for use in Europe.

The open-market approach of marketing uncertified seed of red fescue involves large fluctuations in the revenues received by Canadian growers, both from year to year and from month to month during a single crop year. The price fluctuations are influenced primarily by the demand-and-supply situation in the USA, both for the red fescue seed itself and for alternative amenity grass seeds, particularly Kentucky bluegrass (*Poa pratensis* L.). Furthermore, the long, cold winters in the Peace region are conducive to the maintenance of a high germination potential, even when seed is stored for several years. Thus, during extended periods when seed prices are low, growers who do not need the immediate cash income will store their seed until the 'price is right'. One occasionally hears that this opportunistic approach to marketing has elevated a few growers to 'millionaire' status! Perhaps this explains the reluctance of herbage-seed growers in the Peace region to embrace the contract production of certified seed and move away from the open-market strategy.

13.5 THE FUTURE OUTLOOK

There has been a great proliferation in the number of cultivars of red fescue, as well as of other grass crops, throughout the world in the past two decades. The global demand for the uncertified seed of creeping red fescue grown in Canada is likely to come under considerable downward pressure as the market-place becomes increasingly dominated by certified seed of proprietary cultivars. Perhaps the contract production of certified seed of a broader range of species and cultivars, with its inherent ability to synchronize supply with demand and to stabilize economic returns for the growers, will eventually prevail in the Peace region of Canada as it has in many other areas where herbage seeds are grown. This may help Canadian producers retain their traditional, large share of the global trade in red fescue seed. The urgency of this transition will become acute if trade with the USA is limited to certified seed, and is made consistent with the trading situation that has been in effect with Europe for many years.

In 1995, some 20 years after Maisonneuve *et al.* (1975) suggested that the forage seed industry in the Peace region of Alberta and British Columbia might benefit from the formation of a producer organization, the Peace Region Forage Seed Association was formed. The major objectives are to foster communication between the growers and the processing and exporting agents, and to chart a course for the forage seed industry to meet the challenges of the emerging global marketplace.

The large areas of relatively low-priced agricultural land, the conducive climate, and the freedom from acute environmental issues faced currently by the other major growing regions in the world, suggest that there is an optimistic future for forage seed production in the red fescue growing region of Canada. It is doubtful, however, whether this production will continue to be dominated by just one cultivar of creeping red fescue as it has been for the past three decades. The current trend towards the niche marketing of grass cultivars with specialized characteristics provides an opportunity for Canadian growers to diversify their production. The major challenge for the industry is to identify the species and cultivars of forage crops for which there is predictable demand in the global market-place, and for which seed of high quality can be grown at a competitive economic advantage as we approach the twenty-first century.

ACKNOWLEDGEMENTS

This history was shaped, to a large extent, by the dedicated efforts of a former scientific colleague, Dr C.R. (Bob) Elliott, who nurtured the development of the forage seed industry in the Peace region of Canada for many years, and who was kind enough to review this chapter; I am most grateful for his contributions.

The statistics assembled in this chapter were provided by individuals from several organizations. I express my sincere thanks to Jean Murphy and Bill Robertson of the Canadian Seed Growers' Association, to Nancy Sharp of the Canadian Agricultural Library and Bill Hanson of the Food Production and Inspection Branch of Agriculture & Agri-Food Canada, to Heather Drybrough of the Canada Grains Council, to Bernard Houle of the Agricultural Census Division of Statistics Canada, and to the staff of the Library and the Book and Record Depository at the University of Alberta, Edmonton, who facilitated my access to the current and archival publications of Statistics Canada.

REFERENCES

Canode, C.L. (1968) Influence of row spacing and nitrogen fertilization on grass seed production. *Agronomy Journal* 60, 263–267.

Chilcote, D.O., Youngberg, H.W., Stanwood, P.C. and Kim, S. (1980) Postharvest residue burning effects on perennial grass development and seed yield. In: Hebblethwaite, P.D. (ed.) *Seed Production*. Butterworths, London, UK, pp. 91–103.

Elliott, C.R. (1966) Floral induction and initiation in three perennial grasses. PhD Thesis, University of Saskatchewan, Saskatoon, Canada.

Elliott, C.R. (1968a) Boreal creeping red fescue. *Canadian Journal of Plant Science* 48, 106.

Elliott, C.R. (1968b) Registration of Boreal red fescue. *Crop Science* 8, 398.

Elliott, C.R. (1974) Northern Agriculture: The Peace River Area. *Agrologist (Agricultural Institute of Canada)* 3, 9–11.

Elliott, C.R. (1975) Creeping red fescue seed. *Agriculture Canada, News and Features* 1631, 10–11.

Elliott, C.R. and Baenziger, H. (1977) *Creeping Red Fescue*. Agriculture Canada publication 1122. Canada Department of Agriculture, Ottawa.

Elliott, C.R. and Bolton, J.L. (1970) *Licensed Varieties of Cultivated Grasses and Legumes.* Canada Department of Agriculture publication 1405. Canada Department of Agriculture, Ottawa.

Elliott, C.R., Stacey, E.C. and Doran, W.J. (1961) *Creeping Red Fescue.* Canada Department of Agriculture publication 1122.

Maisonneuve, M., Gareau, L. and Zacharias, J. (1975) *The Production and Marketing of Creeping Red Fescue Seed.* Alberta Agriculture and BC Department of Agriculture.

Smith, J.D. and Elliott, C.R. (1970) Stem eyespot on introduced *Festuca* spp. in Alberta and British Columbia. *Canadian Plant Disease Survey* 50, 84–87.

Smith, J.D. and Elliott, C.R. (1972) *Didymella* stem eyespot of *Festuca* spp. in northern Alberta and British Columbia in 1970 and 1971. *Canadian Plant Disease Survey* 52, 39–41.

Smith, J.D., Elliott, C.R. and Shoemaker, R.A. (1968) A stem eyespot of red fescue in northern Alberta. *Canadian Plant Disease Survey* 48, 115–119.

Stacey, E.C. (1974) *Peace Country Heritage.* Western Producer Book Service, Saskatoon, Saskatchewan, Canada.

Youngberg, H. (1980) Techniques of seed production in Oregon. In: Hebblethwaite, P.D. (ed.) *Seed Production.* Butterworths, London, pp. 203–213.

Lolium multiflorum Lam. (Italian Ryegrass) in Germany

W. Schöberlein[1] and E. Lütke Entrup[2]
[1]*Landwirtschaftliche Fakultät, Martin-Luther-Universität, D-06108 Halle, Germany;* [2]*Deutsche Saatveredelung Breeding Station, D-33154 Thüle, Germany*

14.1 A RETROSPECTIVE VIEW OF ITALIAN RYEGRASS AS A FORAGE CROP, CULTIVAR DEVELOPMENT, AND SEED PRODUCTION

The non-perennial grass species *Lolium multiflorum* Lam. is represented by two forms: a summer annual (annual ryegrass; *Lolium multiflorum* Lam. subsp. *multiflorum*, syn.: *Lolium multiflorum* subsp. *westerwoldicum*) and a winter annual or short-lived perennial (Italian ryegrass; *Lolium multiflorum* Lam. subsp. *italicum* [A. Br.] Volkart).

Summer annual ryegrass requires no vernalization. In Germany it is grown mainly as a catch crop, in mixtures with short-living clover species or as a cover crop in early spring on land that will be used to establish permanent grasslands. When grown as the main crop, it is harvested later than Italian ryegrass. However, if sufficient water is available, the total yield will be similar. Italian ryegrass, with its cultivar-specific vernalization requirements, belongs to the most productive and valuable grass species in field forage cropping, provided that nutrient and water supply are optimal. This species has gained importance as a pure stand or as a companion crop in short rotations when used in mixtures of grass–clover blends, vetch–ryegrass blends or as Landsberg blends with only one overwintering.

The first publications praising the benefits of Italian ryegrass appeared more than 100 years ago and, at that time, farmers were given recommendations for seed propagation (Hannemann, 1870). The origin of Italian ryegrass is in the Mediterranian region. In Germany it attracted interest as a forage crop only at the beginning of the twentieth century. The first ecotypes revealed poor winter hardiness, which allowed them to be grown only in the warmer regions of the Rhineland. After numerous generations in this region, Italian ryegrass became better adapted to the climate of northern Germany. Around 1914, selections began and 14 years later the first cultivars were officially released.

The range of Italian ryegrass became wider after the introduction of the blend 'Landsberger Gemenge', a mixture containing (per hectare) 20 kg Italian ryegrass + 20 kg crimson clover (*Trifolium incarnatum* L.) + 20 kg winter vetch (*Vicia villosa* subsp. *varia* L.). The blend was recommended by the seed company Deutsche Saatveredelung (DSV) in Landsberg/Warthe in 1929 after the typical grass–clover mixtures had fallen victim to the preceding cold winter and the subsequent summer drought in eastern Germany. The mixture used to be sown at the beginning of August and produced excellent yields after overwintering. Thus, it obtained quick popularity especially in northern Germany. With rising inputs of mineral N-fertilizers farmers omitted one or both legume partners and sowed a mixture of winter vetch and Italian ryegrass or Italian ryegrass as the main crop sown at a rate of approximately 40 kg ha^{-1}.

This development pushed forward the demand for seed and gave impetus to intense breeding activities with Italian ryegrass, oriented on productivity, better winter hardiness and disease resistance. These efforts made it possible to enter six new cultivars in the official list of cultivars between 1932 and 1934. In addition, as early as 1928, a hybrid of *Lolium perenne* × *Lolium multiflorum* had been officially registered. It stood out for improved persistence and did not differ from Italian ryegrass in productivity or utilization. The first cultivars of annual ryegrass were released in 1935.

14.2 SEED AVAILABILITY, SEED TESTING AND MULTIPLICATION BEFORE 1945

In the first decades of this century the availability of Italian ryegrass seed in Germany was largely dependent on imports, mainly from England, Ireland and Scotland. After the introduction of official seed testing by Nobbe in 1869, which laid the foundation of the world's first seed testing station in Tharandt, Saxony, commercial seed lots had to be subjected to official analyses of purity and germination capacity. The network of the new seed testing stations, which after 1895 carried out seed analyses in accordance with unified legal rules, offered the farmer warranty for the most important utility values of the purchased seed. This represented tremendous progress, particularly for seed of forage crops. Nevertheless, in those days not much attention was paid to genetically based traits like yield performance and seed quality. The situation changed only gradually after the German Grassland Union (Deutscher Grünlandbund) was founded at the beginning of this century.

According to Weller (1947) the average demand for grass seed in Germany between 1924 and 1943 amounted to approximately 15,000 t year^{-1}. This demand was satisfied by seed imports of about 6550 t and German production of 8450 t. The proportion of Italian ryegrass is unknown; however, estimates of about 30% of the average requirements, i.e. 4500 t, might be realistic. This quantity of Italian ryegrass seed would have been produced from some 4000–4500 ha.

Release and breeding of domestic cultivars promoted seed multiplication in Germany itself. By 1940 Italian ryegrass growers in Germany could choose from eight German-bred cultivars. As a result of national economic policy and the disastrous consequences of the war, seed was almost exclusively produced in Germany. The seed farms owned by the breeding companies used to sow in late summer (from the end of August to the beginning of September) and take one seed harvest in the following summer. Prebasic and basic seed was produced under supervision of the breeder. The subsequent generation, i.e. certified seed, however, was mostly produced on cropping farms on a contract basis via distributing and marketing firms. This organizational set up became well established and is still functioning (prebasic seed → basic seed → certified seed). First year seed crops usually produce around 1.4–1.8 t ha^{-1}.

14.3 SEED PRODUCTION AND DEMAND FOR ITALIAN RYEGRASS IN GERMANY BEFORE REUNIFICATION

The ability of Italian ryegrass to develop reproductive tillers after a forage cut, and to produce mature seeds 8–10 weeks later, enabled growers to produce certified seed and forage. Combining forage and seed cropping with Italian ryegrass offers a chance to simplify the seed growing technology and to produce a higher profit. A first cut for forage (silage or haymaking) yields between 7 and 8 t of dry matter ha^{-1} (Fig. 14.1). In addition to this economic advantage, any off-types can be culled without extra labour.

Seed yield after a forage cut is slightly lower than the seed harvest without a forage cut. Seed yields are further reduced with delays in the first cut for forage; inclement weather is the usual cause of this delay.

Fig. 14.1. Italian ryegrass – forage harvest before a seed crop. (Werks-Foto KRONE, DSV-Archiv.)

Fig. 14.2. Italian ryegrass – seed crop after the first forage cut. (Foto DSV-Archiv.)

The first forage harvest is followed by a nitrogen input of 40–60 kg ha^{-1} to support the regrowth of reproductive tillers, and seed is expected to be mature by mid to the end of August (Fig. 14.2). The seed is direct combine harvested at 35–40% seed moisture content, and therefore requires immediate drying. The conditioned seed yield averages 1 t ha^{-1}. Seed production is subject to the provisions of the Seed Trading Act of 20 August 1985 (BGBl, 1985), last amended by the Act of 25 October 1994 (BGBl, 1994).

In Germany, Italian ryegrass seed can be produced on all arable sites provided they are not too dry, and this is usually combined with cattle farming. Table 14.1 demonstrates the dynamics of the seed-production area for Italian ryegrass in West Germany from 1950 to 1990; since 1986 the area has remained rather stable at 3000–4000 ha, whereas the area of annual ryegrass increased in the same period from 1600 to about 2800 ha. To ensure stable seed yields (1.5–2.0 t ha^{-1}), seed harvests from annual ryegrass are generally taken without a forage cut.

14.4 EXTENDED SEED PRODUCTION OF *LOLIUM MULTIFLORUM* IN THE FORMER EAST GERMANY (GDR)

After the division of Germany at the close of the Second World War, several significant seed companies and their cultivars remained in the territory of the former GDR. In the case of Italian ryegrass three cultivars were available in 1949. Between 1969 and 1987 the range was supplemented by four new cultivars from East German breeders plus one Czech cultivar. There was always one cultivar of annual ryegrass on the market.

Until the end of the 1950s, seed production was mainly on private farms and on about 100 selected state-owned tenant farms which were later transformed into

Table 14.1. *Lolium multiflorum* seed production area in West Germany from 1950 to 1990. (Source: 'Bundessortenamt Hannover'.)

Year	Production area (ha)[1]		
	Italian ryegrass	Annual ryegrass	*Lolium multiflorum* total
1950	1521	173	1694
1955	2089	309	2398
1960	2531	1004	3535
1965	2395	1013	3408
1970	2000	1235	3235
1975	2243	595	2938
1980	1869	1120	2809
1985	2825	2280	5105
1986	3545	1687	5232
1987	3842	1573	5415
1988	4014	1882	5896
1989	3033	1975	5008
1990	3214	2847	6061

[1]Field inspection successful.

nationally owned seed production farms (Gäde, 1993). To promote herbage seed cropping, seed production farms were granted additional nitrogen supplies for their seed growing fields in years with limited fertilizer allocation, and bonuses for surplus production beyond the plan. In the case of Italian ryegrass the order was given to use the first cut for seed production to ensure high and reliable seed yields.

After the extensive transformation of private farms to collective farms (LPGs) beginning in 1960, a number (about 400) of selected LPGs were developed into specialized seed-producing cooperative farms with an area of between 400 and 1200 ha. They produced seed and/or planting material on a large scale. Later, mainly between 1970 and 1975, collective farms of neighbouring villages were amalgamated into huge crop production enterprises with a mean agricultural area of 4000–5000 ha. In these giant farms specialized seed multiplication was an important side-line. Perennial grasses and Italian ryegrass were grown on large fields covering 80–120 ha, which allowed investments in expensive buildings (seed stores of 24 × 80 m with total-floor ventilation and a storage capacity for 430 t of Italian ryegrass) to be economical.

Contracting, consultation and field inspections were under the responsibility of nationally owned enterprises ('VEB Saat- und Pflanzgut') under central management, united in the monopolistic seed trust 'VVB Saat- und Pflanzgut' which was later succeeded by the breeders' and propagators' trust 'Volkseigenes Kombinat für Pflanzenzüchtung und Saatgutwirtschaft' with headquarters in Quedlinburg (Gäde, 1993). The nationally owned enterprises bought up the harvested material from the seed growers, dried it if necessary and supplied the conditioned seed to crop producers or export customers. Prior to delivery, the seed material was certified in government seed testing laboratories in compliance with International Seed Testing Association (ISTA) rules.

In some regions with a high concentration of grass seed crops, adequately equipped seed processing plants for small-grain species were erected. Here, directly after harvesting and precleaning, the moist seed was dried in fully mechanized and ventilated ducts (length: 40 m, width: 3.5 m, height: 2.75 m). If necessary, heated air was used for drying the large homogeneous lots prior to the main cleaning operation. For the purpose of saving energy for drying, dehumidified cold air was used in the 1980s, generated by a mobile dehumidifier pump and installed before the fan. This technology enabled very careful drying and a four- to six-fold increase in the efficiency of electrical energy.

Soon the goal of meeting East Germany's demand for Italian ryegrass seed was reached and even overfulfilled, thanks to improved technical equipment and extension work. This resulted in increased seed exports. There were three major reasons for the endeavours to increase the seed production of grasses including Italian ryegrass.

1. To safeguard complete self-sufficiency in the seed sector of the former GDR.
2. To guarantee delivery of the contracted quantities within COMECON (exports to other socialist countries).
3. To extend exports to 'non-socialist' countries in order to help meet the former GDR's increasing demand for foreign currency (exports at dumping prices).

Grass seed cropping had become an attractive enterprise for experienced and well-managed farms, as the subsidized producer prices promised high financial benefit. Certain western European seed growers began to show rising interest for the multiplication of their cultivars in the GDR. The large and very homogeneous seed lots produced here were available as conditioned and low-priced seed a few weeks after harvesting. In 1989, for example, seed of 21 different cultivars of Italian ryegrass was grown for contractors in western Europe. The yields were from 1.0 to 1.3 t ha^{-1}.

In the 1960s the seed production area for Italian ryegrass was adequate for the requirements of East Germany. In the following decade it increased to 7000 and even 12,000 ha due to the growing export market. Table 14.2 demonstrates the annual area under Italian ryegrass from 1985 to 1990. In 1990, the year of the political changes, a peak of 15,206 ha was reached. At that time grass seed cropping in East Germany covered 39,688 ha or about 73% of the total German grass seed area.

14.5 THE DEVELOPMENT OF ITALIAN RYEGRASS MULTIPLICATION AREAS AFTER THE REUNIFICATION OF GERMANY

After Germany's reunification, drastic changes affected the extent of seed production in the new federal states (area of the former GDR). The planned economy in the former GDR with its subsidized prices for numerous agricultural products including seed had to be replaced by the system of an open market economy. The entire agricultural structure was freed from former centralistic constraints. In this free expression of will the large agricultural enterprises, formerly under government control, changed their status and became agricultural cooperatives with limited liability (GmbH). Additionally, many medium-sized private farms were established. The seed production area and especially the annual percentage of grass seed crops were reduced for the following reasons:

1. The highly subsidized prices and special promotion for grass seed growers granted in the former GDR had ceased.
2. The eastern European (COMECON) markets lapsed after the introduction of the monetary union. Due to their currency deficiency these countries could not afford to import seed from the new federal states of Germany and pay for it in Deutsche Marks.
3. Decreasing interest of western European contractors to produce their cultivars in the former GDR because of the breakdown of price advantages.
4. Reduced demand for forage seed owing to reduced livestock numbers and forage cropping areas in the reunified Germany.
5. Inappropriately low producer prices, that barely met the cost of grass seed production.

Consequently, from 1990 to 1994 the area of grass seed crops in the new federal states of Germany decreased by more than one-half (Table 14.3). However, the grass seed cropping area also declined slightly in the old federal states due to

Table 14.2. *Lolium multiflorum* seed production area in the former East Germany from 1985 to 1990. (Source: 'Saatzucht AG Quedlinburg'.)

	Production area (ha)[1]		
Year	Italian ryegrass	Annual ryegrass	*Lolium multiflorum* total area
1985	9789	3026	12,815
1986	8752	1577	10,329
1987	7936	2848	10,784
1988	4926	2625	7551
1989	8790	4675	13,465
1990	9401	6066	15,206

[1]Field inspection successful.

Table 14.3. Grass seed production area (ha) in the new federal states (former East Germany) from 1990 to 1994 compared to the old federal states (former West Germany) and total Germany.

	New states		Old states		Total Germany
Year	ha	%	ha	%	(ha)
1990	39,688	72.6	14,975	27.4	54,663
1991	38,311	71.5	15,290	28.5	53,601
1992	29,834	68.7	13,591	31.3	43,425
1993	22,717	60.8	14,623	39.2	37,340
1994	16,166	55.7	12,858	44.3	29,024

Table 14.4. *Lolium multiflorum* seed production area in Germany from 1990 to 1995. (Source: 'Bundessortenamt Hannover'.)

Year	Seed production area (ha)[1]			
			Lolium multiflorum	
	Italian ryegrass	Annual ryegrass	Total	%[2]
1990	12,354	8913	21,267	38.9
1991	10,660	10,565	21,225	39.6
1992	9811	4604	14,415	33.2
1993	8780	3198	11,978	32.1
1994	6290	2557	8847	30.5
1995	6681	3802	10,483	34.7

[1] Field inspection successful.
[2] Percentage of the total seed production area in Germany.

the reduced demand for forage seed. Table 14.4 demonstrates the development of the Italian ryegrass seed cropping area in Germany from 1990 to 1995 and the proportion of the total grass seed area. Italian and annual ryegrass rank first, followed by perennial ryegrass (*Lolium perenne* L.) and red fescue (*Festuca rubra* L.).

The extent of Italian ryegrass multiplication is linked with the extent of field forage cropping and catch crop growing, e.g. it is closely related to cattle. According to Kley (1995), Germany's mean Italian ryegrass seed usage is 8500 t year^{-1} (about 22.4% of the total grass seed consumption) and for annual ryegrass, it is 6500 t year^{-1} or 17.1% of the total sown. With an average seed yield of 1 t ha^{-1} from Italian ryegrass and 1.5–2.0 t ha^{-1} from annual ryegrass, 11,700–12,800 ha under Italian ryegrass are required in Germany every year to cover the demand. This estimate, however, is largely dependent on the situation in the European market and the demand for specific cultivars in different regions.

There is a wide spectrum of Italian ryegrass cultivars in Germany. The 'Describing List of Varieties 1995' (Anon., 1995) published by the federal variety board ('Bundessortenamt') contains 32 cultivars of Italian ryegrass (among them 16 tetraploid types) and 27 cultivars of annual ryegrass (17 tetraploids). These cultivars are well adapted to the climate, have been tested for at least 3 years for their value for cultivation and use (yield, resistance to diseases, etc.) and meet the differing demands of various growing regions. Therefore, sufficient seed of these cultivars has to be available for both the present and future needs of agriculture. In the next few years, seed production of Italian ryegrass in Germany should remain at a level comparable with that of recent years (see Table 14.4 for the period 1993–1995).

ACKNOWLEDGEMENTS

The authors thank Mrs Ute Schlosser, Agricultural Faculty, Martin Luther University, Halle-Wittenberg, Germany for translating their manuscript.

REFERENCES

Anon. (1995) In: Gräser, Klee and Luzerne (eds) *Beschreibende Sortenliste 1995.* Bundessortenamt Hannover, Landbuch Verlagsgesellschaft mbH, Hannover, pp. 29–31.

BGBI (1985) *Saatgut Verkehrsgesetz.* Bundesgesetzblatt I vom 20 August 1985, 5, 1633 pp. Last amended by the Bundesgesetzblatt I vom 25 Oktober 1994, 1, 3082 pp. Bonn, Germany.

Gäde, H. (1993) *Beiträge zur Geschichte der Pflanzenzüchtung und Saatgutwirtschaft in den fünf neuen Bundesländern Deutschlands.* Vorträge Pflanzenzüchtung, Heft 23. Verlag Paul Parey, Berlin und Hamburg.

Hannemann, F. (1870) Über Gräser und Weidebau. In: Toussaint, F.W. (ed) *Anleitung zum rationellen Grasbau.* J.U. Kern's Verlag, Breslau, pp. 235–269.

Kley, G. (1995) Bedarfsstruktur und Marktversorgung mit Futterpflanzensaatgut in Deutschland. *Tagungsbericht der 39. Jahrestagung vom 31.8.–2.9.95 der AG Grünland und Futterbau der Gesellschaft für Pflanzenwissenschaften.* TU München-Weihenstephan, Lehrstuhl Grünlandlehre, pp. 28–33.

Weller, K. (1947) *Der Samenbau der Gräser,* Bd. 2. Landbuch-Verlagsgesellschaft mbH, Hannover.

Lolium perenne L. (Perennial Ryegrass) in Denmark

15

K. Svensson[1] and B. Boelt[2]

[1] The Federation of Danish Seed Grower's Associations, Roskilde, Denmark;
[2] Danish Institute of Plant and Soil Science, Department of Cereals, Seeds and Industrial Crops, Roskilde, Denmark

15.1 INTRODUCTION

In Denmark, grass and clover seed production utilizes 2% (50,000 ha) of the 2,500,000 ha of land in agricultural production. Perennial ryegrass (*Lolium perenne* L.) is grown on approximately half of the seed production area, which is equivalent to 1% of the total Danish land area in agricultural use. For land use perennial ryegrass seed crops must compete with cereals (mainly winter wheat (*Triticum aestivum* L.)) which occupy 53% of the agricultural production area, industrial seeds/oilseed rape (*Brassica* spp.) (6%), beets (*Beta* spp.) (5%), pulses (e.g. *Pisum* and *Vicia*) (4%), potatoes (*Solanum tuberosum* L.) (2%), vegetables (1%), set-aside (7%), permanent pasture (6%) and short-term pasture (14%).

Denmark has a long history of forage seed production, and at the beginning of this century around 10,000 ha, mainly cocksfoot (*Dactylis glomerata* L.), was grown. By 1960 the seed production area had grown to 50,000 ha, and there has been little change since that time. However the distribution of species grown has changed markedly. In the 1960s white and red clover (*Trifolium repens* L. and *T. pratense* L.) seed was produced on 15,000 ha, and cocksfoot and perennial ryegrass were each around 9000 ha. In 1972 Denmark joined the European Community, following which the Danish production of perennial ryegrass seed increased and the production of clover seed decreased. Perennial ryegrass seed production reached its peak in 1989 when more than 30,000 ha was grown.

Denmark exports around 90% of its perennial ryegrass seed, mainly to members of the European Union (EU) (Table 15.1). In 1995 non-Danish cultivars occupied approximately 50% of the perennial ryegrass seed crop, although the percentage varies from year to year. Up to 150 different cultivars have been in production in any one year, but in 1994 the total was only 103. On average, 10–20 new cultivars enter production each year. This means that some growers have to frequently change cultivars, and this can pose problems for crop rotation and seed certification requirements.

Table 15.1. Destination and weight (t) of Danish-produced perennial ryegrass seed exported in 1994[1].

Destination	Weight (t)
France	3805
Belgium	1666
Holland	2518
Germany	8442
Italy	1152
UK	3162
Ireland	1496
Spain	479
Greece	55
Portugal	37
Total EU	22,811
Norway	351
Sweden	396
Switzerland	269
Austria	641
Other countries	327
Total	24,795

[1]The Danish Seed Council.

Over the past 10 years (1985–1994) Danish perennial ryegrass seed yields have averaged nearly 1100 kg ha^{-1}, fluctuating from a low of 815 kg ha^{-1} in 1992 to a high of 1360 kg ha^{-1} in 1994 (Table 15.2). Yields have been comparable with those from the UK and greater than those in Germany, but not as high as the Netherlands.

15.2 THE DANISH ENVIRONMENT

15.2.1 Climate

Denmark is located 54–58°N and 8–15°E. No part of the country is more than 50 km from the coast. The climate is described as a temperate coastal climate, which means mild winters (average coldest month (January) temperature of 0°C) and cool summers (average warmest month (August) temperature of 15–17°C). The past 5 years have had very mild winters. Normally perennial ryegrass growth would be expected to stop in early October and start again in early April.

15.2.2 Rainfall

Although Denmark is a small country, there is a large difference in the amount of rainfall in different parts of the country. While rainfall is highest in the autumn, the harvest months of July and August can each receive around 70 mm rain.

Table 15.2. Average perennial ryegrass seed yields (kg ha^{-1}) in four European countries (1987–1995)[1].

Country	Year								
	1987	1988	1989	1990	1991	1992	1993	1994	1995
UK	1033	1000	1183	1119	1062	1012	1264	1193	1136
The Netherlands	1207	1159	1532	1393	1645	1619	1438	1466	1500
Germany	679	773	814	781	772	549	518	962	870
Denmark	1066	1073	1193	1084	1201	815	958	1360	1373

[1]The Danish Seed Council.

15.2.3 Soils

Danish soils are very variable, because of the influence of past glacial periods. Because of the glacial movements the variation in soil type can be very large even within short distances. In general the soils on the islands, and the northern and eastern part of Jutland are composed of mixed deposits, with a clay content of up to 25%. The western part of Jutland has more sandy soils with a low water-holding capacity.

15.2.4 Production Area

Perennial ryegrass seed production is concentrated in Jutland where rainfall is higher than in the rest of the country. Soils in eastern and northern Jutland which have a high water-holding capacity are preferred, but providing irrigation is available, high seed yields have also been produced in western Jutland on lighter soils. The main production areas are therefore characterized by soils with medium clay content in areas with high rainfall.

Recent seasons (1992–1995) have shown that early drought can have a severe influence on seed yield. The seed yields presented in Table 15.2 illustrate the effect that lack of rain in May and June (1992 and 1993) can have on seed yield in Denmark.

15.3 WHY DO FARMERS PRODUCE PERENNIAL RYEGRASS SEED?

15.3.1 Economics

In Denmark the average gross margin for perennial ryegrass seed in the period 1987–1994 has been exceeded only by that for winter wheat (Table 15.3). However, returns to farmers have fluctuated widely depending on season, and prices have been considered low. The poor harvests of 1992 and 1993 discouraged many growers, but a seed shortage and hence increasing prices have encouraged farmers back into perennial ryegrass seed production.

Table 15.3. Gross margin (in relation to perennial ryegrass seed = 100) for different Danish cash crops, 1987–1994[1].

Winter barley	78
Winter wheat	112
Winter rape	100
Peas	76
Spring barley	61

[1]Seed Production Data and Statistical Control, 1994.

15.3.2 System of Payment

In 1994, 103 different cultivars were grown. These ranged from early, tetraploid types to late, amenity types. The mean difference in 1994 between the lowest and highest yielding cultivar was over 1000 kg ha^{-1}. To keep low-yielding cultivars in production, seed companies have a compensatory payment system, which theoretically means that gross income for a low-yielding cultivar does not differ from that for an average-yielding cultivar. However if the price on the international market or from the foreign plant breeder is too low compared to the compensatory price paid to the farmers, seed production of the cultivar stops very quickly.

15.3.3 Distribution of Work

A typical Danish farmer specializes in either dairy production, pig production or in cash-crop production. It is mainly the specialized cash-crop producers who grow perennial ryegrass for seed. It may be in combination with pig production, but very seldom in combination with dairy production.

A normal crop rotation for a farmer with perennial ryegrass for seed production is: spring barley with undersown perennial ryegrass → perennial ryegrass → winter oilseed rape → winter wheat → winter wheat.

It is very important to the farmers that work is distributed evenly over the year. Seed production helps this distribution because the perennial ryegrass is established in spring, and is harvested earlier than the grain crops.

15.3.4 Basics of Management

Perennial ryegrass is considered to be an easy seed crop to grow. In most years there are no problems with diseases and pests in Denmark. In a dense crop, with early lodging, mildew (*Erysiphe graminis* DC.) can occur. One fungicide application in early June will be sufficient. Occasionally crown rust (*Puccinia coronata* Corda) can appear in the spring. Stem rust (*Puccinia graminis* Pers.) is unknown in Denmark. After the establishment of a ryegrass crop, one herbicide treatment against broadleaved weeds (dicotyledons) is normally sufficient. Nitrogen is applied in the spring. Growth regulators are not used, since Cycocel (which is the only one registered) does not have any significant effect on the seed yield. Due to large differences between cultivars, the harvest period begins in mid July for early cultivars and ends up to 4 weeks later for the later cultivars. In general, farmers prefer the early cultivars which are harvested before the grain crops. Ryegrass is direct harvested, because a standing crop dries more quickly after rain. All growers have drying facilities on the farm, and farmers often contract with the seed company to keep the seed at the farm until the seed company is ready to clean it. The farmer will be paid for the storage. It is very unusual that the farmer cleans the seed himself. There is no tradition to do so in Denmark.

A second harvest of the same crop is becoming more common, but it is still only a minority of the growers who use this practice. Danish certification rules allow 3 years of seed harvest on the same field.

A small proportion (10%) of growers establish perennial ryegrass in a pure stand in the autumn. However it is more expensive than undersowing and it can from time to time be difficult to establish the seed because of dry conditions. In order to allow adequate plant development before the winter, a general rule is that perennial ryegrass must be established before 25 August. It is not a question of winterhardiness, but rather a question of the ability to produce a high seed yield in the following year. Over the last few years there has not been any winterkill in perennial ryegrass.

15.4 SEED COMPANIES

15.4.1 Capacity

Normally 50% of Danish seed production is perennial ryegrass. The total Danish seed crop is cleaned in 15 cleaning plants. Perennial ryegrass is relatively easy to clean, and gives the seed cleaners a high cleaning capacity compared to the 'finer' seed of red fescue (*Festuca rubra* L.) and smooth-stalked meadowgrass (*Poa pratensis* L.). The treatment of the seed, the cleaning, and the drying if necessary, is paid by the farmer at a fixed price. The price is calculated from the amount of field-cleaned material delivered.

15.4.2 Contracts

All seed produced in Denmark is certified seed, and is grown under contract. Mostly growers and companies are on a participation contract. This means that the grower delivers all the seed produced to the company, and that the company commits itself to sell at the highest possible price. In some contracts a minimum price is included if certain quality standards can be met.

In order to fully utilize production facilities a company will be interested in high volumes of product. Companies are of course interested in the highest possible price for the seed but they are not as vulnerable as the farmer for low prices. The company's problem, when the price is low, will be to find growers who are willing to grow the seed and take the risks.

15.4.3 Company Structure

The biggest Danish company with about 70% of the production is DLF-TRIFOLIUM. It was established in 1989 through an amalgamation of the four big Danish seed companies, DLF, Trifolium-Silo, SN-seed and the grass section of Daehnfeldt. DLF-TRIFOLIUM is the only Danish-owned seed company which carries out plant breeding. Hunsballe Frø, Anton Nielsen, Morsø Frø and Chr. Kehlet are Danish-owned companies that have production of both Danish and foreign cultivars. Two companies, Wiboltt Frø and Østergaard Frøavl, are owned by the Dutch seed companies, Cebeco and Van der Have.

In order to secure cultivars for multiplication, the Danish-owned companies have contacts with foreign companies. For instance, Hunsballe Frö have a close connection to DSV of Lippstad in Germany.

The 1989 merger between the four big seed companies was a way to prevent competition on the home market and on the international market between the Danish companies. This competition was harmful when they were selling to a foreign market and when foreign breeders tried to contract seed production in Denmark.

15.5 WHY DO BREEDERS FROM EUROPE CONTRACT FOR SEED MULTIPLICATION OF PERENNIAL RYEGRASS IN DENMARK?

Around 50% of the perennial ryegrass production area in Denmark is multiplication of foreign cultivars, mainly from Germany, The Netherlands, France and Belgium.

When a company or a breeder want to multiply their basic seed they demand:

- High quality.
- Reliability, meaning that the seed will be produced and can be delivered in the quantity and at the time it is needed.
- That the payment price is as low as possible.

15.5.1 Quality and Reliability

Of fundamental value for good seed quality is a good extension service, good farm management, good company facilities for cleaning and storage and a tradition for multiplication of quality products. Danish farmers and the Danish seed industry fulfil these demands.

Two weed problems which affect high quality are couchgrass (*Elytrigia repens* L. Beauv.) and annual meadowgrass (*P. annua* L.). The problem with *Poa annua* has increased since production has moved from mainly fodder types to an increasing amount of amenity types. In most cases the *P. annua* is controlled chemically with Tribunil. Control of couchgrass must be during crop rotation, as there is no chemical available in the seed crop. With the decreasing price of glyphosate it should be possible. Also 95% of the perennial ryegrass is direct combined without swathing. The difference in development between the perennial ryegrass and the couchgrass means that the couchgrass often will not be threshed out, and the seed lot can be kept free from couchgrass seed.

The average quality for Danish perennial ryegrass is very high (99.0% pure seed; 0.05% weed seeds; 0.10% other crop seed; 0.83% inert matter; 91% germination (Danish Plant Directorate, 1994)). It is very unusual for cleaned seed not to meet EU standards for subsidy. The main problem in order to meet the EU standard can be low germination, caused by mistreatment during harvest and drying.

15.5.2 Price

Despite competition from the former Comecon members in Eastern Europe, the Danish companies obviously remain competitive. After the change in Eastern Europe, the production in these countries has dropped, and it seems that production will return to Denmark and other West European countries.

15.6 OUTLOOK FOR THE FUTURE

Can seed production of perennial ryegrass in Denmark still remain economically competitive in the future, when faced with an increasing world market price for grain?

In favour of continued production of perennial ryegrass is the fact that the yield in Denmark is relatively high compared to grain yield. The payment area for grain in Denmark is furthermore low compared to Germany, the Netherlands and the UK. Denmark has traditionally grown a higher percentage of spring barley, which has a lower yield potential compared to winter wheat, but the area of winter wheat production increased over the last couple of years.

The higher world market price for grain does not seem to directly influence the price in Denmark. The experience from the 1995 harvest indicated that the EU Commission will keep the grain price low in the EU. Export from the EU has stopped, which means that the price for grain in the EU has not increased.

Environmental issues are of increasing concern, and restrictions on the use of fertilizers and chemicals could affect both yield and quality of perennial ryegrass seed.

15.6.1 Fertilizers

Since 1994 there has been a restriction on the amount of nitrogen (N) fertilizer the Danish farmer is allowed to apply. At present the N permitted for perennial ryegrass is set at the amount normally used (120–130 kg ha^{-1}) so, at this point, there is no acute problem for seed production. However the N level allowed could be reduced, or there could be a levy placed on nitrogen fertilizers, which would reduce the economically optimal application rate, and as a consequence of this, also the seed yield.

15.6.2 Chemicals

Since 1996 there has been a levy on the use of pesticides – 37% on insecticides, and 15% on fungicides and herbicides. Politically it has been decided that the use of pesticides in 1997 should be half the amount used in the reference period from 1981 to 1985. If this decision cannot be fulfilled, the consequence may be other further restrictions, for example higher levies. Some active ingredients may disappear from the agricultural chemicals presently available in Denmark.

15.7 CONCLUSION

If the restrictions and levies only apply in Denmark, the biggest problem the Danish seed industry will have to solve is how to remain competitive for quality and price. Initiatives have already been taken to face that situation.

REFERENCE

Danish Plant Directorate (1994) *Beretring 1993*. 104 pp.

Poa pratensis L. (Smooth-stalked Meadowgrass/ Kentucky Bluegrass) in The Netherlands

16

D.A. Donner[1] and G.E.L. Borm[2]
[1]*ECAF, Blaauwe Kamer 12, 6702 PA, Wageningen, The Netherlands;*
[2]*Applied Research for Arable Farming and Field Production of Vegetables (PAV) PO Box 430, 8200 AK, Lelystad, The Netherlands*

16.1. HISTORY

16.1.1 General

The history of grass seed production in the Netherlands barely covers 50 years. Traditionally, grass seed was collected by hand from the leftovers of haystacks and lofts. Grass seed production before the Second World War was very limited. Only Westerwolds ryegrass (*Lolium multiflorum* subsp. *westerwoldicum*) was grown commercially. Seed from some 500 ha of this very old Dutch species was harvested for both the local market and export. Most seed of perennial ryegrass (*Lolium perenne* L.) was imported from abroad. Seeds of bent grasses (*Agrostis* spp.) and fescue (*Festuca* spp.) were collected from natural stands in open places in the woods, and from heath fields on sandy soils.

After the Second World War the Netherlands was short of seed, especially of fodder grasses. As there was no foreign currency available to pay for the importation of these seeds, commercial seed firms and cooperative organizations took the first steps towards their own breeding programmes. In 1947, 1000 ha of grass seed was grown, 75% of which was Westerwolds ryegrass. By 1957, the Netherlands was self-sufficient for grass seed, and in 1962, seed exports of smooth-stalked meadowgrass began.

Grass seed in the Netherlands is generally considered as a field crop comparable to wheat (*Triticum aestivum* L.) and barley (*Hordeum vulgare* L.), and is grown in rotation with other arable crops. Grass seed producers are interested in the seed more than in a cut for hay. Reasons for the popularity of grass seed and for the rapid increase in area since the 1940s include:

1. The Dutch climate is suitable for the production of grass seed: it is moderate, there are no severe droughts, and winters are seldom severe, resulting in relatively stable yields.

2. Early species like smooth-stalked meadowgrass fit well into the working schedule of arable farms.
3. Generally no additional investment in machinery is needed.
4. Grass seed is a crop that has a positive effect on other crops in the rotation system.
5. Satisfactory weed control with chemicals has been possible.

16.1.2 Smooth-stalked Meadowgrass

The first considerable area of smooth-stalked meadowgrass in the Netherlands was in 1954 when 50 ha were grown. In 1960 the area had increased to 2000 ha, in 1961 it had doubled to 4200 ha, and since then it has fluctuated around this figure.

From 1950 onwards an increasing number of selections appeared on the market. As their use increased rapidly, production of the required quantities of seed needed more attention. Initially this only concerned seed of pasture types, but gradually turf types were introduced and became increasingly sought after. While both types have their specific quality parameters, purity was always the topic for any grass seed production programme. The importance of this characteristic became even more clear in the early-1960s when the export of smooth-stalked meadowgrass to the USA stagnated due to very strict requirements on purity of seed lots. *Poa annua* L. was declared a 'noxious weed' resulting in greater emphasis on cleaning techniques.

In order to improve the differentiation of seeds of the two species, seed size became a selection criterion in smooth-stalked meadowgrass breeding programmes (G.E. van Dijk, Bennekom, 1995, personal communication).

Nowadays the main challenge for the producer of smooth-stalked meadowgrass seed still lies in delivering a high-quality product. Weed control is one of the major constraints to reaching this aim, especially when one has to work within a tight framework of rules and regulations, and strict limits of economic and environmental feasibility.

16.2. ORGANIZATION AND MARKETING

The first grass seed lots were imported from Eastern Europe by merchants at the beginning of this century. These experienced merchants laid a sound basis for the Dutch grass seed trade, and today their firms still play an important role in grass seed production in the Netherlands.

As a major part of grassland in the Netherlands is permanent, the local market for grass seed was relatively small, and seed companies began both importing and also exporting seed. The seed firms had their base in the middle of the country, near Vlijmen, and some of the larger seed companies are still there today.

The local market for grass seed has increased since the need for turf types arose. Grass seed prices were traditionally related to the price of meat, milk and wool, but have gradually come under the influence of a changing standard of living: airstrips, gardens, sports fields and golf courses are luxury products in need of large quantities of grass seed.

16.2.1 Organization

Production of grass seed in the Netherlands is undertaken on a contract basis between farmers and seed companies. Contracts can be made with any of the seed firms; they work on a competitive basis. Although farmers were initially suspicious of any contracts, the system concerning grass seed worked well from the start. This is most likely because grass seed companies have enhanced the seed growers' interest by contracting on a participatory basis, and by delivering very intensive production guidance.

The seed firm provides the farmers with basic seed and advises them from the choice of the cover crop and preparation of the seed bed through to storage of the harvested seed. This is done by well-trained production consultants who visit the farmer at critical times, such as for sowing, fertilizing, weed control and harvesting. The consultants advise and also check whether the advice is carried out. The most common type of contract for grass seed production is the so-called 75–25 participation contract, where 75% of the net wholesale trade price goes to the grower and 25% to the seed company.

Samples taken from the harvested seed before cleaning are analysed by the Seed Testing Service for percentage of clean seed, cultivar purity, moisture content and germination capacity. Prices are fixed according to the results of these analyses. The farmer receives an advance payment in September and sometimes in December. At the end of the selling season, usually May or June in the year after harvest, the seed company settles the final bill. Some firms have a confidential committee of seed growers authorized to verify the books of the firm. This committee is consulted when the farmer's prices are fixed. The system works satifactorily and firms have the confidence of their farmers.

Field inspections for seed certification were, until recently, always carried out by the Netherlands Inspection Service (NAK). In order to reduce production costs of certified seed, seed companies have now been allowed to carry out field inspection themselves, under the supervision of the NAK.

16.3 PRODUCTION

16.3.1 Crop Establishment

Smooth-stalked meadowgrass, a slow starting grass, needs a build-up period to produce a crop sufficiently developed for the vernalization in winter necessary for seed production in the following year. It is commonly grown under a cover crop. Crops suitable as a cover crop should not be too robust and should be harvested in time for the grass seed crop to recover. Traditional cover crops are flax (*Linum usitatissimum* L.), peas (*Pisum sativum* L.), beans (*Vicia* spp.), caraway (*Carum carvi* L.), oilseeds (*Brassica* spp.) and cereals. Apart from the latter, most of these crops have gradually disappeared from rotations, mainly for economic reasons. This leaves the present-day farmer with cereals and especially winter wheat as the most suitable cover crop for smooth-stalked meadowgrass.

Winter wheat can be sown simultaneously or 3–6 weeks before smooth-stalked meadowgrass. The National List classifies wheat cultivars according to their suitability as cover crops. A high score stands for retarded canopy closing in spring, stiff straw, early maturity and little shattering.

In the 1962 cultural guide, growers of smooth-stalked meadowgrass were seriously advised against the use of cereals as a cover crop, mainly due to problems with volunteer plants. With the appearance of TCA (natriumtrichloroacetate) in the early 1970s, the control of wheat volunteers became relatively easy. This lasted until the use of TCA was stopped for environmental reasons in 1991. In spite of an intensive research effort, no adequate alternative for TCA has since been found. Tight regulations for admission, high registration fees and strict environmental laws do not facilitate the legislation of new herbicides for a small area crop like grass seed.

16.3.2 Crop Management

After harvest of the cover crop in summer, the grass needs stimulation to develop sufficiently before winter. Cutting the stubble of the cover crop, nitrogen application, weed control and disease control (if necessary) are common practices in late summer. A dressing of nitrogen very early in spring stimulates reproductive growth. The first harvest of grass seed can take place at the end of June or in the first half of July. Most fields are harvested for 2 years, followed occasionally by a third year if this fits into the rotation, and if the weed situation is sufficiently under control.

After harvest of the grass seed, the grass regrowth is cut two or three times to have a shortcut winter crop. Good results have also been obtained with grazing by sheep. Weed and disease control is carried out if necessary.

16.3.3 Harvest

The traditional way of harvesting grass seed on sheaves, putting the sheaves together in hocks and threshing the seed at complete maturity, has disappeared. Economy of labour, mechanization and the need for increased efficiency in the late-1950s resulted in the system of harvesting the crop by swathing with a windrower and threshing by combine harvester (Evers and Wolfert, 1959; Vreeke, 1989).

Shattering is not a major problem with smooth-stalked meadowgrass, and time of harvesting is less critical than for other grass seeds. However, as the seeds are tightly attached to the glume, threshing may cause problems. It is important that the crop is mature and dry when threshing. Nowadays the harvest in early summer is generally carried out by well-equipped contractors.

Average yields have not increased over the last 20 years (Table 16.1), although yields of 2000 kg ha^{-1} may occur. Seed quality has, however, improved substantially over this period. A range of cultivars of different types of smooth-stalked meadowgrass is available on the market.

Table 16.1. Areas and seed yields of smooth-stalked meadowgrass in the Netherlands. (Source: NAK/Produktschap GZP.)

Period	Total (ha)	Average yield (kg ha^{-1})
1955	106	1971–1975: 1080
1960	2768	
1965	5465	1971–1980: 1110
1970	1782	
1975	3746	1973–1987: 1110
1980	5262	
1985	2263	1978–1989: 1030
1990	5141	
1995	4506	1981–1992: 1000

16.4 BREEDING

The earliest breeding of smooth-stalked meadowgrass in the Netherlands mainly focused on its suitability for grassland. Turf types were introduced gradually and breeding aims have developed consequently. Whereas feeding quality, taste and productivity are major parameters for pasture types, colour, slow growth rate and resistance to treading are more important for turf types.

The need of introgression of winterhardiness into the existing material was a characteristic of the first Dutch smooth-stalked meadowgrass breeding activities. Collecting expeditions were organized to Scandinavian countries and Iceland, resulting in cultivars with names such as Blue Bell and Ice Lady. Saltings along the coast were a source for salt tolerance, an important trait for grass grown along highways where a high amount of salt may be applied during cold winters (Tj. Veenstra, Vlijmen, 1995, personal communication).

A look at the earliest editions of the Dutch 'Descriptive List of Varieties of Arable Crops' gives an idea of how breeding in the Netherlands started. The 8th edition (1931) stated that *Poa pratensis* L. in the Netherlands was to be found in pastures as 'a tasty, tillering undergrass'. The USA was mentioned as the main production area and it was stated (for no specific reason) that 'seed from Missouri (sold under the name: Kentucky bluegrass, Missouri grown) is to be preferred above seed from Kentucky'. Amongst several growers in the Netherlands, A. van der Have in Kapelle who 'multiplies American material under selection and two varieties of Prof. Mayer Gmelin' and B. van Engelen in Vlijmen who 'multiplies local material under selection' were mentioned. These names are still connected to some of the main grass seed companies in the Netherlands. Furthermore, this 1931 edition referred to names in Poland and Germany (Weihenstephan) 'who are trying to set up a breeding programme, maybe with success'.

The first results of Dutch breeding activities were seen in 1937 when the Association of Dutch Breeders appeared with a collective cultivar named Neerlandia I (hooi/weide). From 1950 onwards, increasing numbers of local selections and foreign selections adapted to local conditions appeared on the market. Accordingly, the descriptions of American material became more impertinent: 'early heading and soon becoming stiff' (1944), 'coarse but usable' and 'susceptible to rust' (1948), which by that time accounted for most of the material mentioned.

Some of the Dutch cultivars which appeared on the list were Brabantia (1954, Van Engelen), Veldbeemdgras C.B. (1958, later called Delft, Cebeco), Prato (1959, Van der Have), Arista (1965, Van Engelen), Baron (1970, Barenbrug) and Parade (1972, Van der Have). The most common foreign cultivars were Merion (USA), Ötofte (Denmark), Steinacher (Germany), Fylking (Sweden) and Julia (Germany).

Poa pratensis L. is an apomictic species, meaning that all the progeny of a plant are identical, even if this plant is heterozygous or aneuploid. Different cultivars have different numbers of chromosomes. Cross-fertilization is very rare in most genotypes. This has influenced breeding methods used for this species. The difficulty in producing new genetic variation limits progress.

The collection of ecotypes in various countries is a common method to obtain new starting material for selection. Even in places where the crop has never been grown, it is possible to find cultivars already existing, due to apomictic spread of the species in the past. Artificial ways to increase variation, such as irradiation, treatment with mutagenic chemicals and more recently somaclonal variation have been applied with limited success. Certain techniques may help to obtain a few percent hybrid seed from crosses between different plants. Quite often, however, those hybrids are not completely apomictic, which causes problems afterwards in obtaining uniformity and for seed certification.

Once the breeder has obtained suitable genetic variability, breeding is relatively straightforward. Seed of each plant is sown in replicated plots to compare for important characteristics, such as early establishment, density, wear tolerance, diseases, colour, fineness of leaves and seed yield. Due to the apomixis, breeder's seed can be multiplied from one individual superior genotype. Careful roguing of non-apomictic offspring needs to be carried out in breeders- and prebasic seed production, as seed certification requirements are strict (L. Beerepoot, Oosterhout, 1995, personal communication).

16.5 RESEARCH

16.5.1 General

Agricultural research structurally made a great leap forward after the Second World War due to large investments made by the Dutch government. The first reports of research on smooth-stalked meadowgrass go back to the early 1950s.

One of the pioneers was A. Evers who initiated and carried out a sound research programme first at the Central Institute of Agricultural Research (CILO) and later at the Research Station for Arable Crops (PAW, presently PAV). The main constraints of production at the time were: slow germination, slow initial growth, the problem of contamination with seeds of *Poa annua* and the difficulties of threshing. A special project for grass seed production on salty soils was initiated in 1953 as a result of the serious flooding of the Southwestern part of The Netherlands, which is the main grass seed production area. Having dealt with the 'basic' cultural problems throughout the years, weed control is still a major research item.

The main topics of research on smooth-stalked meadowgrass will be reviewed. The larger part of this research is still carried out at the PAV in Lelystad.

16.5.2 Cover Crops

Research in the 1970s and 1980s (Meijer, 1984) made clear which conditions favour the production of inflorescences in seed crops of smooth-stalked meadowgrass and red fescue. Based on this knowledge, it was possible to determine a crop management system for the most important cover crop, winter wheat, in order to optimize the development of the undersown grass seed crop (Meijer, 1987). This includes selecting a late-closing winter wheat cultivar, decreasing the sowing rate of winter wheat, and postponing part of the earliest nitrogen (N) application for winter wheat to combine with an early autumn sowing of smooth-stalked meadowgrass.

The production of protein crops (peas and field beans) was stimulated in the late 1980s by the European Community. The combination of the expected withdrawal of TCA to control volunteer plants of winter wheat, and a change of the winter wheat cultivars to more leafy types, led to an increased interest in peas and field beans as potential cover crops for slow developing grass seed crops. Seed yield of smooth-stalked meadowgrass undersown in these cover crops can be as good as, or even higher, than that of the crop undersown in winter wheat (Borm and Vreeke, 1991). However the establishment of the grass in spring in the protein crops is more risky than in winter wheat, resulting sometimes in a complete failure of the grass seed crop and/or greater contamination of the harvested seed with *Poa annua*.

After another change in EC policy, protein crops have now practically disappeared from the Dutch crop rotation, and this once more leaves the grower of smooth-stalked meadowgrass with winter wheat as the most acceptable cover crop.

16.5.3 Autumn Treatment

Apart from manipulating the crop management of the winter-wheat cover crop, it is also possible to stimulate inflorescence production by close cutting the grass immediately after harvest of the wheat. Better light penetration after cutting allows the tillers to remain compact, lessens the severe competition in early spring between the elongated tillers, and increases the number of fertile tillers (Meijer and Vreeke, 1988). This has resulted in an almost proportional increase in seed yield. In most experiments, seed yield was highest with the application of N immediately after the wheat harvest (Meijer, 1988).

Smooth-stalked meadowgrass is usually harvested for 2 years. Mowing two or three times in the late summer and autumn after the first seed harvest has resulted in higher seed yields (Borm, 1993). The positive effect of mowing before the second or later seed harvest also seems to be partly connected with the increase in shoot density and the reduction in the length of the leaf sheath. A late application of N (first half of October) gave a higher seed yield compared to an early application of N (Borm, 1993).

Grazing by sheep for some time in the autumn can give the same positive effects as mowing. The intensity of grazing has to be above 1000 animal-days (number of grazing days × number of sheep) per hectare (Wander, 1993).

16.5.4 Fertilizer

Nitrogen fertilizer received much research attention at the end of the 1970s and in the early 1980s. It was concluded that highest seed yields were achieved when 60 kg N ha^{-1} was applied in the autumn, although if the crop was grazed, the autumn N rate had to be increased to 90–125 kg ha^{-1} (Wander, 1993). No relationship was found between the optimum spring rate and the soil-N status. Highest seed yields were obtained at a spring N rate of 110 kg ha^{-1} (Meijer, 1988). Smooth-stalked meadowgrass did not respond to phosphate in the autumn or in spring (Wander et al., 1996).

16.5.5 Diseases and Pests

Diseases and pests do not cause major problems in smooth-stalked meadowgrass seed crops in The Netherlands, but may lead incidentally to critical situations. In 1977 some plots in the southwestern part of the Netherlands were seriously infected with gall midges (*Mayetiola* spp.) causing yield reductions of up to 60%. This then led to research on control possibilities for these insects, and on the economic feasibility of these control methods. Gall midges can be controlled with insecticides (permethrin and parathion) but this is only profitable above certain thresholds (Horeman, 1990a).

The main diseases which occur in smooth-stalked meadowgrass in the Netherlands are brown rust (*Puccinia brachypodii* Otth.), mildew (*Erisyphe graminis* DC.), Drechslera leafspots (*Drechslera* spp.) and ergot (*Claviceps purpurea* (Fr.) Tul) (Horeman, 1990b). In a 3-year trial (1986–1988) no significant effect of spring rust infection on crop yield was found. Treatment with fungicides increased yields in some cases but was never economically feasible (Horeman, 1989a,b). Severe infections of mildew in autumn can reduce the first harvest seed yield by 50%, depending on the cultivar and development stage of the crop (Bor, 1977). Damage due to this disease in autumn is more serious than damage caused in spring. Only in very well-developed crops does autumn mildew infection not cause any damage. Trials have been started recently to reassess the effects of rust following a severe outbreak in 1995 which significantly reduced seed yield.

16.5.6 Weed Control

As for other seed crops, weed control has been a major research item for smooth-stalked meadowgrass over the years. Research carried out today is not basically different from that in the 1950s. Throughout the years, however, it has adapted to changes in cultural practices, weed population, grass seed cultivars, available chemicals, and last but not least, to political changes including laws concerning the environment. Screening of herbicides on an experimental basis started in 1956 and has continued on a regular basis since.

Dicots like coltsfoot (*Tussilago farfara* L.), creeping thistle (*Cirsium arvensis* (L.) Scop.) and common chickweed (*Stellaria media* (L.) Vill.) are controlled with herbicides acting as growth regulators. In the past calcium cyanamide was used as a corrective measure during winter; the efficiency is influenced to a large extent by weather conditions.

Control of monocots in grass seed obviously causes more problems. *Poa annua*, black grass (*Alopecurus myosuroides* Huds), volunteers from cover crops like flax, later winter wheat and other cereals have always needed special attention. In the 1960s, when *P. annua* was declared a 'noxious weed' by the USA, methods for its control were emphasized, as mechanical cleaning of this weed is almost impossible.

The Ministry of Agriculture initiated a 10-year national programme (MJPG) in 1990, aimed at reducing the amount of active ingredients used for crop protection in the year 2000 by 50%. Since TCA was withdrawn in 1991, the grower of smooth-stalked meadowgrass is facing a serious problem of how to control volunteer plants of winter wheat, which is nowadays the most feasible cover crop. Diuron is now the only product registered for the control of volunteer wheat plants. However, the use of diuron is complicated by the relationship between the effectiveness and selectivity of the product. Mechanical control of volunteer wheat plants in smooth-stalked meadowgrass by means of harrowing, hoeing and brushing was not very successful (Baltus and Zweep, 1994). Looking for alternative herbicides against volunteer wheat plants resulted in some interesting products like quizalofob and propaquizafob, which however are not yet registered for use in grass seed. Recent stopping of the production of methabenzthiazuron adds a serious new problem to solve in controlling *Poa annua* and *P. trivialis* L.

If problems with the legislation of new products for weed control cannot be solved reasonably quickly, and if sowing in open land appears to not be economically feasible, smooth-stalked meadowgrass may soon turn into an endangered species in the Netherlands.

REFERENCES

Baltus, P.C.W. and Zweep, A.T. (1994) Mechanical control of volunteer wheat in Kentucky bluegrass grown for seed. In: *PAGV Jaarboek 1993/1994* pp. 144–146.

Bor, N.A. (1977) Meeldauwbestrijding in veldbeemd. *Bedrijfsontwikkeling* 8, 826–828.

Borm, G.E.L. (1993) Autumn treatment of perennial ryegrass intended for the first and second seed-harvest and of smooth-stalked meadowgrass and red fescue intended for the second or later seed-harvest on clay soils. In: *PAGV Report* 162. Research Station for Arable Farming, Lelystad, 157 pp.

Borm, G.E.L. and Vreeke, S.(1991) Effects of several cover crops on undersown smooth stalked meadow grass and red fescue grass seed crops. *Journal of Applied Seed Production* 9 (suppl.), 45–46.

Evers, A. and Wolfert, J.E. (1959) *De Zaadteelt van Veldbeemd*. PAW, Wageningen.

Horeman, G.H. (1989a) Effects of fungicides on perennial ryegrass (*Lolium perenne*) seed production. In: *Proceedings of the XVIth International Grassland Congress*, Vol. 3, pp. 667–668.

Horeman, G.H. (1989b) Noodzaak van roestbestrijding in Engels raai- en veldbeemdgras. In: *PAGV Report* 94. Research Station for Arable Farming, Lelystad, 134 pp.

Horeman, G.H. (1990a) Graszaadstengelgalmuggen (*Mayetiola* spp.) in veldbeemdgras. In: *PAGV Report 118*. Research Station for Arable Farming, Lelystad, 51 pp.

Horeman, G.H. (1990b) Inventarisatie van ziekten en plagen in veldbeemdgras en Engels raaigras voor zaadproduktie in 1989. In: *PAGV Report 119*. Research Station for Arable Farming, Lelystad, 29 pp.

Meijer, W.J.M. (1984) Inflorescence production in plants and in seed crops of *Poa pratensis* L. and *Festuca rubra* L. as affected by juvenility of tillers and tiller density. *Netherlands Journal of Agricultural Science* 32, 119–136.

Meijer, W.J.M. (1987) The influence of winter-wheat cover crop management of first-year *Poa pratensis* L. and *Festuca rubra* L. seed crops. *Netherlands Journal of Agricultural Science* 35, 529–532.

Meijer, W.J.M. (1988) Nitrogen fertilization of grass seed crops as related to soil mineral nitrogen. *Netherlands Journal of Agricultural Science* 36, 375–385.

Meijer, W.J.M. and Vreeke, S. (1988) The influence of autumn cutting treatments on canopy structure and seed production of first year crops of *Poa pratensis* L. and *Festuca rubra* L. *Netherlands Journal of Agricultural Science* 36, 315–325.

Vreeke, S. (1989) Time of swathing of grass crops for seed production. In: *PAGV Jaarboek 1997/1988*, 64–71.

Wander, J.G.N. (1993) Grazing sheep on smooth-stalked meadow grass (*Poa pratensis* L.) before the second seed harvest. In: *PAGV Jaarboek 1992/1993*, pp. 134–143.

Wander, J.G.N., Steenhuizen, J.W. and Ehlert, P.A.I. (1996) Response of perennial ryegrass, red fescue and smooth-stalked meadowgrass grown for seed to phosphorus fertilization and soil P level. In: Schoberlein, W. and Forster, K. (eds) *Proceedings of the Third International Herbage Seed Conference*, 18–23 June, Martin-Luther-Universitat, Halle-Wittenberg, Germany, pp. 261–295.

Dactylis glomerata L. (Cocksfoot) in New Zealand

17

M.J. Hill
Seed Technology Centre, Department of Plant Science, Massey University, Palmerston North, New Zealand

17.1 INTRODUCTION

Cocksfoot (*Dactylis glomerata* L.) is a constituent perennial in most of New Zealand's dairy pastures, and also finds a place on much of the land devoted to cattle and sheep production. It is well adapted to moderate fertility and low soil moisture and, after perennial ryegrass (*Lolium perenne* L.), is the most important of New Zealand's certified forage seed crops, with a production of 515 t of seed from 1000 ha in 1993 (MAF, 1993).

The fluctuating and sometimes low yields obtained from commercial cocksfoot stands strongly suggest that cocksfoot seed production is a specialist operation. This is partly true, as evidenced by seed yields of over 1000 kg ha^{-1} by specialist growers (Rolston, 1991). However, climate is important; a heavy frost at anthesis, lack of spring and early summer rain and bad weather at harvest all having major effects in reducing seed yields. Management of the crop is of course also important, and in particular, establishment, weed control, grazing, harvesting and fertilizer use as well as pests and diseases can all play a major role in the success or otherwise of cocksfoot seed production.

17.2 SEED PRODUCTION – HISTORY AND CULTIVARS

Herbage seed production in New Zealand dates back more than 100 years with exports in the 1880s of 1400 t of grass seed, mainly cocksfoot, ryegrass (*Lolium* spp.) and tall fescue (*Festuca arundinacea* Schreb.) (MacKay, 1987).

Cocksfoot has occupied a prominent position in New Zealand for many years, the first bred cultivar, Grasslands Apanui, having been released in 1953, followed by Grasslands Kara and Grasslands Wana in 1981. However the history of cocksfoot seed production in New Zealand goes back to the 1850s using seed imported from England, initially grown on Bank's Peninsula near Christchurch, and later elsewhere in Canterbury (e.g. Ashburton County) and in Southland and Otago provinces (McPherson, 1948).

It was with the early settlers of Bank's Peninsula that the history of cocksfoot had its origins in New Zealand. From its humble beginnings in homestead gardens it developed into a highly organized industry. Although no longer continuing, the cocksfoot industry of Bank's Peninsula was responsible not only for making New Zealand famous for its seed and the name 'Akaroa' known throughout the world, but also for pasture development in both the North and South Islands, which was important in establishing New Zealand as a major agricultural nation (Coulson, 1979).

Bank's Peninsula is situated approximately in the middle of the east coast of the South Island of New Zealand, on the margin of the Canterbury Plains. It is only 580 km^2 in area, dominated by the craters of two volcanoes which now form Lyttleton and Akaroa harbours. The Peninsula was the first part of Canterbury Province to be settled by Europeans and the first cocksfoot seed crop was produced in 1852. By the 1870s settlers throughout the Peninsula were harvesting cocksfoot seed on a large scale as 'Akaroa' cocksfoot and from the late 1880s to the mid 1900s many Peninsula farmers relied on the production of cocksfoot seed as their main source of income. During the late 1880s an area of nearly 10,000 ha was cut for seed.

Harvesting usually commenced in the first or second week of January and continued until the end of February or early March. Crops were cut when the seeds appeared 'doughy' (about 5 weeks from first flowering) using a reap hook or sickle. Cut sheaves were left on the cut stubble to dry for a period of 10–12 days before being carried in to be threshed by flail and sieved by hand or on riddling machines.

The success of the cocksfoot seed industry on Bank's Peninsula was unquestionably climatic, with rich volcanic soil, and the fact that, unlike the Danish, French and American cocksfoot cultivars available, which were characteristically short-lived (although often heavy seed producers), 'Akaroa' cocksfoot was more persistent, had a high density crown, was multi-tillered with high leaf production, and was more suitable for grazing purposes.

While the area cut for cocksfoot seed increased in Canterbury between 1929 and 1931, the amount of land used for this purpose on the Peninsula decreased. This was attributed to the development of cocksfoot pastures on the Canterbury Plains and the use of reapers and binders to harvest the seed. Large threshing machines were easily moved from field to field on the Plains, an operation quite impossible on the rugged hills of Bank's Peninsula.

In 1933, Government Certification of cocksfoot seed cultivar Akaroa, and the establishment of a registered Association of Seedgrowers, gave a further boost to the industry. However by 1939 labour cost increases created doubt as to the future viability of the cocksfoot seed business, a situation which worsened during the Second World War. By 1945 a series of bad harvests, competition overseas from Danish cocksfoot (available at 5 cents kg^{-1} compared with 30 cents for New Zealand cocksfoot) and the cessation of exports to Australia and the UK created a situation where seed production on the Peninsula was uneconomic. Many growers abandoned cocksfoot seed production as a farming enterprise.

In the 1950s and 1960s cocksfoot seed production began in earnest on the Canterbury Plains, often on very large individual farmer areas. At that time the crop was cut with a binder and stooked in the field before threshing with a stationary mill. Stand management in the 1960s involved no grazing of the seed production area between harvests, although techniques to remove crop residue by baling the straw after harvest were often practised.

The practice of heavy fertilizer (NPK) application in autumn (March) and subsequent topping (cutting) of stands in late April was increasingly popular from the mid 1960s. The rejuvenation of cocksfoot stands resulted in more even rows of plants, improved light penetration, and the opportunity for 'the plants to feast on everything they needed in autumn and early winter' (Angus McKay, Methven, personal communication, 1996). The early strong seedheads which appeared in the spring were seen as a 'sure sign of high yield'.

The major emphasis on the development of the New Zealand cocksfoot industry began in the 1970s when, with an increasing scarcity and cost of farm labour, many growers became more aware of the advantages of mechanical harvesting and seed drying in an attempt to produce higher yields and better seed quality (particularly better germination).

Even today, old-fashioned binders are still used on some Canterbury farms. This labour-intensive, 'out-of-date' method of harvesting cocksfoot has continued to be used with excellent results. After the binder has mown the cocksfoot, the sheaves are stooked in the field by hand and left for about 10 days before being threshed in a stationary thresher. John McKay of Methven has recorded seed yields of up to 1050 kg ha^{-1} using this method (Lucas, 1984).

Despite this, the self-propelled combine harvester is still predominant on most cocksfoot seed crops, whether the crop is direct harvested at about 35% seed moisture content and artificially dried, or swath harvested and threshed after varying times of natural drying. Perhaps the pioneers of direct harvesting were J.D. Marr and Sons of Rakaia who in the 1970s investigated direct harvesting in association with an artificial (supplemental heat) drying system. While often successful in terms of yield, some problems were encountered with seed germination and high cleaning losses. As a consequence of the appearance of rotary hay mowers Marr and Sons developed a 'double cut' swath harvesting system. This involved the cocksfoot crop being cut on a high stubble with a windrower. The swath was then immediately undercut with a rotary mower to drop the swath a further 25–30 cm onto a short firm stubble base. The importance of this system was immediately obvious, since it protected the cut swath from disturbance and damage by the strong northwesterly winds which pervade the Canterbury Plains. Seed yields and quality were increased dramatically.

Currently, two cultivars, Grasslands Wana (390 t) and Grasslands Kara (97 t) comprise the bulk of the New Zealand cocksfoot crop, with other cultivars such as Grasslands Tekapo, Saborto, Condor and Grasslands Apanui being produced in much smaller amounts. Grasslands Wana was bred for use in sheep pastures (Rumball, 1982b) due to its low crown and dense tiller formation. Grasslands Kara was bred for use in dairy pastures due to its more erect habit and sparser tillering morphology (Rumball, 1982a). Although bred for New Zealand conditions, these cultivars also have potential for use overseas, and are in demand in Australia and

South America (Lolicato and Rumball, 1994). While seed exports of New Zealand cultivars continue to increase, there is also the opportunity for multiplication and re-export of seeds of overseas cultivars in New Zealand, particularly for the Japanese market, which uses cocksfoot as a component of dairy pastures, but does not produce seed (Rowarth *et al.*, 1991).

17.3 ENVIRONMENTAL REQUIREMENTS FOR SEED PRODUCTION

Cocksfoot generally requires a minimum annual rainfall of 600 mm and a growing season of at least 8 months. It is adaptable to a wide range of soils but does better on rich loams. It is better suited than ryegrass to medium moisture and shade and moderate fertility conditions, but will not thrive where drainage is poor.

Following winter, the crop requires fine weather at pollination followed by rain during seed fill and a dry sunny period for ripening (Hare, 1990). Pollen dispersal by wind is restricted to several brief periods during the day and is positively affected by both temperature and light intensity (Langer, 1972). Cocksfoot exerts anthers slowly and this is influenced by weather conditions on the previous day, e.g. warm sunny weather during one day initiates the exertion of the anthers which are slowly released over the period of that day and night. By the following morning the anthers are exerted and release pollen over a period of about 20 min (Langer, 1972). Temperatures below 12.8°C at night followed by a cloudy day inhibit pollen release, but high day temperatures due to lack of cloud give normal pollen release (Hill, 1980). Subsequent seed development is greatly enhanced by warm temperatures, high light intensity and adequate water.

There is no precise evidence to indicate which factors are important in affecting seed set. In general, environmental factors which delay or prevent the opening of florets reduce the chances of pollination, but low temperatures and wet or dull conditions (Jones and Brown, 1951) have also been observed to have this effect. In New Zealand, even under favourable circumstances, the proportion of florets which set seed seldom exceeds 70% and may often be much less. For example, Johnston (1960) showed that as many as 10% of the florets in cultivar Grasslands Apanui cocksfoot may be morphologically sterile and thus incapable of developing a seed.

Between-cultivar differences also occur. Although such differences are reflected in seed yield, the yield components, particularly fertile tiller numbers, spikelets per head and to a lesser extent florets per spikelet and seed weight are likely to vary considerably (Table 17.1). Of these, seedhead numbers and panicle structure are likely to be of most significance. Certainly in cultivar Grasslands Wana higher seed yield has been shown to occur mainly as a result of greater numbers of fertile tillers. This dominant yield component has been previously highlighted in cocksfoot (Langer, 1980).

Cocksfoot is known to be particularly susceptible to late-season frosts. Frost destroys pollen viability and as a result no seed is formed when flowering plants undergo frost conditions of −3 to −6°C for 3.5–5 h (Ede, 1968). Lighter frosts also

Table 17.1. Cultivar differences in seed yield, purity and yield components in cultivars Grasslands Wana and Grasslands Kara. (After Zahid, 1996.)

Cultivar	Seed yield (kg ha^{-1})	% Pure seed	Fertile tillers (no. m^{-2})	Spikelets (no. head^{-1})	Florets (no. spikelet^{-1})	1000 seed wt (g)
Wana	1135[a]	69.6[a]	682.0[a]	410.6[a]	4.5[b]	0.95[b]
Kara	826[b]	72.5[a]	555.5[b]	358.4[b]	4.7[a]	1.21[a]

Cultivar data with common letters do not differ significantly at $P < 0.05$.

influence pollen viability, delay seed development and reduce panicle numbers. In extreme conditions panicle production in cocksfoot may fail completely. Cocksfoot seed crops in Canterbury are considered to be more sensitive to frost injury at the late stage of development (e.g. booting) than in the early development stages.

Factors influencing often large differences between potential and actual seed yield (e.g. 15–20% floret site utilization) include fertilization, seed set and maturation of seed (Brown, 1981). This compares with perennial ryegrass where the maximum percentage of florets setting seed is about 65% (Anslow, 1963).

Despite these variables, the climate in the major cocksfoot seed production area in New Zealand (Ashburton county) is dominated by mean monthly maximum air temperatures during harvest of 31–32°C and a relative humidity which may often fall to 20–30% on days when the warm, dry, northwesterly winds prevail. These conditions give the area a reputation for generally reliable weather at harvest, and the opportunity to recover consistently high seed yields.

17.4 FACTORS AFFECTING SEED PRODUCTION

17.4.1 Sowing and Establishment

Soil type

Cocksfoot in New Zealand is grown successfully on a wide range of soil types, but the most satisfactory seed yields are usually obtained from river silts, free-working loams or peaty soils containing an ad-mixture of clay. Stiff clayey soils which are inclined to crack during dry weather, or soils with an ironstone pan, do not suit cocksfoot. In other words, the crop demands good drainage and facilities for deep rooting. Good wheat land will usually produce good yields of cocksfoot if the soil has an open texture (Nixon, 1962). It is doubtful if an attempt should be made to establish cocksfoot for seed on lighter or drier soils types, as seasonal conditions cause too great a fluctuation in yield.

Place in rotation

Cocksfoot for seed has no definite place in the cropping rotation, particularly where the land is reasonably fertile. However, cocksfoot should be part of a rotation which allows for good weed control. Contamination with hair grass (*Vulpia* spp.), goose grass (*Bromus mollis* L.), yellow gromwell (*Amsinckia calycina* (Moris) Chater), wild oats (*Avena fatua* L.) or twitch (*Agropyron repens* (L.) Beauv.) should be avoided, and crops should either not be sown into areas with known contamination (Brown and Lill, 1990), or ground preparation should aim at reducing these to a minimum before sowing. Because of the difficulty of separating other grass seeds from cocksfoot during machine cleaning, cocksfoot crops should not follow another grass seed crop. (e.g. *Lolium* spp.). If fertility has been reduced as a result of preceding fertility depleting crops, it may be advisable to improve the physical condition of the soil and its fertility with greenfeed crops assisted by suitable fallowing.

Sowing rates

Opinions vary on the best quantity of cocksfoot seed to sow to establish a suitable stand for seed production. It is generally accepted that the best-yielding stands, and also those which produce the best-quality seed, are those in which the plants are fairly wide apart. Given sufficient room, individual plants develop robust growth with bold heads, whereas in a thickly sown stand the plants do not seem to have the same vitality. The difficulty, of course, is to estimate the necessary sowing rate to give the optimum thickness. This will vary with the soil type and the condition of the field. The amount of annual weed growth likely to compete with the young cocksfoot is also a factor. Hare (1990) has suggested an optimum density of about 200 plants m^{-2} for cocksfoot. In New Zealand it is common to sow cocksfoot in 45-cm rows at 3–5 kg ha^{-1} (Brown and Lill, 1990), although 2 kg ha^{-1} can be used if a good seedbed has been prepared. Cocksfoot is not sown with a companion crop.

Time of sowing

Cocksfoot is usually spring sown (end of September/early October) and harvested 15 months later (Brown and Lill, 1990), although it is possible with some cultivars to produce an acceptable seed yield from an autumn sowing (mid March) (Wilson *et al.*, 1994). The success of autumn sowing (and harvesting 9 months later) depends on the time of sowing and the tillering capacity of the cultivar. The tillers produced must have time to reach a size where they can respond to inductive conditions (i.e. complete the juvenile stage). Wilson *et al.* (1994) found that only one of five cultivars trialled produced an acceptable seed yield (680 kg ha^{-1}) following autumn sowing, and that the same cultivar when spring sown at the same site yielded 916 kg ha^{-1} because more reproductive tillers were produced. Spring-sown crops produce more vegetative tillers than autumn-sown crops by mid July, and as few tillers formed after mid July become reproductive (Wilson, 1959), it is not surprising spring-sown crops outyield autumn-sown crops.

17.4.2 Weed Control

Weeds reduce yield and profits and the quality of cocksfoot seed lots for subsequent sowing. Yield losses result from weed competition in the growing crop and from separation losses at harvest and during cleaning. Economic losses result from the cost of weed control, increased seed cleaning costs and loss in the value of seed lots which fail to meet weed contamination standards set by the New Zealand certification scheme (Rolston *et al.*, 1985). Major weed seed impurities found in machine cleaned cocksfoot lots include ryegrass (*Lolium* spp.), soft brome (*Bromus mollis* L.), Yorkshire fog (*Holcus lanatus* L.), *Poa annua* L., docks (*Rumex* spp.), Vulpia hairgrass (*Vulpia* spp.) and field pansy (*Viola* spp.) (Rowarth *et al.*, 1991).

Weed control is largely achieved with herbicides. Despite this, the full range of physical, cultural and perhaps biological practices can all have a place in obtaining the high level of physical purity required to achieve certified seed standards (Rolston *et al.*, 1985).

The availability of suitable herbicides for cocksfoot seed crops is an increasing problem for seed producers. Generally, unless a herbicide is already registered for use on pasture or large volume crops, it is unlikely to become available for recommended use on seed crops (Rolston, 1994). Some herbicides, however, have been shown to provide the crop tolerance and specific weed control needed in cocksfoot seed crops (Rolston, 1991).

Ethofumesate (Nortron) is active on wild oat and a number of other grass species and can be used in cocksfoot crops that are not undersown with legumes (Rolston *et al.*, 1985). *Poa annua* can be a serious weed after autumn sowings. Ethofumesate at 1.5 kg ha^{-1} has been shown to control *P. annua* and increase seed yields of autumn-sown Wana cocksfoot seven-fold (Brown *et al.*, 1983).

Most New Zealand cocksfoot cultivars, once established for 12 months, will tolerate a range of soil active herbicides. These include atrazine, simazine and diuron at 1.0–2.5 kg a.i. ha^{-1}. Ryegrass as a weed in established cocksfoot crops can be controlled with a dichofop-methyl (Hoegrass) plus atrazine mixture (1.0 + 1.0 kg a.i. ha^{-1}). Similarly terbicil (Sinbar) at 0.5 kg a.i. ha^{-1} plus diuron gives excellent weed control in cocksfoot seed crops (Rolston, 1994). Non-herbicide control of Californian thistle (*Cirsium arvense* L.) has been achieved by the use of cocksfoot as a 'smother crop' against this weed (Musgrave, 1949).

17.4.3 Crop Management

The optimum management of cocksfoot stands for seed production is a subject on which very little reliable information is available. The effect of grazing, burning or fertilizer treatments on seed yields and quality is largely a subject of conjecture.

It had been the practice in the past with many New Zealand seed-production stands – perhaps most of them – to be harvested for seed with no further attention until the following harvest. In some cases stands treated in this way continued to produce profitable yields of seed for 10–20 years or more. Over the years such a stand gradually builds up a mass of decaying leafage, which prevents the crown of the cocksfoot, a shade-loving plant, from becoming unduly exposed to direct sunlight. If the dead material is not disturbed, a cocksfoot stand will remain relatively pure for quite a few years (McPherson, 1948).

While this mass of decaying leaf assists in maintaining a reasonably pure stand, it is considered to also have a detrimental effect on the yield and the quality of the seed. It is well known that if a cocksfoot stand is subjected to indiscriminate grazing, particularly by sheep, plants gradually die out and are replaced by foreign growth. New Zealand and overseas experience strongly suggests that grazing of cocksfoot crops must be carefully managed if seed yields are to be maintained. Though severe grazing of an area is recognized as detrimental to seed production, light and controlled grazing with cattle may have a beneficial effect on both seed yield and quality. It is the opinion of many cocksfoot seed producers that some grazing of stands in autumn is desirable. The stock trample a certain amount of the decaying material into the ground, allowing some sunlight into ground level, and a much more vigorous type of plant results. However first-year crops are not grazed (Brown and Lill, 1990).

The burning of stands to remove a portion of the decaying material has been attempted with indifferent results. In some cases good yields of seed have followed burning, but in other cases the reverse has happened. There is evidence that burning to remove straw and stubble after harvest is useful in controlling residual litter and also pests such as army worm (*Persectania aversa* Walker) and cocksfoot midge (*Stenodiplosis geniculata* Reuter var. *dactylidis* (Westend)) infestation. However burning, particularly in the autumn, frequently results in an opening up of the sward, which can give rise to increased weed density and a reduction in cocksfoot plant population. Grazing is considered to be a better way of reducing excess growth and litter (Nixon, 1962).

New Zealand results on the use of autumn-nitrogen (N) on cocksfoot are conflicting, some workers reporting a response to autumn-N, while others finding no response (Lambert, 1956). It is usual, however, to apply 20 kg N ha^{-1} in the autumn of the establishment year.

A cocksfoot crop loses relatively high amounts of nutrients (particularly N, P and K) in the straw and to a lesser extent in the seed (Table 17.2), and 80–100 kg N ha^{-1} in early spring as a split application is usual (Brown and Lill, 1990). Phosphorus and potassium are only applied if soil tests show deficiencies.

Cycocel (chlormequat; CCC) shortens internode length and reduces lodging in cocksfoot (Hampton, 1988), and New Zealand growers apply this growth regulator at stem elongation and 10 days later (two × 1 litre ha^{-1} applications) to prevent lodging and improve ease of harvesting (Brown and Lill, 1990).

17.4.4 Diseases and Pests of Cocksfoot

Plant diseases generally affect seed crops more than grazed pastures, as grazing removes much of the diseased plant. In a forage seed crop, diseased material is generally not removed and therefore the disease is allowed to incubate (Latch, 1980).

In New Zealand, 36 different diseases have been isolated from cocksfoot, most due to fungal pathogens (Pennycook, 1989). Of these, stem rust (*Puccinia graminis*

Table 17.2. Nutrient loss (kg ha^{-1}) in cocksfoot seed and straw (Davies, 1939).

	Nutrient loss			
	N	P	K	Ca
Seed (500 kg ha^{-1})	13	2	5	1
Straw (5500 kg ha^{-1})	30	6	81	17
Total	43	8	86	18

Pers.), stripe rust (*Puccinia striiformis* var. *dactylides* Westend), leaf scald (*Rynchosporium orthosporium* Caldwell) and eyespot (*Mastigosporium rubricosum* (Dearn & Barth) Nannf.) are considered to be most important. Bacterial diseases are occasionally important, particularly bacterial wilt (*Pseudomonas syringae* (van Hall)) and Rathay's or yellow slime disease (*Clavibactor rathayi* (E.F. Smith)) (Johnston, 1956).

Rolston *et al.* (1989) have used propiconazole (125 g a.i. ha^{-1}) and mancozeb (1.5 kg a.i. ha^{-1}) in 300 l water ha^{-1} to successfully control eyespot in cocksfoot. Both fungicides had a marked effect in maintaining green leaf area (GLA) by delaying senescence. Seed yield increases of 15–21% were obtained from early fungicide application where pathogen infection was low. Yield increases were attributed to maintaining GLA during seed development and maturation, resulting in an increase in seeds per spikelet. Similarly, propiconazole (125 g a.i. ha^{-1}) applied at anthesis and during seed development has been shown to significantly increase seed yields, by 72% and 98% in cultivars Grasslands Wana and Grasslands Tekapo respectively (Wilson *et al.*, 1994). Most growers apply a fungicide just before flowering, and may follow this up with a further application 3–4 weeks later to ensure control of rust.

Cocksfoot stands are ideal for harbouring insects, and many different types are found associated with the plants. However, in New Zealand, only four are considered to be important pests – cocksfoot stem borer (*Glyphipteryx achyloessa* (Meyrick)) cocksfoot midge, cocksfoot thrip (*Chirothrips pallidicornis* (Haliday)) and army worm.

17.4.5 Harvesting, Threshing and Seed Cleaning

The two major problems with harvesting cocksfoot seed are that the seed tends to shatter easily once mature and also that there is uneven ripening within the inflorescences. The most common traditional method of harvesting cocksfoot in New Zealand has been to windrow the crop, allow it to mature on the cut straw, and then to thresh it with a combine harvester fitted with crop lifters. Cocksfoot is ready for seed harvest when the stems turn yellow for 15 cm below the head, or shatter when rubbed in the palm of the hand, even though culms and leaves may still be green. Seed shattering is common in cocksfoot and severe yield losses may occur if the crop is not harvested at the appropriate time.

Cocksfoot seed requires more care, more elaborate equipment and more experienced operators for cleaning than most other grass seeds. It is also extremely difficult to estimate the percentage of clean seed in thresher-run cocksfoot seed, so that buyers and sellers are likely to miscalculate the percentage and estimate either much more or less clean seed than is actually present in a threshed but uncleaned seed lot.

17.5 SEED QUALITY

In New Zealand the germination percentage of cocksfoot is often poorer than that of ryegrass, with 15–20% of seed lots having a germination of less than 80%, often as a result of high levels of immature seed (K.A. Hill, Palmerston North, 1996, personal communication). Seed weight, and particularly a reduction in the number of multiple seed units in cleaned cocksfoot seed, are likely to be important factors in improving seed germination and seed establishment potential (Scott and Hampton, 1985). Perhaps as an appreciation of the value of improved quality of cocksfoot seed in New Zealand the establishment of the New Zealand Cocksfoot Growers Association in 1988 is of particular interest. This Association was formed for two reasons. First to exercise more control by growers over the marketing and promotion of cocksfoot seed through its agents, Cropmark New Zealand Society Ltd, and second, to ensure, for the first time, that growers were recompensed for quality. Based on a 90% purity, 85% germination quality standard, growers are paid a premium price for their seed. Below-grade seed attracts a lower price by approximately 10 cents kg^{-1} (Gunn, 1986).

17.6 CONCLUSION

Cocksfoot seed production in New Zealand has a long history. Many growers consistently produce high yields of high quality cocksfoot seed. This, along with the development of new and improved cultivars, will ensure the demand for seed continues in the future. Although average yields of cleaned seed of approximately 500 kg ha^{-1} occur nationally, specialist growers consistently produce high quality seed crops of well over 1000 kg ha^{-1}. There is no doubt that cocksfoot seed production will continue to be recognized for its quality, and that seed growers will continue to grow the crop profitably by the use of controlled management and the development of innovative methods which have been a continuing feature of cocksfoot seed production in New Zealand.

REFERENCES

Anslow, R.C. (1963) Seed formation in perennial ryegrass. 1. Anther exsertion and seed set. *Journal of the British Grassland Society* 18, 90–96.

Brown, K.R. (1981) Inefficient conversion of floret populations to actual seed harvested in grass-seed crops. *Proceedings of the XIV International Grassland Congress*, pp. 2666–2668.

Brown, K.R. and Lill, C. (1990) Cocksfoot. In: Rowarth, J.S. (ed.) *Management of Grass Seed Crops*. Grasslands Research and Practice Series No. 5. New Zealand Grassland Association, Palmerston North, pp. 56–57.

Brown, K.R., Rolston, M.P. and Archie, W.J. (1983) 'Grasslands Wana' cocksfoot seed production. *Proceedings of the New Zealand Grassland Association* 44, 24–29.

Coulson, J. (1979) *Golden Harvest: Grass Seeding Days on Bank's Peninsula*. Dunmore Press, Palmerston North, New Zealand.

Davies, R.P. (1939) The chemical composition of the seed and straw of various grasses. *Welsh Journal of Agriculture* 15, 250–260.

Ede, R. (1968) *Grass and Clover Crops for Seed*. Bulletin 204. Ministry of Agriculture, Fisheries & Food, London.

Gunn, J. (1986) *Unified Approach Paying off for Cocksfoot Growers*. Farm Review, The Timaru Herald, Timaru, New Zealand, 7 October 1986.

Hampton, J.G. (1988) Herbage seed production. *Advances in Research and Technology of Seeds* 11, 1–28.

Hare, M.D. (1990) Establishment, spacing density & grazing effect. In: Rowarth, J.S. (ed.) *Management of Grass Seed Crops*. Grassland Research & Practice Series No. 5. New Zealand Grassland Association, Palmerston North, pp. 9–20.

Hill, M.J. (1980) Temperate pasture grass-seed crops: formative factors. In: Hebblethwaite, P.D. (ed.) *Seed Production*. Butterworths, London, pp. 137–149.

Johnston, M.E.H. (1956) Bacteriosis, a disease of cocksfoot. *New Zealand Journal of Agriculture* 93, 443.

Johnston, M.E.H. (1960) Investigation into seed setting in cocksfoot seed crops in New Zealand. *New Zealand Journal of Agricultural Research* 3, 345–357.

Jones, M.D. and Brown, J.G. (1951) Pollination cycles of some grasses in Oklahoma. *Agronomy Journal* 43, 218–222.

Lambert, J.P. (1956) Seed production studies. *New Zealand Journal of Science and Technology* 37, 467–477.

Langer, R.H.M. (1972) *How Grasses Grow*. Studies in Biology 34. William Clowes and Sons, London.

Langer, R.H.M. (1980) *Pasture plants*. In: Langer, R.H.M. (ed.) *Pastures. Their Ecology and Management*. Oxford University Press, Auckland. 39–74.

Latch, G.C.M. (1980) Importance of diseases in herbage seed production. In: Lancashire, J.A. (ed.) *Herbage Seed Production*. Grassland Research and Practice Series. No. 1. New Zealand Grassland Association, Palmerston North, pp. 36–40.

Lolicato, S. and Rumball, W. (1994) Past and present improvement of cocksfoot (*Dactylis glomerata* L.) in Australia and New Zealand. *New Zealand Journal of Agricultural Research* 37, 379–390.

Lucas, D. (1984) *Binders Best for Cocksfoot on Modern Methven Cropping Farm*. Farm and Station Section. The Christchurch Press, Christchurch, New Zealand.

MacKay, T. (1987) *A Manual of the Grass and Forage Plants Useful to New Zealand*. Crown Land Development, Wellington.

McPherson, G.M. (1948) Seed production in New Zealand: cocksfoot. *New Zealand Journal of Agriculture* 78, 33–41.

MAF (1993) *Seed Certification Statistics (1992/93)*. Seed Certification Bureau, Ministry of Agriculture & Fisheries, Christchurch, New Zealand.

Musgrave, M.M. (1949) Control of Californian thistle with cocksfoot. *New Zealand Journal of Agriculture* 79, 23.

Nixon, G.W. (1962) Cocksfoot seed production in New Zealand. *New Zealand Journal of Agriculture* 92, 293–298.

Pennycook, S.R. (1989) *Plant Diseases Recorded in New Zealand*, vol. 1. Plant Disease Division, DSIR, Auckland, New Zealand.

Rolston, M.P. (1991) Cocksfoot seed crop tolerance to herbicides applied to seedling and established stands. *Journal of Applied Seed Production* 9, 63–68.

Rolston, M.P. (1994) Weed control for herbage seed crops. Seed Industry Training Course, SeedComm, Ashburton, New Zealand.

Rolston, M.P., Brown, K.R., Hare, M.D. and Young, K.A. (1985) Grass seed production: weeds, herbicides and fertilisers. In: Hare, M.D. and Brock, J.L. (eds) *Producing Herbage Seeds*. Grassland Research and Practice Series No. 2, New Zealand Grassland Association, Palmerston North, pp. 15–22.

Rolston, M.P., Hampton, J.G., Hare, M.D. and Falloon, R.E. (1989) Fungicide effects on seed yield of temperate forage grasses. *Proceedings of the XVI International Grassland Congress*, pp. 669–670.

Rowarth, J.S., Rolston, M.P. and Johnson, A.A. (1991) Weed seed occurrence in cocksfoot seedlots. *Proceedings of the 44th New Zealand Weed and Pest Control Conference* 296–299.

Rumball, W. (1982a) 'Grasslands Kara' Cocksfoot (*Dactylis glomerata* L.). *New Zealand Journal of Experimental Agriculture* 10, 49–50.

Rumball, W. (1982b) 'Grasslands Wana' Cocksfoot (*Dactylis glomerata* L.). *New Zealand Journal of Experimental Agriculture* 10, 51–52.

Scott, D.J. and Hampton, J.G. (1985) Aspects of seed quality. In: Hare, M.D. and Brock, J.L. (eds) *Producing Herbage Seeds*. Grassland Research and Practice Series No. 2, New Zealand Grassland Association, Palmerston North, pp. 43–52.

Wilson, J.R. (1959) The influence of time of tiller origin and nitrogen level on the floral initiation and ear emergence of four pasture grasses. *New Zealand Journal of Agricultural Research* 2, 915–932.

Wilson, S.M., Hampton, J.G., Hill, M.J. and Rolston, M.P. (1994) Effect of cultivar, time of sowing and fungicide application on seed yield in cocksfoot (*Dactylis glomerata* L.). *Proceedings of the Agronomy Society of New Zealand* 24, 103–108.

Zahid, M.I. (1996) Cocksfoot (*Dactylis glomerata* L.) seed production. MAgrSc Thesis, Department of Plant Science, Massey University, New Zealand.

Lotus corniculatus L. (Birdsfoot Trefoil) in North America

18

P.R. Beuselinck
USDA-ARS, Plant Genetics Research Unit, Columbia, Missouri 65211, USA

18.1 INTRODUCTION

Floras of the eastern USA first listed broadleafed birdsfoot trefoil (*Lotus corniculatus* L.) in 1900. In the northwest USA birdsfoot trefoil was first mentioned as occurring in ballast near Portland, Oregon in 1917. McKee and Schoth (1949) indicated that specimens in herbaria showed that plants were collected in New Jersey as early as 1876, in North Carolina and New York (NY) in 1885, and in Alabama in 1888. Beginning in 1905, the United States Department of Agriculture (USDA) initiated experimental trials with birdsfoot trefoil, often in cooperation with state Agricultural Experiment Stations (AES). Thus, between 1905 and 1908 the Washington AES at Pullman, Washington, found that birdsfoot trefoil provided a dense vigorous growth, but concluded it was inferior to lucerne (*Medicago sativa* L.); in 1909 birdsfoot trefoil grew well in Arlington, Virginia; seed provided to the Illinois AES proved birdsfoot trefoil to be reasonably well adapted to southern Illinois in the 1927–1929 trials; the North Carolina AES maintained plantings of birdsfoot trefoil at Statesville, North Carolina from 1931 to 1935. In 1934, Cornell University surveyed pastures in New York and recognized birdsfoot trefoil as a plant of forage value in meadows and pastures (Johnstone-Wallace, 1938). Birdsfoot trefoil as a forage crop was popular among farmers for 20–25 years because of its ability to grow and survive in soils of low fertility.

18.2 BIRTH OF AN INDUSTRY

Favourable reports from New York about the attributes of birdsfoot trefoil created excitement among producers and seedsmen to establish birdsfoot trefoil as the 'poor man's lucerne'. Reports of a persistent and high-producing legume that could be hayed or directly grazed on soils with poor drainage, low fertility or acid conditions were hard to resist. Birdsfoot trefoil was purported to grow on a wide range of soils and conditions including droughty, infertile, acidic or mildly alkaline soils, mine spoils, and under saline and waterlogged conditions. The seed from New York birdsfoot trefoil threshed from naturalized meadows cut for hay generated enough seed for the early purposes of testing and making small plantings. However, shattering losses during field curing were considerable, making it difficult to meet an increased demand for seed. Early yields in New York between 1937 and 1942 ranged from 20 to 50 kg ha^{-1}, demonstrating the difficulty in obtaining seed. Those yields were much lower than the seed yields of 280–390 kg ha^{-1} reported from Europe (MacDonald, 1946). The early work of MacDonald (1946) demonstrated that with improved fertility and handling to reduce seed loss, birdsfoot trefoil was capable of producing over 225 kg ha^{-1}.

Although birdsfoot trefoil has been grown for forage in other countries for centuries, the difficulty in obtaining seed has been the major obstacle to the increased use of this crop. MacDonald (1946) quoted an 1892 statement made by Vilmorin-Andrieux concerning the multiplication of birdsfoot trefoil in France, 'Lotus seed is scarce and so difficult to harvest that it is doubtful if its culture can ever be extended.' One hundred years later there still remains unanimous agreement that seed production is the most difficult challenge to widespread dissemination of this crop.

18.3 A GROWING INDUSTRY

The first record of New York birdsfoot trefoil seed production is from 1937 when three growers produced 1192 kg on 40.5 ha (MacDonald, 1946). From 1934 to 1947, sales of birdsfoot trefoil seed only identified the seed as being of New York origin. Eventually, the name 'Empire' was given to a strain of birdsfoot trefoil found growing in Preston Hollow, New York (MacDonald, 1957). Throughout the remainder of this chapter, Empire will be used to refer to the original unnamed naturalized population and the cultivar. During the early years of Empire production, seed was purchased directly from the producer. As New York seed supplies increased, the Champlain Valley producers organized a cooperative in 1942 to assist its members in the marketing of their seed. However, the annual New York seed production vacillated between 2300 and 7300 kg between 1940 and 1946 (MacDonald, 1946). Production in those years may have been hampered by resource shortages during the Second World War.

The end of the Second World War in 1945 saw demand for increased agricultural production. Meat was scarce in the USA, and better pastures meant improved grazing for livestock and greater production of meat and milk. The producers of the Champlain Valley Cooperative, aided by agricultural researchers at Cornell University, the USDA-Agricultural Research Service, and agents of the USDA-Soil Conservation Service (SCS), started to make dramatic increases in 1947 with approximately 16,000 kg of seed produced from 100 ha (Fig. 18.1); production in 1949 reached 50,000 kg and 185,000 kg was produced in 1950.

In 1948, Empire sold for \$4.50–\$6.00 kg^{-1} and by 1950 other states in the USA as well as Canada became increasingly interested in growing birdsfoot trefoil as a forage crop. Demand for seed exceeded supplies, resulting in high seed prices

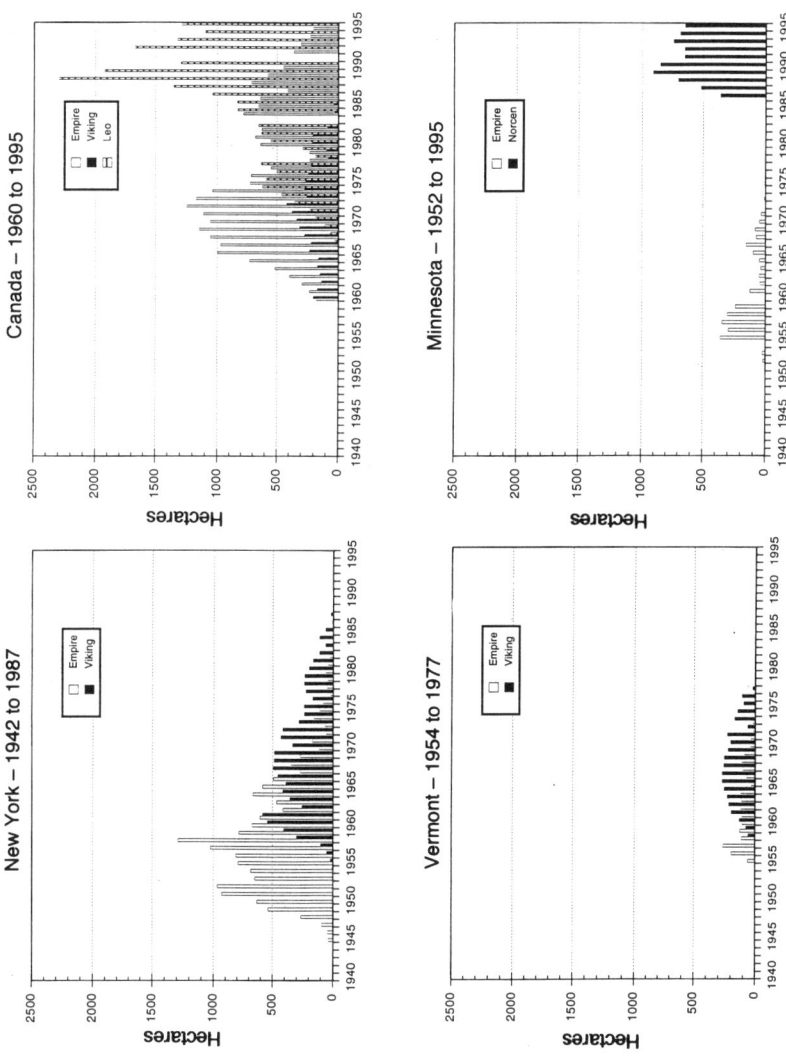

Fig. 18.1. Certified seed production of predominant cultivars of birdsfoot trefoil in major North American production areas.

which stimulated the importation of seed from Europe. Italian birdsfoot trefoil sold for as low as $2.50 kg^{-1}. Researchers later demonstrated that Italian birdsfoot trefoil was less winterhardy than Empire and discouraged its planting, but the recommendations did little to deter sales of imported seed. The high demand for Empire and limited seed supplies resulted in the importation of over 445,000 kg of birdsfoot trefoil seed into the USA in 1950. By 1954, Empire was still selling at $5.00 kg^{-1} and Italian for $1.50 kg^{-1}. Imported birdsfoot trefoil was later simply termed European birdsfoot trefoil as seed from Europe did not retain a country-of-origin identity.

As demand for birdsfoot trefoil seed for pasture plantings continued to grow, other areas tried to develop seed production. Small areas were harvested in Oregon and California (McKee, 1949), but major production remained centred in states of the northcentral and eastern USA including Iowa, Michigan, Minnesota, Missouri, Wisconsin, South Dakota and Vermont, and in Canada. Production records are incomplete, as early production was typically used to supply local needs for seed before the development of certified programmes in the late 1940s and early 1950s. There was demand for Empire and producers in many states were encouraged by SCS and AES agents to produce seed. Income levels of about $40–60 ha^{-1} were anticipated by producers as yields of 110 kg ha^{-1} were predicted for seed bringing $2.75–3.30 kg^{-1}. Occasionally, yields as high as 560 kg ha^{-1} were reported, although producers were cautioned to expect difficulty in obtaining a 110 kg ha^{-1} yield. Still, the prospect of producing 560 kg ha^{-1} was an attractive lure. Production in Iowa and Missouri in the 1950s was small and supplied some local demand; major commercial production continued to be centred in New York.

18.4 A MATURE INDUSTRY

Production data were obtained from seed certifying agencies with a history of certified production of birdsfoot trefoil (Figs 18.1, 18.2). New York certified production continued to increase through the 1950s (Fig. 18.1). Between 1948 and 1975, New York had at least 400 ha in production annually. Production peaked in 1958 at about 1300 ha and remained above 400 ha until 1975. The introduction of the cultivar Viking in 1957 began a shift in production away from Empire. Viking, being more upright, yielded 15–20% more herbage than Empire in New York. The production of Viking peaked at about 570 ha in 1961 and surpassed the New York production of Empire in 1967. The number of New York hectares devoted to certified seed production continued to decline after 1961. By 1987, production of certified birdsfoot trefoil seed had ceased in New York. Seed production was initiated in Vermont and Minnesota between 1953 and 1954, but their contributions to the total seed supply were minor compared to New York (Fig. 18.1).

Relative to USA production, Canadian seed production did not start until 1960, but increased rapidly to fill the void created by declining supplies from New York (Fig. 18.2). The overall area of production of Empire reached its highest levels between 1966 and 1970 because of increased production from Canada (Fig. 18.2).

Canadian production of certified seed of cultivar Viking ended in 1985, at about the same time that it ended in the USA. But in 1995, Canada continued to maintain about 200 ha of Empire seed production. Besides producing the cultivars Empire and Viking, Canada initiated the production of the cultivar Leo in 1967 (Fig. 18.1). Certified seed production of Leo peaked at over 2200 ha in 1988 and continues to account for the majority of Canadian birdsfoot trefoil seed production.

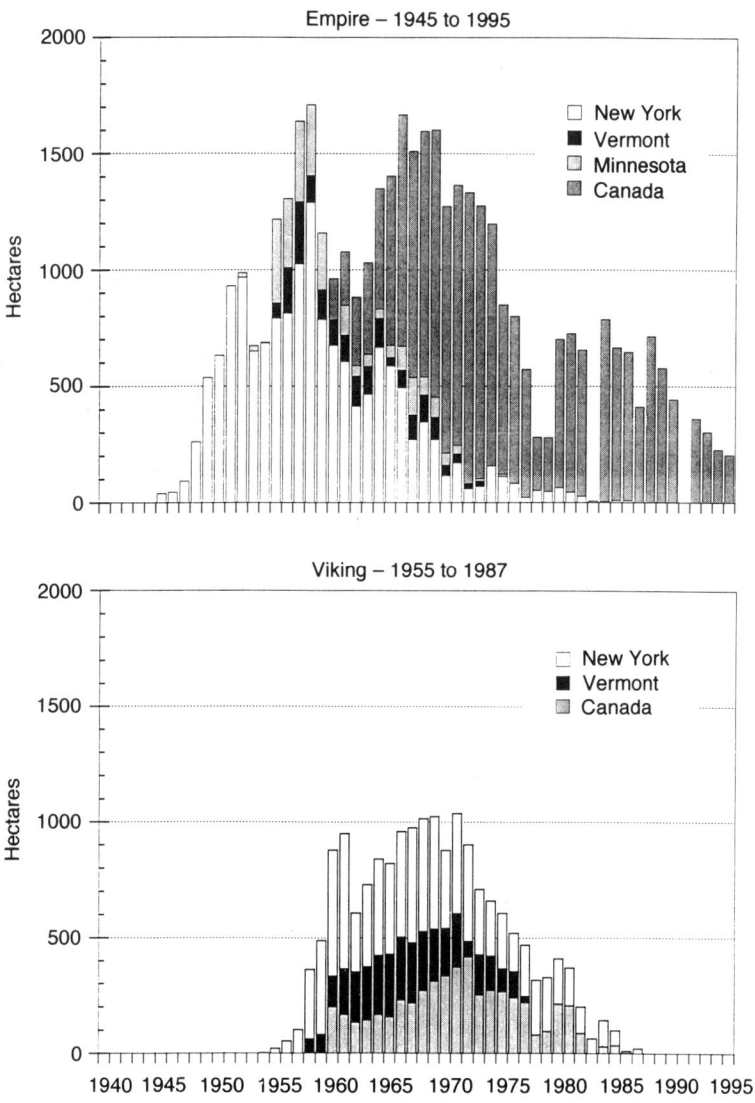

Fig. 18.2. Certified seed production of cultivars Empire and Viking in North America.

18.5 CONSTRAINTS TO SEED PRODUCTION

18.5.1 Physiological Considerations

Birdsfoot trefoil is a qualitative, long-day plant requiring at least 16 h of daylight for full flowering; a shorter daylength restricts flower production. Its indeterminate flowering habit results in seed set over an extended period of time. These crop characteristics require special consideration for commercial seed production. In North America, long days at latitudes above 40°N favour a contracted period of intense blooming that is advantageous for seed production (Beuselinck and McGraw, 1988) and this illustrates why birdsfoot trefoil seed production started in New York (42.5°N lat.) and could be conducted in Oregon (45.5°N lat.). In North America, there is a corridor of agricultural land between 40 and 50°N lat. that is well suited to birdsfoot trefoil seed production. This corridor contains the production areas of Minnesota (49°N lat.), Wisconsin (46.5°N lat.), Michigan (47°N lat.) and Canada (50°N lat.). Above 50°N lat., birdsfoot trefoil blooms well, but the growing season is too short to consistently set and harvest a crop of seed. Below 40°N lat., culture of birdsfoot trefoil is not conducive to good bloom, and yield and quality of seed is inferior to more northern areas.

Although commercial seed yields can approach 600 kg ha^{-1}, average yields range from about 50 to 175 kg ha^{-1}, and typically are 100 kg ha^{-1} or less (McGraw *et al.*, 1986). Indeterminate flowering, the limited distribution of photosynthate to reproductive growth, flower and pod abortion, and pod dehiscence lead to low seed yield (McGraw and Beuselinck, 1983). Pods within a field mature unevenly, and any delay in cutting after the pods are mature results in significant seed loss from pod dehiscence and seed shattering. Pod dehiscence and subsequent seed loss due to shattering can be high when relative humidity drops below 40%, and it may be more than coincidental that commercial seed production areas are generally located near large bodies of water.

18.5.2 Pests and Diseases

Several pests and diseases of birdsfoot trefoil seed have contributed to the decline of production in the eastern USA and increased production in the Upper Midwest and Canada. Early reports claimed that birdsfoot trefoil was a perennial with a longevity comparable to lucerne (MacDonald, 1946). As birdsfoot trefoil spread beyond New York, especially south, reports of disease problems affecting stand longevity were encountered (Krenzin, 1953; Miller *et al.*, 1964). The production of seed was important for the commercial use of this crop to new areas, yet even within areas of seed production, the incidence of premature stand decline became more common.

It is likely that changes in seed production practices inadvertently promoted the development and spread of disease organisms harmful to stand longevity and productivity. In the early 1940s seed was harvested from birdsfoot trefoil grown with companion grasses such as poverty grass (*Danthonia spicata* L.), Canada bluegrass (*Poa compressa* L.) and the escaped introductions of Kentucky bluegrass (*Poa pratensis* L.), meadow fescue (*Festuca pratensis* Huds.) and timothy (*Phleum pratense* L.) (MacDonald, 1946). Seed yields were low, usually averaging less than 50 kg ha^{-1}, even though reports of grass–birdsfoot trefoil fields of 20+ years old were common. Later, birdsfoot trefoil culture for seed production was modified to a monoculture. Complete farming areas in the Champlain Valley of New York and Vermont were grown as near monocultures, and these fields remained in production for many years without the benefits of rotation.

Crown and root rots have been the most important diseases of birdsfoot trefoil (Seaney and Henson, 1970). Organisms implicated in this disease complex are species of *Fusarium*, *Verticillium*, *Macrophomina*, *Mycoleptodiscus*, *Rhizoctonia* and *Sclerotinia* (Ostazeski, 1967). An especially virulent isolate of *F. oxysporum* Schlect. is considered to have destroyed the New York and Vermont production areas. Forms of *Fusarium* have been found in production fields in Minnesota, Wisconsin, Michigan and Iowa. Most of the fungal organisms listed can be spread as seed-borne contaminants (Beuselinck *et al.*, 1989). Birdsfoot trefoil is also susceptible to parasitism by root knot (*Meloidogyne hapla* Chitwood, *M. javanica* (Treub) Chitwood, *M. arenaria* (Neal) Chitwood and *M. incognita* (Kofoid & White) Chitwood) and root lesion nematodes (*Pratylenchus penetrans* Cobb). Synergism between *Fusarium* spp. and *R. solani* Kuehn with root lesion nematodes lowers birdsfoot trefoil productivity and increases plant mortality (Thompson and Willis, 1975). Although one organism may never be determined as *the* culprit, it is apparent that disease organisms have contributed to the demise of birdsfoot trefoil seed production in New York and Vermont, and contribute to low and variable yields in current production areas.

It does not appear that insects have eliminated the production of seed in any areas, but contribute to lower yield, which can make seed production uneconomical. A number of insects cause forage and seed losses (Seaney and Henson, 1970). The meadow spittlebug (*Philaenus spumarius* L.) causes plant stunting and abortion of flower buds. Three plant bugs (*Adelphocoris lineolaris* Goeze, *Lygus lineolaris* Palisot de Beavois and *Plagiognathus chrysanthemi* Wolff) and the potato leafhopper (*Empoasca fabae* Harris) reduce yield by destroying stem terminals and flowers (Wipfli *et al.*, 1990). The trefoil seed chalcid (*Bruchophagus platypterus* Walker) is a host-specific insect that parasitizes seeds (Kamm, 1992) and reduces yield by 40% or more.

Weed problems have not eliminated the production of seed in any areas, but like insects, contribute to lower yields, complicate harvesting and cleaning, and increase inputs required to pass certification. During the early years of seed production in the USA, when seed was harvested from birdsfoot trefoil–grass mixtures, weeds were controlled by mowing or haying. Many early birdsfoot trefoil seed producers fed livestock the early cut hay or residue after the seeds were threshed from the windrowed or baled crop (MacDonald, 1946). When birdsfoot trefoil was grown in monoculture, weeds became a more widespread problem. Seed certification standards dictated low weed presence as the industry matured, necessitating the use of herbicides for good weed control. As herbicides were applied, restrictions on feeding hay or residue from treated fields eliminated one option for producers, but several chemicals provided necessary broadleaf and grass control. The relatively small number of hectares of seed production are not sufficient to encourage chemical companies to commit their resources to the testing and documentation required for labelling their herbicide for birdsfoot trefoil. The development of cultivars resistant to herbicides may simplify weed control in birdsfoot trefoil, leading to enhanced production, but commercial cultivars with resistance to herbicides have not been developed.

18.6 THE FUTURE OF SEED PRODUCTION

Today, the birdsfoot trefoil seed industry is healthy. In 1995 there were about 1500 ha in production in Canada. Certified production in the USA is estimated at about 650–735 ha in northern Minnesota and a few certified hectares in northern Michigan (approximately 100) and Wisconsin (approximately 10). Yet, the presentation of certified seed hectares in Figs 18.1 and 18.2 may create the misleading impression that seed production of birdsfoot trefoil is terminating in the USA and future production will depend upon Canada. The USA continues to produce about the same level as Canada, so the discrepancy between reported and actual hectares in production requires explanation.

To date, certified seed production has been well documented in Canada. Further, in Canada, seed of named cultivars can only be produced under the certification programme; uncertified production is minor. In contrast, in the USA there is little production of certified seed and considerable uncertified production that is not reported. Producers in some states like Minnesota can instruct their state certification agency to withhold information on their level of production, adding to the misimpression that very little seed is produced in the USA.

Annual USA production of birdsfoot trefoil seed is estimated to be approximately 270,000 kg. Using an average yield of about 110 kg ha^{-1}, it is estimated that 2475 ha are in production, indicating a discrepancy of approximately 1640 ha of production between reported certified hectares and estimated total production. Another consideration is that although some 2400+ ha are harvested for seed production, a doubling of that figure to 4800 ha is considered a more plausible estimate. This is because many hectares considered for possible production of seed are not harvested because of: (i) poor seed production related to poor weather at bloom, during pod fill, or at harvest time; (ii) high weed seed content making it difficult to harvest and clean; (iii) the herbage was used for a hay or pasture crop at the expense of seed production.

The differences between the USA and Canada production of certified versus uncertified seed can partially be understood by examining the marketing of birdsfoot trefoil seed. All birdsfoot trefoil seed marketed in Canada by a cultivar name must be certified, whereas all uncertified seed is marketed as common seed. Seed destined for countries in the European Union must be certified and currently Canada has the only North American cultivar (Leo) on the registered list of cultivars. In contrast, USA sales of uncertified seed of named cultivars is a common practice, with the only exception to this practice for cultivars protected under Article V of the Plant Variety Protection Act. The practice of producing and selling uncertified cultivars has undermined the pricing of birdsfoot trefoil seed in the USA because uncertified seed is less expensive than certified seed. Thus, it is difficult for the producer to justify the added labour and cost of certified production, and for the seed broker to pay a premium to the producer. The result has been that uncertified production has dominated in the USA, and the market readily absorbs uncertified seed production from Canada and South America.

It is a sad note that the identities of individual cultivars, and their respective attributes over other cultivars have been mongrelized in the USA into a generic commodity by an indifferent industry. Unless conditions change, it will become increasingly difficult to justify public or private investments in the development of new cultivars. To increase profit margins, seed companies may have to develop or license proprietary cultivars to gain exclusive control of production and marketing.

The level of birdsfoot trefoil production is neither static nor predictable. It will become increasingly difficult to assess birdsfoot trefoil seed production, at least in the USA, as fewer records are maintained and the industry moves toward proprietary cultivars and increased secrecy. For the near future uncertified seed of public cultivars will continue to comprise the majority of production in the USA, and certified seed will comprise the majority of production in Canada. New cultivar releases are destined to be handled with more stringent production control as seed companies try to increase their profit margin after investing in proprietary cultivars. Breeding efforts to produce cultivars that are self-pollinating, have reduced shattering or non-shattering pods, or reduced photoperiod requirements could have positive effects on stabilizing seed production. However, until the vagaries of weather and pests can be controlled, the production of birdsfoot trefoil seed will remain somewhat unpredictable.

REFERENCES

Beuselinck, P.R. and McGraw, R.L. (1988) Indeterminate flowering and reproductive success in birdsfoot trefoil. *Crop Science* 28, 842–844.

Beuselinck, P.R., Kremer, R.J. and McGraw, R.L. (1989) A survey of microorganisms associated with birdsfoot trefoil seed. *Journal of Applied Seed Production* 7, 32–37.

Johnstone-Wallace, D.B. (1938) Pasture improvement and management. *Cornell University Extension Bulletin* 393, 1–42.

Kamm, J.A. (1992) Influence of celestial light on visual and olfactory behavior of seed chalcids (Hymenoptera : Eurytomidae). *Journal of Insect Behavior* 5, 273–287.

Krenzin, R.E. (1953) Trefoil – opinions of birdsfoot trefoil in several states. *Forage Notes* 8, 37–38.

MacDonald, H.A. (1946) *Birdsfoot Trefoil (Lotus corniculatus L.): Its Characteristics and Potentialities as a Forage Legume*. Cornell University Agricultural Experiment Station Memo 261.

MacDonald, H.A. (1957) Viking birdsfoot trefoil. *Forage Notes* 12, 143.

McGraw, R.L. and Beuselinck, P.R. (1983) Growth and yield characteristics of birdsfoot trefoil. *Agronomy Journal* 75, 443–446.

McGraw, R.L., Beuselinck, P.R. and Ingram, K.T. (1986) Plant population density effects on seed yield of birdsfoot trefoil. *Agronomy Journal* 78, 201–205.

McKee, R. and Schoth, H.A. (1949) *Birdsfoot Trefoil and Big Trefoil*. Circular 625. United States Government Printing Office, Washington, DC.

Miller, J.D., Kreitlow, K.W., Drake, C.R. and Henson, P.R. (1964) Stand longevity studies with birdsfoot trefoil. *Agronomy Journal* 56, 137–139.

Ostazeski, S.A. (1967) An undescribed fungus associated with a root and crown rot of birdsfoot trefoil (*Lotus corniculatus*). *Mycologia* 59, 970–975.

Seaney, R.R. and Henson, P.R. (1970) Birdsfoot trefoil. *Advances in Agronomy* 22, 119–157.

Thompson, L.S. and Willis, C.B. (1975) Influence of fensulfothion and fenamiphos on root lesion nematode numbers and yield of forage legumes. *Canadian Journal of Plant Science* 55, 727–735.

Wipfli, M.S., Wedberg, J.L. and Hogg, D.B. (1990) Damage potentials of three plant bug (Hemiptera : Heteroptera : Miridae) species to birdsfoot trefoil grown for seed in Wisconsin USA. *Journal of Economic Entomology* 83, 580–584.

Medicago sativa L. (Lucerne/Alfalfa) in Canada

19

D.T. Fairey and N.A. Fairey
Northern Agriculture Research Centre, Agriculture and Agri-Food Canada, Beaverlodge, Alberta T0H 0C0, Canada

19.1 LUCERNE AND LUCERNE SEED IN CANADA

Lucerne (*Medicago sativa* L.) is the most widely used forage legume in Canada. It is a predominant component in mixtures for hay, silage and pasture, and is grown extensively in monoculture with varied end-uses such as dehydrated pellets, pharmaceutical products or seed. This crop was first introduced into Canada in 1871 with an importation of sheep from France; the shepherd who was responsible for the delivery of the sheep to a farm in Ontario also delivered a kilogram of lucerne seed (Armstrong *et al.*, 1948). The progeny of this seed became known as Ontario Variegated, and this strain was widely grown in eastern Canada (Bolton, 1962). However, Ontario Variegated was not hardy enough for the prairie provinces of Manitoba, Saskatchewan and Alberta and, in 1908, Grimm and Baltic were imported from Minnesota, USA, into Suffield, Alberta (Wheeler, 1951). The fate of the Baltic seed is unknown but, by 1914, the Grimm selection provided the ancestry of a large part of the Grimm lucerne grown in western Canada. The cultivar Grimm was developed from Grimm 666 which was formulated in 1912 from the best 1300 single-plant selections based on both winter hardiness and yield. Cultivar Grimm continues to be licensed for use today; however, the date of licensing is not recorded, as seed of this cultivar was first distributed in 1923 before licensing was started (Elliott and Bolton, 1970).

Lucerne seed production in Canada is predominantly of winter-hardy cultivars. Lucerne stands are sown in the early spring, some time after mid May, and a first seed crop is harvested the following year. Establishment with companion cereal (e.g. wheat, *Triticum aestivum* L., barley, *Hordeum vulgare* L. and oats *Avena sativa* L.) crops, often to suppress weeds, is not uncommon; however, competition from the cereal reduces the yield of the first seed crop. A maximum of eight seed crops can be harvested from a certified stand and most stands are maintained for 3–5 years. Seed yields are usually highest for the first crop and substantial reductions often occur with successive harvests (Fairey and Lefkovitch, 1992).

Increased inputs (herbicides, insecticides, cultivation) are necessary to help offset these reductions in yield in older seed stands.

19.1.1 Cultivars

Of all the forage legumes in Canada, lucerne has consistently held the unique distinction of being the species that has the largest number of cultivars registered for use. Agriculture and Agri-Food Canada (AAFC), Food Production & Inspection Branch, publishes a list of registered cultivars of all field crops in Canada in the autumn of each year, with updates on new additions and deletions on a quarterly basis. In the recent lists (Agriculture and Agri-Food Canada, 1995, 1996), there are about 125 registered cultivars. Most of these are relatively non-hardy, bred in the private sector in the USA, and tested in the eastern provinces of Canada where they are recommended for use. In contrast to this, most of the winter hardy cultivars have been bred in the public sector within Canada and have been tested and used in the western provinces. The Research Branch of AAFC has been a prominent participant in both the breeding and performance evaluation of the winter hardy cultivars, and has had a significant impact on shaping the lucerne seed industry in Canada.

Until 1990, an annual publication from AAFC, Research Branch, provided a listing of the recommended forage cultivars for each province in Canada. The registration of a cultivar is based on the production of herbage in one or more provinces in Canada; once registered, a cultivar can be sold in any Canadian province and local performance data are not required. Recommendations, on the other hand, are based on local agronomic performance. In the final publication on recommended forage crop cultivars (Goplen, 1990), only 17 (Table 19.1) of the 82 listed lucerne cultivars were recommended for two or three of the prairie provinces of Alberta, Saskatchewan and Manitoba where most of Canada's lucerne seed is produced. A common trait of these 17 cultivars was winter hardiness with an acceptable yield and resistance to prevailing diseases. A majority (13 of the 17) were bred at AAFC Research Stations. In contrast to this, 49 cultivars were recommended for Ontario, 23 for Quebec and 10 for the Atlantic provinces. Many of the cultivars recommended in eastern Canada, particularly Ontario, were bred by the private sector (52 private sector USA; three private sector Canada; six public sector USA; four public sector Canada).

Starting in the late 1960s a network of testing sites was established across the country at most research stations, and promising plant material and cultivars from both the public and private sector were tested on a regular basis. For lucerne, testing at the western locations included seed performance, and these data, which were made available to farmers and the seed trade at no cost, generally laid the framework for seed production.

In addition to having performance data of the hardy lucerne cultivars, both the farmers and seed trade had ready access to the distribution rights. Breeders' seed of cultivars bred by AAFC was distributed between the western Canadian provinces in accordance with the 'Forage Seed Project', whereby each cultivar was allotted, in almost equal proportions, to farmers and the Canadian seed trade applicants who in turn multiplied and sold the seed. In 1976, SeCan, a

Table 19.1. Lucerne cultivars recommended for use in the Canadian prairie provinces. (From Goplen, 1990.)

Cultivar	Breeder	Canadian distributor
1. Admiral	Private USA[1]	Private, Canada[2]
2. Algonquin	Public, Canada[3]	Public, Canada[4]
3. Anchor	Private USA	Public, Canada
4. Angus	Public, Canada	Public, Canada
5. Anik	Public, Canada	Public, Canada
6. Apica	Public, Canada	Public, Canada
7. Barrier	Public, Canada	Public, Canada
8. Beaver	Public, Canada	Private, Canada
9. Drylander	Public, Canada	Public, Canada
10. Heinrichs	Public, Canada	Public, Canada
11. Peace	Public, Canada	Public, Canada
12. Rambler	Public, Canada	Private, Canada
13. Rangelander	Public, Canada	Public, Canada
14. Roamer	Public, Canada	Public, Canada
15. Spreador II	Private, USA	Private, Canada/USA[5]
16. Vernal	Public, USA[6]	Private, Canada
17. 120	Private, USA	Private, Canada/USA

[1] Private company in the USA.
[2] Private company in Canada.
[3] Agriculture Canada, Research Station, Research Branch.
[4] Canadian Forage Seed Project and/or SeCan that provided equal access of distribution rights to farmers and Canadian seed companies.
[5] The Canadian subsidiary of a USA Company.
[6] University, USA.

non-profit-making corporation, was formed to promote and distribute cultivars bred at publicly funded institutes. The membership of SeCan is open to both farmers and the seed trade who have equal access to the rights of cultivars given to SeCan. Until recently, all AAFC varieties were given to SeCan on a preferential basis; this policy has recently been changed to permit competitive open bids from any interested party.

Most of the lucerne seed grown in Canada was, and still is, of public, winter hardy cultivars. While a large proportion is used in the domestic market in the west, there is often a surplus that is exported and much more that is stored for extended periods, only to surface when a demand, such as the conservation, government-funded programmes in the USA, becomes great enough. Much of this seed is traded as 'common' (i.e. uncertified) with large fluctuations in price based on demand.

19.1.2 The Pollinator

Armstrong and White (1935) proved that the stigmatic surface effectively blocks pollen tube penetration in untripped lucerne flowers. While pollen grains could germinate on the surface of the membrane, the pollen tubes did not penetrate the stigma until the stigmatic membrane was ruptured, usually by tripping. In a lucerne flower, the anthers are slightly longer than the pistil. As soon as the flower is tripped, the surface of the stigma is held firmly against the standard petal. Therefore, self- or cross-pollination can only occur during the few seconds after the flower has been tripped before the stigma contacts the standard petal. When the flower is tripped by a bee, the body of the bee that is interposed between the staminal column and the standard petal acts as a rupturing surface for the stigmatic membrane, and also as a carrier of foreign pollen that is transferred to the ruptured stigmatic surface only seconds before the bee's body is showered with the pollen of the flower being visited. The bee therefore provides the tripping and cross-pollination required for good seed set.

By the mid-1950s, it became increasingly obvious that there was a decrease in the populations of native bees in western Canada (predominantly *Bombus*, *Megachile* and *Melitta*) that nested in the margins of seed fields and made a substantial contribution to pollination (Hobbs, 1973; Pankiw *et al.*, 1978). Many realized that the role of the honey bee in the pollination of lucerne in Canada was relatively unimportant and less than ideal (Bolton, 1962; Pankiw and Bolton, 1965). In response to this limitation, programmes on the introduction and management of *Megachile rotundata* Fab., the alfalfa leafcutting bee, were initiated at AAFC Research Stations at Beaverlodge and Lethbridge (in Alberta), and at Melfort (in Saskatchewan) in the 1960s. The introduction of the alfalfa leafcutting bee, indigenous to Eurasia, and the development of management techniques for this bee in the different lucerne seed-growing regions (Pankiw and Siemens, 1974; Pankiw *et al.*, 1978), resulted in the revitalization of the lucerne seed industry in Canada in the 1980s as is evident from both the quantity (Fig. 19.1) and land area (Fig. 19.2) of production.

A significant and often overlooked by-product of the lucerne seed industry in Canada is leafcutting bee cells. This insect pollinator overwinters as a dormant prepupa in a cell made of leaf pieces; the cells are incubated in the spring so that the emergence of the adult bees is synchronized with flowering of the crop. The adult bee pollinates the crop and reproduces before dying at the end of the season. On average about 2.0–2.5 times the number of cells taken out are brought back in to winter storage. Most of the excess cells are sold to markets in the USA and form a significant part of the farmer's income from lucerne seed production. While many seed growers have acquired the skills to manage the bee, others choose to hire the services of a 'custom pollinator' to look after the pollination operation. Many custom pollinators offer their services in both the Canadian prairie provinces and in the Pacific northwest, USA. In Canada, the services of a custom pollinator are usually provided in exchange for one-third of the seed of the crop being pollinated in addition to the dormant bee cells that are taken out at the end of the season. In the USA, however, a flat fee per hectare of crop (approximately $150–200) is charged and all the dormant larvae and equipment are disposed of at the end of the growing season because of problems with chalkbrood disease (*Ascosphaera* spp.) in the Pacific northwest states.

The success that Canada has achieved as a producer of lucerne seed is, in large part, a reflection of the enormous success of public research in providing for 'the public good'. Most of the research on both the management of the lucerne crop and

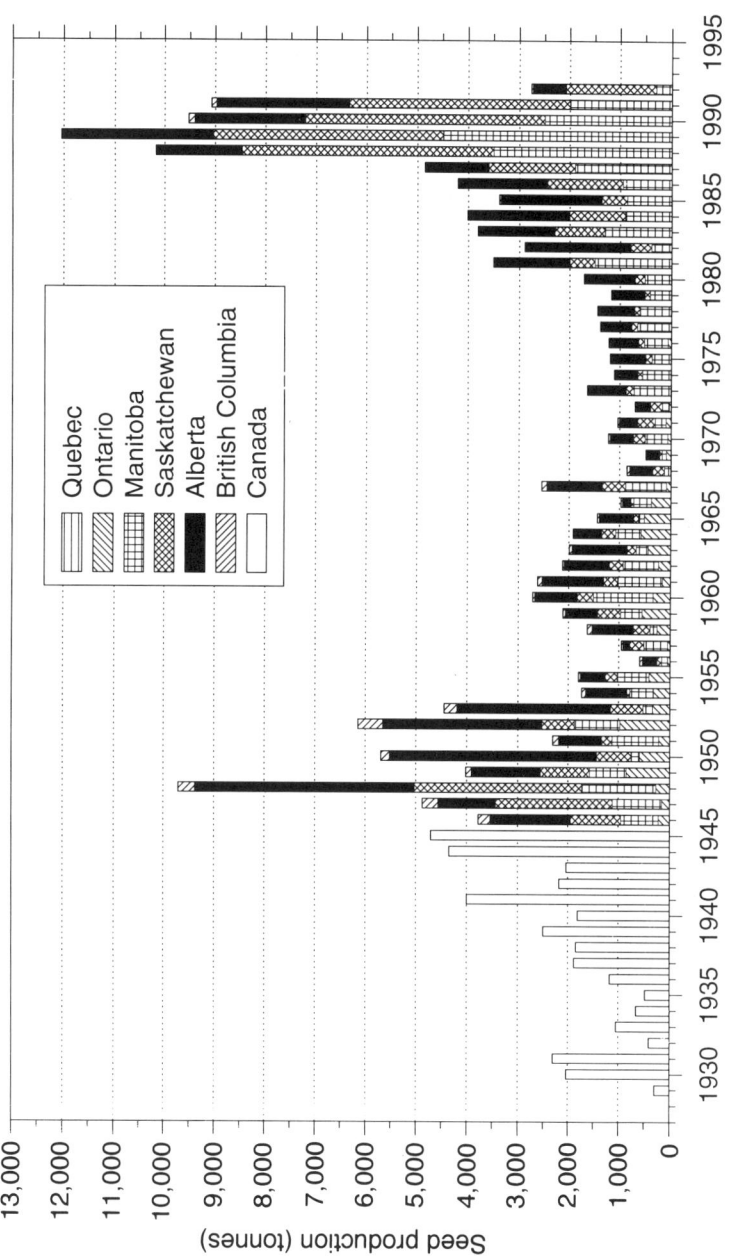

Fig. 19.1. Lucerne seed (certified and uncertified) production in Canada. (Sources: Dominion Bureau of Statistics, Canada, Agricultural Division; Agriculture Canada, Food Production & Inspection.)

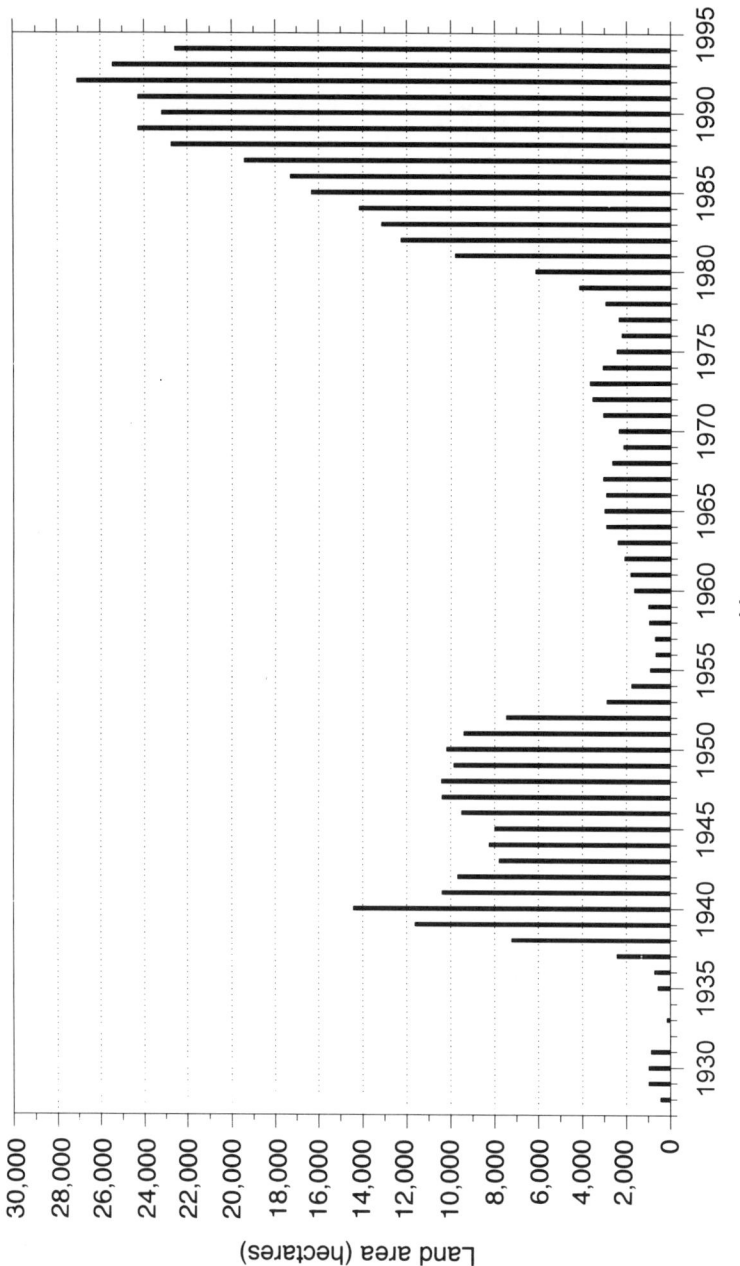

Fig. 19.2. Annual certified lucerne seed area (ha) in Canada. (Source: Canadian Seed Growers' Association.)

the insect pollinator that is required for successful seed production has been at AAFC Research Stations.

19.1.3 Commodity Associations

The Canadian farmer has played a prominent role in promoting opportunities for lucerne seed production. In the early-1970s, when large bee populations were gradually being established on many farms in western Canada, farmers in the prairie provinces formed provincial 'leafcutting bee associations' to serve as an information network. By the mid-1980s, these associations changed their names and emphasis to encompass all the aspects of lucerne seed production that related to the crop, the pollinator and marketing. The Manitoba Alfalfa Seed Producers' Association (MASPA), the Saskatchewan Alfalfa Seed Producers' Association (SASPA) and the Alberta Alfalfa Seed Producers' Association (AASPA) replaced these leafcutting bee associations. In addition, these new provincial associations were coordinated by a national body, the Canadian Alfalfa Seed Council (CASC) that consists of two voting representatives from each of the prairie provinces and the non-voting positions of an Executive Secretary/Manager, and a Technical Advisory Committee of three provincial forage specialists and two Research Scientists from AAFC. Today, MASPA has metamorphosed into the Manitoba Forage Seed Association (MFSA) with an interest in all forage seed crops; SASPA continues unchanged; AASPA operates as two Branches, the Irrigated Branch in southern Alberta and the Peace Branch (with grower-members from the Peace River region of both Alberta and British Columbia) who work in conjunction with the newly formed Peace Region Forage Seed Association (PRFSA) whose mandate includes all forage seed crops. CASC is responsible for the operation of the Cocoon Testing Centre, a laboratory where bee cell quality is determined on a fee-for-service basis. The Council also coordinates the activities of the provincial associations and disseminates information on production technology and marketing, via an annual seminar that is held on a rotational basis in each of the prairie provinces.

19.2 SEED PRODUCTION STATISTICS

19.2.1 Regions of Production

A record of annual lucerne seed production has been available from 1929 to 1992 (Fig. 19.1). The data in this figure for the years 1929 to 1945 (Agricultural Statistics, 1946) are for the total Canadian production. Unfortunately, details on precisely where in Canada this production originated were not documented. However, this deficiency was remedied from 1946 to the present, when production in relation to province of origin was delineated by AAFC, Food Production & Inspection Branch, Ottawa, Ontario. In most years, the seed grown in Canada came from the western Canadian provinces of British Columbia, Alberta, Saskatchewan and Manitoba. In some years, small quantities of seed were produced outside the west in Ontario and Quebec. To date, lucerne seed has never been produced in the Atlantic provinces.

19.2.2 Volumes of Production

Between 1929 and 1945, annual seed production was under 5000 t and ranged from a low of less than 500 t in 1929 to about 4700 t in 1945 (Fig. 19.1). In the next 8-year period, between 1946 and 1953, annual production was generally higher albeit variable from year to year; a high of almost 10,000 t of seed was produced in 1948. From 1954 to 1980 production levels rarely exceeded 2000 t and were often less than 1000 t. After 1981 a marked increase in production occurred, with a maximum of more than 12,000 t in 1989.

While there has been some speculation and discussion on the factors that influence the large annual fluctuations in lucerne seed production through the years, Bolton (1962) assessed the Canadian situation accurately with his observation that bees which trip and cross-pollinate lucerne flowers play a primary role in seed production, while other factors such as climate, soil, crop management, disease and injurious insects are of secondary importance because their influence comes into play primarily after the flowers have been tripped and cross-pollinated. According to Hobbs (1973) and Pankiw *et al.* (1978), prior to 1954, adequate populations of native bees in the margins of seed fields were instrumental in the production of abundant seed yields (Fig. 19.3), and the low yields between 1954 and 1980 were attributed to a decrease in the populations of these bees. The widespread use of the leafcutting bee in the 1980s, after populations of this insect were gradually increased in western Canada from the initial introductions of small numbers from the USA in the late-1960s, resulted in the revitalization of the lucerne seed industry in Canada (see Figs 19.1, 19.2).

19.2.3 Crop Area

Information on the total area in lucerne seed crops, and the number of farms where lucerne seed was grown, is available from the census data that are taken at 10-year intervals. While the records on both the amount of seed and province of production have been available since 1929, area data that include both certified and uncertified seed have been collected on four occasions to date, i.e. 1961, 1971, 1981 and 1991 (Table 19.2). In 1961, when the first lucerne seed census was taken, lucerne was grown for seed on 6511 farms; however, subsequent data show a dramatic drop in the number of farms. A comparison of the 1981 and 1991 data would indicate that while farm numbers remained below 1300 there was an almost threefold increase in land area in production during this 10-year period, perhaps indicative of a large increase in the size of lucerne seed fields on individual family farms situated in western Canada.

Annual data on the number of hectares of certified lucerne seed crops inspected each year (Fig. 19.2) are maintained by the Canadian Seed Growers' Association (CSGA). Between 1929 and 1937 the area was generally less than 1000 ha. There was an appreciable increase from 1938 to 1952 when between 7000 and 14,000 ha of lucerne were harvested for seed. From 1953 to the late-1970s a dramatic decrease was observed with less than 3000 ha in most years. According to Pankiw *et al.* (1978) this period was characterized by extensive land clearing, an increase in field size, and a decrease in the native vegetation that

(a)

(b)

Fig. 19.3. A lucerne seed crop on the banks of the Peace River at Taylor, British Columbia a. The J.R. Ardill farm in 1950: bees were not provided for pollination; crop yielded 336 kg ha^{-1}. (Photograph credit: AAFC, Beaverlodge, Alberta.) b. Don & Hazel Pedersen's farm in 1982: note 'seed rings' around shelters for leafcutting bees that were introduced during crop bloom; crop yielded 890 kg ha^{-1}. (Photograph credit: Daphne Fairey, AAFC, Beaverlodge, Alberta. Sponsors: Peace River Agriculture Strategic Planning Society & Peace Region Forage Seed Association.)

Table 19.2. Agricultural census information on lucerne seed production in Canada.

Region of Canada	1961 Farms (no.)	1961 Land area (ha)	1971 Farms (no.)	1971 Land area (ha)	1981 Farms (no.)	1981 Land area (ha)	1991 Farms (no.)	1991 Land area (ha)
Canada	6511	na	3105	32,749	1123	22,843	1256	60,284
Newfoundland	0	na	1	1	0	0	0	0
Prince Edward Island	5	na	6	28	0	0	0	0
Nova Scotia	8	na	9	67	0	0	0	0
New Brunswick	3	na	4	31	0	0	0	0
Quebec	141	na	234	2390	0	0	10	70
Ontario	3188	na	734	3322	0	0	79	1146
Manitoba	1354	na	561	6903	428	7844	475	22,516
Saskatchewan	707	na	701	7458	300	5732	412	24,703
Alberta	1062	na	785	11,641	353	8772	262	10,700
British Columbia	43	na	69	899	42	596	18	1148

na, not available.

surrounded seed fields and provided a nesting habitat for native bees. After 1980, when the use of the leafcutting bee for pollination commenced, there was a steady increase in the area of lucerne grown for seed; a high exceeding 26,000 ha was recorded in 1992.

It should be noted that the area for certified seed production (Fig. 19.2) and the total amounts of seed produced in Canada (Fig. 19.1) do not always follow a similar trend because the total amount of seed represents the harvest of both certified and uncertified stands. The price of seed usually dictates whether an uncertified stand is harvested for hay or seed, or whether some or all of the required inputs for optimum seed production such as herbicides, insecticides and/or pollinators are justifiable.

19.2.4 Exports and Imports

Exports of lucerne seed exceeded 500 t from 1926 to 1928, 1936 to 1954 (excluding 1944), 1961 to 1963, and from 1987 to the present. The periods of greatest export activity were from 1937 to 1954 during which an all-time high of 7300 t was exported in 1948, and from 1987 to the present with a high of 6600 tonnes being attained in 1990 (Fig. 19.4). A majority of the exported Canadian seed is from winter hardy cultivars that can be assigned to an autumn (fall) dormancy rating or FDR (Barnes et al., 1978; Tueber et al., 1984) of 1, 2 and 3. These types of cultivars were, and are still, predominant in the stands that are grown in western Canada.

Significant importation of lucerne seed occurred in the early 1950s, and this has continued; generally between 1000 and 3500 t of seed have been imported annually (Fig. 19.4). A majority of this seed is of cultivars that are of the less winter

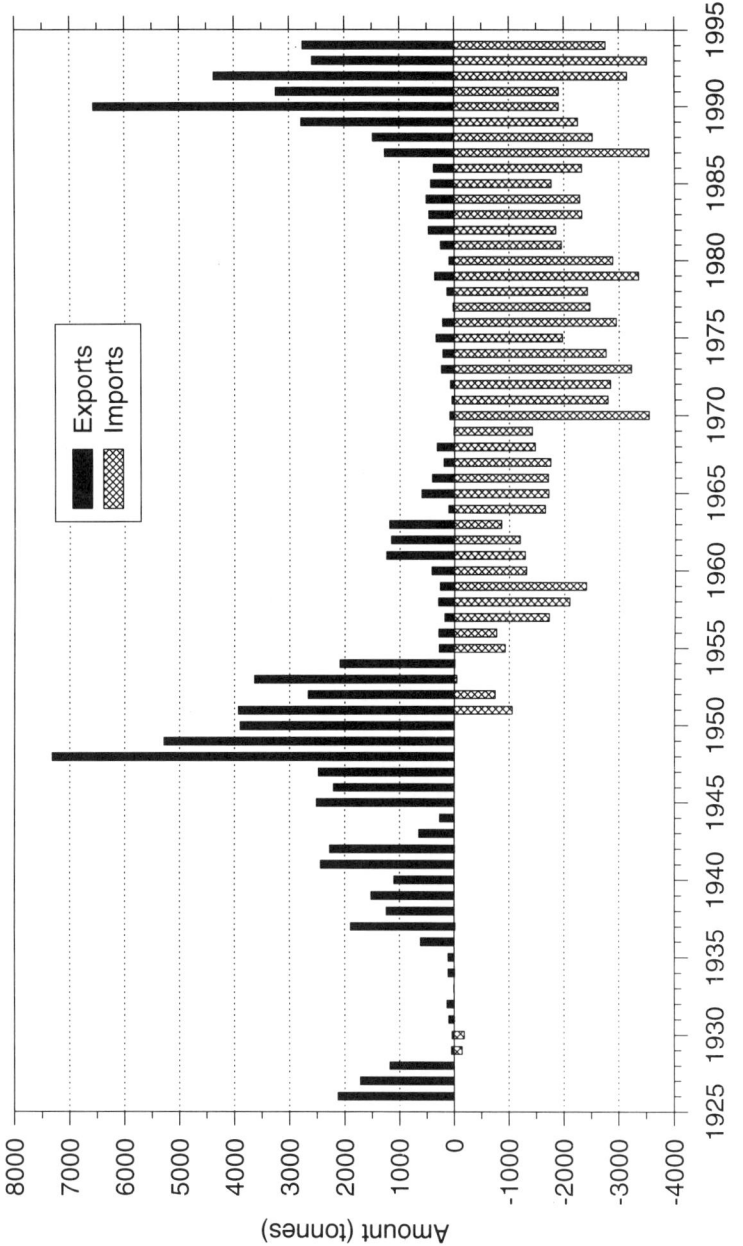

Fig. 19.4. Lucerne seed exported from and imported into Canada annually. Note: import amounts for 1926–1928 and for 1944–1950 are not available because they were grouped together with those for clovers; import amounts for 1929–1943 are included but are generally too low to register on the bar chart. (Sources: Dominion Bureau of Statistics, Canada; Agricultural Division, Statistics Canada; Agriculture Canada, Food Production & Inspection Branch.)

hardy types with FDRs of 4 or more. Generally, the certified seed imported is of cultivars bred in the USA and subsequently registered and recommended for use somewhere in Canada, either in one or more of the Atlantic provinces, or in Ontario or Quebec where the higher yielding, non-falcata types of lucerne are utilized for herbage production.

In most years there is a negative balance of payments between the Canadian export and import of lucerne seed (Fig. 19.5). For approximately 50 of the 70 years of annual data on record, Canadian exports were worth less than $1 million. The value of exports was strong from 1945 to 1953 and since 1987, and attained an all-time high in 1990 of $14 million. The value of lucerne seed imports has shown an increasing trend since the early 1960s, reaching $7 to $11 million annually in the 1990s.

19.3 CHALLENGES FACING THE INDUSTRY

The 1990s are unfolding as the watershed years for the lucerne seed industry and, for that matter, for the entire forage seed industry in Canada. To date, research in the public sector has formed the nucleus for seed production technology and cultivar development for both the farmer and seed trade. Recently, however, funding for public sector research has been in a downward spiral. As we approach the twenty-first century, the industry has arrived at a point where its direction and priorities must be redefined, and new alliances and/or arrangements will have to be made to acquire the technology the industry needs to remain competitive.

Opportunities exist for contract production of the less hardy lucerne cultivars. According to Squires and Casavant (1993), lucerne seed production in the USA is shifting northwards into the Pacific Northwest states because of urban and industrial encroachment, and because environmental concerns have begun to limit the use of intensive inputs in California and some of the southern states. These authors have also observed a trend towards private companies contracting with growers in the USA and Canada to grow proprietary seed rather than public cultivars. For the Canadian farmers a market opportunity of $8–10 million exists if the challenge of import substitution for seed of less hardy cultivars for eastern Canada is considered. To date, studies in Canada's northern-most lucerne seed production areas in the Peace region of northwest Alberta and northeast British Columbia have shown that, with the use of suitable production technology, it is feasible to grow one or two seed crops of many of the less hardy lucerne cultivars, even those with FDRs of 7, 8 and 9 (D.T. Fairey, AAFC, Beaverlodge, Alberta, 1996, personal communication).

It is becoming increasingly obvious that if Canadian farmers are to continue to be able to afford to use leafcutting bees for pollination of lucerne, and hope to be able to have surplus bee cells to sell, then they have to make the move toward producing a higher-value seed crop of lucerne, or to pollinate other high-value crops such as hybrid canola (*Brassica* spp.), fruit or berry crops. To date, the technology to use leafcutting bees to pollinate clovers and canola (*Brassica campestris* L.) has been developed (Fairey and Lefkovitch, 1990) but there is a

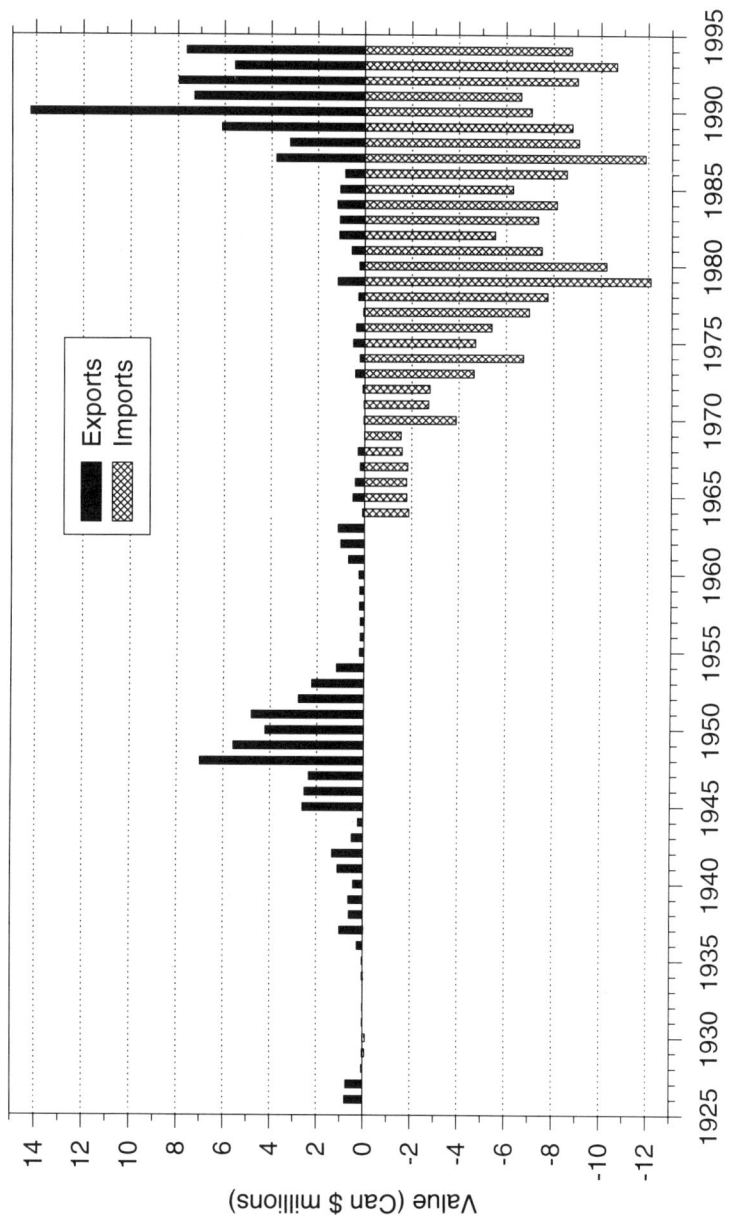

Fig. 19.5. The value of Canada's exports and imports of lucerne seed. Note: import values for 1926–1928 and for 1944–1950 are not available because they are grouped together with those for clovers; import values for 1929–1943 are included but are generally too low to register on the bar chart; import values for 1951–1963 were unavailable. (Sources: Dominion Bureau of Statistics, Canada; Agriculture Division, Statistics Canada.)

clear need for more research and development on these alternative pollination opportunities.

There are encouraging signs that the many participants in the industry have accepted the fact that publicly funded research will not provide for all their needs. Producer groups are establishing a framework that allows for the collection of a levy on the seed marketed; these levies will be used to fund whatever activities advance the good of the industry. This has been initiated already for all forage seeds by the MFSA in Manitoba, and for lucerne seed by the SASPA in Saskatchewan.

Canada has the basic technology to produce lucerne seed. The challenge ahead is to build a network that will provide both the market opportunities and technology to stay competitive with all the partners sharing the costs and benefits.

ACKNOWLEDGEMENTS

The statistics assembled in this chapter were provided by individuals from several organizations. We thank Jean Murphy and Bill Robertson of the Canadian Seed Growers' Association; Nancy Sharp of the Canadian Agricultural Library; Bill Hanson of the Food Production and Inspection Branch of Agriculture & AgriFood Canada; Heather Drybrough of the Canada Grains Council; Bernard Houle of the Agricultural Census Division of Statistics Canada; the staff of the Library and the Book and Record Depository at the University of Alberta, Edmonton.

REFERENCES

Agriculture and Agri-Food Canada (1995) *List of Varieties which are Registered in Canada.* Publication T-1-07-95, Variety section. Plant Products Division, Agriculture and Agri-Food Canada, Nepean, Ontario.

Agriculture and Agri-Food Canada (1996) *Quarterly Supplement to List of Varieties which are Registered in Canada.* Publication T-1-07-96, Variety section. Plant Products Division, Agriculture and Agri-Food Canada, Nepean, Ontario.

Agricultural Statistics (1946) *Quarterly Bulletin of Agricultural Statistics.* July to September. Dominion Bureau of Statistics, Canada, Agricultural Division, Ottawa, Ontario.

Armstrong, G.M. and White, W.J. (1935) Factors influencing seed-setting in alfalfa. *Journal of Agricultural Science* 25, 161–179.

Armstrong, J.M., Nowosad, F.S., Ripley, P.O. and Kalbfleisch, W. (1948) *Alfalfa for Hay, Silage and Pasture.* Publication 735. Canada Department of Agriculture, Ottawa, Ontario.

Barnes, D.K., Smith, D.M., Stucker, R.E. and Elling, L.J. (1978). Fall dormancy in alfalfa: a valuable predictive tool. In: Barnes, D.K. (ed.) *Report of the 26th Alfalfa Improvement Conference,* Brookings, SD. 6–8 June 1978. US Government Printing Office, Washington, DC, 34 pp.

Bolton, J.L. (1962) *Alfalfa: Botany, Cultivation and Utilization.* Leonard Hill (Books), London.

Elliott, C.R. and Bolton, J.L. (1970) *Licensed Varieties of Cultivated Grasses and Legumes.* Publication 1405. Information Division, Canada Department of Agriculture, Ottawa, Ontario.

Fairey, D.T. and Lefkovitch, L.P. (1990) Reproduction of *Megachile rotundata* Fab. foraging on *Trifolium* spp. and *Brassica campestris*. *Acta Horticulturae* 288, 185–189.

Fairey, D.T. and Lefkovitch, L.P. (1992) Seed yields of consecutive harvests from forage legume stands. *Journal of Applied Seed Production* 10, 25–30.

Goplen, B.P. (1990) *Recommended Forage Crop Varieties for 1990*. Canadex, Publication 120.30. Agriculture Canada, Communications Branch, Ottawa, Canada.

Hobbs, G.A. (1973) *Alfalfa Leafcutter Bees for Pollinating Alfalfa in Western Canada*. Publication 1495. Information Division, Canada Department of Agriculture, Ottawa, Ontario.

Pankiw, P. and Bolton, J.L. (1965) Characteristics of alfalfa flowers and their effects on seed production. *Canadian Journal of Plant Science* 45, 333–342.

Pankiw, P. and Siemens, B. (1974) Management of *Megachile rotundata* in northwestern Canada for population increase. *Canadian Entomologist* 106, 1003–1008.

Pankiw, P., Siemens, B. and Lieverse, J.A.C. (1978) Breeding and management of *Megachile rotundata* for alfalfa seed production in northwestern Canada (lat. 55–58°N). In: *Proceedings of the IVth International Symposium on Pollination*; Maryland Agricultural Experimental Station Special Miscellaneous Publication 1, pp. 273–277.

Squires, G.W. and Casavant, K.L. (1993) *Alfalfa Seed: A Market Analysis*. Report to the Northwest Alfalfa Seed Growers' Association. Washington State University, Pullman, Washington.

Teuber, L.R., Marble, V.L., Lehman, W.F., Kawaguchi, I.I., Miller, M.K., Hartman, B.J., Hunt, O.J., Barnes, D.K., Burrows, B., Lancaster, D.L. and Gripp, R.H. (1984) Climatic and dormancy data reduces need for many regional alfalfa trials. *California Agriculture* 38, 12–14.

Wheeler, W.A. (1951) *Beginnings of Hardy Alfalfa in North America*. Reprinted from Seed World, Northrup King & Co., Minneapolis, Berkeley & Boise, 31 pp.

Trifolium pratense L. (Red Clover) in France

20

S. Bouet and G. Sicard
FNAMS, Le Verger, 49800 Brain sur l'Authion, France

20.1 HISTORICAL DEVELOPMENT AND DISTRIBUTION

20.1.1 Distribution and Adaptation Around the World

Red clover (*Trifolium pratense* L.) is a temperate region crop which originated in southeastern Europe. Approximately 20 million ha of this important forage crop are grown in the world (Taylor, 1985). It is a major crop in France, Germany, Italy and Austria and its production extends across the continent into Poland, Yugoslavia, Hungary, Russia, and even into Japan. It is also grown in the USA, Canada, South Africa, Chile, New Zealand and Australia (Taylor, 1985). Red clover can be grown either as a pure sward or in a mixture with grasses.

20.1.2 Importance of Red Clover in France

Introduction and development

Red clover was cultivated in Europe during the third and fourth centuries. It has long been an important crop in France; the first record of red clover cultivation in France appeared in 1583 (Taylor, 1985). Over recent decades, the area of red clover in France has decreased significantly, from 1,200,000 ha in 1950 to 230,000 ha in 1980, mainly as a result of increased use of cereals for animal feed. Recently, however, there has been a renewed interest in red clover, because the crop allows farmers to reduce their use of nitrogen fertilizer, and to reduce their dependence on cereal-based animal feeds.

France was the largest user of red clover seeds in the European Union (EU), 2000 t in 1994. Germany was the next largest user, 1000 t in 1994 (GNIS, 1995).

Main uses

Red clover is a short-lived perennial that produces for approximately two or three seasons. It is generally grown in mixtures with grasses. In a French survey completed in 1982, red clover was grown in association with Italian ryegrass (*Lolium multiflorum* Lam.) on 235,000 ha, and only on 70,000 ha as a pure stand (Arnaud *et al.*, 1993). The main areas where red clover is used as a forage crop in France are in the Massif Central, Bretagne and the Vendée. More recently, farmers in cereal-growing regions have started sowing red clover in set-aside lands, in order to take advantage of the quality forage and nitrogen-fixing ability.

Different management systems are used in France; rotational grazing in the case of red clover/grass associations, or hay or silage for pure swards. The first crop of hay or silage is taken at the prebloom stage, to optimize the yield of digestible nutrients. Two or three annual forage harvests are possible.

20.1.3 Statistics of Seed Production in the EU and France

Seed production in the EU

Red clover seed production is concentrated in a few countries in the EU. France has always been the main producer with 2000 t in 1993, more than 75% of EU production. The rest of the production (around 500 t) is scattered across Germany, Italy, Denmark, Great Britain and Austria.

Seed production in France

During the 1950s, red clover seed production was just over 6000 t per year. From 1970 to 1980, production decreased slowly reaching an average of about 3800 t per year with the exception of 1976, a particularly bad drought year (Fig. 20.1). Seed production in 1976 was very low and imports had to compensate for the lack of seeds. Since 1980, seed production has generally continued to decrease as a result of the decrease in demand in Europe.

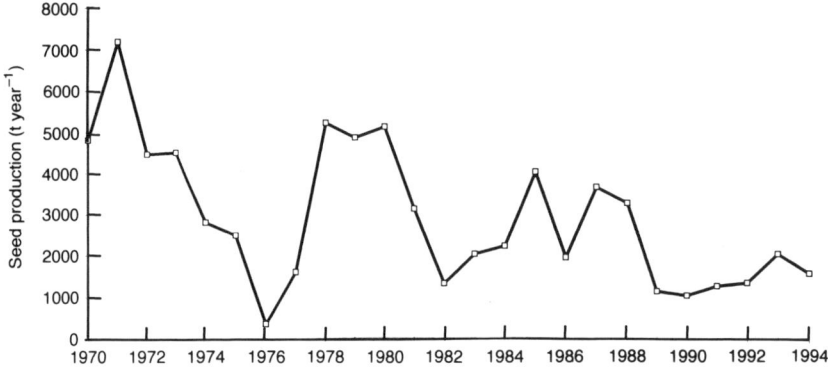

Fig. 20.1. French red clover seed production, 1970–1994. (From GNIS, 1995.)

20.2 AREAS OF SEED PRODUCTION IN FRANCE

20.2.1 Characteristics of these Areas

Regions of production

More than 60% of French red clover seed production comes from central France (Region Centre). The 'department' (district) which produces the most clover in this region is Indre. Additional production is scattered across the west, southwest and east of France.

Climate

Region Centre is a continental area with an average of 700–750 mm of rain each year (1951–1980). The winter is cold and humid, with 30–35 days under 0°C. The last frost generally occurs at the end of March or the beginning of April. In spring, the months of April and May are often dry. Summer is mild with an average temperature of 19°C and generally dry with only a few storms during July and August. Rainfall in July improves the flowering stage of red clover seed production. August rains usually allow growers to grow red clover without irrigating.

Soil

Red clover performs best on fertile well-drained soils with a high moisture-holding capacity. Heavy textured soil is preferred to light sandy or gravelly soil which is sensitive to drought. Contrary to lucerne (*Medicago sativa* L.), red clover can produce satisfactory yields even at a low pH (5 to 6).

In Region Centre, red clover is grown on two types of soil: clay soil based on calcareous rock, and silt loam. The former is characteristic of the cereal area. This soil is well drained, containing 30–45% clay with a high pH (7–8) and a good level of organic matter. However, deep soil is preferred, particularly when irrigation is not available. The latter is characteristic of the mixed farming regions. This deep silt loam soil contains 8–15% clay, has a lower pH (5–7) and low organic matter. The structure is often poor. In addition, rapid drying of the soil sometimes leaves a crust on the surface.

20.2.2 Historical Development in Region Centre

Region Centre has always been a traditional red clover seed production area. Before 1950, farms were 20–30 ha in size and focused on animal husbandry (sheep, cattle, and dairy farming). Red clover was a very common crop for forage on all soil types. It was undersown by hand in spring barley, 1 or 2 days after sowing the barley. Sometimes, however, the red clover density was too high, and barley yield was very low. Forage production was always the priority, and the first cut of hay around the prebloom stage occurred too late for subsequent seed production from the regrowth. Cultivars were short-lived perennials and the red clover crop had to be renewed every year.

For the seed harvest, red clover was originally hand-cut and dried in the sun. It remained in the field in haystacks covered by wheat straw. Threshing was completed in autumn or winter on dry days. With the introduction of mechanization, special steam-powered threshers for clover were sold in the region. These machines were equipped with two cylinders; the first to thresh the pods and the second to hull the seeds. Seed yield was variable, from 0 to 400 kg ha^{-1}, and averaging 150 kg ha^{-1}. Sometimes, seed was severely damaged by insects in the absence of pesticides. After threshing, seed was cleaned with a suitable round-holed sieve to remove larger objects. The introduction of the combine harvester completed the process of mechanization.

The seed market was organized in Issoudun, a small town in the Indre department. Each grower had to bring a sample of his harvest and sell it according to its quality. Several merchants operated in this forage crop seed market.

20.3 EVOLUTION TO AN INTENSIVE CROP

20.3.1 Evolution of Techniques

Progressively, techniques of seed production have improved. The use of pesticides (carbetamide, MCPP, toxaphène) began after 1960. In 1961 seed certification became compulsory. During the same period, livestock disappeared in the most fertile areas in order to make room for cash crops, mainly cereals, rape (*Brassica* spp.) and sunflower (*Helianthus* spp.). Farms became larger, with an average of 85 ha, and the largest up to around 300 ha. In this system, red clover seed production was kept as an interesting first crop in the rotation where it reached up to 10% of the total farm area. On other soils, animal farming remains an important activity; farms are generally smaller, but farmers continue to grow red clover.

The farmers of the Region Centre have developed their own techniques for growing red clover. After a wheat (*Triticum aestivum* L.) crop, the seedbed is prepared either with a cover crop or for direct sowing. Red clover is sown at 3–5 kg ha^{-1} in August. Rainfall is generally adequate to encourage germination, but irrigation may also be used to encourage growth. At sowing, approximately 40–50 kg ha^{-1} of potassium and phosphorus are applied.

Weed control is essential to the production of clean clover seed, so sowing the crop on relatively weed-free land is important. Clover seed has to be free of certain weeds such as sheep's sorrel (*Rumex acetosella* L.). One or two herbicides are generally applied; bentazone is applied at the seedling stage in autumn and carbetamide is applied in the winter to control grasses. Clipping or cutting the crop in spring can have a positive effect in controlling some weeds, e.g. broom rape (*Orobanche minor* Sm.) and broadleaf weeds. The clipping is completed around the 20 May in Region Centre. The goal is to delay flowering and to encourage it in July when pollinator activity is highest, and so that seed harvest can occur before the first rainfalls of September. Clipping also reduces the vegetation and lodging, allowing better pollination.

In cereal-growing areas, the forage is not needed but must be removed or at least cut and finely chopped to allow good regrowth of the crop. In the livestock area, some farmers still prefer to delay the clipping and take a hay crop: in this case the seed yield is generally reduced.

The major disease affecting red clover is *Sclerotinia* crown rot caused by *Sclerotinia trifoliorum* Erikss. Wet weather in autumn encourages disease development in first-year crops. Since 1990, a warning system for this disease has been operating locally. The 'Institut National de la Recherche Agronomique' (Raynal, 1987) is able to produce artificial sclerotia in the laboratory. Every year, these are sent out to 15 seed growers throughout the region. The sclerotia are buried in the soil at red clover sowing time in an identified spot. Under favourable climatic conditions, these sclerotia produce apothecia; the presence of apothecia is an indication of a possible disease outbreak and farmers are advised to apply fungicide (carbendazim; vinchlozoline) in the autumn.

Several insects damage red clover. The clover seed weevil (*Apion trifolii* L.) is the most serious and is well known by the red clover seed growers. It deposits its eggs in flower buds where its larvae feed on the seeds. Severe attacks in the past have contributed to crop losses in the region. Currently, insecticides are sprayed to control this insect, two applications of parathion between bud stage and the beginning of flowering, and one application of bifenthrine or phozalone at the beginning of flowering.

Another pest, the clover root borer (*Hylastinus obscurus* Marsham), was reported in Region Centre in 1983. Adult and larvae feed on red clover roots causing serious damage, particularly on second-year crops. As no chemicals are registered for this pest, seed growers were forced to grow red clover for 1 year only in order to avoid damage. Severe infestations have now been reported in the traditional red clover areas. Consequently, production over the past 2 years has decreased, in spite of new, more perennial, cultivars being introduced.

In the cereal-growing area, an environment composed of large fields does not favour good pollination, because red clover is often surrounded by crops more attractive to pollinators, such as sunflowers. Bumblebees (*Bombus* spp.) are especially effective pollinators of red clover, but often are inadequate in number. In addition, wild bees make a limited contribution. Consequently, some seed growers provide two or three hives of honeybees (*Apis* spp.) per hectare.

Trials completed in this area since 1986 have shown the interesting effect of irrigation which allows growers to regulate yield in dry years. In the cereal area, farmers already equipped to irrigate maize (*Zea mays* L.) and peas (*Pisum sativum* L.) irrigate the red clover seed crop in July to stimulate flower production and seed development. One or two irrigations, depending upon the weather, can be applied, but irrigation of the maize always has priority.

At least 4 days before harvest, a chemical desiccant (diquat) is sprayed on the crop to dry the vegetation quickly, allowing the standing crop to be harvested by direct combining. Harvest occurs at the end of August or the beginning of September. To increase the yield, seed growers carefully adjust the threshing surface, closing the concave portion by three-quarters in order to increase the rubbing surface area. This guarantees the best threshing results.

20.3.2 Evolution of Seed Yield

Progressively, seed yield has increased with changes in techniques. The ability to control insects has improved significantly. The use of herbicides and fungicides has improved weed and disease control. Today from seven to ten different pesticides may be applied to red clover seed crops, and with irrigation, yield can be regulated. However yield always depends on certain factors that cannot be controlled, particularly the weather. High moisture-holding capacity in the soil, limited growth of the crop after clipping, dry conditions to ensure good pollination, and sufficient water during the flowering stage often lead to a successful seed crop. Today, seed yield averages around 700 kg ha^{-1} in Region Centre. Under the best conditions, yield reaches 1000 kg ha^{-1}.

20.4 RED CLOVER SEED ORGANIZATION IN FRANCE

20.4.1 Seed Production

Seed growers in France mainly produce for significant local cooperatives such as Semagra, Episem and La Franciade in Region Centre, or for small local organizations which manage an important area. In the cooperatives, seed production is often better organized with technical staff who advise seeds growers. These two types of local production units are involved with bigger cooperative groups like Semences de France, or with seed firms such as Jouffray, Carneau and Tourneur which are situated outside the region. These groups and firms sell across the French market or export seed. About 25 diploid cultivars and five tetraploid cultivars are produced in France (GNIS, 1995). The grower must contract with a registered seed company. The agreement defines the cultivar, the area and the minimum price.

20.4.2 Breeding Research

Red clover breeding has decreased in France over the past 20 years, reflecting the decreasing market. Today breeding is carried out in France at the public research institute (INRA), and also by one private company GIE AMFO (a RAGT group subsidiary).

20.4.3 Registration in the Catalogue of Species and Cultivars

The official body responsible for examining cultivars is the Comité Technique Permanent de la Sélection (CTPS). It sets the conditions for studying cultivars within each species. The Groupement d'Etude et de Contrôle des Variétés et des Semences (GEVES) of the Institute National de la Recherche Agronomique makes all necessary observations and tests. The agricultural value of each new cultivar is tested over 3 years in a multilocated trial network. New cultivars are protected under plant variety rights (PVR) legislation.

20.4.4 Official Controls

Seeds must be produced in accordance with technical regulations published by the Ministry of Agriculture. All French certification regulations comply with European guidelines relating to production and marketing of seed. The Service Officiel de

Contrôle et de Certification (SOC) under the auspices of the Ministry of Agriculture is the organization in charge of enforcing the technical regulations and issuing official certificates for seeds.

20.4.5 Seed Growers' Organization

French seed growers belong to a union called the FNAMS – Fedération Nationale des Agriculteurs Multiplicateurs de Semences. This union has a technical staff who work to improve all seed production techniques and who have been involved with red clover seed production since 1955. Trials are conducted in Region Centre and in the southwest; herbicide, fungicide, irrigation and harvest techniques are tested for their contribution to yield improvement. FNAMS also directly advises seed growers about techniques for their fields.

In 1961, a laboratory (LABOSEM), based in Angers, was created by the seed growers to analyse the quality of their seeds (germination and purity). This analysis is recognized by most of the seed firms, which pay the seed growers upon these analyses. Each year, minimum prices are discussed by seed growers (FNAMS) and seed firms at the national level.

20.5 CONCLUSION

Red clover seed production in France started from a traditional farming enterprise and has evolved into an intensive crop. In Region Centre, the birthplace of red clover seed production, the experience of the farmers allows them to succeed in the production of a high-yielding first crop in a mainly cereal growing system. The expertise of seed growers, supported by a modern seed industry infrastructure, ensures a quality product.

REFERENCES

Arnaud, J.D., Legall, A. and Pflumlur, A. (1993) *Fourrages no. 134, Evolution des surfaces en légumineuses fourragéres en France*. Association Francaise pour la production fourragére, Versailles, pp. 145–154.

GNIS (1995) *Données Statistiques*, Groupement National Interprofessional des Semences, Paris.

Raynal, G. (1987) Facteurs agissant sur la formation des apothècies de *Sclerotinia trifoliorum* Eriks en conditions contrôlèes. *Agronomie* 9, 715–725.

Taylor, N.L. (1985) Red clover. In: Taylor, N.L. (ed.) *Clover Science and Technology*. Number 25. American Society of Agronomy, Madison, Wisconsin, pp. 457–564.

21 *Trifolium repens* L. (White Clover) in New Zealand

P.T.P. Clifford
AgResearch, PO Box 60, Lincoln, New Zealand

21.1 INTRODUCTION

New Zealand's white clover (*Trifolium repens* L.) seed production has evolved from supplying the modest local, turn-of-the century, land development needs to the present-day requirements of successfully competing in an international market. This has relied upon: high inputs into breeding, seed science, and technology transfer, to ensure high product performance; efficient seed production systems; quality control; competitive pricing; and market alignment.

21.2 HISTORICAL BACKGROUND

21.2.1 Phosphate Commission 1920

The real beginnings of New Zealand's seed industry date back to the period when superphosphate fertilizers gained general acceptance for farm use (Brock *et al.*, 1989). Clovers were recognized as the key species, providing nitrogen (N) through N fixation to sustain pasture productivity. Fuelled by phosphate, the enhanced pasture production was utilized through increased stocking rates to the benefit of nutrient recycling and thereby improved soil fertility status (Brock *et al.*, 1989). The move to improved, clover-based, pasture management systems in the 1920s highlighted existing limitations arising from variations in clover and grass strain stability and poor seed quality. In response, the newly formed government plant breeding station at Palmerston North released three pasture cultivars in the mid-1930s. One of these was the white clover now known as cultivar Grasslands Huia. This cultivar still remains the world's largest commodity trader at about 35–40% of total white clover sales (Mather *et al.*, 1995).

21.2.2 First Seed Certification System, 1929

A voluntary seed certification system was developed to maintain strain advantage and provide quality seed over many generations. The system also curtailed the spread of undesirable weed species both within and across arable and pastoral regions. This same system still prevails today, conforming to all international standards (Hampton and Scott, 1990). Thus, the stage was set for the development of a fledgling seed industry, initially based on long-term government investment in a pastoral export economy. Since these beginnings, clover-based pastoral management systems now embrace most of the world's temperate zones, thereby greatly enhancing the potential to extend New Zealand's seed export opportunities.

21.3 THE EARLY ERA

21.3.1 1920s to 1940s

Even from the late 1920s, the Canterbury province in the South Island predominated, with the area ranging from 40 to 70% of the national total (Leith, 1949). Canterbury's advantages are a well-distributed annual rainfall of 500–750 mm with cool winters and warm summers, and predominantly flat-to-undulating alluvial-based soils, rising from sea level to 300 m. Up to the beginning of the Second World War only about 2000 ha were used annually for seed multiplication. This was attributed to the high labour inputs needed for either stacking or carting the crop to a stationary clover huller for threshing, and price-competitive imported seed. The lack of European competition during the war, coupled with the introduction of header-harvesting efficiencies, produced a fourfold increase in area. Postwar, the area further doubled to 18,500 ha, based on the need to replace war-torn pastures in Europe.

The major seed production system was from ryegrass–clover sowings taken for white clover seed in the second season. Summer fallow presowing, and grazing, were the only weed control measures. Seed yields as high as 400 kg ha^{-1} were recorded but more than 100 kg ha^{-1} was considered satisfactory. Lime was used to balance soil acidity and phosphate was applied in autumn to promote growth. Pest problems included larvae of the endemic insects, grassgrub (*Costelytra zealandica* White) and porina moth (*Wiseana* spp.), which could both cause high plant mortality. Clover casebearer moths (*Coleophera spissicormis* Haworth and *Frischella* L.) were noted as pests of the future.

21.3.2 1950s and 1960s

The conversion of dryland pastoral areas by way of overhead irrigation to arable use, gave a significant area increase where 'controlled' management could be practised.

Soils used for clover seed production are commonly classified on a basis of additional summer moisture requirements. Light soils need irrigation every year, medium soils need it in some years and heavy soils do not require irrigation. For the more traditional medium and heavy land, the crop rotations were (Garrett, 1957):

1. Medium soils
old grass → partridge peas (*Pisum sativum* L.) or rape (*Brassica napus* L.) → wheat (*Triticum aestivum* L.) → barley (*Hordeum vulgare* L.) or oats (*Avena sativa* L.) → green feed → summer fallow → new grass → ryegrass seed (*Lolium* spp.) → 4–5 years with one crop of white clover seed usually in the first year.

2. Heavy soils
old grass → garden peas → wheat → barley → greenfeed → summer fallow → new grass → ryegrass seed → 4–5 years grazing with two crops of white clover seed, one in the first year.

Rotation length indicates that sheep were still an important part of the arable economy. In this period, spring oversowings of cereal crops were the first attempt at a 'specialist' system. The then low sowing costs in relation to about a 40% success rate of achieving 300 kg ha^{-1} of clover seed were considered worthwhile. This system evolved from the taking of 'volunteer' crops establishing from earlier seed losses; this practice is no longer permitted within seed certification. The development of a counter-rotating beater pickup for the combine harvester greatly enhanced retrieval of these low-bulk pure stands. Clover casebearer had now become a pest causing significant economic losses, limiting the taking of a second crop on many farms.

21.4 PRESENT DAY PRACTICE – 1970s ONWARDS

21.4.1 Research Role

Initial reproductive physiology research was carried out by Dr R.G. Thomas in the early 1960s and has been a base for the last 30 years of research (Thomas, 1961). The DSIR Grasslands (AgResearch since 1992) formed a specialist Seed Production Unit in 1972, which provided the major science requirements from which modern growing, and seed certification technology, has evolved.

The science embraced four key elements:

1. Form (morphology) – how the plant grows.
2. Function (physiology) – plant response to environment constraints; both vegetative and reproductive.
3. Seed quality – as related to genetic stability and certification requirements to ensure purity standards.
4. Yield consistency – growing systems based on cultivar interrelationships with environment and management (Clifford 1980, 1985, 1987; Clifford *et al.*, 1985, 1990, 1995).

Paralleling the plant studies was research on weed and pest management. The evolution of pure, non-competitive sowings placed a far greater emphasis on herbicide control of weeds. Paraquat, dinoseb acetate and 2, 4-D ester were the key early chemicals. Health hazard problems led to the removal of dinoseb in the mid-1980s. A lack of effective alternatives then started the move to total-rotation weed control practices. Not until 1995, with the registration of one new chemical (flumetsulam) and the possibility of a further two, has within-crop control again become a prospect. Chemical controls have been formulated for grassgrub and porina caterpillar, while biological control via two wasp parisitoids (*Bracon variegator* Nees and *Neochrysocharis* spp.) has virtually eliminated the casebearer. Clover damage by slugs (*Derocerus* spp. and *Arion* spp.), particularly in direct-drilled establishments, is of concern where not checked.

21.4.2 Technology Transfer

The first real inputs associated with technology development and transfer commenced in the mid-1970s. Over the next 10 years, close collaboration between the AgResearch Seed Production Unit and Lincoln Agricultural College (now Lincoln University) Mixed Cropping Unit staff, took science from the plot to field-scale reality. An important aspect of this era was the annual and special field days to ensure full exposure to growers over this crucial development phase. At the same time, technology transfer in herbicide use became important, with interest from private consultants and the formation of farm discussion groups run predominantly by Ministry of Agriculture and Fisheries staff (now a private consultancy – Agriculture New Zealand).

A government-assisted, technology for business growth 3-year programme in the early 1990s was the major approach that lifted seed yield consistency over the environmental diversity of the Canterbury–North Otago growing region. The programme involved an AgResearch staff member becoming responsible for the overall management of Challenge Seeds (now Wrightsons Limited) white clover seed crops of cultivars Grasslands Kopu and Grasslands Tahora on over 300 farms. The rationale was to improve grower and company representative confidence, understanding and response to the issues confronted, particularly those associated with cultivar change. In terms of success, over 1000 ha were taken through cultivar change in the final year with an overall yield increase of 20% once change conditions had been met. This programme greatly enhanced the level of technology, particularly in relation to cultivar placement. Most companies now have their own clover seed specialists, with AgResearch giving inputs to the head licensee companies* of Grasslands-bred cultivars.

Over the last 10 years the Herbage Seed Subsection of Federated Farmers has played an important role in funding field days.

21.4.3 Seed Certification

The major areas of difference between New Zealand and OECD requirements revolve around those associated with white clover cultivar-change crops (Anon., 1984). Clifford *et al.* (1990) demonstrated the necessity for a 5-year break period between cultivars (compared to three for most overseas countries) coupled with a minimum row spacing of 30 cm to allow for adequate crop inspection. These

---

*Company licensed by AgResearch (the breeder) to multiply and market seed of a particular cultivar.

cultivar change requirements denote a rigid approach to quality control. The success of these stringent requirements is acknowledged by the fact that in New Zealand there is only a 2% rejection rate for cultivar-change crops (Clifford et al., 1990). International recognition of these standards is evident in the fact that seed of 27 overseas cultivars is now being multiplied for re-export. Once field-change crop certification requirements are met, future sowing technology reverts to grower preference.

21.4.4 Systems Alternatives

The systems outlined in Table 21.1 and discussed below are those used or developed on the Lincoln University Mixed Cropping Unit (McCartin, 1985):

1. Traditional ryegrass–white clover sowings are now less than 10% of their peak, and are mainly with cultivar Grasslands Huia. More astute grazing management has promoted seed yields of up to 800 kg ha^{-1} in good seasons.

2–5. The limitations to high seed yields within these systems relate to competition for light and weed control, with autumn sowings being more affected. Some growers have reduced cereal sowing and applied nitrogen rates, with lowered cereal yields being more than offset by better clover returns in the following season. Grasslands Huia is the major cultivar used. In general, low inputs are traded off against a lower but acceptable seed yield to the grower. The advent of new broad-spectrum chemicals for within-crop weed control may significantly increase the area in non-change proprietary cultivars established in this manner.

6. Reduced timespan of a pea crop in the ground leads to less establishment competition. Pre- and within-crop herbicide control is also an advantage for white clover.

7–8. Autumn clover sowings following a cereal, either direct drilled or after minimum cultivation, have become the dominant crop establishment method. Presowing soil incorporation of trifluralin is a standard practice, with minimum tillage to offset any reactivation of the soil buried clover seed burden.

Table 21.1. White clover seed yields from different establishment strategies. (From McCartin, 1985.)

	Establishment strategy	White clover seed yields (kg ha^{-1})
1	Ryegrass and clover Ryegrass seed 1st year White clover seed 2nd year	350–450
2	Autumn sowing with wheat in 15-cm row spacings	300–400
3	Autumn sowing with wheat in 30-cm row spacings	250–400
4	Spring sowing with barley in 30-cm row spacings	250–400
5	Oversowing wheat in the spring with white clover	250–500
6	Undersowing peas with white clover	300–600
7	Direct drilling clover seed into burnt wheat stubble in late summer (February)	700–935
8	Direct drilling clover seed into barley stubble	700–935

Row-spaced sowings are frequently used, even in cultivar non-change crops. Seeding rate is standardized at 3 kg ha^{-1}. Up to 2 kg ha^{-1} of ryegrass is often sown with the clover as a protectant. All crops are heavy rolled in late winter/early spring to ensure land uniformity for harvest. Recently there has been a move to more direct-combining of crops.

Direct-drilling of ryegrass stubble has become a favoured option, particularly for cultivar change. This practice extends the autumn establishment period. Over 90% of the change crops are autumn sown, to facilitate adequate removal of any contaminants establishing between the rows.

21.4.5 Government Subsidy Transition

In the early 1980s the Government ceased funding cultivar development, maintenance, and administrative costs associated with breeders (Prebasic) and basic seed multiplication of public cultivars. Over the last 5 years these costs to AgResearch have been reimbursed through a grower levy. At the same time science funding eligibility and its allocation were reviewed. Public Good Science was redefined, and now means that it has no perceived commercial advantage. Appropriateness of individual programmes is currently assessed by independent peer review, in competition for 1–3-year funding with universities and the private sector. As such, guaranteed funding of government institutions is now measured in terms of value to the New Zealand and the international science base.

In 1992, government science was further rationalized by the formation of ten autonomous Crown Research Institutes (CRIs), of which AgResearch is one. The price of this autonomy was a 10% annual dividend to the government based on taxpayer's investment. This redirection has placed a far greater incentive on CRIs to take science from the shelf to the market-place. All profits realized are now available for 'self-determined' re-investment. This 10-year transition in science funding has significantly modified AgResearch's association with the industry.

21.4.6 Agresearch–Industry Relationships

Commercialization of plant improvement science programmes in the form of transposing elite germplasm collections into proprietary cultivars is now carried out by the Cultivar Development and Maintenance Unit. This unit is totally funded commercially through royalties and joint-industry ventures. As a result there has been a rapid increase in cultivars released through companies interested in maintaining and developing local market share. With the removal of cultivar Grasslands Huia from the United Kingdom Recommended Cultivar List, offshore breeding programmes now cover most countries of perceived market opportunity. A key feature of these programmes is long-term market sustainability. As such, acceptable returns to the seed multiplier have now been placed on an equal footing with increased pastoral advantage.

Commerce now funds all evaluations, including pesticide screenings. Of note have been the substantial financial inputs into technology by the Herbage Seed Subsection of Federated Farmers from seed grower levies. These levies have now

21.5 FUTURE DIRECTIONS

These relate directly to world seed requirements, area availability for clover multiplication, and cultivar numbers to be accommodated. Currently, the annual world trade is about 9500 t, with New Zealand providing 50–55% of that seed (Mather *et al.*, 1995). Over the last 15 years, the area in clover seed production has diminished by almost half, due to competition from more lucrative land-use alternatives. It is now static at about 15000 ha (Clifford *et al.*, 1995) (Table 21.2). Of this area, 77% remains in cultivar Grasslands Huia which accounts for 80% of New Zealand's export volume. The slow decline in the Grasslands Huia area and the only gradual increase of 'preferred' proprietary cultivars is indicative of a prolonged world seed shortage, with price per kilogram of Grasslands Huia seed being almost similar to that of the proprietary cultivars.

Of future concern is the existence of a diverse range of New Zealand breeding programmes, let alone the retention of New Zealand's open-door policy on overseas cultivar multiplication. Cultivar numbers in certification (38), rather than area, are the present major problem. Current cultivar-change certification requirements have proven to be justified. The only modification potential is through the development of a consistent buried-seed-count technology which would eliminate or reduce the definitive 5-year requirement. More effective area use is related to two factors. Greater Head Licensee inputs into technology transfer could raise yields by 25% to the advantage of both grower and area requirements. Similarly, rotation modification could enhance the existing area. Currently the timespan between

Table 21.2. Seed production area for Public, New Zealand and Overseas Proprietary cultivars for seasons 1990–1991 to 1994–1995 (MAF Seed Certification Statistics). (From Clifford *et al.*, 1995.)

	Cultivar type									Total area (ha)
	New Zealand Public			New Zealand Proprietary			Overseas Proprietary			
Seasons	Area (ha)	% of total	Yield (kg ha^{-1})	Area (ha)	% of total	Yield (kg ha^{-1})	Area (ha)	% of total	Yield (kg ha^{-1})	
1990–1991	9898	82	385	722	6	343	1396	11[17]	327	12016
1991–1992	12,283	73	381	1492	9	194	2930	18[27]	420	16705
1992–1993	12,271	77	284	867	5	438	2803	18[23]	214	15941
1993–1994	12,607	81	144	1071	7	234	1921	12[19]	180	15599
1994–1995	10,149	75	N/A	11,540	11	N/A	1856	14[25]	N/A	13545
Means	11442	77	324	1138	8	302	2181	15[23]	285	14761

N/A, not available.
[1] Superscript figures in parentheses are total proprietaries.

crops on the same area is 3–4 years. At 15,000 ha grown annually, the total clover-contaminated area is 60,000 ha.

A reduction of the area in cultivar Grasslands Huia, to the advantage of proprietary cultivars, is beginning to occur. The current release of one broad-spectrum within-crop herbicide, coupled with the possible release of a further two, will alleviate the perceived change-crop weed control problems for many potential growers. Acceptance by present Grasslands Huia growers of these advances could render a further 5000 ha immediately available for proprietary multiplication. Of equal concern to these impending growers is cultivar-change choice. With 38 cultivars available, choice in relation to seed yield and consistency remains the greatest imponderable for most potential growers. Many growers who were at the forefront of change have now diversified into more than one cultivar per farm, depending on total area. A key factor for modern-generation growers is to minimize the past constraints of a one-cultivar only system of unknown market lifespan.

REFERENCES

Anon. (1984) *Seed Certification 1984*. Agricultural Data Series. Ministry of Agriculture and Fisheries, Wellington.

Brock, J.L., Caradus, J.R. and Hay, M.J.M. (1989) Fifty years of white clover research in New Zealand. *Proceedings of the New Zealand Grassland Association* 50, 25–39.

Clifford, P.T.P. (1980) Research in white clover seed production. In: Lancashire, J.A. (ed.) *Herbage Seed Production*. Grassland Research and Practice Series No. 1. New Zealand Grassland Association, Palmerston North, pp. 64–67.

Clifford, P.T.P. (1985) Effect of leaf area on white clover seed production. In: Hare, M.D. and Brock, J.L. (eds) *Producing Herbage Seeds*. Grassland Research and Practice Series No. 2. New Zealand Grassland Association, Palmerston North, pp. 25–31.

Clifford, P.T.P. (1987) Producing high seed yields from high forage producing white clovers cultivars. *Journal of Applied Seed Production* 5, 1–9.

Clifford, P.T.P., Rolston, M.P. and Williams, W.M. (1985) Possible solutions to contamination of white clover seed crops by buried seed. In: Hare, M.D. and Brock, J.L. (eds) *Producing Herbage Seeds*. Grassland Research and Practice Series No. 2. New Zealand Grassland Association, Palmerston North, pp. 67–73.

Clifford, P.T.P., Baird, J.J., Grbavac, N. and Sparks, G.A. (1990) White clover soil seed load; effect on requirements and resultant success of cultivar change crops. *Proceedings of the New Zealand Grassland Association* 52, 95–98.

Clifford, P.T.P., Sparks, G.A. and Woodfield, D.R. (1995) The intensifying requirements for white clover cultivar change. In: Woodfield, D.R. (ed.) *White Clover: New Zealand's Competitive Edge*. Agronomy Society of New Zealand Special Publication No. 11; Grassland Research and Practice Series No. 6. New Zealand Grassland Association, Palmerston North, pp. 19–24.

Garrett, H.E. (1957) *Small Seeds in New Zealand Farm Management*. Canterbury Agricultural College publication. Canterbury, New Zealand, 78 pp.

Hampton, J.G. and Scott, D.J. (1990) New Zealand seed certification. *Plant Varieties and Seeds* 3, 173–180.

Leith, C.C. (1949) Seed production in New Zealand: white clover. *New Zealand Journal of Agriculture* 79, 59–64.

Mather, R.D.J., Melhuish, D.T. and Herlihy, M. (1995) Trends in the global marketing of white clover cultivars. In: Woodfield, D.R. (ed.) *White Clover: New Zealand's Competitive Edge*. Agronomy Society of New Zealand Special Publication No. 11, Grassland Research and Practice Series No. 6. New Zealand Grassland Association, Palmerston North, pp. 7–14.

McCartin, J. (1985) Alternative establishment strategies for white clover seed production. In: Hare, M.D. and Brock, J.L. (eds) *Producing Herbage Seeds*. Grassland Research and Practice Series No. 2. New Zealand Grassland Association, Palmerston North, pp. 33–35.

Thomas, R.G. (1961) The influence of environment on seed production capacity in white clover (*Trifolium repens* L.). I controlled environment studies. *Australian Journal of Agricultural Research* 12, 227–238.

Trifolium subterraneum L. (Subterranean Clover) in Australia

K.G. Boyce
*Seed Services Centre, Primary Industries, South Australia,
GPO Box 1671, Adelaide, South Australia 5001*

22.1 INTRODUCTION

Subterranean clover (*Trifolium subterraneum* L.), also known as sub clover, is an annual forage legume species so named because of its characteristic of active burial of its burr (fruiting head) and seed. The species is a native of the Mediterranean basin and the Atlantic seaboard of western Europe (Gladstones and Collins, 1983). It was accidentally introduced into Australia with early European settlement and became widely naturalized over southern areas of the continent. It has now been sown for pasture, or has become naturalized as a self-regenerating species in pastures over at least 20 million hectares of temperate climate areas of southern Australia (Cocks *et al.*, 1980).

Sub clover became generally recognized as a valuable pasture species in Australia in the last two decades of the nineteenth century. Its main characteristics include an ability to tolerate heavy grazing, particularly by sheep, its adaption by natural selection (and latterly by targeted breeding) to a wide range of flowering and seed maturity times, and its production of large quantities of impermeable seed effectively buried away from grazing influence.

In addition to its value as a pasture plant, sub clover plays a key role in providing nitrogen in the mixed cereal cropping and grazing areas typical of the 250–800 mm annual rainfall belt of southern Australia. Soils in these areas are chronically deficient in macronutrients in their native state. As cereal cropping intensity in this zone increased with no nutrient input in the early decades of this century, crop yields fell markedly. In the 1930s phosphatic fertilizer was introduced to the cropping phase, and self-regenerating annual legumes with their nitrogen-fixing capacity were introduced to the pasture phase of the rotation; the result was remarkably increased crop yields. Sub clover was the species introduced in the acidic and neutral soils and annual medics (*Medicago* spp.) in the alkaline soils. This pattern of agricultural production has continued essentially unchanged for five decades.

The species *T. subterraneum* is divided into three subspecies (Katznelson and Morley, 1965), each of which occupy more or less specific ecological niches in the winter rainfall agricultural areas. The subspecies are *subterraneum*, *yanninicum* and *brachycalycinum*, of which the former has the largest number of cultivars and is the most important in agricultural terms.

Since the value of the species as a pasture plant for stock feed and a source of nitrogen for cropping soils became evident at the turn of the twentieth century, increasing efforts have been made to identify superior, naturalized strains in Australia, and to collect germplasm from original sources for plant breeding. The Australian sub clover plant improvement programme and the national germplasm bank for the species are located at the Western Australian Department of Agriculture in Perth. The programme, which has been coordinated in operation across southern Australia, has concentrated on selection (and in latter years, breeding) of cultivars with a range of characteristics. These include flowering time and maturity, relative hardseededness, oestrogenicity, resistance to clover scorch (*Aureobasidium caulivorum* (Kirch.) W.B. Cooke) and morphological uniqueness for certification purposes.

22.2 ENVIRONMENTAL ADAPTION

22.2.1 Climate

Sub clover seed production is undertaken in southern areas of Australia which have a Mediterranean-type climate characterized by cool, wet winters and hot, dry summers. For dryland crops, the lower annual rainfall limit for economic production is 350–375 mm; most production takes place in the 400–700 mm rainfall belt. Both irrigated and dryland production is undertaken in concentrated areas of New South Wales, Victoria and South Australia, but production in Western Australia is limited to dryland situations.

Summer rainfall incidence limits seed production in the southeast of the country where precipitation over the harvest period from mid December to the end of February exceeds 150 mm. Ecologically, the subspecies are adapted to both differing rainfall areas and soil types:

- Subspecies *subterraneum*: well-drained acidic soils, 250–750 mm annual rainfall.
- Subspecies *yanninicum*: poorly drained acidic soils and areas subject to periodic waterlogging, over 600 mm annual rainfall.
- Subspecies *brachycalycinum*: neutral to alkaline clay loam or self-mulching soils, 500–750 mm annual rainfall.

22.2.2 Soils

The most suitable soils for seed production are neutral to slightly acidic loams or sandy loams. Hard setting clays or clay loams are not ideal, as many cultivars must bury their burrs for seed to be formed. Light has been found to inhibit the development of seed (Taylor, 1979). Subspecies *subterraneum* and *yanninicum* employ active burial mechanisms by positive geotropism of the peduncle, a feature not so prevalent in subspecies *brachycalycinum*. Clay soils also tend to set hard in summer making harvest recovery of the seed very difficult.

Soils containing small stones are avoided, as these are recovered along with the seed at harvest and cause significant wear to harvesting machinery as well as being difficult to remove in the seed cleaning process.

Australian soils are generally deficient in the major, and many of the trace elements, for adequate growth of sub clover. Phosphatic fertilizer is generally applied with each crop, and trace elements, particularly molybdenum, are applied periodically where subsequent seed crops are taken on the same land. For dryland crops the fertilizer rate should be that used for cereal crops: in irrigated areas this should be at least doubled.

22.3 SEED PRODUCTION IN AUSTRALIA

Commercial seed production of sub clovers in Australia began in the hills east of Adelaide in South Australia in the 1890s. Burrs were gathered by hand raking the soil. They were then threshed through stationary, cereal-threshing machines, and the seed winnowed and sieved from the trash.

In the 1930s with the need for larger amounts of seed in new areas of land being brought into cultivation in the cereal belt, burrs were collected on gangs of large rollers covered with sheep skins drawn by animals or mechanical means across the fields. The burrs collected were subsequently threshed and cleaned in stationary cleaning machines in the field.

In 1962 the process of harvest was significantly simplified by the introduction into farming of mechanically drawn suction harvesting machinery. This was the single most important advance in sub clover seed production technology. It facilitated the supply of large quantities of cheap seed in the late 1960s which in turn allowed the economic development from native scrub of large tracts of new cropping land, particularly in Western Australia. This harvest technology, with some modification, is still used generally to this date.

The first commercially available cultivars of sub clover were Mt Barker (1906), Dwalganup (1929), Tallarook (1935), Bacchus Marsh (1937) and Yarloop (1939). These have been added to significantly over the years, particularly from the early 1960s through plant improvement programmes, so that there are at present 31 cultivars registered. Of these, 24 are from the subspecies *subterraneum*, four from the subspecies *yanninicum* and three from the subspecies *brachycalicinum*. Some 85–90% of seed produced, however, comes from only 10 cultivars.

Annual production of seed is volatile, due to the preponderance of rainfed production, particularly in New South Wales and Western Australia. Production under the official seed certification schemes in each State during the decade of the 1980s averaged 4400 t, with a high of 6880 t in 1988/89 and a low of 2700 t in the drought year of 1982/83. Production during the 1990s has averaged 3400 t, generally declining from the peak of 1988/89 to 1560 t in 1994/95.

In years of average or below-average rainfall, the majority of seed is produced under seed certification schemes, as most seed comes from specialized areas grown by professional seed producers. In high-rainfall years, considerable quantities of uncertified seed may be produced from sub clover pasture by contractors; the quantity harvested may reach 50–60% of the certified production, causing difficulties for orderly marketing. However, the economics of seed production increasingly restrict production to irrigated areas operated by professional growers.

Seed yields vary considerably, being mainly dependent on the availability of soil moisture: low-input dryland production areas have average yields of 100–300 kg ha^{-1}, high-input dryland areas have yields of 400–600 kg ha^{-1} and irrigated areas from 1000 to 1200 kg ha^{-1}. Most of the seed produced until the early 1970s was used domestically. Export marketing increased from that time, and in some years now constitutes up to 30% of the total production, averaging just over 800 t annually over the last 5 years. Most of the export markets are in countries of the European Union (EU).

22.4 FACTORS IN PRODUCTION SUCCESS

22.4.1 Sowing Technology

Methods used for successful seed crop establishment vary. The cheapest method is to establish a sub clover stand under a cereal crop. Although the sub clover seed is sown at the same time as the cereal in the autumn, it must not be sown deeper than 15 mm (vs. 45–50 mm for the cereal seed), necessitating the use of a specialized small-seed box attached to the rear of the combine seeder. Seed is metered onto the soil surface at 8–10 kg ha^{-1} and buried with light trailing harrows. The seed set in the year of the cereal crop is allowed to regenerate in the following year, and the seed crop is harvested at the end of that season. This method of establishment presents difficulty for weed control in the cereal crop, and if there is poor seed set in the establishment year, there will be insufficient seed to regenerate to form a dense sub clover stand in the subsequent year.

A stale seedbed technique may give more reliable establishment results. After autumn rains have fallen and initial weed seed germination in the field has begun, the crop area is sprayed twice, about 10 days apart, with a desiccant herbicide, killing seedlings that have established. The clover seed is then sown into this seedbed with sowing shoes (drill openers) that cause minimal soil disturbance. This method is very effective, provided thought has been given to weeds that need to be controlled by postemergence herbicides.

The most effective, but most costly, establishment method is to sow into a normally prepared crop seedbed. Seedbed preparation generally includes removal of any cereal stubble by heavy grazing, stubble mulching or fire harrowing, followed by normal ploughing and subsequent cultivation to remove germinating weed seeds. Herbicides such as trifluralin or pendimethalin incorporated during cultivation, generally give good weed control. Weed control in the crop has a significant bearing on the effectiveness of seed and burr pickup at harvest.

If the area has not grown sub clover before, the seed should be inoculated with an effective strain of *Rhizobium* bacteria and lime coated to ensure adequate plant nodulation and nitrogen fixation.

22.4.2 Pest and Disease Management

Sub clover seed crops, particularly in the establishment phase and at flowering, need protection from red-legged earthmite (*Halotydeus destructor* Tucker) and lucerne flea (*Sminthurus viridis* Linnaeus). Native budworm (*Heliocoverpa punctigera* Wallengren) can also destroy flowers and developing burrs. Diseases are generally not a problem except for clover scorch disease in some susceptible cultivars. The severity of the disease can often be controlled by minimizing excessive forage material by judicious grazing with livestock in spring.

22.4.3 Grazing Management

The seed crop should be kept grazed by sheep to a height of 4–5 cm from mid winter until the plants begin to flower, after which only light grazing should occur. At full flower all grazing should cease. Lush forage growth restricts flower production and leads to poor seed set on flowers under the leaf canopy. Grazing also assists with efficient insect, weed and clover scorch control. The actual time of removal of stock will vary with the cultivar grown, being from early August for early cultivars to late September for late cultivars. For dryland crops the relative availability of moisture for finishing the growth and seed set will also influence the decision.

22.4.4 Irrigation

As there is more control over water output, spray irrigation is better than flood irrigation for sub clover seed production. In an average year in Australia, three or four irrigations of 50–75 mm of water will be necessary. The first irrigation should be applied before the crop first wilts in spring and subsequently soil moisture needs to be maintained to prevent excessive wilting. Extra care must be taken with cultivars of the subspecies *yanninicum* which quickly collapse irretrievably even under mild moisture stress.

Once burr development begins, minimal irrigation should be given, as overwatering will cause rotting of the burrs.

22.4.5 Harvest Technology

As the burr is buried beneath the ground, the soil surface must be prepared for harvest.

When the burr has fully matured and the plants are dry, the vine material is cleared from the area with a finger-wheel side-delivery rake. The vine is either left in windrows or completely removed from the field with a forage harvester or baler. The vine remaining on the surface of the soil is broken up in the heat of the day with heavy harrows or lengths of chain wire mesh, weighted and dragged across the field at high speed.

Buried burrs are best brought to the soil surface with a broad-footed tyned cultivator set to a depth of about 2 cm, or at the depth where the majority of burrs have accumulated. Burr burial depth will depend on the soil type, cultivar and soil moisture conditions during seed setting. Following the cultivation unit, the ground is lightly harrowed and meshed with chain wire mesh to break down soil clods, even out the surface, and bring the burrs to the surface (Fig. 22.1). A number of passes of the preparatory implements may be required to bring the burr to the harvest position.

When prepared, the burr and loose soil are sucked from the soil surface with a suction harvesting machine (Fig. 22.2). The harvesting operation is slow (1–1.5 km h^{-1}) and requires power of at least 70 kW for each harvesting machine. It is common to attach several harvesting machines together to get over the ground more quickly to minimize the risk of crop loss due to unseasonable rainfall. Harvesting is most efficient in very hot weather. In cooler weather, the vine becomes toughened and plant material is not able to be forced through the wire mesh of the harvester's drum. This leads to overloading of the harvester, and the subsequent clogging of the system can lead to costly harvester breakdowns.

Due to the high level of inert matter and weed seeds in the harvested sample, sub clover seed must be cleaned before commercial sale. This is generally done in a stationary cleaning plant mainly using indent cylinders and a gravity table. Despite the harvesting methodology, the seed sample may have a high level of hardseededness, which is generally unacceptable to the commercial seed trade and farmers. A scarifier is often used in line in the cleaning operation to facilitate the reduction of this problem.

22.4.6 Genetic Quality Management

Official seed certification programmes for sub clover have been available in Australia since 1935, to ensure cultivar purity of seed entering agriculture.

Fig. 22.1. Subterranean clover burrs on the soil surface ready for the harvesting operation.

Fig. 22.2. Subterranean clover suction harvester in operation.

Standards for cultivar purity have been set at 95% for certified grade. This is generally adequate for most cultivars, as the seed is normally sown for pasture as a mixture of cultivars.

Cultivars that are distinguishable from each other in field inspection are certified by a modified scheme which results in mixed generations in the seed lot when production is carried out on the same field over subsequent years. This is the normal practice, as it is difficult to change cultivars from year to year, due to the significant contamination from hard seed left in the field after harvest.

Where cultivars may be indistinguishable by morphological markers in the field, a pedigree system of certification is employed, utilizing special methods in the field and in postcontrol tests to identify genetic contaminants.

Both seed certification methodologies have been accepted into the OECD schemes for seed moving in international trade (Boyce, 1990), and documented in the rules of the OECD subterranean clover and similar species scheme.

REFERENCES

Boyce, K.G. (1990) The OECD Seed Certification Scheme for subterrenean clover and similar species. *Plant Varieties and Seeds* 3, 195–199.

Cocks, P.S., Mathison, M.J. and Crawford, E.J. (1980) From wild plants to pasture cultivars: annual medics and subterranean clover in southern Australia. In: Summerfield, R.J. and Bunting, A.H. (eds) *Advances in Legume Science*. Royal Botanic Gardens, Kew, UK, pp. 569–596.

Gladstones, J.S. and Collins, W.J. (1983) Subterranean clover as a naturalised plant in Australia. *Journal of the Australian Institute of Agricultural Science* 49, 191–202.

Katznelson, J. and Morley,F.H.W. (1965) A taxonomic revision of sect. *Calycomorphum* of the Genus *Trifolium*. I. The geocorpic species. *Israel Journal of Botany* 14, 112–134.

Taylor, G.B. (1979) The inhibitory effect of light on seed and burr development in several species of *Trifolium*. *Australian Journal of Agricultural Research* 30, 845–907.

Index

Note: page numbers in *italics* refer to figures and tables

abscisic acid 85, 98
abscission
    floret 51
    process 85
ageing, seed 235
Agriculture and Agri-Food Canada 362
air screen cleaners 192–193
    aspiration 192–193
    screens 193
aleurone layer 80
alfalfa *see* lucerne
alfalfa leafcutting bee 164–166, *167*, 168, 171–172
    disease control 166, 168
    life history 164
    management for pollination 165–166, *167*
    parasites 168
    use in alfalfa production 364, 368, 372
alfalfa mosaic virus 146
alkali bee 168–169
amenity use 1
analytical purity standards 222
ancymidol 116
anemophilous flowers 153, 154
anther dehiscence 154
anthesis
    days after 183–184
    number of ovules per unit area 56
    reproductive system efficiency 248–250
    temperature 117–118
anthocyanin loss from grass seed heads 83
anti-quality factors 208
apical development 27
apical dominance 12
apomixis 155, *156*
    breeding methods 263–264
    gametophytic 79
    smooth-stalked meadowgrass 334
asteosclereids 96
atrazine 210
auxins 12
    abscission process 85
awns, grass seed crop losses 112–113
axillary bud 28

bacteria
    biological control
        pests 120
        weeds 114
    diseases of legume crops 147

403

bacterial wilt   147, 347
　　seed quality   228
basic seed   265
bee
　　alkali   168–169
　　bumble   160, 169–170, *171*, 381
　　honey   160, 161–163, *164*
　　mason   170–172
　　nectar robbing   161
　　pollination   157, 158, 160,
　　　　161–166, *167*, 168–172
　　wild species   172
　　see also alfalfa leafcutting bee
bee fly   169
biological control
　　insect pests   120, 146
　　weeds   114
　　white clover pests   388
biotic stress resistance, seed yields   253
blind seed disease   111, 118
　　climatic conditions   253
　　seed quality   227
bloat, ruminant   208
boron   132
break crops for legume seed   129
breeders seed   264–265
　　Syn generations   264–265
breeding methods   260–264
　　apomictic species   263–264
　　cross-fertilizing species   261–263
　　higher seed yields   243–244
bromegrass, smooth   285–286
brown patch disease, tall fescue   295
brown rust   119
bumble bee *see* bee, bumble
burning
　　postharvest management   205–206, 207
　　postharvest residue removal   121

candivars   245
　　apomictic species   264
carbohydrate status, floral induction   24
carbon-banding   107
caryopsis   73
　　developmental
　　　　morphology/physiology   76–83
　　embryogenesis   76–79
　　endosperm development   79–81
　　seed coat development   81–83

　　isolation from parent plant   84
cereals, photosynthetic activity of ears 81–82
certification schemes   105, 106
　　forage seed trade   271–272
certified seed   265
　　birdsfoot trefoil   354–355, 358–359
　　Canadian red fescue   306
　　lucerne   368, 370
　　perennial ryegrass   325
　　quality verification   196
chaff removal   192
chalkbrood disease   166, 168, 364
chlormequat   116, 141
chlorophyll loss from grass seed heads 82–83
cleaning   347
　　seed loss   54
climate
　　legume seed   126–127
　　red clover   379
　　subterranean clover   396
clipping, weed seed control   121
clonal evaluation   261
closing
　　grass seed crop management 115–116
　　legume seed crop management 137–141
clover
　　red   54, 377–383
　　　　breeding research   382
　　　　climate   379
　　　　controls   382–383
　　　　diseases   380
　　　　flowers   56
　　　　forage removal   380
　　　　France   377
　　　　growing with Italian ryegrass   378
　　　　harvest   379–380, 381
　　　　insect pests   381
　　　　irrigation   381
　　　　pollination   381
　　　　regions of production   379
　　　　registration   382
　　　　seed development   86, *87*, 88, *90–91*
　　　　seed growers' organization   383
　　　　seed market   380
　　　　seed production   378–380, 382
　　　　seed yield   381–382

soil 379
stem elongation 16
techniques of production 380–381
uses 378
weed control 380
world distribution 377
subterranean 54, 395–401
burrs 399, 400
climate 396
cultivars 401
diseases 399
environmental adaptation 396–397
flowers 56
genetic quality management 400–401
grazing 399
harvesting 397, 399–400, *401*
irrigation 399
nitrogen 395, 396
pests 399
seed certification 398–399
seed production 397–398
seed yield 398
soil 396–397
sowing technology 398
subspecies 396
white 54, 385–392
buds 25–26
cultivars 385
development in New Zealand 386–387
floret development *29*
flowers 56
inflorescence initiation 30
pests 387
research role 387–388
rotation 387
seed certification 386, 388–389
systems alternatives 391–392
technology transfer 388
weeds 387, 392
clover casebearer 387, 388
clover phyllody 147
clover rot, seed quality 227–228
clover scorch disease 399
cocksfoot 284, 339–348
bacterial disease 347
burning after harvest 346
crop management 345–346
crop rotation 343–344
cultivars 340, 341–342
diseases 346–347
environmental requirements 342–343
fertilizers 341
frost susceptibility 342–343
fungal pathogens 346–347
grazing 345–346
harvesting 341, 347
herbicides 345
insect pests 347
pests 346–347
control 346
production in New Zealand 340–341
rainfall 342
seed
cleaning 347
production 339–342, 343–347
quality 347–348
yield 339, 343
soil type 343
sowing 343–344
temperature 342
threshing 347
weed control 343–344, 344–345
Coleoptera, pests 143–144
coleoptile 79
coleorhiza 79
colour sorters 195
combining, direct 187, 188
COMECON markets 5, 317
Common Agricultural Policy (EU) 5
Danish production 273
companion cereal 130
lucerne 361
cotyledonary outgrowths 94
cotyledons
grass 78–79
legumes 94–95
respiratory activity 94–95
cover crops, legume seed 129–130
crop colour 184
crop establishment
grass seed 106–108
legume seed crop management 128–132
crop growth, seed loss 182–183
crop residues
livestock use 207
removal from grass seed 121

crop ripeness, seed moisture   84
cropping system, legume seed   127
cross-fertilizing species   261–263
cross-pollination   155–158
    lucerne   364
crown rot   357
crown rust   119, 324
crowns   14
cultivars
    birdsfoot trefoil   358–359
    development   262–263
    DUS requirements   262
    endophytic status   120
    evaluation systems   243
    fertile tillers at harvest   49
    floret site utilization   53
    florets per spikelet   51
    identification with protein
        electrophoresis   224
    improvement programmes   3
    lucerne   361, 362–363
    maintenance   264–266
      basic seed   265
      breeders seed   264–265
      certified seed   265
      cultivar stability   265–266
      prebasic seed   265
    purity   223–225
    recommended list   275
    reproductive system   263
    seed viability   198
    seed yield   243–244
      differences   257–258
    selection for grass seed crop
        management   105–106
    specialist in Netherlands   277
    spikelets per tiller at harvest   50
    stability   265–266
    thousand seed weight descriptor
      226
    verification   221, 224–225
    white clover   385
cultivation
    inter-row   208–209
    postharvest management   208–209
cutting
    forage removal in legume crops
      137–138
    grasses   187–188
    legumes   185
    postharvest management   205–206

cylinder separator, indented   193–194
cytokinins   98
cytoplasmic male sterility   260–261

daminozide
    biosynthesis inhibition   140–141
    pollination effects   62
damping off   118, 146
daylength
    birdsfoot trefoil   356
    critical   247
    floral induction   19–20
    seed yield   247
    tillering   11
decapitation, tillering   11–12
defoliation
    fertile tillers at harvest   48
    grass seed crop management
      115–116
    legume seed crop management
      137–141
    spikelets per tiller at harvest
      50
    tillering   11
dehiscence, birdsfoot trefoil   252
dehullers   195
Denmark
    climate   322
    domestic consumption   275
    exports   275
    forage seed trade   273–275
    foreign cultivars   274
    herbage seed production   321
    rainfall   322
    seed companies   274–275
dessicant, chemical   381
developmental seed
      morphology/physiology
    embryogenesis   93–95
    legumes   89, *90*–*91*, 92–97
    temperature   94
Diptera, pests   144
direct-combining   187, 188
diseases   236
    birdsfoot trefoil   356–357
    burning for control   206
    cocksfoot   346–347
    foliar   119–120
    grass seed crop management
      118–120

legume seed crop management 146–147
    seed/seedhead 118–119
    stem 119–120
    subterranean clover 399
    see also fungal disease
distinctness, uniformity and stability (DUS) 262, 272
dormancy 98
drapers 187
drought, tillering 11
dryers 191
drying process 190–192
    seeds 200
    systems 191–192
    temperature 191

ecological repair 1
embryo
    cell division 94
    chlorophyll retention 94
    cotyledonary outgrowths 94
    grasses 77
        abortion 79
        axial symmetry 77, 78
        bilateral symmetry 78
        formation without fertilization 79
        size 78
    nutrient sources in legumes 92–93
    sac 93
        aposporous 264
        suspensor 93
embryogenesis, legumes 93–95
emergence
    field 233–234
    seed vigour 234
endophytes
    animal health problems 208
    content of seed lot 292
    novel 230, 293
    seed quality 230
    seed viability 199
    tall fescue 284, 292–293, 294
    turfgrass value 293, 295
endosperm 79
    aleurone layer 80
    cell division 79–80
    consistency 184
    development in grasses 79–81
    embryo nutrition in legumes 89
    hardness 80–81
    legume 89, *90–91*, 92
    meristematic layer 80
    protein reserves 80
    starch reserves 80
    water arrival 84
entomophilous flowers 153–154
environmental stress 236
equilibrium moisture content 190
ergot 118–119
    seed quality 227, 228
ethofumesate 107
European Union (EU) 271
exports, Denmark 275
eyespot 347

female-sterile plants, male-sterile plant pollination 261
fertilization 155
    seed yields 250–251
fertilizers
    cocksfoot 341
    grass seed 106
    perennial ryegrass 327
    requirements for legume crops 132–133
    smooth-stalked meadowgrass 336
    see also nitrogen
fescue 281
    red 297–298
        Boreal cultivar 300
        crop statistics *299*, 300, 302, *303–305*, 306
        establishment 307
        evolution as seed crop 297–298, 300
        exports 302, *304–305*
        land area *299*
        marketing 307–308
        Olds cultivar 297–298
        rejuvenation of stands 307
        seed production 300–301, 302, *303*, 306, 308–309
        seed storage 307–308
        stem eyespot disease 306–307
        uncertified 306, 308
    tall 284, 287
        breeding developments 290–292
        brown patch disease 295
        diseases 295

fescue *contd*
   endophytes   292–293, 294
   history   287, 289
   Kentucky   31, 289
   Oregon growers   291–292
   production amounts   289–290
   seed certification   292–293
   seed crop   *291*
   seed production areas   289–290
   turf potential   290–292
   turfgrass   293
   uses   *288*
field heating, dry seed/wet seed   189
field performance of stored seed   204
film coating   196
fire
   seed germination   121
   see also burning
floral development   246
floral differentiation   26–32
   floret number   30–32
   ovules per inflorescence   30–32
   spikelet number   30–32
   timing   28–30
floral induction   246
   carbohydrate status   24
   classification according to
      requirement   19–24
   juvenility   17–19
   light intensity   24
   long-day requirement   35
   primary   17
   requirement   19–24
     dual   20–24
     single   19
   secondary   17, 22–24
   shoot contribution to seed yield
      24–26
   short-day primary   34
floral initiation   26–32, 246
   double ridge stage   26, 28
   main axis elongation   27
floral structure, potential seed yield   56
floret site utilization
   biological   52
   cultivar   53
   economic   52
   factors affecting   52–54
   increase   53
   loss during harvesting/cleaning   54
   nitrogen application   53
   seed abortion   52
   seed yield   52–54
   water deficit   117
florets
   abortion   60
   abscission   51
   development in white clover   29
   differentiation   26–27
   per inflorescence   59–60
   per spikelet   51
   size utilization   52–54
   spikelet   26–27
   temperature effects   31–32
flower bud development, legume crops
     142
flower tripping   157, 158
   lucerne   363–364
flowering, legume crops   142
foliar diseases   119–120
forage removal, legume seed crop
     management   137–138
forage seed
   production location   34–35
   uses   1–2
forage seed trade   271–272
   certification schemes   271–272
   Denmark   273–275
   Netherlands   275–278
   New Zealand   278–281
   United Kingdom   281–283
   United States of America
     283–286
   weed contamination   272
forage species, temperate   2
fungal disease   118–119
   burning crop residues for control
     205
fungi
   legume crops   146
   storage   228–229
   see also endophytes
fungicides
   cocksfoot   347
   damping off   118
   postharvest seed treatment   196
   rust   119
   seed treatment   200
   seed weight   54
   seedborne infection prevention
     199
fusarium wilt   146

galactomannan 92
gamete fusion 155
gapping 208
genetic shift, cultivar stability 265–266
genetic variation, seed yield 244–246
genotypes, screening 255
germinability
    acquisition 74
    water loss 84
germination
    capacity 75–76
    field emergence 233–234
    potential and seed quality 231–232
    seed deterioration 232
    testing 236
    vigour loss 233
gibberellic acid biosynthesis inhibitors 51
gibberellin biosynthesis inhibitors 116, 139
glume opening 154
government subsidies 5
grain weevil 229
grass seed
    coat development 81–83
    Danish production 273–274
    development 73–76
    maturation 71–86
    mixtures 277
    ripening 83–86
    shedding 83–86
    yield components 45–54
grass seed crop management 105
    clipping 121
    closing 115–116
    crop establishment 106–108
    cultivar selection 105–106
    defoliation 115–116
    diseases 118–120
    endophyte 120–121
    environmental constraints 106
    fertilizer application 106
    grazing 121
    growth regulators 116
    herbicides 113
    hygiene 114–115
    nitrogen 108–111
    nutrients 108–111
    pests 120
    physical constraints 106
    postharvest management 121
    residue removal 121
    rotation 274
    row spacing 108
    seedbed 106
    site preparation 106–107
    site selection 105–106
    sowing rate 108
    species selection 105–106
    techniques 113–114
    temperature 117–118
    time of sowing 107
    water 116–117
    weed control 112–115
grasses 2
    companion for legumes 130
    developmental morphology/physiology 76–83
    double ridge 30
    dual floral induction requirement 20–21
    embryo 77
    embryogenesis 76–79
    endosperm development 79–81
    fertile tillers 45–47
    floral induction 18
    floral initiation 26–27
    inflorescence initiation/differentiation timing 30
    juvenility phase 18
    polyploidization 251
    production regions 4–6
    secondary floral induction 22–23
    shoot density 32–34
    shoot formation 9–13
    single floral induction requirement 19
    tiller seed production capacity 24, 25
    timing of floral differentiation 28–30
    vivipary 23
    water loss in maturing seeds 83–85
grazing
    biological weed control 114
    cocksfoot 345–346
    forage removal in legume crops 137–138
    postharvest management 208
    smooth-stalked meadowgrass 336
    subterranean clover 399
    weed seed control 121
green leaf area, maintenance 119–120
grey leaf spot 295

growth regulators   116
   florets per spikelet   51
   inflorescence numbers   59
   legume crops   139–141
   seed setting   61
   seed weight   54, 62

hard seeds   232
hardseededness   89, 198
   development   95–97
harvest
   index   243
   mass maturity   84
harvest methods
   cutting   185
   legumes   185, 187
   pre-cutting   185
harvest time   184
   optimal   185, *186, 187*
harvesting
   cocksfoot   347
   floret site utilization   54
   mechanical damage   199
   red clover   379–380, 381
   seed yields   252–253
   subterranean clover   397, 399–400, *401*
hat test   184
haustorium, endosperm   89, *90, 92*
heat
   damage   189–190
     detection   229
   seedborne infection prevention   199
herbicides   209–210
   birdsfoot trefoil   357–358
   cocksfoot   345
   grass seed   113
   inter-row spraying   209
   legume seed   133, *134–135*, 136
   postemergence   *135*
   postharvest management   209–210
   safety margins   113
   smooth-stalked meadowgrass   336–337
   strip spraying   209
   tolerance   113
     seed yields   253–254
   white clover   387–388, 392
   see also pesticides
heritability
   broad-sense   245
   estimates   245
   seed yield   244–246
Heteroptera pests   144
High Voluntary Standards (UK)   271
hilar fissure   96
Homoptera pests   144
honey bee   160, 161–163, *164*
   brood rearing   162
   flower recognition   163
   life history   161–162
   management for pollination   162–163
   parasite control   163, *164*
hot spots, fungally heated   228–229
humidity, pollination   155
hydration, seed performance   235
hygiene
   grass seed   114–115
   legume seed crop management   133, *134–135*, 136
Hymenoptera, pests   144

inbreeding, allelic interaction loss   251
industrial uses   1
inflorescence size
   light effects   32, *33*
   nitrogen   32
inflorescences per shoot   58–59
inflorescences per unit area
   reproductive potential   247
   seed yield   59
insect pests
   biological control   146
   birdsfoot trefoil   357
   cocksfoot   347
   control   145
     integrated   145–146
   infestation effects on seed quality   229–230
   legume seed crop management   143–146
   plant selection for resistance   146
   red clover   381
   seed treatment   146
   white clover   387
insect pollination   157, 158, 160–161
   management   161–166, *167*, 168–172
insecticides   145–146

postharvest seed treatment 196
white clover 388
see also pesticides
International Rules for Seed Testing 224
International Seed Testing Association 272
international trade 271, 272
irrigation
    fertile tillers at harvest 49
    moisture stress 117
    postharvest management 210
    red clover 381
    subterranean clover 399

juvenility 17–19
    length 18
    mechanisms 18–19
    termination 18

Kentucky 31, 289
Kentucky bluegrass *see* meadowgrass, smooth-stalked

land reclamation 1
latitude, birdsfoot trefoil 356
leaf development 9
leaf primordium, bud development 28
leafcutting bees *see* alfalfa leafcutting bee
legume seed chalcid 229
legume seed crop management 127
    chlormequat 141
    closing 137–141
    companion cereal/grass 130, 361
    cover crops 129–130
    crop establishment 128–132
    crop site 127–128
    cropping system 128
    daminozide 140–141
    defoliation 137–141
    diseases 146–147
    fertilizer requirements 132–133
    forage removal 137–138
    growth regulators 139–141
    herbicides 133, *134–135*, 136
    hygiene 133, *134–135*, 136
    irrigation 142
    optimal density 131–132
    paclobutrazol 139–140

    pests 143–146
    row spacing 131–132
    seedbed preparation 127–128
    site selection 127–128
    sowing date/rate 130
    triapenthol 141
    water management 141–143
    water stress 142–143
    weed control 133, *134–135*, 136
legumes 2
    axillary bud 28
    buds 25–26
    cotyledons 94–95
    developmental seed morphology/physiology 89, *90–91*, 92–97
    floral induction requirement
      dual 21–22
      single 19
    floral initiation 27–28
    flowering pattern 25
    hardseededness 89, 95–97
    harvest methods 185, 187
    latitude effects on seed yield 55
    ovules 92
    perennation 16
    production regions 4–6
    seed
      loss 92–93
      maturation 71–86, 86, *87*, 88, 89, *90–91*, 92–97
      yield 54–62
    shoot
      categories 25
      density 32–34
      formation 13–14, *15*, 16–17
    storage protein 89
    suspensor 93
lemma, non-hairy 253
Lepidoptera, pests 144
light
    inflorescence size 32, *33*
    intensity and floral induction 24
    line 95
lipid peroxidation, seed ageing 235
livestock, endophyte effects 208
lodging
    fertile tillers at harvest 48–49
    genetic variation for resistance 249
    nitrogen 111, 116
    seed loss 182–183

lodging contd
    seed yield loss   116
lodicule enlargement   154
long-day reaction, critical day length   247
long-day requirement   35
lucerne   54, 285, 361–374
    certified seed   368, 370
    commodity associations   367
    companion cereal   361
    crop area   366, 368, 370
    cross-pollination   364
    cultivars   361, 362–363
    distribution rights   362–363
    exports   370, 371, 372, 373
    flower tripping   363–364
    imports   370, 372, 373
    pollinators   363–364, 367
    research funding   372
    seed area   366
    seed development   88
    seed production   361–362, 365
        region   365, 367
        volume   365–366, 368
    seed yields   361–362
    self-pollination   364
    stem development stages   29
    see also alfalfa leafcutting bee
lucerne rust   146

machine vision, cultivar purity   224
macrosclereids   95
maize, reproductive development phases   73
market demands   6–7
marketing, Danish production   274–275
mason bees   170–172
    life history   170
    management for pollination   170–172
mass maturity   74
    cotyledon respiratory activity   95
    harvest   84
matric priming   235
matriconditioning   197
meadowgrass
    smooth-stalked   330
        autumn treatment   335–336
        breeding   333–334
        crop establishment   331–332
        crop management   332
        cultivars   334
        diseases   336
        fertilizer   336
        harvest   332
        marketing   330–331
        pests   336
        production   331–332
            organization   330–331
        research   334–337
        weed control   336–337
meristematic layer of endosperm   80
methyl bromide, legume seed fumigation   147
micropyle   96
mildew   119
    perennial ryegrass   324
moisture stress   236
    fertile tillers at harvest   49
    florets per inflorescence   59–60
    inflorescence density increase   59
    irrigation   117
mowers   185
mycoplasma, legume crops   147

nectar secretion   161
nematodes
    legume crops   147
    seed quality   228
net blotch   295
Netherlands
    climate   329
    cover crop research   335
    forage seed trade   275–278
    grass seed   329–337
        mixtures   277
    legumes   335
    protein crops   335
    recommended cultivars   275
    research   334–337
    seed companies   278
    specialist cultivars   277
neutron probes   117
New Zealand
    forage seed trade   278–281
    marketing   279–280
    organizations   280–281
    Plant Variety Rights (PVR)   279
    seed certification   280–281
    seed production   278–279

New Zealand Grain and Seed Trade
    Association   280
nitrogen
    blind seed disease   111
    fixation
        legume crops   129
        Rhizobium inoculation   197
    grass seed   108–111
    inflorescence size   32
    legume crop requirements   132
    lodging   111, 116
    seed yield component response
        110–111
    subterranean clover   395, 396
nitrogen fertilizer
    fertile tillers at harvest   48
    floret site utilization   53
    optimum requirements   109–110
    perennial ryegrass   327
    seed weight   54
    smooth-stalked meadowgrass   336
    tillering   11
    timing of application   110
non-shattering mutants   252
nutrients, grass seed   108–111

oats, cover crop for legume seed   130
open-pollination   261
orchardgrass   284
Oregon Forage Grass Seed Endophyte
        Test   292
Organization for Economic Cooperation
        and Development (OECD)   271,
        272
osmoconditioning   197
osmotic priming   235
osteosclereids   95
outcrossing rate   155
ovary position   88
overinduction   35
ovules
    abortion   92, 236
        fertility   60
        self-pollination   157
    fertility   92
    legumes   92
    number   256
    seed yield   56
    sterility   60
    successful fertilization   251

packaging of seeds   202–204
paclobutrazol   48–49, 53, 116
    inflorescence numbers   59
    legume crops   139–140
    raceme number   58
panicle number, seed production
        environment   35
pasture, clover-based management
        systems   385
pathogens
    avoidance   226
    elimination   226
    seed hygiene   226–228
Peace Region Forage Seed Association
        308
pelleting   196
pericarp   73
    evaporation   84
    testa relationship   81
pest control
    burning   206
    cocksfoot   346
pesticides
    perennial ryegrass   327
    red clover   381
    see also herbicides; insecticides
pests   120
    birdsfoot trefoil   356–357
    cocksfoot   346–347
    infestation   236
    legume seed crop management
        143–146
    postharvest seeds   201
    subterranean clover   399
pharmaceutical products   1
phosphates, white clover   385
phosphoenolypyruvate carboxylase   94
phosphorus
    grass seed   112
    legume crop requirements   132
photoperiod
    floral induction   20–21, 22
    primary induction   34
photosynthesis inhibitors   113
phytochrome pigments   10–11
phytoplasma   147
pickup   185, 187, 188
plant breeding, selection   245
plant density, seed yield   58
Plant Variety Protection Act (USA;
        1970)   292

Plant Variety Rights (PVR)
    legislation 230
    New Zealand 279
ploidy level 244
pods per raceme
    abortion rate 60
    seed yield 60
pollen
    germination 157
    grains and temperature effects 249
    production 154–155
    receptivity of stigma 157
    sterility 154
    transport distance 159
    viability 154–155
pollen tube
    bursting 154
    elongation 62
    entry delay to zygote division 77
    formation 154
    growth 157, 251
      inhibition 157
    temperature effects on growth 249
pollinating mechanisms 159–161
pollination 153–155
    daminozide effects 62
    inter-row cultivation 209
    legumes 249–250
    red clover 381
    seed setting 256–257
    tetraploid red clover 60, 249–250
    type in grasses *156*
    visitation rate 160
    wind 159, 249
    see also insect pollination
pollinators, custom 364
polycross 261
polygenes 245
polyploidization of grasses 251
polyvinyl alcohol 235
postharvest management
    burning 205–206, 207
    cultivation 208–209
    cutting 206–207
    drying process 190–192
    equipment
      primary 192–195
      secondary 195
    fertile tillers at harvest 49
    grass seed 121
    grazing 208

    heat damage 189–190
    herbicides 209–210
    irrigation 210
    packaging of seeds 202–204
    pests 201
    processing effects 199–200
    quality verification 196
    seed
      crop 204–210
      drying 189–192
      processing 192–197
      storage 197–204
      treatment 196–197
    storage environment 200–201
    weed control 205
potassium, grass seed 112
prebasic seed 265
pregermination 235
processing, mechanical damage 199
production regions 4–6
proembryo 77
progeny testing 245
protein crops, Netherlands 335

quality control, technological
    development 235–236
quantitative trait loci 245
quarantine regulations 226

raceme number 58
rachilla 26
rachis 26
rainfall
    cocksfoot 342
    subterranean clover 396
relative humidity, atmospheric for seeds
    201
relative storability index 197–198
remuneration 5
reproductive development phases of
    grasses 73
reproductive potential
    inflorescences per unit area 247
    potential seed sites 247
reproductive stage 246–247
reproductive system
    efficiency 248–254
    size determination 247
    size and seed yield 247–248

Index

respiration, seed development 95
revegetation 1
Rhizobium inoculation 197
Rhizoctonia stem blight 146
rhizomes 10, 14
    genotype evaluation 255
    shoot numbers 57
rhynchosporium leaf blotch 119
ribulose biphosphate carboxylase 94
roller mill, velvet 194
root and crown rot disease 143
root rot
    birdsfoot trefoil 357
    seed quality 228
rotation, legume seed crop place 128–129
row spacing, grass seed 108
rust 119
ryegrass 281
    annual 284, 311
    Italian 311–318
        blend 312
        cultivars 314–315
        forage and seed cropping combination 313
        in GDR 314–316
        German reunification effects 317–318
        red clover growing 378
        seed availability 312–313
        seed multiplication 312–313, 316–318
        seed processing plants 316
        seed production 313–314
        tillering 313, 314
    perennial 284–285, 321–328
        capacity 325
        contracts 325
        diseases 324
        exports 321
        fertilizers 327
        harvesting 325
        management 324–325
        payment system 324
        price 327
        production 322
        quality 326–327
        reliability 326–327
        seed companies 325–326
        seed multiplication 326–327
        work distribution 324

salinity, pollination 155
Sclerotinia crown rot 381
scutellum 78–79
seed
    abortion 52, 60
        nutrient supply 92
    ageing 235
    air-screen cleaner 113
    analysis 221, 272
    artificial drying 189–190
    basic 265
    buried 114–115
    cleaning 54, 347
    coating 196–197
        polyvinyl alcohol 235
    companies
        Denmark 274–275
        Netherlands 278
    crops
        management 6
        seed yield 257–260
    deterioration 201
    dormancy 198
        weed species 114
    drying 189–192, 200
        rate 84
    field heating 189
    field performance of stored 204
    fill 251, 254
    growth rate 254
    handling 200
    hygiene 226–230
    international collection 4
    maturity 198
    number and seed weight relationship 254
    packaging 202–204
    polisher 195
    potential sites 247, 248
    pre-basic 265
    priming 197
    production problems 3–4
    purity 6
    quality 219–220
    ripening in grasses 83–86
    sampling intensity 221
    setting 256–257
    sprouting 183
    storage 197–204
    temperature effects on viability 201
    treatment 196–197, 200

seed *contd*
    insect pests 146
    vigour 97–98
    water loss 84
    see also certified seed
seed-borne disease 272
    control 226–227
seed certification 236
    cultivar maintenance 264
    cultivar purity 223, 224
    legume seed 128
    New Zealand 280–281
    purity requirements 225
    schemes 222
    seed hygiene 226
    subterranean clover 398–399
    tall fescue 292–293
    varietal 271
    white clover 386, 388–389
seed coat development 73
    grasses 81–83
    hyaline layer 81
    pigment changes *82*, 83
    testa cell layers 81
seed damage
    drying 231
    extent 231
    germination effects 230
    mechanical injury 230
    seed quality 230–231
    viability 230
seed deterioration
    low germination 232
    mechanisms 231
    seed quality 230–231
seed development
    floret site utilization 52–53
    legumes 86, *87*, 88–89
      timing 86
    red clover *90–91*
    reserve accumulation 84
    respiration 95
    seed yields 251
seed development in grasses 73–76
    germinability 74–75
    germination capacity 75–76
    rate 75
    ripening stage 76
    stages 73–76
seed harvest management 181–185, *186*, 187–188

crop colour 184
crop growth 183
days after anthesis 183–184
endosperm consistency 184
harvest methods 185, *186*, 187–188
harvest time 184, 185, *186*, *187*
seed loss 181
seed moisture content 184, 185
seed shattering 182
stage of development determination for crop 183–184
stripping ripeness 184
seed loss
    crop growth 182–183
    harvesting 181
    legumes 92–93
    lodging 182–183
    sprouting 183
    uneven ripening 183
    vegetative growth 183
seed lot
    description 221
    endophyte content 292
    variation 221
    weed contamination 222
seed maturation
    grasses 71–86
    hardseededness development 95–97
    legumes 71–86, 86, *87*, 88, 89, *90–91*, 92–97
    seedcoat development 95–97
    shedding in grasses 85–86
    water loss 95–97
      grass seed 83–85
seed moisture content 184, 185, 189, 198–199
    crop ripeness 84
    equilibrium 190
seed quality
    analytical purity 221–222
    assessment 220
    cocksfoot 347–348
    components 220, 221–234
    control 220
      schemes 236
    cultivar purity 223–225
    damage 230–231
    definitions 220
    deterioration 230–231
    dimensions 220

Index

endophytes 230
  field performance 233–234
  germination potential 231–232
  insect infestation 229–230
  management factors 236
  mite infestation 229–230
  potential storability 234
  seed hygiene 226–230
  seed lot description/variation 221
  seed size 225–226
  storage fungi 228–229
  tests 219
  vigour 233–234
seed retention 182
  non-shattering mutants 252
  seed yields 252
seed set 158–159
  per flower 256
  success 157–158
seed shattering 182, 184, 185
  birdsfoot trefoil 352
  harvest 252
  index 182
  resistance 252
  seed loss at harvest 85
seed shedding
  florets per spikelet 51
  grasses 83–86
    physiology 85–86
  seed loss at harvest 85
seed size 225–226
  quality indicator 225–226
  seed yield 61–62
seed viability 74
  endophytes 199
  hardseededness 198
  inherent factors 197–199
  moisture content 198–199
seed weight
  fungicide 54
  growth regulators 54
  nitrogen application 54
  seed number relationship 254
  seed yield 54, 60–62
seed yield 3, 243–244
  actual 248
  biotic stress resistance 253
  birdsfoot trefoil 352, 356
  breeding methods 260–264
  candivars 245
  cocksfoot 339, 343
  components 55, 254–260
  cultivar 243–244
    differences 257–258
  dynamic aspects 246
  fertile tillers 45–47
  fertilization 250–251
  florets
    abortion 60
    per inflorescence 59–60
    per spikelet 51
    site utilization 52–54
  genetic variation 244–246
  grasses 158
    components 45–54
  harvesting 252–253
  herbicide tolerance 253–254
  heritability 244–246
  inflorescence type 56
  inflorescences
    per shoot 58–59
    per unit area 59
  legumes 54–62
  lucerne 361–362
  number of shoots 58
  per plant 255, 256
  perennial ryegrass 72
  plant density 58
  plants per unit area 57
  pods per raceme 60
  potential 247, 248
    and floral structure 56
  processing 252–253
  production environment 35
  realized 247, 248, 254–260
  reproductive stage 246–247
  reproductive system size 247–248
  seed
    crops 257–260
    development 251
    fill duration 254
    retention 252
    size 61–62
    total growth rate 254
    weight 54, 60–62
  seeds per pod 60–62
  shoots per plant 57–58
  sowing rate 57
  spaced plants 255–257, 258–260
  spikelets per tiller 49–50
  subterranean clover 398
seed-yielding capacity 261

seedbed
    grass seed   106
    preparation for legume seed   127–128
seedcoat
    development   95–97
    impermeability   97
seeds per pod   60–62
selection
    ovule number   256
    plant breeding   245
    seed setting   256
self-compatibility   155, 156
self-fertility   157
self-incompatibility   155–156
    mechanisms   157–158
self-pollination   155–158
    lucerne   364
separation mechanisms   188
separators, spiral   194–195
shading, tiller development   205
shoots
    aerial   14
    density   32–34
    environmental control of growth   14, 16
    fertile per unit area   57–58
    flower-carrying ability   60
    formation   9–14, *15*, 16–17
    legumes   13–14, *15*, 16–17
      demography   16–17
      internal control of development   16
      longevity   16–17
    per plant and seed yield   57–58
    subsidiary   10
    types   10
short-day primary induction   34
site selection, legume seed   127–128
slime disease, yellow   347
soil
    amelioration   1
    erosion control   1
    pH   132–133
    red clover   379
    subterranean clover   396–397
    type
      cocksfoot   343
      legume seed   127
    value of forage crop culture   5
sooty blotch   146
sowing date
    fertile tiller number at harvest   47–48
    legume seed   130
sowing rate
    fertile tiller number at harvest   48
    grass seed   108
    legume seed   130
    seed yield   57
sowing technology for subterranean clover   398
sowing time for grass seed   107
species, seed viability   197–198
specific gravity table   194
spikelets   26
    florets   26–27
    per tiller   49–50
stem cutters   3, 275
stem diseases   119–120
stem eyespot disease   306–307
stem rust   119
    breeding for resistance   253
    cocksfoot   346–347
stigma, receptivity   249
stolon   10, 14
    shoot numbers   57
    tips   142
storability potential   234
storage environment   200–201
storage proteins   72
straw
    burning   205
    removal   205
stripe rust   119, 347
stripping ripeness   184
strophiolar region   96, 97
stubble management   205
subsidies   5
sulphur, grass seed   112
suspensor, embryo   93

tannins   84
technology transfer   388
temperature
    anthesis   117–118
    cocksfoot   342
    developmental seed morphology/physiology   94
    effects on seed viability   201
    floral induction   19–20, 21–22, 23, 24

floret number effects 31
grass pollen viability 154–155
grass seed crop management 117–118
legume shoots 16
pollen tube growth 157
primary induction 34
tillering 11
testa, pericarp relationship 81
tetraploids, pollination 249–250
thatch build-up prevention 205
thick-headed fly 169
thinning 209
thousand seed weight
  cultivar descriptor 226
  cultivar differences 257
  seed number relationship 254
  seed size 225
threshing 188
  cocksfoot 347
  smooth-stalked meadowgrass 332
tiller 10
  age
    florets per spikelet 51
    spikelets at harvest 50
  demography 13
  elongation 13
  fertile 45–49
    cultivars 49
    defoliation 48
    florets per spikelet 51
    irrigation 49
    lodging 48–49
    nitrogen fertilizer 48
    number at harvest 47–49
    postharvest management 49
    seed yield 45–47
    sowing date/rate 47–48
    weeds 49
  floret number 31
  longevity 13
  number
    florets per spikelet 51
    spikelets at harvest 50
  reproductive 204
  seed production capacity 24, 25
  vegetative 117
tillering
  after burning 205–206
  decapitation 11–12
  defoliation 11

environmental control 10–11
internal control 11–12
Italian ryegrass 313, 314
seasonal effects 13
temperature 117
timothy 285
trefoil, birdsfoot 54, 351–352, 353, 354–359
  certified seed 354–355, 358–359
  constraints to seed production 356–358
  cultivars 352, 353, 354, 355, 358–359
  day length 356
  diseases 356–357
  flowers 56
  herbicides 357–358
  latitude 356
  pests 356–357
  seed
    production 352, 353, 354–355, 358–359
    scarcity 352
    yield 352, 356
  shattering 352
  uncertified seed 358–359
  weeds 357–358
triapenthol, legume crops 141
tripping *see* flower tripping
turf 1
turfgrass, tall fescue 293, 295

ultrasonic treatment, seeds 200
underinduction 34
United Kingdom
  agricultural market 282
  amenity market 282–283
  fescue production 281
  forage seed trade 281–283
  marketing 282–283
  organizations 283
  ryegrass production 281
  seed production 281–282
United States of America
  annual ryegrass 284
  cocksfoot 284
  forage seed trade 283–286
  lucerne 285
  perennial ryegrass 284–285
  smooth bromegrass 285–286

United States of America *contd*
  tall fescue 283–284, 284, 287, *288*, 289–293, *294*, 295
  timothy 285
urea 109

value for cultivation and use (VCU) 272
varroa mite 163
vernalization 204
  annual ryegrass 311
  floral induction 17
  seed yield 246–247
Verticillium wilt 146
vigour, seed quality 233–234
virus disease, legume crops 146
vivipary 23

wasps, parasitic 120
water
  grass seed crop management 116–117
  loss
    seed germinability 84
    seed maturation 95–97
  management in legumes 141–143
  stress 142–143
weathering damage 236
weed control
  biological 114
  cocksfoot 343–344, 344–345
  cultural methods 114
  grass seed crop management 112–115
  herbicides 113, 133, *134–135*, 136
  legume seed crop management 133, *134–135*, 136
  postharvest management 205
  red clover 380
  smooth-stalked meadowgrass 336–337
weeds
  birdsfoot trefoil 357–358
  clean field preparation 133
  contamination in forage seed trade 272
  fertile tillers at harvest 49
  grass seed crop losses 112–113
  perennial ryegrass 324
  seed
    dispersal 115
    dormancy 114
    in seed lots 222
    separation 194
  white clover 388, 392
wind pollination 159, 249

Zadok's scale 80
zygote
  cell division 77, 93
  dormancy 93